HANDBOOK OF NEUROCHEMISTRY

VOLUME V

METABOLIC TURNOVER
IN THE NERVOUS SYSTEM

PART A

HANDBOOK OF NEUROCHEMISTRY
Edited by Abel Lajtha

HANDBOOK OF NEUROCHEMISTRY

Edited by Abel Lajtha

New York State Research Institute
for Neurochemistry and Drug Addiction
Ward's Island
New York, New York

VOLUME V

METABOLIC TURNOVER
IN THE NERVOUS SYSTEM

PART A

PLENUM PRESS • NEW YORK–LONDON • 1971

ISBN 978-1-4615-7168-1   ISBN 978-1-4615-7166-7 (eBook)
DOI 10.1007/978-1-4615-7166-7
Library of Congress Catalog Card Number 68-28097

SBN 306-37705-5

© 1971 Plenum Press, New York
Softcover reprint of the hardcover 1st edition 1971
A Division of Plenum Publishing Corporation
227 West 17th Street, New York, N.Y. 10011

United Kingdom edition published by Plenum Press, London
A Division of Plenum Publishing Company, Ltd.
Davis House (4th Floor), 8 Scrubs Lane, Harlesden, NW10 6SE, London, England

Contributors to this volume:

M. M. Brand   The Division of Neurochemistry, Department of Neurology, The Mount Sinai School of Medicine, New York, New York (page 337)

H. C. Buniatian   Institute of Biochemistry, Academy of Sciences of Armenia S.S.R., Yerevan, Armenia, S.S.R. (page 235)

Robert Main Burton   Department of Pharmacology and the Beaumont–May Institute of Neurology, Washington University School of Medicine, St. Louis, Missouri (page 199)

Sze-Chuh Cheng   New York State Research Institute for Neurochemistry and Drug Addiction, Ward's Island, New York (page 283)

P. Greengard   Department of Pharmacology, Yale University School of Medicine, New Haven, Connecticut (page 317)

Monique Jacob   Centre de Neurochimie du Centre National de la Recherche Scientifique, Strasbourg, France (page 165)

A. Lajtha   New York State Research Institute for Neurochemistry and Drug Addiction, Ward's Island, New York (pages 49 and 551)

G. M. Lehrer   The Division of Neurochemistry, Department of Neurology, The Mount Sinai School of Medicine, New York, New York (page 337)

Paul Mandel   Centre de Neurochimie du Centre National de la Recherche Scientifique, Strasbourg, France (pages 165 and 249)

N. Marks   New York State Research Institute for Neurochemistry and Drug Addiction, Ward's Island, New York (pages 49 and 551)

Margaret R. Murray   Departments of Anatomy and Surgery, College of Physicians and Surgeons, Columbia University, New York, New York (page 373)

J. M. Ritchie   Department of Pharmacology, Yale University
                     School of Medicine, New Haven, Connecticut
                     (page 317)

Sidney Roberts   Department of Biological Chemistry, School
                     of Medicine and the Brain Research Institute,
                     University of California Center for the Health
                     Sciences, Los Angeles, California (page 1)

R. Rodnight   Department of Biochemistry, Institute of Psychia-
                     try, (British Postgraduate Medical Federation,
                     University of London) London, England (page
                     141)

Yasuzo Tsukada   Department of Physiology, School of Medicine,
                     Keio University, Tokyo, Japan (page 215)

PREFACE

Volume V deals with the problems of turnover in the nervous system. "Turnover" is defined in different ways, and the term is used in different contexts. It is used rather broadly in the present volume, and intentionally so. The turnover of macromolecules is only one aspect; here "turnover" indicates the simultaneous and coordinated formation and breakdown of macromolecular species. The complexities of cerebral protein turnover are shown in a separate chapter dealing with the synthesis of proteins, in another on breakdown, and in still another on the relationship of these two (showing how the two halves of turnover are controlled). The fact that most likely the two halves of protein turnover, synthesis and breakdown, are separated spatially and the mechanisms involved are different further emphasizes the complexity of macromolecular turnover.

"Turnover" is used in a different context when the turnover of a cycle is discussed; but here again a number of complex metabolic reactions have to be interrelated and controlled; some such cycles are discussed briefly in this volume, additional cycles have been discussed with metabolism, and some cycles still await elucidation or discovery.

A different type of reaction which can also be interpreted as "turnover" is the exchange of a compound in the brain for another identical or similar one from outside this tissue, with no prior metabolic transformation of the compound to be replaced. Replacement via exchange involves processes of diffusion and transport; these are also treated in the present volume, along with a fairly detailed theoretical background for them, since they have special importance in the nervous system, which represents perhaps the highest complexity of membranes and is perhaps the richest in membrane structures among all the organs.

Other subjects discussed in Volume V are processes that often are not strictly considered part of the processes of turnover, such as hormonal factors affecting metabolism, growth, and development.

Some years ago the question could be asked whether there is turnover of the major components of the nervous system, especially because this organ functions as storage of memory. Since it is a depository for information of a permanent nature, it would not be unreasonable to assume that most of its structures are permanent. More recent studies, some of which are reviewed in the present volume, have shown, however, that not only does metabolic activity of all kinds proceed with great rapidity and intensity in the nervous

vii

system, but the turnover of most compounds occurs at fairly high rates, so that one can assume that only a very small portion of this organ is permanent.

What portion of this activity—turnover, metabolism, exchange, etc.—is specific for the brain and not present in other organs is a difficult question to answer at the present time, although there are a number of compounds and perhaps specific metabolic reactions that are present only in nervous structures and no other elements in the organism. Certainly most components and most reactions present in the brain are also present in other organs; this perhaps shows the unity of life processes. A somewhat more pertinent question, and one that is easier to answer, is whether the processes that one can study in nervous tissues are the ones responsible for the function of the brain. The more we learn about the nervous system, the more we find that its functions, its control processes, and its malfunctions can be understood in terms of biology and that the reactions that are described in the present volume have a great deal to do with the requirements and functions of the nervous system. As perhaps also in other areas, the more we know, the better we understand, the more remains to be found out and understood. This, however, should not—and, I am sure, does not—in any way discourage present and future workers in this field. What is really shown here is that even formidable problems are not unapproachable and their solution will not escape us forever. The rapid growth of investigations in this area clearly emphasizes that the hope of understanding and the importance of acquiring knowledge in this area are recognized increasingly by many.

Abel Lajtha

New York, New York
July 1971

CONTENTS OF PART A

Chapter 4

RNA Metabolism   165
by Paul Mandel and Monique Jacob

Chapter 5

The Turnover of Lipids   199
by Robert Main Burton

Chapter 9

The Tricarboxylic Acid Cycle 283
by Sze-Chuh Cheng

Chapter 10

Metabolism and Function in Nerve Fibers. 317
by P. Greengard and J. M. Ritchie

CONTENTS OF PART B

Chapter 18

Protein Turnover   551
by A. Lajtha and N. Marks

Chapter 1

PROTEIN SYNTHESIS

Sidney Roberts

Department of Biological Chemistry
School of Medicine and the Brain Research Institute
University of California Center for the Health Sciences
Los Angeles, California

I. INTRODUCTION

The basic mechanisms of protein synthesis are presumed to be identical in all cells. However, as detailed information accumulates on the comparative aspects of this process in the highly differentiated and diverse cells which comprise complex organisms, striking points of difference emerge. These differences appear to be derived principally from highly selective adaptations of the cell and its various protein-synthesizing systems to specific, but often changing, environments. The cellular and subcellular elements of the nervous system exhibit a remarkable degree of structural and functional specialization which may not be duplicated in other living systems. The mature neuron is generally isolated from the systemic environment by various barrier mechanisms and is intimately dependent for its sustenance upon neighboring nonneural elements. Moreover, neurons assume a variety of unique forms related to their specific functions. In some cases, the neuronal cell body is located at a great distance from its axonal extensions. These conditions place unusual demands and limitations upon the metabolic activities of both neurons and glia. In addition, the special requirements for information processing and storage undoubtedly contribute to the development of unique characteristics in the protein-synthesizing systems of these cells. The discussion which follows will summarize available information on the properties of protein-synthesizing systems in nervous tissues and will attempt to relate the findings to the special functions of these tissues. Certain pertinent topics will be dealt with in greater detail in other chapters of this Handbook. These include RNA synthesis and metabolism, mitochondrial structure and function, protein turnover, axoplasmic flow, axonal protein synthesis, and physiological and pathological alterations in protein synthesis.

1

II. COMPONENTS OF PROTEIN-SYNTHESIZING SYSTEMS

Most of the detailed information available on the properties of neural protein-synthesizing systems is based upon investigations of whole brain or cerebral cortex of common laboratory animals. These studies suffer from the limitation that the systems analyzed were derived from several cell types. Moreoever, the data obtained from one area or phase of development in the nervous system may not be applicable to other regions or developmental stages. In recent years, attempts have been made to characterize neural protein-synthesizing systems in animals of different ages, in different areas of the central nervous system, in fractionated neuronal and glial preparations from brain, and in neuronal and nonneuronal elements outside of the central nervous system. The results of these investigations may be expected to contribute significantly to understanding of the relationship between protein synthesis and specialized function in the nervous system.

A. Physical and Chemical Properties of Microsomal and Ribosomal Systems

1. General Aspects

The primary site of protein synthesis in the nervous system, as in all cells, is the cytoplasmic ribosome. This conclusion was heralded by the finding that microsomal proteins in the mammalian brain became labeled more rapidly and more extensively than proteins of other subcellular fractions after administration of radioactive amino acids *in vivo*.[1-3] Protein synthesis may also occur on ribosomes or other RNA-containing structures in nuclei, mitochondria, and additional subcellular sites of neural tissue. Quantitatively, the latter processes are probably relatively minor, but qualitatively they may be of very great significance to the economy of the nervous system.

Cytoplasmic ribosomes exist in several physical forms which vary with cell type and functional state.[4] These organelles occur either as single ribosomes (monoribosomes) and subunits thereof, or as dimers, trimers, tetramers, etc., of these basic particles. *In situ*, a large proportion of ribosomes appears as aggregates associated with strands of messenger RNA (polyribosomes). The various ribosomal forms may exist "free" in the cytoplasm or bound to the microsomal membranes (endoplasmic reticulum).

Cytoplasmic ribosomes of the neuron are localized mainly in the perikaryal region.[5] Under normal circumstances, these ribosomes occur in close proximity to the endoplasmic reticulum, thus comprising the Nissl substance. However, neuronal ribosomes apparently exist largely unattached to these membranes[6,7] (Fig. 1). In contrast, cytoplasmic ribosomes in glial cells may be mostly membrane bound.[8] Neuronal damage resulted in dispersion of the Nissl substance (chromatolysis) and possibly in partial disruption of polyribosomes.[9] Cytoplasmic ribosomes associated with the endoplasmic reticulum (ergastoplasm) were also described at the cell membrane of proximal dendrites in the primate spinal cord.[10] Morphological

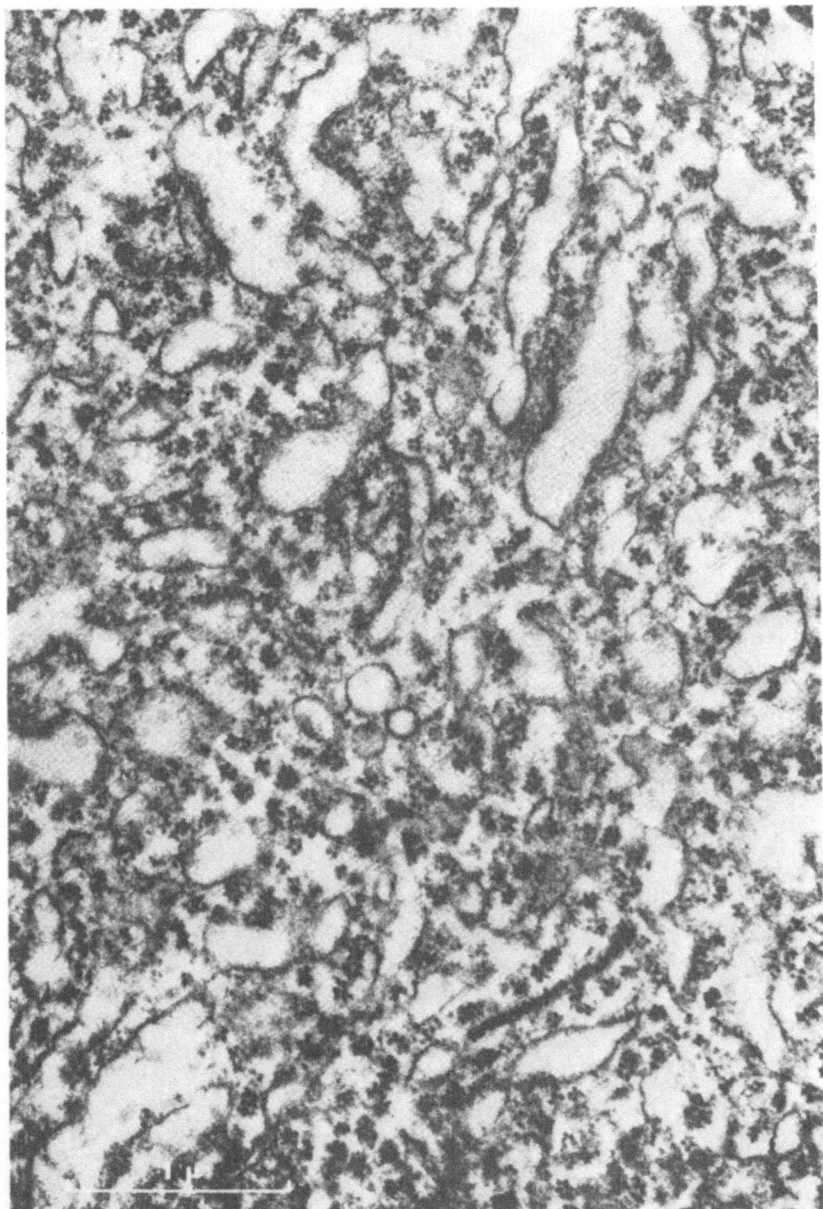

Fig. 1. Survey micrograph of a section from Deiters' neuron of a normal rabbit. The cytoplasm is extremely rich in ribosomes, most of which are found in groups in the cytoplasmic matrix. Only a few ribosomes are seen along the membranes of the endoplasmic reticulum. × 30,000. (Reprinted with permission from Ekholm and Hydén.[6])

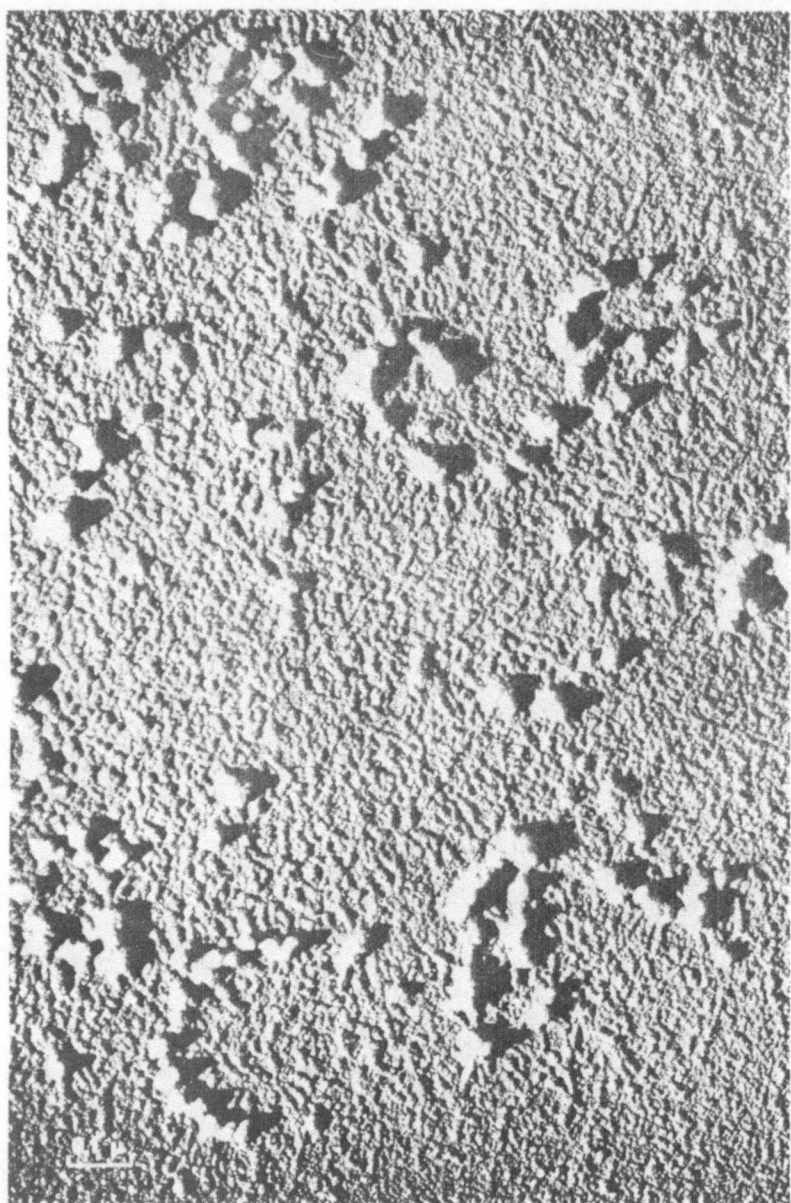

Fig. 2. Platinum-shadowed preparation from Deiters' cells of a normal rabbit.
Several polyribosomes of various sizes as well as single ribosomes are seen. × 90,000.
(Reprinted with permission from Ekholm and Hydén.[6])

evidence is lacking for the occurrence of ribosomes in the distal portions of nerve axons. However, RNA of the ribosomal type has been found in this location (Section II,A,2).

Ekholm and Hydén[6] examined the morphological properties of ribosomes which were carefully extruded from the large Deiters' neurons of the lateral vestibular nucleus in the rabbit. Three basic ribosomal species were revealed by electron microscopy (Fig. 2). The majority of these particles had a diameter which approximated 230 Å (monoribosomes). The other two basic particles (presumably subunits) were about half this size and increased in numbers when Mg^{2+} ions were omitted from the preparation medium. Aggregates of each of these particles existed, sometimes attached to thin filaments of messenger RNA (10–20 Å diameter) from which they could be released in a nonaggregated state by the action of ribonuclease. Many of the aggregates (30%) were tetramers, but dimers, trimers, pentamers, and hexamers were also common. In addition, assemblies of 12 to 15 ribosomes were observed. Monoribosomes constituted a fairly large percentage of the ribonucleoprotein particles. Reflecting the situation in situ, most of the ribosomes in these preparations were not attached to membranes. However, the proportions of polyribosomes, their size distribution, and relationship to endoplasmic reticulum varied within different neuronal populations.[7] Indirect evidence was obtained for the presence of a relatively high proportion of "free" ribosomes (presumably mainly of neuronal origin) in homogenates of rat brain compared to analogous preparations from other tissues. For most tissues, a detergent (e.g., sodium deoxycholate) must be added to the postmitochondrial supernatant or to the isolated microsomes to disrupt the microsomal membrane and release the bulk of cytoplasmic ribosomes. In contrast, omission of detergent in cerebral preparations only slightly decreased the yield of ribosomes.[11,12]

Polyribosomes have been successfully isolated from homogenates of rat cerebral cortex, as from other tissues, by sedimentation through sucrose solutions of high density.[11,13] This procedure resulted in selection of the larger aggregates present in the original homogenates. Relatively high concentrations of Mg^{2+} ions (10–12 mM) were required to protect against the action of endogenous ribonucleases and to produce the maximum yield of heavy polyribosomes (Fig. 3). Zonal centrifugation of cerebral polyribosomes through sucrose layers of increasing density also removed most of the adsorbed contaminants, as shown by the high RNA concentration ($> 60\%$ of total RNA + protein), the relative absence of ribonuclease and other enzymes, and the high absorbance at 260 mμ.[12,13] The yield of polyribosomes was equivalent to approximately 0.2 mg RNA per gram tissue. Polyribosomes have also been obtained in relatively high proportions from brain homogenized in the presence of bentonite[17] or an endogenous ribonuclease inhibitor.[18] Each of these methods resulted in a greater total yield of ribosomes, but in a smaller proportion of aggregates and a lesser degree of purity than the zonal centrifugation procedure. Polyribosomes isolated from cerebral cortex of adult rats in the presence or absence of detergent evidenced great

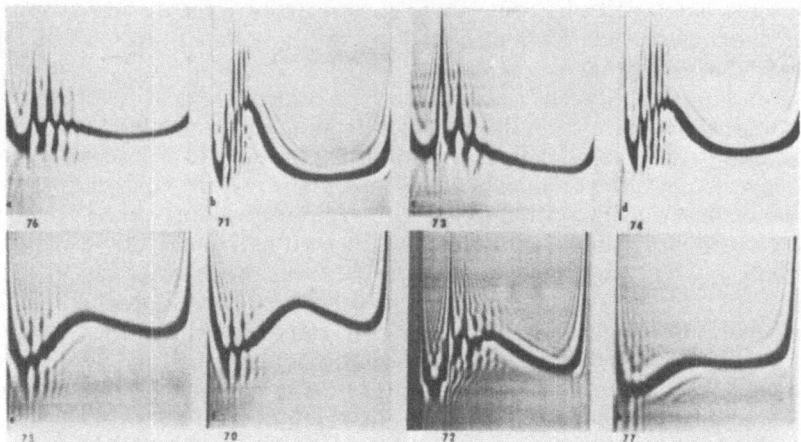

Fig. 3. Sedimentation patterns of cerebral and hepatic polyribosome preparations. Analyses were carried out in a 12-mm cell in the Spinco model E analytical ultracentrifuge at 20°C. Photographs were obtained using schlieren optics at the following intervals after rotor reached full speed of 29,500 rev/min : (a) 7 min ; (b) 2 min ; (c) 6 min ; (d) 2 min ; (e) 5 min ; (f) 5 min ; (g) 6 min ; (h) 3 min. Concentrations of ribosomal protein varied from 2 to 4 mg/ml. $S_{20,w}$ values are shown for the monoribosomal components. (a) Cerebral ribosomes prepared by method of Wettstein, Staehelin, and Noll[14] and analyzed in 5 mM $MgCl_2$, 25 mM KCl and 50 mM Tris–HCl buffer (pH 7.4). (b) Cerebral ribosomes prepared by method of Munro, Jackson, and Korner[15] and analyzed in 10 mM magnesium acetate, 40 mM NaCl, 100 mM KCl and 20 mM Tris–HCl buffer (pH 7.6). (c) Cerebral ribosomes prepared by method of Wilson and Hoagland[16] and analyzed in 8 mM $MgCl_2$, 25 mM KCl and 100 mM Tris–HCl buffer (pH 7.5). (d) Cerebral ribosomes prepared by method of Zomzely et al.[13] and analyzed in 12 mM $MgCl_2$, 100 mM KCl and 50 mM Tris–HCl buffer (pH 7.4). (e) Hepatic ribosomes prepared by method of Wettstein et al.[14] and analyzed in 5 mM $MgCl_2$, 25 mM KCl and 50 mM Tris–HCl buffer (pH 7.4). (f) Hepatic ribosomes prepared by method of Munro et al.[15] and analyzed in 10 mM magnesium acetate, 40 mM NaCl, 100 mM KCl and 20 mM Tris–HCl buffer (pH 7.6). (g) Hepatic ribosomes prepared by method of Wilson and Hoagland[16] and analyzed in 8 mM $MgCl_2$, 25 mM KCl and 100 mM Tris–HCl buffer (pH 7.5). (h) Hepatic ribosomes prepared by method of Zomzely et al.[13] and analyzed in 12 mM $MgCl_2$, 100 mM KCl and 50 mM Tris–HCl buffer (pH 7.4). (Reprinted with permission from Zomzely, Roberts, Brown and Provost.[13])

instability when exposed to reduced concentrations of Mg^{2+} ions (1 mM or less) or pancreatic ribonuclease.[12,13] Similar preparations from rat liver were relatively insensitive to these alterations in the medium. The high proportions of "free" and relatively small ribosomal aggregates in the mature neuron may be a consequence of polyribosome instability. Several physiological and pathological phenomena support the concept that cerebral polyribosomes may be unusually sensitive to alterations in the environment. These include the process of chromatolysis following nerve injury,[9] the disaggregation of brain polyribosomes induced by electroconvulsive treatment in the rabbit,[19] and the reaggregation of brain polyribosomes in rats exposed to ambient light and sound for 15 min after 3 days in the dark.[20]

Cerebral ribosomes have been isolated largely as monoribosomes and small aggregates ("mixed ribosomes") by treatment of the microsomal pellet with sodium deoxycholate.[21–24] This type of preparation has been useful for investigation of the properties of ribosomal subunits and the response of cerebral ribosomes to added messenger RNA. Cerebral mixed ribosomes were obtained in much higher yield (about 0.6 mg RNA per gram tissue) than cerebral polyribosomes.[24] They also had a lower content of RNA (about 40 to 45 %), even when highly purified. Preparations of cerebral mixed ribosomes consisted primarily of monomeric and dimeric ribosomes (Fig. 4). However, a small proportion of polyribosomes (10 to 15 %) was present. These aggregates exhibited the same sensitivity to disruption in the presence of low concentrations of Mg^{2+} or ribonuclease observed in polyribosomal preparations.[12,13] The dimers were also readily dissociated when the Mg^{2+} concentration of the suspending medium was very low. However, complete dissociation to subunits did not occur with highly purified cerebral ribosomes depleted of endogenous ribonuclease unless ribosomal-bound Mg^{2+} was first removed by addition of EDTA.[13] The corrected sedimentation coefficients ($S_{20,w}$) of the cerebral monoribosome and its subunits were approximately 80S, 60S, and 40S. Comparable values have been reported for the analogous ribosomal species obtained from other eukaryotic cells.[4] Upon dissociation of the 80S ribosome, both subunits appeared to be released at the same rate although the smaller subunit broke down more rapidly.[13,25] In large part, the association–dissociation behavior of cerebral ribosomes seemed to be similar to that of other mammalian ribosomes.[4] The presence of the synthetic messenger, polyuridylic acid, resulted in aggregation or delay of breakdown of cerebral mixed ribosomes during incubation.[12] This finding supported the evidence from studies of amino acid incorporation that cerebral mixed ribosomes were partially depleted of natural messenger RNA.

The existence of metabolically active ribosomal precursors, membrane-bound RNA and messenger RNA-protein complexes ("informosomes") in preparations of cerebral microsomes has been postulated.[26] Evidence was obtained for the occurrence of light ribonucleoprotein particles (20–60S) which incorporated radioactivity from precursors of RNA and protein more rapidly than ribosomal subunits. Ribosomelike moieties earlier isolated from preparations of cerebral mixed ribosomes exhibited sedimentation constants intermediate between those of the two ribosomal subunits.[23] These 50–55S particles were active in amino acid incorporation. However, during incubation in the presence of 12 mM Mg^{2+} 80S ribosomes were formed. The possibility exists that the 50–55S preparation may have consisted of a hybrid mixture of the two subunits rather than a single discrete ribosomal species. The sedimentation properties and base compositions of the RNA moieties comprising this particle were consistent with this conclusion (D. M. Schneider and S. Roberts, unpublished observations).

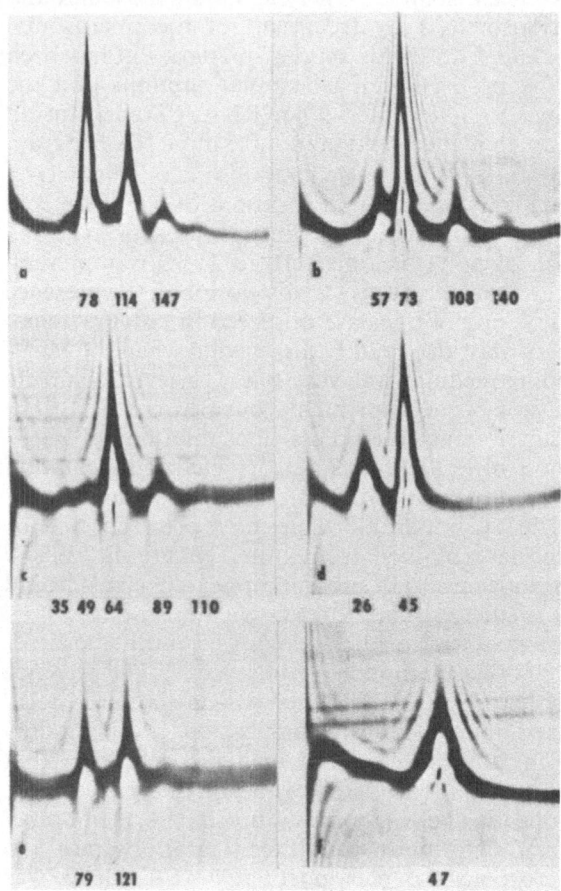

Fig. 4. Sedimentation patterns of cerebral mixed ribosomes. Analyses were carried out in a 12-mm cell in the Spinco model E analytical ultracentrifuge. Temperature of analysis was 20°C except in (c) (13°C) and (d) (6°C). Photographs were obtained using schlieren optics at the following intervals after rotor reached full speed of 29,500 rev/min: (a) 17 min; (b) 14 min; (c) 19 min; (d) 33 min; (e) 17 min; (f) 20 min. Concentrations of ribosomal protein varied from 2 to 3 mg/ml. $S_{20,w}$ values are shown: (a) Analyzed in 4 mM $MgCl_2$, 25 mM KCl and 50 mM Tris–HCl buffer (pH 7.4). (b) Analyzed in 100 mM KCl, 50 mM Tris–HCl buffer (pH 7.4). (c) Dialyzed 16 hr against 1 mM Tris–HCl (pH 7.6); analyzed in the same medium. (d) Analyzed in 50 mM KCl, 1 mM Tris–HCl buffer (pH 7.6), containing 3 μmoles EDTA/mg ribosomal protein. (e) Duplicate of sample (d) dialyzed 5 hr at 4°C against 12 mM $MgCl_2$, 100 mM KCl, 1 mM Tris–HCl buffer (pH 7.6) and analyzed in the same medium. (f) Dialyzed 40 hr at 4°C against 100 mM KCl, 50 mM Tris–HCl buffer (pH 7.4) and analyzed in the same medium. (Reprinted with permission from Zomzely et al.[13])

Available information indicates that many of the physicochemical properties of cerebral monoribosomes and ribosomal subunits are comparable to those of similar entities obtained from other sources.[27] Sedimentation analyses of purified RNA extracted in the presence of phenol from cerebral ribosomes of the adult rat revealed two main peaks.[22,28] These components exhibited sedimentation coefficients of approximately 28S and 18S and were present in the weight ratio of approximately 2:1 as anticipated on the basis of an equal distribution of the 60S and 40S subunits from which the RNA was derived[28] (Fig. 5). A small peak, presumably corresponding to 5S ribosomal RNA and residual 4S transfer RNA, was also present. The base compositions

Fig. 5. Sucrose density-gradient pattern of ribosomal RNA from adult rat cerebral cortex. Ribosomes were prepared by the method of Zomzely *et al.*[24] Ribosomal RNA was extracted in the presence of cold phenol and purified essentially as described by DiGirolamo, Henshaw, and Hiatt.[29] The purified RNA was layered on a linear sucrose density-gradient (20 to 5%) containing 8 mM EDTA, 100 mM NaCl and 100 mM Na acetate (*p*H 5.1). Centrifugation was carried out in a Spinco SW25 rotor at 0°C and 22,500 rev/min for 18 hr. The effluent from the top of the gradient was monitored continuously for absorbance at 254 mμ. (Reprinted with permission from Schneider and Roberts.[28])

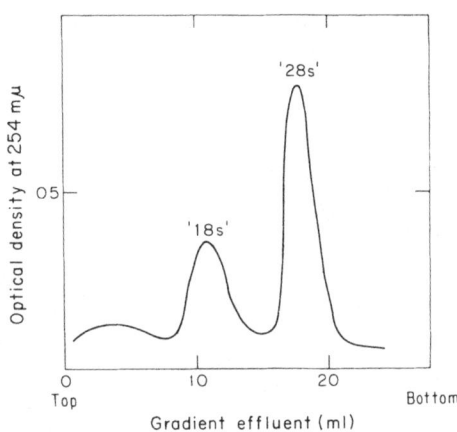

of the fractionated RNA species were similar to those reported for 28*S* and 18*S* ribosomal RNA from other mammalian cells[28] (Table I). In less purified brain ribosomal preparations, including microsomes and polyribosomes, the slow-moving peak of absorbance in gradient profiles may be considerably larger. In addition, a diffuse background of RNA molecules which became rapidly labeled with radioactive orotate has been noted in ribosomal preparations[31,32] (Fig. 6). The species of polyribosomal RNA with template activity in an *E. coli* ribosomal system (presumably messenger RNA) exhibited sedimentation values varying from 7*S* to >28*S*.[32] RNA isolated

TABLE I

Base Compositions of 28*S* and 18*S* Ribosomal RNA from Adult Rat Liver and Cerebral Cortex[a]

	18*S*		28*S*	
	Liver	Brain	Liver	Brain
UMP	22.5 ± 0.2	22.9 ± 0.2	18.8 ± 0.1	19.0 ± 0.3
GMP	31.0 ± 0.6	31.4 ± 0.4	35.5 ± 0.3	36.0 ± 0.5
AMP	20.1 ± 0.3	19.8 ± 0.2	16.8 ± 0.2	16.6 ± 0.3
CMP	26.6 ± 0.4	25.8 ± 0.4	28.9 ± 0.2	28.9 ± 0.4
$\dfrac{GMP + CMP}{AMP + UMP}$	1.35	1.34	1.81	1.82

[a] Reprinted with permission from Schneider and Roberts.[28]
    Each value, expressed as moles %, represents the average ± S.E.M. of six determinations. Ribosomal RNA was isolated, purified and fractionated as outlined in the legend to Fig. 1. Analyses of base compositions were carried out by the method of Katz and Comb.[30]

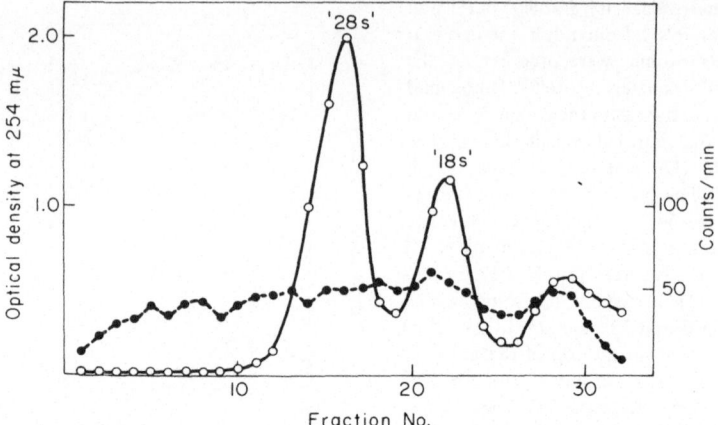

Fig. 6. Sucrose density-gradient pattern of microsomal RNA from adult rabbit brain 30 min after injection of orotic acid-^{14}C into the subarachnoidal space. RNA was extracted, purified, and analyzed essentially as described in the legend to Fig. 5. Centrifugation was carried out at 0°C and 20,000 rev/min for 13 hr. O——O, absorbance; ●–––●, counts/min. (Reprinted with permission from Vesco and Giuditta.[32])

from polyribosomes of cerebral cortical gray matter and hindbrain-medullary white matter of the adult rat exhibited properties of messenger RNA, including DNA-like base composition, high template activity in an homologous ribosomal system, and great sensitivity to ribonuclease (see, e.g., Zomzely et al.[199]). The major species of RNA in these preparations had sedimentation values of approximately 8S and 15S.

Ribosomal RNA in its native state and RNA extracted from brain ribosomes both exhibited similar hyperchromic effects following hydrolysis with alkali or ribonuclease and after treatment with urea.[27] Exposure to EDTA or urea was capable of accentuating ribonuclease activity in cerebral ribosomal preparations. The conclusions were drawn that electrostatic and hydrogen bonds were both involved in maintaining ribosomal stability and that latent ribonucleases were released when cerebral ribosomes were disrupted. Ribosomal ribonucleases exhibited pH optima of 5.4 and 7.9, with higher activity in the acid region. They catalyzed the release principally of 3'-mononucleotides from yeast RNA. These ribonucleases also hydrolyzed transfer RNA and ribosomal RNA. They were inhibited by ions (especially divalent cations) and detergents. The occurrence of alkaline phosphomonoesterases (hydrolyzing both 3'- and 5'-nucleotides) and 5'-phosphodiesterase in ribosomal preparations from goat brain has also been reported.[27] These enzymes were firmly associated with the ribosomal structure since they generally resisted all attempts at removal, including overnight dialysis. The latter finding recalled the observation that the 30S subunit of E. coli ribosomes contained ribonuclease so firmly bound that it behaved as though it were an integral constituent of the ribosomal proteins.[33] The relative instability of the small subunit of cerebral ribosomes suggests that an analogous localization of ribonucleases may occur in this preparation.

Analyses of the amino acid composition of proteins extracted from purified rat cerebral ribosomes revealed an excess of basic residues (including amide N) over acidic amino acids (Table II). Ribosomal protein was extracted by the acetic acid procedure described by Waller and Harris.[34] Amino acid analyses were carried out essentially as described by Roberts.[35] Half-cystine was determined as cysteic acid.[36] Ammonia values were estimated by extrapolation to zero time of hydrolysis and were not included in the total, but were used as a measure of amide nitrogen. These results closely resembled the data obtained for ribosomal proteins derived from other sources.[4] Cerebral microsomal preparations contained relatively more protein of an acidic nature (Table II) and less RNA than ribosomes from the same source.[24,37] Phospholipids and other lipids are also present in microsomal

TABLE II

Amino Acid Composition of Total Protein, Microsomes and Purified Ribosomes from Rat Cerebral Cortex[a]

Amino acid	Total protein[b]	Microsomes[b]	Ribosomes[b]
Lysine	7.20	6.37	9.16
Histidine	2.38	2.44	2.27
Arginine	5.59	6.22	8.22
Aspartic acid	9.29	9.54	8.57
Glutamic acid	11.67	12.54	10.57
Glycine	7.45	7.33	7.22
Alanine	8.02	7.64	7.90
Valine	6.36	6.42	6.69
Leucine	9.24	8.96	8.17
Isoleucine	4.94	4.93	5.13
Threonine	5.33	5.21	4.94
Serine	6.04	5.91	5.10
Proline	5.43	5.30	5.43
Tyrosine	3.12	2.94	2.62
Phenylalanine	4.32	4.09	3.71
Methionine	2.52	2.43	2.06
Half cystine	1.10	1.73	2.24
Ammonia	(7.63)	(9.08)	(9.90)
	100.00	100.00	100.00

[a] From R. Di Bernardo, C. E. Zomzely, and S. Roberts, unpublished data.
[b] Expressed as micromoles per 100 μmoles.

membranes and may serve to stabilize or protect the attached ribosomes.[38] Although the endoplasmic reticulum of neurons may possess unique physico-chemical properties which are related to the peculiar and changeable morphology of the Nissl substance, no evidence is available to substantiate this hypothesis.

2. Cytological and Regional Variations

A large proportion of the total RNA present in all areas of the central nervous system appears to be localized in the microsomal fraction.[39–41] This fact, coupled with analyses of the RNA/DNA ratio in different regions of the central nervous system,[40,42,43] as well as morphological evidence,[44] indicated that the ribosomal population of neurons may be several times that of the glia in specified areas of the adult nervous system. Analyses performed on isolated cells also revealed that the RNA content of neurons was con-siderably higher than that of the associated glia.[45] This ratio may reach 10:1 in areas of large neurons, e.g., Deiters' and hypoglossal nuclei,[9,46] but appears to be only a few times that of the glia in regions where small neurons predominate, e.g., fornix or corpus callosum.[43] In addition, the relative numbers of neurons may vary widely in different areas of the brain cortex; e.g., from 16 % in the motor cortex to 46 % in the calcarine cortex of the adult human.[43]

Base compositions of RNA extracted with the aid of ribonuclease from rabbit hypoglossal neurons and glia supported the conclusion that differences in RNA content of these cells were primarily a function of the ribosomal population.[47] Base ratios (GMP + CMP)/(AMP + UMP) of neurons and glia were 1.3 and 0.8, respectively. Thus, neuronal RNA resembled cytoplasmic RNA more than the corresponding glial fraction, whereas the base composi-tion of the latter was more of the nuclear (DNA) type. Similar, but generally less profound, differences in other areas of the central nervous system have been described by the Göteborg group, employing phenol as well as ribo-nuclease extraction techniques.[45] Conclusions derived from these data may be limited by the occurrence of RNA species which exhibit differential extractability in the presence of phenol or ribonuclease.[48]

Morphologically demonstrable ribosomes have not been found in the distal portions of axons. However, RNA molecules with base compositions and density-gradient profiles characteristic of ribosomal RNA were isolated from the giant axons of Mauthner neurons in the goldfish,[49] myelin-free cranial nerve roots in the cat,[50] and hypoglossal and vagal nerves in the rabbit.[51] Transfer and messenger RNA also appeared to be present in gradient-fractionated extracts of vagal and hypoglossal nerves.[51] The origin of ribosomallike RNA in the axon has variously been attributed to axoplasmic flow, migration from the supporting cells, and local synthesis.[52] Evidence which points to local synthesis of axonal protein raises the possibility that this RNA may be associated with protein in a functional ribosomal state *in situ* (Section III,B,2).

3. Developmental Changes

Development of the central nervous system is associated with striking alterations in relative proportions of neurons and glia and in the morphological characteristics of these cells. Thus, the glial population which was very small in the cortex of immature rats far exceeded the neuronal complement in adult animals.[53] Concomitantly, maturing neurons exhibited a marked outgrowth of cytoplasmic processes. Ribosomal transformations accompanied these phenomena. Ribosomes probably occur in a dispersed form in cytoplasm of the embryonic neuron (including dendritic processes), but then tend to accumulate in the perikaryal region during early development.[54] This migration may be correlated with alterations in the relative protein-synthesizing capacities of the cell body and its extensions. Studies carried out on isolated hippocampal neurons from rat brain revealed that the total RNA content per cell increased during development and then declined in old age.[55] During the period of active RNA synthesis, the $(GMP + CMP)/(AMP + UMP)$ ratio rose about 25% (from 1.33 to 1.66). These findings suggested that neuronal formation of cytoplasmic RNA, relative to that of DNAlike RNA, was increased during development. During maturation, ribosomal and microsomal RNA in rat cerebral cortex increased up to about 18 days of age and then declined[41] (Fig. 7). Since the cytosine–guanine ratio of RNA extracted from these organelles was greater in adult animals than in 4-day-old rats, the suggestion was advanced that a selective loss of certain cerebral ribosomes may have occurred. If this finding is valid, these ribosomes

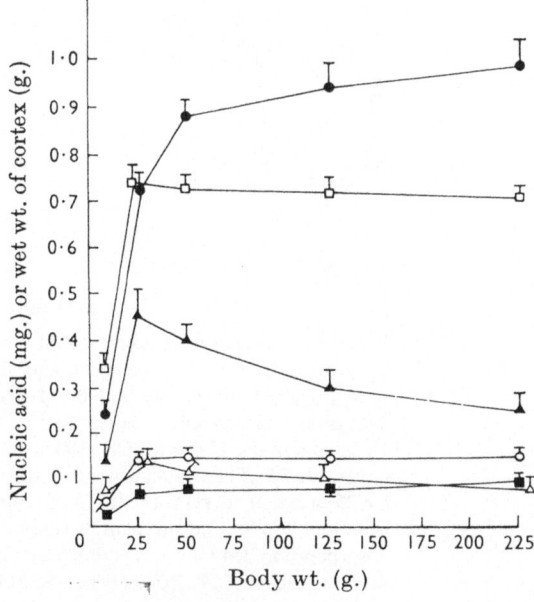

Fig. 7. RNA and DNA content of subcellular fractions of rat cerebral cortex during development. Each point represents the average ± S.E.M. of six to eight determinations, each on one whole cortex. ●, wet wt (g); □, DNA (mg); ▲, ribosomal RNA (mg); △, microsomal RNA (mg); ○, nuclear RNA (mg); ■, transfer RNA (mg). (Reprinted with permission from Adams.[41])

may be concerned with the synthesis of proteins with special significance for developing cerebral cells; e.g., proteins involved in cell division.

Evidence has been obtained which indicates that alterations in the proportions of diverse messenger RNA molecules associated with cerebral ribosomes occur with aging. A higher proportion of relatively heavy ribosomal aggregates was found in polyribosomal preparations isolated from brain[17] and cerebral cortex[56] (Fig. 8) of newborn rats compared to adult animals. This pattern may derive from the greater stability of certain of the larger messenger RNA-ribosome complexes in the immature brain,[56] as well as the more rapid synthesis of larger species of messenger RNA. Thus, the turnover rate of hybridizable RNA in brain cytoplasm (presumably messenger RNA) was higher in the immature rat than in the adult animal, even though the proportion of the total RNA which hybridized was similar in both age groups[57] (Sections II,C,1 and III,A,3). Differences in turnover rates of hybridizable RNA in whole brain may reflect the greater neuronal contribution to overall formation of messenger RNA in the immature brain compared to the adult, as well as developmental differences in the properties of these molecules synthesized by both neurons and glia. Transfer RNA in cerebral

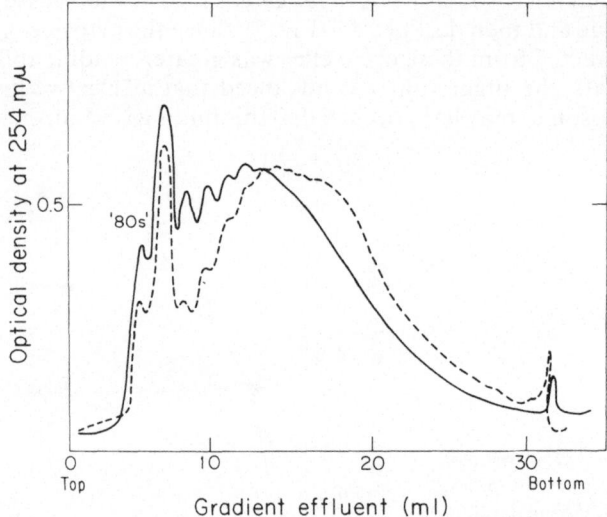

Fig. 8. Sucrose density-gradient profiles of polyribosomes isolated from cerebral cortices of newborn and adult rats. Polyribosomes were prepared by the method of Munro et al.,[15] but without detergent, and layered on linear sucrose density gradients (35 to 15%) containing 12 mM $MgCl_2$, 100 mM KCl and 50 mM Tris–HCl buffer (pH 7.6). The gradients were centrifuged for 1.5 hr in a Spinco SW25 swinging bucket rotor at 25,000 rev/min and 0°C. The effluent from the top of the gradient was monitored continuously for absorbance at 254 mμ. ____, adult cerebral cortex; _ _ _, newborn cerebral cortex. (C. E. Zomzely and S. Roberts, unpublished data.)

tissue, like ribosomal RNA, exhibited an early increase during development but not a later decline[41] (Fig. 7). The various amino acid-activating enzymes in acetone powders prepared from white matter of guinea pig brain generally revealed only minor alterations in activity with age.[58] However, the cysteine-activating enzyme could not be detected in cerebral white matter of animals which were nine days of age or older, and the activity of the glutamate-activating enzyme increased with age in the guinea pig.

Changes which occurred in cytoplasmic RNA of neurons following nerve injury[9] were somewhat similar to those observed during development. Within 24 hr after the hypoglossal nerve of the rabbit had been crushed, dispersion of cytoplasmic RNA-containing aggregates had occurred. This chromatolytic process was followed within a few days by axonal outgrowth. During the subsequent period of axonal maturation, a transient but marked enhancement in total RNA content of the nerve occurred and was accompanied by reassembly of the RNA-containing particles to give the normal Nissl structure. These regenerative changes may have been associated with variations in ribosomal aggregation. Alterations in base compositions of RNA extracted from the Mauthner nerve axon and its myelin sheath have been described following spinal cord transection in the goldfish.[59] The major change was a transient increase in the adenine–guanine ratio. Stimulation of total RNA synthesis, without any obvious change in properties of the RNA, has been noted within a few hours after addition of nerve growth factor to tissue cultures of chick embryo ganglia.[60,61]

During development, protein-synthesizing systems from mammalian tissues other than brain revealed patterns which were at variance with those outlined above. For example, microsomal protein of liver, which was lower than that of brain in the immature rat, rose to about four times the cerebral value (on a DNA basis) in the adult.[62] Unlike the situation in brain, the proportion of polyribosomes in hepatic ribosomal preparations was greater in the adult rat than in the newborn animal.[17]

B. Physical and Chemical Properties of Mitochondrial and Synaptosomal Systems

1. General Aspects

Following the parenteral administration of isotopically labeled amino acids to experimental animals, radioactivity appeared most rapidly in the microsomal fraction of most tissues. However, in certain tissues (e.g., muscle) the mitochondrial fraction may become labeled first[63] and, in others (e.g., immature brain), radioactivity may appear almost as rapidly in this fraction as in the microsomes.[64] Mitochondria probably contribute to the overall rate of protein synthesis to varying extents in different tissues. In nervous tissue, mitochondrial protein synthesis may be relatively high, especially during early development.

Significant quantities of DNA and RNA were demonstrated in mitochondrial fractions from a number of cell types, including brain cells. The

RNA/protein ratio of purified mitochondria obtained from either cell bodies or synaptosomes of rat cerebral cortex[65,66] was about one-third that of microsomes from this source (Table III). The membrane fractions from osmotically shocked synaptosomes contained even more RNA (on a protein basis) than the mitochondria. DNA was also present in cerebral mitochondria in significant concentration. Mitochondrial DNA occurred as circular filaments about 5 μ in length in cells of most or all tissues, including brain.[67]

TABLE III

Nucleic Acid Content of Subcellular Fractions from Rat Cerebral Cortex[a]

Fraction	RNA[b]	DNA[b]
Nuclei	30	77
Microsomes	40	0
Soluble	20	0
Crude mitochondrial fraction	16	3.0
Cell body mitochondria	12	6.3
Synaptosomal fraction	12	0.4
Vesicles	—	—
Membranes and ghosts	14–21	0–0.2
Synaptosomes	11	0.2
Mitochondria	13	6.3

[a] Reprinted in modified version with permission from Austin and Morgan.[65]
[b] Values given as micrograms/milligram protein.

At least in certain cells, the density of mitochondrial DNA differed from that of nuclear DNA, resulting in its appearance as a "satellite" band in sedimentation patterns obtained after centrifugation in CsCl density-gradients.[68] However, mitochondrial DNA in mouse brain exhibited the same buoyant density (1.701) as the major nuclear DNA species.[69] Nevertheless, differences between mitochondrial DNA and nuclear DNA in brain were indicated by the finding that the former preparation renatured more readily after heating and was presumably less heterogeneous. In addition, hybridization experiments revealed that base sequences of these two DNA species were different.[70] These studies further suggested that mitochondria from mouse brain contained significant quantities of messenger RNA. However, a considerable portion of this RNA appeared to be coded for by nuclear DNA. The relative homogeneity of cerebral mitochondrial DNA also indicated that its informational content for messenger RNA synthesis must be quite limited.

Additional components of mitochondrial protein-synthesizing systems have not yet been isolated from cells of the nervous system. However, purified mitochondria from other cells have been shown to contain ribosomes, transfer RNA, amino acid-activating enzymes, RNA polymerase, and DNA polymerase.[71] Certain of these materials may have properties different from those of microsomal protein-synthesizing systems. For example, ribosomes isolated from rat liver mitochondria exhibited a sedimentation coefficient of only 55S and an average diameter of 145 Å.[72] These ribosomes were active in amino acid incorporation. Relatively low sedimentation values were reported for one or more of the ribosomal RNA species associated with mammalian mitochondria.[73] The lighter of the two major species (17S) was less methylated than its microsomal counterpart (18S) and may have originated within the mitochondrion. Finally, the (GMP + CMP)/(AMP + UMP) ratio of cerebral mitochondrial RNA may be slightly lower than that of ribosomal RNA.[66,74]

2. *Cytological and Regional Variations*

Mitochondrial structure and function vary significantly in different cell types, within different parts of the same cell, and in different areas of the nervous system.[5,7,8] Mitochondria of oligodendroglia and Schwann cells may be selectively involved in synthesis of the major proteins of the myelin sheath (proteolipids), whereas mitochondria or other membranous structures of neuronal synaptosomes may participate in formation of axolemmal or synaptic membrane proteins, as well as polypeptides which function in neurotransmission (Section III,D). However, functional variations of this nature have not yet been demonstrated in diverse mitochondrial populations.

3. *Developmental Changes*

The population of mitochondria in rat brain cells increased rapidly during development, reaching a value at 21 days of age which was four times that at 1 day.[75] Concomitantly, mitochondrial protein per gram brain was elevated almost 75%. This increase accounted for two-thirds of the total elevation in brain protein over the same time period. Brain cells contained twice as many mitochondria as liver cells in the adult rat.[76] These observations support the concept that mitochondrial synthesis of protein may be of special significance in the nervous system.

C. Physical and Chemical Properties of Nuclear Systems

1. *General Aspects*

In large measure, protein synthesis in all eukaryotic cells is directed from the nucleus. Thus, the primary site of RNA synthesis is on DNA templates in this organelle.[77] Two types of nuclear DNA have been reported. The major component (chromatin), which is the more heterogeneous, is formed in the nucleus proper and serves as a template for the synthesis of most of the cytoplasmic messenger RNA. Transfer RNA may also be formed in the nucleus proper.[78] The minor or "satellite" nuclear DNA component has a slightly different density and is presumed to arise in the nucleolus.[79] Nucleolar DNA codes for the synthesis of ribosomal RNA.[80] The properties of cerebral nuclear DNA fit into this pattern. Thus, DNA isolated from mouse brain nuclei exhibited major and minor components with densities of 1.702 and 1.691, respectively.[69] Total nuclear DNA revealed poor renaturation properties in its melting profile, indicating considerable heterogeneity.

The multienzyme systems responsible for nucleotide polymerization and cyclization are found in highest concentrations in the nucleus. Some properties of cerebral RNA polymerases have been described.[81–83] RNA isolated from cerebral nuclei by phenol extraction procedures revealed characteristics which were distinct from those of the analogous cytoplasmic preparation and provided direct evidence for the formation of relatively large amounts of messenger RNA in the brain of adult as well as immature animals.[57,84–86] These distinguishing characteristics included more rapid turnover rate, base composition more closely approaching that of DNA, greater template activity in ribosomal amino acid-incorporating systems, more heterogeneous sedimentation behavior, and greater capacity for hybridization with homologous DNA.

RNA synthesis in rat brain appeared to be directed toward the formation of messenger RNA to a greater extent than in liver[57] (Table IV). This conclusion was consistent with the findings that RNA polymerase activity was higher in brain than in liver of the adult rabbit and monkey.[82,83] Cerebral nuclei may produce an unusually high proportion of messenger RNA species of relatively small molecular size. Thus, an appreciable fraction

TABLE IV

Hybridizable RNA in Rat Brain and Liver[a]

Tissue	Hybridizable RNA[b]	
	Nucleus	Cytoplasm
Adult brain	2.1	0.65
Adult liver	1.2	0.48
Newborn brain	2.2	0.88
Newborn liver	1.5	0.40

[a] Reprinted with permission from Bondy and Roberts.[57]
[b] Values are expressed as percent of total radioactivity hybridizing to homologous DNA. RNA was labeled by exposure to dimethylsulfate-^{3}H. The labeled RNA (8–12 μg) was incubated for 18 hr with 25–40 μg of denatured rat liver DNA. The average of 3 separate experiments is shown.

of the total template activity in cerebral nuclear RNA preparations was associated with species of molecular size equivalent to 16–18S.[84,86] In more recent investigations, evidence has been obtained that even smaller messenger RNA molecules (8-12S) are present in cerebral nuclei and cytoplasm[32,87] (Zomzely *et al.*[199]). Moreover, cerebral polyribosomal preparations contained a higher proportion of relatively small aggregates than similar preparations from other mammalian tissues.[12] In certain instances, at least, polyribosome size appears to be related to the length of attached messenger RNA.[88] The prominence of small messenger RNA molecules in cerebral preparations may be partially a consequence of unique lability of certain of the larger polyribosomes present.[12]

2. *Cytological and Regional Variations*

Several investigations have dealt with gross variations in DNA and RNA concentrations in different areas of the nervous system.[40,42,43] Relatively high levels of these nucleic acids were found in cerebral and cerebellar cortices compared to those measured in white matter and in lower regions of the neuroaxis. These variations reflected differences in the absolute and relative proportions of neuronal and glial nuclei and their activity in nucleic acid synthesis. Consistent with these data was the finding that RNA polymerase activity was greater in cerebral cortex than in white matter or lower brain areas.[81–83]

Experiments on separated cells revealed that turnover of RNA molecules may be more active in glia obtained from rabbit hypoglossal tissue[47] or brain cortex[89] than in neurons from the same sources. This conclusion was consistent with data from earlier autoradiographic studies.[90] However, in view of the much greater quantity of RNA in the neuron,[9,46] its overall rate of synthesis was undoubtedly higher in this cell type. The base compositions of total neuronal and glial RNA[47] suggested that relatively more of the RNA synthesized by glia was of the nuclear type, compared to the neuron (see also Ref. 91).

3. *Developmental Changes*

DNA content of rat cerebral cortex appeared to reach a maximum during the period of neuronal maturation and active myelination[41] (Fig. 7). However, even higher levels of DNA have been recorded in brain of the newborn rat.[92] Variations in the association of cerebral DNA and nuclear protein may occur with aging. Thus, the ratio of protein to DNA in nucleoprotein preparations from mouse brain was reported to decrease markedly from 3 days of age to 6 months.[93] Concomitantly, the mean temperature of denaturation of the nucleoprotein also declined. These findings suggested that stabilization of the DNA helix may vary with age and be reflected in alterations in RNA synthesis. Several observations indicate that RNA formation is more active in cerebral nuclei of young animals than in adults.

Total nuclear RNA in rat cerebral cortex reached a maximum at 18 days of age and then plateaued[41] (Fig. 7). During the latter time period, RNA species with sedimentation constants approximating 17S and 27S gradually decreased in nuclear preparations from rat brain.[85] Simultaneously, the concentration of 5S RNA molecules increased. These findings, suggesting a decline in messenger (and ribosomal) RNA synthesis with age, were supported by the observations that template activity[85] and messenger RNA turnover[57] were higher in cerebral nuclear RNA from the newborn rat than the adult (Section III,A,3). However, the proportions of hybridizable RNA in nuclear and cytoplasmic RNA preparations from rat brain were not greatly altered during development[57] (Table IV). Moreover, the apparent changes in distribution of nuclear RNA species during development were not accompanied by alterations in base composition.[41] Cerebral RNA polymerase ("aggregate-enzyme") obtained from disrupted crude nuclei was more active in preparations from young rats (12 days) than adults.[81] In contrast, the opposite picture was obtained when RNA polymerase activity was assayed directly in purified preparations of cerebral nuclei from newborn and adult rats.[83] The physiological significance of these RNA polymerase determinations may have been limited by the presence of ribonuclease and other contaminants in the nuclear preparations and the use of high salt concentrations to enhance enzyme activity.

III. MECHANISMS OF PROTEIN SYNTHESIS

A. Protein Synthesis in Microsomal and Ribosomal Systems

1. General Aspects

Certain distinctive features have been noted in the protein-synthesizing characteristics of microsomal and ribosomal preparations from nervous tissue. These may be related to unique aspects of ribosomal organization in this tissue, as well as to specialized properties of the attached messenger RNA molecules. The relative roles of genetic and environmental factors (including brain-barrier mechanisms) in the establishment of these differences are unknown.

Microsomal preparations from various tissues generally exhibit lower amino acid-incorporating activity, expressed as specific activity of the labeled proteins, than the corresponding ribosomal fractions. However, when gross contamination of the microsomal fraction with membrane proteins was taken into account, incorporation of uniformly labeled L-leucine-^{14}C by rat cerebral microsomes, on an RNA basis, was found to be twice that of the analogous "mixed ribosome" preparation during the first 30 minutes of incubation.[94] Initial incorporating activity of the microsomal fraction was comparable to that observed with a highly active polyribosomal preparation from the same source.[12] These findings were consistent with the observation that ribosomes bound to the cerebral microsomal membrane were largely aggregated.[95] Nevertheless, incorporation of radioactive amino acids into protein of cerebral microsomal fractions plateaued after 15 to 30 min of incubation,[24,62,96,97] whereas the analogous ribosomal systems continued to incorporate actively for 60 to 90 min.[12,18,21,24,98] The apparent loss of incorporating activity in the microsomal systems during prolonged incubation may be related to release of certain fatty acids or other inhibitory materials from the microsomal membrane[99] or to dilution of the radioactive amino acid pool as a result of breakdown of membrane protein during incubation. The existence of a substantial amino acid pool in incubation mixtures containing cerebral microsomes was supported by the observation that these preparations were not stimulated by addition of amino acid mixtures.[24,62,96,100] Certain of these mixtures enhanced amino acid incorporation in brain ribosomal systems (see below).

Basic requirements for protein synthesis in microsomal and ribosomal systems from nervous tissue were similar to those for comparable preparations from other cell types.[11,21,24,62,96,98,101–105] These requirements included a source of amino acid-activating enzymes (synthetases), transfer RNA molecules, and amino acid transferases, usually provided as crude microsomal supernatant or as the fraction precipitated from this supernatant at pH 5. ATP, GTP, and appropriate concentrations of certain monobasic and dibasic ions were also necessary for optimal activity. Amino acids incorporated were found in peptide linkages within the interior of the polypeptide molecules formed.[96]

Microsomal and ribosomal systems from brain tissue were unusually sensitive to variations in factors essential for optimal protein synthesis.[94] Amino acid-incorporating activity in these systems exhibited striking alterations in response to changes in concentration of microsomal supernatant or pH 5 enzymes.[21,24] In contrast, analogous hepatic systems were saturated with low ratios of the activating fraction to ribonucleoprotein particles. To some extent these results may be due to a deficiency of transferases in cerebral ribosomal systems.[104] However, Murthy[106] reported that amino acid-incorporating activity of brain ribosomes was not increased by addition of the supernatant remaining after precipitation of pH 5 enzymes from cerebral or hepatic postmicrosomal preparations. These supernatants were rich in transferase enzymes.[104] Moreover, hepatic pH 5 enzymes were less effective with cerebral ribosomes than the corresponding cerebral preparation, even though both pH 5 fractions were equally active with hepatic ribosomes.[106] Interpretation of these results is complicated by the fact that messenger RNA molecules attached to cerebral ribosomes are uniquely sensitive to the degradative actions of ribonuclease.[12] Hepatic subcellular fractions may contain higher concentrations of endogenous nucleases than the corresponding cerebral preparations.[12]

The requirement of ATP for amino acid activation in cell-free protein-synthesizing systems can usually be satisfied by provision of adequate concentrations of preformed cofactor. However, cerebral microsomal and ribosomal preparations were dependent upon the presence of an ATP-generating system for optimal activity.[24] An excess of preformed cofactor markedly inhibited the activity of cerebral ribosomes, but had little effect on the analogous hepatic system.

Ribosomal systems from a variety of cells generally require the addition of an amino acid mixture for optimal amino acid-incorporating activity. Contradictory data have been presented for cerebral ribosomes in this regard. Thus, when L-leucine was present at a saturating level (0.16 mM), its incorporation into protein of rat cerebral ribosomes was significantly inhibited[24] by the simultaneous presence of a mixture of 20 nonradioactive amino acids in proportions based on the free amino acid composition of rat cerebral cortex.[35] Other investigators, using different conditions, have variously observed inhibition or stimulation of incorporation of one amino acid in the presence of others.[11,21,98,100,103,105,107] To some extent, these variations may have been due to differences in pool sizes of amino acids. Nevertheless, the data suggest that cerebral protein-synthesizing systems are unusually sensitive to relatively slight variations in the amino acid composition of the medium. Thus, even in those experiments where certain amino acid mixtures stimulated amino acid incorporation, other combinations were inhibitory.[11] A significant portion of the stimulatory activity was attributable to dicarboxylic amino acids and certain other amino acids[11,105] which are present in the free amino acid pool of the brain in high concentrations.[35] The postulated inhibitory neurotransmittor, γ-aminobutyric acid, had unique stimulatory effects upon cerebral protein-synthesizing systems.[11,103] Stimulatory and inhibitory effects of certain amino acids on incorporation of others into protein have also been described for brain slices in vitro and cerebral tissue in vivo.[94,107–114]

Protein-synthesizing activities of brain microsomal and ribosomal preparations were markedly affected by ion concentrations. Endogenous amino acid-incorporating activity was maximal in the presence of 10–12 mM-Mg^{2+} and 100 mM-K^+.[24] NH_4^+, Na^+, and K^+ could be substituted for one another provided that the total concentration of these ions remained constant.[11,103,115] Analogous systems from other sources generally exhibited maximal amino acid-incorporating activity in the presence of lower ion levels and were less sensitive to relatively minor alterations in these levels. High concentrations of Mg^{2+} appeared to maintain polyribosomal structure in the cerebral systems in part by inhibiting ribonuclease action.[12] Even higher levels of Mg^{2+} were required for maximal stimulation of amino acid incorporation by preparations of cerebral nuclear RNA[87] or by polyuridylic acid.[12] In these instances, Mg^{2+} presumably exerted its effects by facilitating attachment of messenger RNA molecules to ribosomes.[116]

In common with preparations from other sources,[117,118] the response of cerebral ribosomes to exogenous messenger RNA was also favored by conditions which resulted in depletion of endogenous messenger RNA and appearance of ribosomal monomers and subunits to which added messenger molecules could readily bind.[23] Cerebral ribosomes obtained from adult rats were unusually responsive to polyuridylic acid compared to similar preparations from other mammalian tissues.[12,17] This responsiveness apparently resulted from the unique sensitivity of endogenous messenger RNA molecules in the cerebral system to the action of nucleases.[12] Nuclear RNA prepared from rodent brains stimulated amino acid incorporation in ribosomal systems.[84–86] The maximum degree of stimulation in homologous ribosomal systems was about threefold[87] (Table V). Cytoplasmic RNA fractions were generally quite inactive. RNA isolated from purified polyribosomes of rat brain exhibited high template activity in the presence of 80S ribosomes from the same source (Zomzely et al.[199]).

TABLE V

Stimulation of Amino Acid Incorporation into Cerebral Ribosomal Protein by Cerebral RNA Preparations[a]

Addition	Amount (μg)	Incorporation (cpm/mg protein)	Stimulation (%)
None	—	5,080	—
Ribosomal RNA	75	5,350	5
	150	5,310	5
Nuclear RNA	75	6,850	35
None	—	6,200	—
Nuclear RNA	34	11,590	87
	68	13,700	121
	102	15,900	157
	136	17,760	187

[a] Reprinted with permission, in modified form, from Samli and Roberts.[87]
 Each sample contained 0.06 mg of ribosomal protein and 0.4 mg of pH 5 enzyme protein, prepared from rat cerebral cortex, in a final volume of 0.1 ml. The RNA preparations were extracted at 65°C in the presence of phenol. The radioactivity incorporated was derived from a mixture (0.02 μC) of uniformly labeled ^{14}C-L-amino acids (Nuclear-Chicago) which was added after a preincubation period of 10 min. Incubation was carried out for 30 min at 37°C.

The protein-synthesizing process in nervous tissue exhibited certain other unique characteristics. "Mixed ribosomes" from rat cerebral cortex initially incorporated amino acids more slowly than the analogous hepatic preparation, but the maximal levels attained were comparable.[24] The cerebral system released only about half as much protein to the incubation medium as the hepatic system. These differences may have been related to the fact that the cerebral system contained a smaller proportion of polyribosomes and these disaggregated more readily during incubation. Comparable results were obtained when amino acid-incorporating activities of polyribosomes from rat cerebral cortex and liver were investigated in vivo and in vitro.[12] For example, radioactivity from injected ^{14}C-amino acids was initially incorporated most actively into the larger polyribosomes (> tetramer) found in postmitochondrial supernatants of either cerebral or hepatic tissue from the rat (Fig. 9). However, radioactivity in the smaller cerebral aggregates was also relatively high and increased markedly within 15 min after injection of labeled precursor, presumably as a consequence of rapid breakdown of the larger polyribosomes. Little breakdown of hepatic polyribosomes occurred. These results also suggested that cerebral cells normally contain a high

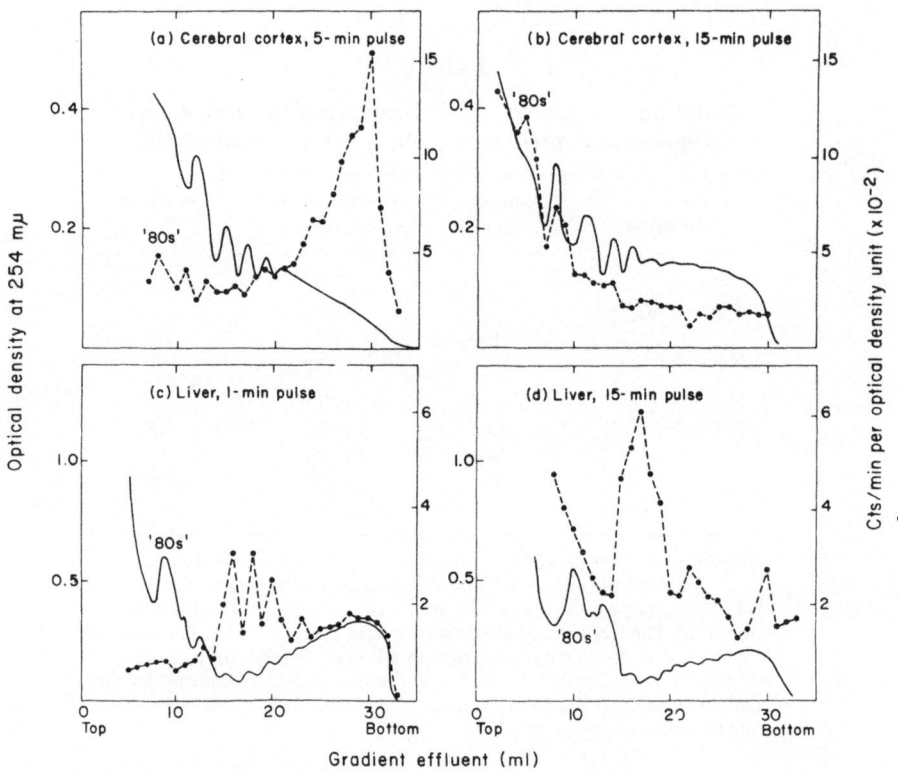

Fig. 9. Protein-synthesizing activities of cerebral and hepatic polyribosomes *in vivo*: relationship to state of aggregation. Cerebral postmitochondrial supernatants were prepared from rats injected intracisternally with uniformly labeled L-leucine-^{14}C (10 μC) 5 or 15 min before autopsy. Hepatic postmitochondrial supernatants were prepared from rats injected via the portal vein with 20 μC of leucine-^{14}C 1 or 15 min earlier. These supernatants were prepared in 12 mM MgCl$_2$, 100 mM KCl, 50 mM Tris–HCl buffer (pH 7.6) and layered on a linear sucrose density-gradient (35 to 15%) containing the same materials. The gradients were centrifuged for 4 hr in a Spinco SW25 rotor at 25,000 rev/min and 0°C. The effluents from the top of the gradient were monitored continuously for absorbance at 254 mμ and collected as 1-ml samples for determination of radioactivity. ——, optical density; –•––•–, radioactivity. (Reprinted with permission from Zomzely et al.[12])

proportion of relatively small messenger RNA molecules. These may be active in the formation of polypeptides which serve specialized functions in the brain.

Effects of a wide variety of physiological and pharmacological agents on cerebral protein synthesis have been investigated. *In vitro*, low concentrations of pancreatic ribonuclease completely inhibited amino acid incorporation by cerebral microsomes or ribosomes[96,98,100,106] and resulted in virtually complete disruption of cerebral polyribosomes.[13] A much lesser effect on ribosomal aggregation was produced in analogous hepatic systems. Puromycin also markedly inhibited amino acid incorporation.[11,98,100,103] Actinomycin D, chloramphenicol, DNase, and cycloheximide had relatively little effect on endogenous incorporation.[11,103,105] Differential actions of various inhibitors on endogenous and poly U-stimulated incorporation of phenylalanine by rat cerebral ribosomes have been described.[106] The effects of a wide variety of other pharmacological agents on incorporation of amino acids by cerebral ribosomes *in vitro* have also been investigated.[103,105,106] The physiological significance of these experiments is uncertain.

2. Cytological and Regional Variations

The majority of studies *in vivo* of regional variations in protein synthesis in the nervous system are difficult to interpret. Incorporation of isotope into protein by different areas of the nervous system after parenteral administration of labeled amino acids depends upon other factors in addition to the rate of protein synthesis. These include blood supply to the area, ease of passage of the amino acid through various cellular barriers, size and rate of change of the precursor pool, and elapsed time after administration. The time factor is important not only because of the existence of proteins with widely different turnover rates,[119] but also because of the movement of preformed proteins from one subcellular site to another, e.g., from ribosome to mitochondrion, from perikaryal region to distal axon. Attempts to relate protein synthesis in different areas of the nervous system to their relative contents of neurons and glia are generally complicated by the fact that those areas in which glia are abundant are also rich in axonal elements. However, autoradiographic studies suggested that neurons, oligodendrocytes, and Schwann cells of the mammalian neuroaxis incorporated radioactive precursors into nucleic acids and proteins, whereas astrocytes, microglia, and satellite cells were normally inactive.[90]

Axoplasmic flow undoubtedly carries preformed proteins and nucleic acids from their site of origin in the neuronal cell body to the distal portions of the axon.[52,102,120–124] These substances may also enter the axon from surrounding Schwann cells[59,125] and may be formed locally as well.[52,123,126–131] In direct support of the latter conclusion, sciatic and spinal nerves *in vitro* incorporated radioactive amino acids into protein, particularly in the caudal portions of axon and myelin sheath.[123,129,132,133] Incorporation was stimulated two- to threefold within a few days after sectioning or crushing the nerve[132] and was markedly inhibited by the addition *in vitro* of puromycin.[123] Conflicting results were obtained with actinomycin D[129,133] and no inhibition was seen with chloramphenicol.[129] Certain of these characteristics of axonal protein synthesis suggested the intervention of nonribosomal mechanisms (Section III,B,2).

Following systemic administration of radioactive amino acids, proteins in higher (cortical) areas of the central nervous system generally incorporated label earlier than white matter.[1,2,134] However, the reverse situation has been noted after intracisternal administration of isotopic leucine.[56] Localized increases in amino acid incorporation were produced by different types of neural stimulation, e.g., in ventral horn motor neurons, motor cortex, and visual cortex of exercised rats,[135] in occipital cortex of dark-adapted rabbits after chronic exposure to light,[136] in hippocampal pyramidal neurons contralateral to the used paw of rats induced to transfer handedness,[137] and in cortex, caudate nucleus, and white matter of rat brain in the presence of elevated circulating levels of Na^+.[138] Selective enhancement of protein synthesis in localized areas of the central nervous system was also suggested by changes in RNA content and base ratios in response to specific stimuli.[45] Differential effects on neuronal and glial elements were noted in the latter experiments.

Conflicting data have been reported which deal with the relative capacities of cellular and subcellular preparations from different areas of the brain to incorporate radioactive amino acids into protein. Johnson and Luttges[139] found that cell suspensions from cerebral cortex of young mice (5 to 11 days old) were slightly more active in this process than the analogous preparations from cerebellar or basal areas. In a comparable fashion, slices of white matter from adult guinea pig brain were only 80 % as active as gray matter in the uptake of glycine-^{14}C into protein.[112] Microsomal fractions derived from cerebral gray matter incorporated DL-lysine-1-^{14}C into protein more rapidly than the analogous preparations from white matter in the adult rat but not in the immature (12-day) animal.[140] The difference in the adult brain appeared to be located in both the microsomal fraction *per se* and the enzyme fraction (postmicrosomal supernatant). However, these results may be explained in part by the fact that microsomal preparations from cerebral white matter were more highly contaminated with membranous structures than the corresponding materials from gray matter, particularly in the adult animal. Purified polyribosomes from cerebral cortical gray matter and hindbrain-medullary white matter exhibited comparable activities in amino acid incorporation in immature as well as adult rats.[56] Although the highest activities of amino acid-activating enzymes were reported in cerebral cortex and other areas of adult rabbit brain which contained a high proportion of neurons,[141] conflicting data have also been obtained.[56]

The turnover rates of messenger RNA molecules undoubtedly vary in different parts of the nervous system. On the basis of experiments with actinomycin D, administered *in vivo* or added to brain slices *in vitro*, Appel[142] calculated that messenger RNA molecules in rat cerebral cortex all possessed half-lives of less than 10-12 hr. The majority of messenger RNA species in the "subcortex" were reported to have shorter half-lives (about 4 hr); however, the existence of molecules with longer half-lives (> 20 hr) also seemed indicated. In contrast, Orrego and Lipmann[114] in similar experiments with slices of whole rat brain reported an average half-life of approximately $2\frac{1}{2}$ hr for all cerebral messenger RNA molecules. Data obtained with actinomycin D in cellular systems are highly suspect due to the demonstrated toxicity of this antibiotic under such conditions.[143,144] More direct studies of messenger RNA turnover in rat brain, using hybridization techniques with radioactive RNA extracted from this organ, suggested the presence of a very heterogeneous population with half-lives varying from a few hours to a

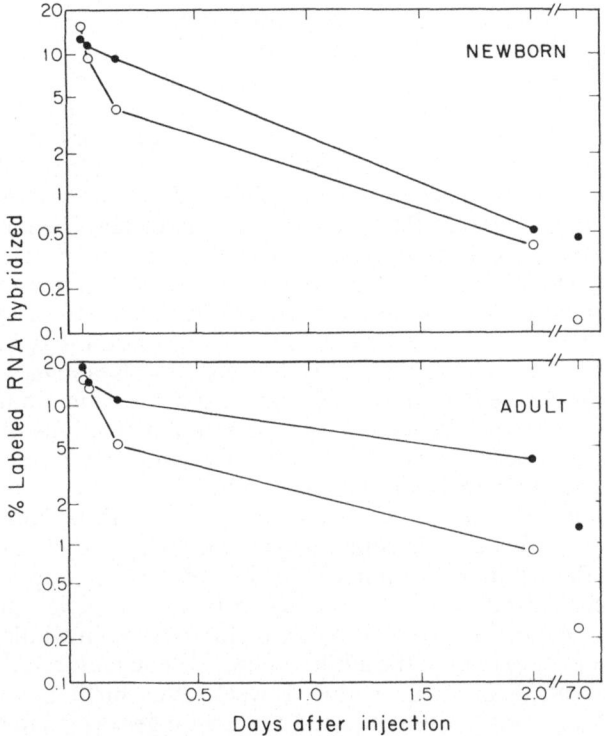

Fig. 10. Hybridization capacity of radioactive nuclear and cytoplasmic RNA from newborn and adult rat brain. The cell fractions were obtained at varying intervals after intracervical injection of uridine-5-^{3}H. RNA was extracted in the presence of phenol, purified, and incubated for 18 hr with denatured rat liver DNA immobilized on nitrocellulose filters. The radioactivity bound was measured by scintillation counting. ●, Nuclear RNA; ○, cytoplasmic RNA. (Reprinted with permission from Bondy and Roberts.[57])

few days[57] (Fig. 10). Differentiation on the basis of brain area has not yet been attempted with this technique.

3. Developmental Aspects

Although numerous investigations support the conclusion that overall protein synthesis in the brain is most active during early development of this organ, conflicting data have also been obtained. Moreover, a clear picture has not yet emerged regarding the relative contributions of neurons and glia and of different subcellular organelles to this phenomenon. At a gross level, Gaitonde and Richter[145] reported that incorporation of intracisternally-administered methionine-^{35}S into brain proteins decreased with age in the rat. This process appeared to be associated with the appearance of proteins which exhibited relatively slow turnover rates.[146] However, when the radio-activity of the amino acid precursor pool was taken into account in similar experiments with leucine-^{3}H, no evidence was obtained for a developmental decrease in cerebral protein synthesis *in vivo*.[56] Soon after birth, a sharp drop in incorporation of L-phenylalanine-^{14}C into cerebral cortical proteins of the miniature piglet has been noted.[147] This phenomenon was followed by a more gradual decrease for the first day or so thereafter. The initial dramatic decline in amino acid incorporation also occurred with premature birth effected by hysterectomy. Since the total concentration of free amino acids in the brain tended to decrease after birth, and was accompanied by an increase in their specific activity, the alteration in incorporation may have reflected a true decrease in protein turnover. In addition, these data suggested that some factor in the uterine environment was responsible for maintaining a high rate of protein synthesis in the fetal cortex. Subcellular fractions obtained from brain tissue after administration of radioactive amino acids *in vivo* revealed age-associated differences in labeling. Thus, 19 hours after parenteral administration of L-serine-^{14}C specific activities of all cerebral subcellular fractions (except the fraction containing myelin) were higher in the 3-day-old rat than in the 12-day-old animal.[148] Similarly, at varying intervals after intraperitoneal injection of tyrosine-^{3}H, proteins of nuclei, mitochondria, microsomes, and cell sap were all more highly labeled in whole brain of the newborn rat (1 day) than in the adult animal.[64] The major decline in amino acid incorporation occurred within 1 week after birth in mitochondria, microsomes, and cell sap, but required a much longer period in nuclei (about 4 weeks) and was not as extensive. The greatest decrease was found in the mitochondrial fraction. Nevertheless, turnover calculations did not support the conclusion that the rate of cerebral protein synthesis declined with age in these studies.[64]

Investigations *in vitro* of developmental influences on cerebral protein synthesis have also produced conflicting data. In one study, brain cortical slices from 16-day-old rats incorporated valine-^{14}C into protein at a rate which was only 16% of the value found in animals which were 7 days old or younger.[114] By 90 days of age, a further decrease to less than 2% of the maximal value was observed. When the protein synthesized by rat brain slices under similar experimental conditions was divided into two fractions on the basis of solubility in a chloroform–methanol mixture, the insoluble fraction exhibited a maximum rate of protein synthesis neonatally, but formation of the soluble fraction (proteolipids) was most rapid at about 8 days of age, i.e., during active myelination.[149] The proteolipids may be of mitochondrial origin (Section III,D). In common with studies *in vivo*, interpretation of experiments with surviving brain tissue was limited by the fact that developmental alterations in amino acid transport and pool size were not properly assessed. Thus, age-associated changes in pool sizes of several amino acids have been described.[150] In addition, transport of valine-^{14}C into cell suspensions from mouse brain has been shown to decrease rapidly during maturation.[139]

Several investigators have reported that amino acid-incorporating activities of microsomal and ribosomal systems isolated from brain decreased during development.[17,62,85,115,140] Changes in both ribosomes and activating factors were implicated in these findings.[85,97,140] However, purified polyribosomes from cerebral cortical gray matter or hindbrain-medullary white matter of the rat exhibited no age-associated differences in protein synthesis.[56] Moreover, the *p*H 5 fractions from these brain regions in the adult rat appeared to be at least as active in amino acid incorporation as the analogous preparations from immature or newborn animals.[56]

Whether or not the overall rate of protein synthesis varies in the brain with age, developmental alterations certainly occur in the nature of the proteins synthesized. These alterations must derive in part from changes in the population of messenger RNA molecules. Thus, the adult brain may contain higher proportions of species which have a smaller molecular size,[17,56] are synthesized less rapidly[57] (Fig. 10), have lesser template activity,[85] and are less firmly associated with cerebral ribosomes. The latter possibility was supported by the observation that polyphenylalanine synthesis in response to polyuridylic acid, relative to endogenous protein synthesis, was greater with cerebral polyribosomes from adult rats than with analogous preparations from newborn animals.[17] The unique nature of alterations in cerebral protein synthesis during development was suggested by the finding that basal amino acid incorporation in *hepatic* ribosomal preparations increased with age, while the relative response to polyuridylic acid decreased. Moreover, elevated rather than reduced polyribosomal populations may appear in hepatic tissue and other extracerebral tissues during development. The relationship of developing brain-barrier mechanisms to changes in cerebral protein synthesis with age has not been clarified. However, the resulting restriction in amino acid uptake may be responsible for reduction in pool size of several essential amino acids.[150] Contraction of this pool could in turn lead to adaptive changes in the synthesis or properties of messenger RNA, polyribosomes, and activating factors which eventuate in alterations in cerebral protein synthesis.[113,151,152]

B. Protein Synthesis in Mitochondrial and Synaptosomal Systems

1. General Aspects

Until recently, evidence for the occurrence of mitochondrial protein synthesis in the nervous system was based primarily upon the observation that certain cerebral enzymes and other proteins were more concentrated in isolated mitochondria than in other subcellular fractions. However, this localization could readily be explained by transport from microsomal sites of formation[153] or by association with mitochondria during isolation.[154] Within the past few years, purified cerebral mitochondria have been shown to contain certain of the major components essential for protein synthesis and actively to incorporate amino acids into protein both *in vivo* and *in vitro*. However, the role of mitochondrial protein synthesis in the specialized functioning of the nervous system has not yet been clearly delineated (see Section III,D).

Klee and Sokoloff[155] demonstrated that reconstituted homogenates of rat brain (cell debris and nuclei discarded) incorporated radioactivity from DL-1-leucine-^{14}C into proteins of the mitochondrial fraction (which was subsequently reisolated) as well as into microsomal protein. The crude mitochondria *per se* also exhibited appreciable amino acid-incorporating activity under appropriate incubation conditions.[156] This fraction was undoubtedly grossly contaminated with myelin and nerve endings, and may have contained microsomal fragments as well. Subsequent investigations in other laboratories revealed that highly purified mitochondrial preparations from brain were active in amino acid incorporation.[103,157] In incubation experiments with crude mitochondria from immature rat brain the proteolipid fraction became more highly labeled than the residual protein which was insoluble in chloroform–methanol (Fig. 11). However, proteolipid synthesis accounted for only a small fraction of the total incorporation of amino acid into protein.[156] Mokrasch[158] reported that "mitochondrial" fractions of rat brain were more active than other cell fractions in incorporating a mixture of ^{14}C amino acids into proteolipid. Labeling of proteolipid occurred primarily in the interior of the phosphatidopeptide moiety of the molecule.[156,158] Chloramphenicol, but not ribonuclease, inhibited amino acid incorporation into proteolipid. Interpretation of the data of Mokrasch[158] as unequivocal evidence for mitochondrial synthesis of proteolipid was marred by the fact that the most active fraction was probably relatively poor in mitochondria and rich in other membranous structures which may be capable of protein synthesis *per se*.[130] However, amino acid incorporation into the latter fractions was resistant to chloramphenicol, a potent inhibitor of protein synthesis in cerebral mitochondrial preparations.[130,157]

Optimal conditions for amino acid incorporation by cerebral mitochondria depended markedly on mitochondrial state.[103] No requirement for exogenous pH 5 enzymes or ATP existed with purified, intact prepara-

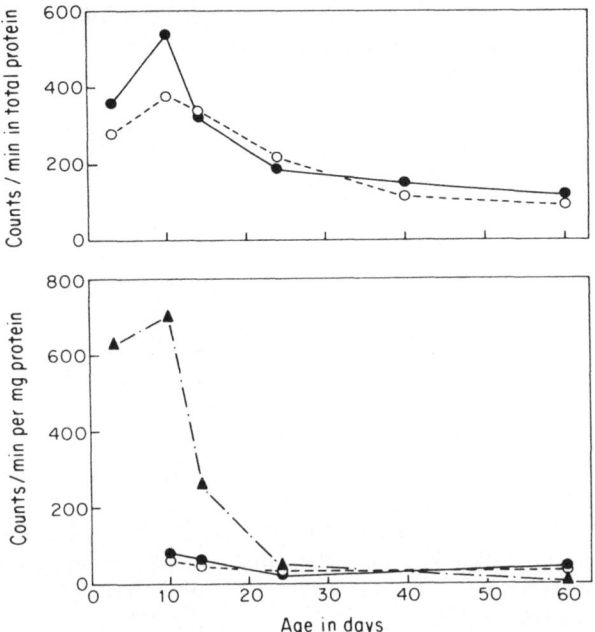

Fig. 11. Influence of age on amino acid incorporation into proteolipid and other proteins by a crude mitochondrial fraction of rat brain *in vitro*.

Top: Incorporation into total protein. The incubation mixture contained (in μmoles): sucrose, 225; AMP, 5; GTP, 0.25; potassium phosphate buffer (*p*H 7.4), 20; MgCl$_2$, 10; Na DL-β-hydroxybutyrate, 50; EDTA, 0.03; DL-leucine-1-^{14}C, 0.8 (5.4 mC/mmole), 0.15 ml of crude mitochondrial suspension. Incubation time at 37°C was 25 min. Incorporation into total protein in medium is shown. $-\bullet-\bullet-$, without ribonuclease; $-\bigcirc-\bigcirc-$, 85 μg ribonuclease added.

Bottom: Incorporation into proteolipid and residual protein. The incubation mixture and its constituents were present in 25 times the amounts indicated above. Specific activity of leucine was 8 mC/mmole. Ribonuclease was present in all flasks. Incubation time at 37°C was 50 min. $-\blacktriangle-\cdot-\blacktriangle-$, proteolipid protein; $-\bigcirc--\bigcirc-$, residual protein (insoluble in chloroform-methanol); $-\bullet-\bullet-$, total protein. (Reprinted with permission from Klee and Sokoloff.[156])

tions.[103,157] In appropriate circumstances, amino acid incorporation was stimulated by ADP, GTP, and substrates of the citric acid cycle (especially α-ketoglutarate). Puromycin, as well as chloramphenicol, inhibited protein synthesis by cerebral mitochondria.[130] Cycloheximide[130] and actinomycin D[157] were relatively ineffective in this regard. The physiological significance which may be attributed to effects of inhibitors and pharmacological agents on mitochondrial protein synthesis is limited. Thus many substances, including ribonuclease and actinomycin D[159] are probably unable to penetrate the outer membrane of intact mitochondria. Moreover, mitochondrial energy metabolism, upon which protein synthesis in this organelle depends, exhibits marked sensitivity to alterations in the environment. Among the neurochemical substances which have been reported to inhibit amino acid incorporation by cerebral mitochondria *in vitro* may be listed chlorpromazine and serotonin.[103] Stimulation was obtained with γ-aminobutyric acid and, in the presence of ADP, with acetylcholine, dopa, epinephrine, and norepinephrine as well. A mixture of 19 amino acids, in addition to L-leucine-^{14}C, had no effect on mitochondrial protein synthesis.[157]

2. *Cytological and Regional Variations*

Synaptosomal, as well as cell body, mitochondria from rat cerebral cortex appeared to be capable of incorporating amino acids into protein under appropriate conditions *in vitro*.[65,130,131] These findings seemed to implicate synaptosomal mitochondria in local axonal synthesis of certain proteins. However, evidence has been presented that synaptosomal mitochondria may normally derive a large proportion of their proteins from the neuronal cell body by axoplasmic flow.[160] Fractionation of synaptosomes after incubation in an amino acid-incorporating system revealed that the membrane and soluble fractions were most highly labeled[130] (Table VI). Unlike synaptosomal mitochondria, incorporation by these fractions was inhibited by cycloheximide, but not by chloramphenicol. Puromycin was inhibitory with all fractions, but ribonuclease was not. These results and the distribution of RNA in synaptosomal fractions (Table III) suggested that membranous structures of the axon and nerve terminal may contain ribosomes which have not yet been visualized in the electron microscope and are protected from ribonuclease action by the enveloping membrane.[129]

No information is available concerning the relative protein-synthesizing activities of mitochondria from different parts of the nervous system. Nor have glia and neurons been compared in this respect.

3. *Developmental Changes*

The mitochondrial population of brain cells increased during the phase of active growth and this increase was reflected in an elevation of total mitochondrial protein.[75] Certain enzymes which are localized in this organelle also increased in concentration or activity during development.[161,162] A crude mitochondrial fraction isolated from rat brain revealed maximal incorporation of DL-leucine-1-^{14}C into total protein at about 10 days after birth[156] (Fig. 11). This peak coincided with the period of most active myelination and was related to rapid proteolipid synthesis (see also Ref. 149). Between the tenth and twenty-fifth day, the specific activity of

TABLE VI

Incorporation of Leucine into Synapto-somal Fractions of Rat Cerebral Cortex[a]

Fraction	Specific activity (cpm/mg protein)
Intact synaptosomes	330 ± 19.4
Soluble fraction	274 ± 12.5
Vesicles	2
Membranes	260 ± 12.2
Mitochondria	170 ± 16.3

[a] Reprinted with permission from Morgan and Austin.[130]

Synaptosomes were isolated on a ficoll gradient and incubated in media containing 0.2 mM EDTA, 10 mM K_2HPO_4, 10 mM $MgCl_2$, 10 mM Tris–HCl (pH 7.4), 200 μg/ml sodium penicillin G, 60 μg/ml streptomycin sulfate and 0.3 M mannitol or 0.25 M sucrose. After 10 min preincubation at 37°C, the reaction was started by addition of 0.5 μC/ml of L-leucine-$U^{14}C$ (150 mC/mmole). After incubation for 30 min, the synaptosomes were pelleted, osmotically shocked by suspension in distilled water, and fractionated on a sucrose gradient.

proteolipid fell precipitously to one-tenth its highest value, whereas the specific activity of the residual protein dropped only about 50 %[156] (Fig. 11). Mitochondria in infant brain may provide energy and other factors required for mitochondrial *and* microsomal protein synthesis more effectively than the analogous structures in the adult brain.[155] Moreover, mitochondria from immature brain were uniquely capable of responding to physiological concentrations of thyroxine with an increased production of factors which appeared to function at a late stage in the assembly of the polypeptide chain. These phenomena may represent an important hormonal control of protein synthesis in the developing brain.

C. Protein Synthesis in Nuclear Systems

Recent investigations support the concept that nuclear protein synthesis may be an active process in the brain of the immature animal. When sub-cellular components were isolated after incubation of chopped cerebral cortex from young rats (10–15 days old), the nuclear fraction revealed the most rapid incorporation of L-[U-^{14}C]leucine into protein[65] (Fig. 12). Moreover, isolated cerebral nuclei from 25–35 day old rats, under incubation conditions which were optimal for proteolipid synthesis by mitochondrial fractions,

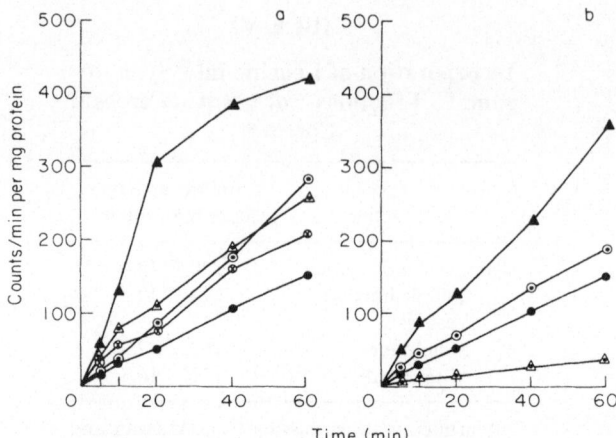

Fig. 12. Incorporation of L-leucine into subcellular components of chopped cerebral cortex from young rats (10–15 days old). Incubation was carried out at 37°C in Krebs-phosphate buffer (pH 7.4) containing 10 mM glucose and 0.5 μC/ml L-[U-^{14}C] leucine (150 mC/mmole). After incubation, the chopped tissue was washed at 0°C, homogenized in 0.32 M sucrose, and fractionated. (a) Fractions obtained by differential centrifugation. $\otimes$, homogenate; ▲, nuclei; ●, crude mitochondrial fraction; △, microsomes; ⊙, soluble fraction (high-speed supernatant). (b) Fractions obtained from the crude mitochondrial pellet after centrifugation on a discontinuous sucrose density gradient (0.8 and 1.4 M). ●, crude mitochondrial fraction; △, myelin; ⊙, synaptosomes; ▲, mitochondria. (Reprinted with permission from Austin and Morgan.[65])

incorporated [^{14}C]amino acids into chloroform–methanol insoluble proteins to a greater extent than the microsomal and mitochondrial fractions.[158] However, the purity of the nuclei was not determined.

Data obtained with other mammalian cells support the general concept of nuclear protein synthesis.[163] Ribosomes have been demonstrated in purified nuclei from mammalian tissues. Hepatic nuclear ribosomes exhibited properties which were similar to those of cytoplasmic ribosomes, including comparable sedimentation characteristics and base compositions for the associated RNA.[164] Nuclear ribosomes may function in the formation of histones,[165] which, by combining with specific sites on nuclear DNA, may regulate messenger RNA synthesis.[166] The observation that histones turn over very slowly in the adult brain[167] supports the hypothesis that they may be formed or replaced only at the time of DNA replication during cell division.[165]

D. Nature of Proteins Synthesized

The spectrum of proteins formed in the nervous system undoubtedly varies with region and state of development. Thus, proteins synthesized by neurons and glia and by diverse populations of these cells appear to differ. Moreover, various subcellular sites probably elaborate characteristic proteins whose nature may change during development.

Certain proteins are unique to the nervous system or to certain parts of this system. The highly acidic, immunologically specific "S-100" protein, first described by Moore and McGregor[168] in aqueous extracts of brain belongs in this category. This protein, which was reported to be free of phosphorus, carbohydrate, and lipid, and rich in dicarboxylic acids, was found in nervous tissue of all species studied.[169] It had a molecular weight of 20,000–25,000.[170] The "S-100" protein may exist in several different conformational states (B. Moore, unpublished observations) rather than as a mixture of two or more closely related proteins.[171,172] In one study, evidence has been presented for the *de novo* formation of "S-100" protein by ribosomes isolated from rabbit brain cortex.[100] A significant proportion of the proteins labeled after incubation of this system with leucine-^{14}C was precipitable by antibody to "S-100" protein. Although rapid turnover of "S-100" proteins in rat brain has been reported by McEwen and Hydén,[172] conflicting data have also been obtained (B. Moore, unpublished observations). Available information indicates that "S-100" protein may constitute about 0.3–0.5% of the total soluble proteins of rat brain.[170] Micro-immunodiffusion and fluorescent antibody studies conducted on Deiters' nucleus of rabbit brain stem suggested that "S-100" protein was primarily concentrated around (but not in) the nuclei of oligodendrocytes and was scattered throughout the glial membrane system.[174] In these studies, "S-100" protein also appeared to be present in the large neuronal nuclei. The suggestion was advanced that this highly acidic protein may interact with basic histones in neuronal nuclei to control genetic activity.[174] However, Dravid and Burdman,[175] employing immunochemical techniques, were unable to demonstrate "S-100" protein in isolated nuclei from rat brain even though large amounts of other soluble, acidic proteins were present. Moore and Perez[170] reported that "S-100" protein was localized mainly in white matter of the central nervous system and was present in the axoplasm of peripheral nerve as well, probably as a result of leakage from surrounding glia (B. Moore, unpublished observations). An additional acidic protein specific to the nervous system (designated 14-3-2) was concentrated in gray matter.[170] Its molecular weight may be about 50,000. A basic protein which appears to be present only in neurons and primarily in the nucleus has been isolated by Tomasi and Kornguth.[176] This protein was present in all species studied and was described as a histone of asymmetrical structure with a molecular weight of about 18,000.

Other proteins specific to the nervous system may be formed in microsomal locations.[84,177] Herriman and Hunter[84] reported that cerebral nuclear RNA from 4-week-old mice, added to an homologous ribosomal system, enhanced the production of the basic encephalitogenic protein associated with brain myelin.[178] The smallest nuclear RNA component with apparent template activity exhibited a sedimentation coefficient which was less than 16S.[84] However, a much smaller messenger RNA unit would be required to code for the encephalitogenic protein which appears to have a probable minimum molecular weight of 4000 or 8000.[179] Immunoprecipitation data indicated that this myelin protein contained about 4% of the total isotope incorporated into protein by the cerebral ribosomal system. Ribosomal formation of membrane proteins was also suggested by the work of Lim and Adams.[95] These investigators noted that microsomal preparations obtained from *immature* rat brain were capable of incorporating L-valine-^{14}C into membrane proteins. This property was lacking in adult animals.

Developmental changes in the nature of proteins synthesized by the nervous system were also indicated by other findings. For example, "S-100" protein did not appear in extracts of rat brain until 10–12 days after birth.[170] Concentrations of this protein in brain attained half the maximum adult level in about 30 days. Addition of nerve growth factor to tissue cultures of sensory ganglia from chick embryos resulted in selective stimulation of the synthesis of certain proteins.[180,181] Wender and Waligóra[182,183] noted that the amino acid composition of brain proteins altered during development in the guinea pig. The change was greater in white matter than in gray matter. Gross changes were seen in the distribution of proteins of varying solubility in the rat brain during development.[184] The ratio of albuminlike proteins to globulins increased markedly between the fifth and tenth days of age.

Mitochondrial synthesis of specific proteins may be of particular significance to development and specialized function in the nervous system. Proteolipids, the major protein constituents of myelin,[185] may be of mitochondrial origin (Section III,B,1). The proteolipid fraction of white matter from beef brain differed in amino acid composition from similar fractions of nonneural origin.[186] Proteolipids of the nervous system are presumably formed by oligodendroglia centrally and Schwann cells peripherally.[90] Synthesis of these proteins normally occurs at a significant rate only during myelination. Turnover studies suggested that certain proteins in the mitochondrial membranes of adult rat brain were made locally and were broken down only with the demise of the mitochondrion, i.e., with a half-life equal to 26–30 days.[153] In contrast, water-soluble proteins from the same source exhibited a half-life of only 18 days and probably were mainly of extra-mitochondrial origin. The limited coding properties of mitochondrial DNA (Section II,B,1) supports the deduction that the mitochondrion may be capable of synthesizing only a small complement of different proteins. Mitochondrial protein synthesis in the liver appears to be directed principally toward the formation of the inner membrane proteins.[187,188] Although the situation may be different in the nervous system, particularly in the immature animal, it seems likely that the major portion of even those enzymes and other proteins found localized in brain mitochondria are manufactured elsewhere.

No direct evidence is available to establish the subcellular site of synthesis for any enzyme of the nervous system. The unreliability of circumstantial evidence is illustrated by experiments on acetylcholinesterase. This enzyme has been employed as a microsomal marker in neuronal preparations.[154]

It also appears to be selectively localized in synaptic membranes[130,189,190] The enzyme may be synthesized locally in the distal axon[126-128] where ribosomes have not been demonstrated.[5] Regeneration of acetylcholinesterase activity in cat hypoglossal nerves after irreversible inactivation or transection was only partially inhibited by puromycin,[128] but was actually stimulated by actinomycin D.[191] Moreover, acetylcholinesterase activity in cultures from embryonic chick brain was unaffected by addition of chicken liver histone, puromycin or polylysine, but was elevated by actinomycin D.[192] Concomitantly, synthesis of proteins presumed to be largely of microsomal origin (lactic dehydrogenase and total brain protein) was stimulated by the histone preparation and markedly inhibited by actinomycin D, puromycin, and polylysine. The results suggested that acetylcholinesterase may be specified by a stable messenger RNA and that the synthesis of this enzyme can be accomplished in sites other than the cytoplasmic ribosome. Since mitochondria and other membranous components of the synaptosome appear to be active in protein synthesis (Section III,B,1,2), one of these structures may be the site of synthesis of acetylcholinesterase in the distal axon.

Cells of the nervous system form a variety of proteins and polypeptides which function in storage and processing of information. In the normal adult animal, where development or repair of nervous tissue is proceeding very slowly, synthesis of these specialized polypeptides may account for a large proportion of the total protein-synthesizing activity. This concept seems borne out by the high capacity of hypothalamic tissue from the guinea pig to incorporate cysteine-^{35}S or L-tyrosine-^{3}H into the arginine-vasopressin molecule *in vitro*.[193] After a lag period of 1–3 hr, the specific activity of this polypeptide exceeded that of total tissue protein. Synthesis of vasopressin was further enhanced in dehydrated guinea pigs. These observations also pointed up the fact that results obtained in investigations of overall protein synthesis in complex cellular preparations from the nervous system may not be applicable to the behavior of selected areas, each of which may produce specific proteins and react differently in various experimental situations.

IV. CONCLUSIONS. PHYSIOLOGICAL REGULATION OF PROTEIN SYNTHESIS

Early investigations of amino acid uptake by brain proteins *in vivo* suggested that protein synthesis in this organ proceeded at a rate which was appreciably lower than that of other metabolically active tissues such as liver,[194] was greater in higher centers than in subcortical regions,[90] and declined progressively with age.[145] These phenomena were linked to knowledge that neuronal renewal was minimal in the adult brain[195] and that development of brain-barrier mechanisms restricted passage of essential metabolites between brain and blood.[196] The latter concepts are still useful in explaining some of the unusual characteristics of protein synthesis in the central nervous system, even though conclusions regarding possible developmental and regional variations in cerebral protein synthesis may have to be significantly modified in the light of new information.

Rapid extension of interest in the relationship between protein synthesis in the neuron and the specialized functions of this cell followed recognition that cerebral protein synthesis was a very active process, even in the adult brain. Detailed studies of turnover of brain proteins, combined with experiments in which brain-barrier mechanisms were partially circumvented by local administration of radioactive precursors, first led to this realization.[119] Nervous tissues, generally, appear to contain a significant proportion of proteins with relatively short half-lives. Certain of these unstable proteins may be involved in processes of neurosecretion and neurotransmission. Other proteins with longer half-lives presumably contribute to brain structure and provide for certain requirements of the distal axon by the process of axoplasmic flow.[120]

Overall synthesis of brain proteins may not be substantially more rapid in young animals than in adults. However, removal from the uterine environment *per se* seems to trigger a decline in cerebral protein synthesis.[147] Among the possible causes for this phenomenon may be mentioned changes in substrate availability and in the control of protein synthesis by humoral substances in the placental circulation.

The profound alterations in the spectrum of proteins synthesized by the brain during development suggest that this process is unusually sensitive to variations in the internal and external environment. Data have accumulated in recent years which strongly support this conclusion. Alterations in systemic levels of amino acids, ions, and other physiological substances, as well as pharmacological agents, undoubtedly have profound acute and chronic effects upon the rate of cerebral protein synthesis, as well as the nature of the proteins elaborated.[94,197] Numerous reports have dealt with changes in RNA and protein synthesis in central and peripheral nerves resulting from specific and nonspecific stimuli, nerve damage, muscular activity, and behavioral experiences.[198] The actions of these effectors were variously ascribed to influences at the transcriptional or translational levels in the regulation of protein synthesis. At the transcriptional level, neural stimuli were reported to effect alterations in the amounts and types of RNA synthesized by the nerve cell. This observation implied that turnover rates of at least certain messenger RNA molecules were changed. Changes in protein synthesis at the translational level may be envisioned as involving principally the capacity of the neural ribosome to function at variable efficiencies or to associate selectively with messenger RNA species which were already present in optimal concentrations. The unique instability of certain messenger RNA-ribosome complexes in the nervous system may provide a mechanism whereby the coding and transmission of information are significantly facilitated.[12,197]

Intelligent appraisal of the functional consequences of environmental alterations in neural protein synthesis requires a detailed knowledge of the properties of the pertinent protein-synthesizing systems. Information relating to these characteristics is actively in the process of being crystallized, but still encompasses many areas of obscurity and relative ignorance. This state of affairs is attributable partly to the complexity of neural tissues and the difficulties in sampling, and partly to the peculiar characteristics of the systems being investigated. Continued and expanded efforts to delineate the differential properties and specific contributions of protein-synthesizing systems derived from the various cell types and subcellular fractions of nervous tissue give promise of providing a rational basis for associating protein synthesis and specialized function in the nervous system.

ACKNOWLEDGMENTS

Supported by research grants from the National Institute of Neurological Diseases and Blindness, U.S. Public Health Service (NB-07869) and the United Cerebral Palsy Research and Educational Foundation. The author is indebted to Mrs. Carole Feingold for capable bibliographic and secretarial assistance.

V. REFERENCES

1. A. Lajtha, S. Furst, and H. Waelsch, The metabolism of the proteins of the brain, *Experientia* **13**:168–172 (1957).
2. S. Furst, A. Lajtha, and H. Waelsch, Amino acid and protein metabolism of the brain—III. Incorporation of lysine into the proteins of various brain areas and their cellular fractions, *J. Neurochem.* **2**:216–225 (1958).
3. D. H. Clouet and D. Richter, The incorporation of [^{35}S]labelled methionine into the proteins of the rat brain, *J. Neurochem.* **3**:219–229 (1959).
4. M. L. Petermann, *The Physical and Chemical Properties of Ribosomes*, Elsevier Publishing Co., Amsterdam (1964).
5. S. L. Palay and G. E. Palade, The fine structure of neurons, *J. Biophys. Biochem. Cytol.* **1**:69–88 (1955).
6. R. Ekholm and H. Hydén, Polysomes from microdissected fresh neurons, *J. Ultrastruct. Res.* **13**:269–280 (1965).
7. C. Sotelo and S. L. Palay, The fine structure of the lateral vestibular nucleus in the rat. I. Neurons and neuroglial cells, *J. Cell Biol.* **36**:151–179 (1968).
8. E. Mugnaini and F. Walberg, Ultrastructure of neuroglia, *Ergeb. Anat. Entwicklung.* **37**:194–236 (1964).
9. S.-O. Brattgård, J.-E. Edström, and H. Hydén, The chemical changes in regenerating neurons, *J. Neurochem.* **1**:316–325 (1957).
10. D. Bodian, A suggestive relationship of nerve cell RNA with specific synaptic sites, *Proc. Nat. Acad. Sci. U.S.* **53**:418–425 (1965).
11. A. T. Campagnoni and H. R. Mahler, Isolation and properties of polyribosomes from cerebral cortex, *Biochemistry* **6**:956–967 (1967).
12. C. E. Zomzely, S. Roberts, C. P. Gruber, and D. M. Brown, Cerebral protein synthesis. II. Instability of cerebral messenger ribonucleic acid-ribosome complexes, *J. Biol. Chem.* **243**:5396–5409 (1968).

13. C. E. Zomzely, S. Roberts, D. M. Brown, and C. Provost, Cerebral protein synthesis. I. Physical properties of cerebral ribosomes and polyribosomes, *J. Mol. Biol.* 20:455–468 (1966).

14. F. O. Wettstein, T. Staehelin, and H. Noll, Ribosomal aggregate engaged in protein synthesis: Characterization of the ergosome, *Nature* 197:430–435 (1963).

15. A. J. Munro, R. J. Jackson, and A. Korner, Studies on the nature of polysomes, *Biochem. J.* 92:289–299 (1964).

16. S. H. Wilson and M. B. Hoagland, Studies on the physiology of rat liver polyribosomes: Quantitation and intracellular distribution of ribosomes, *Proc. Nat. Acad. Sci. U.S.* 54:600–607 (1965).

17. M. R. V. Murthy, Protein synthesis in growing-rat tissues. II. Polyribosome concentration of brain and liver as a function of age, *Biochim. Biophys. Acta* 119:599–613 (1966).

18. Y. Takahashi, K. Mase, and H. Sugano, Preparation of polysomes from rat brain tissue, *Biochim. Biophys. Acta* 119:627–629 (1966).

19. C. Vesco and A. Giuditta, Disaggregation of brain polysomes induced by electroconvulsive treatment, *J. Neurochem.* 15:81–85 (1968).

20. S. H. Appel, W. Davis, and S. Scott, Brain polysomes: Response to environmental stimulation, *Science* 157:836–838 (1967).

21. M. R. V. Murthy and D. A. Rappoport, Biochemistry of the developing rat brain. VI. Preparation and properties of ribosomes, *Biochim. Biophys. Acta* 95:132–145 (1965).

22. S. Yamagami, Y. Kawakita, and S. Naka, Some physical, chemical and biochemical properties of ribosomal RNA from guinea pig brain cortex, *J. Neurochem.* 12:607–611 (1965).

23. C. E. Zomzely, S. Roberts, D. M. Brown, and D. Rapaport, Isolation of 55-S ribonucleoprotein particles with amino acid-incorporating activity, *Biochim. Biophys. Acta* 103:529–531 (1965).

24. C. E. Zomzely, S. Roberts, and D. Rapaport, Regulation of cerebral metabolism of amino acids—III. Characteristics of amino acid incorporation into protein of microsomal and ribosomal preparations of rat cerebral cortex, *J. Neurochem.* 11:567–582 (1964).

25. J. M. Azcurra and O. Z. Sellinger, Cerebral microsomes. IV. On the attachment of ribosomes to the microsomal membranes of rat cerebral cortex, *Brain Res.* 6:359–362 (1967).

26. J. Samec, P. Mandel, and M. Jacob, Occurrence of light ribonucleoprotein (RNP) particles in the microsomal fraction of adult rat brain, *J. Neurochem.* 14:887–892 (1967).

27. R. K. Datta, Brain ribosomes, *Brain Res.* 2:301–322 (1966).

28. D. M. Schneider and S. Roberts, Base compositions of 18 S and 28 S RNA fractions from rat cerebral ribosomes, *J. Neurochem.* 15:1469–1471 (1968).

29. A. DiGirolamo, E. C. Henshaw, and H. H. Hiatt, Messenger ribonucleic acid in rat liver nuclei and cytoplasm, *J. Mol. Biol.* 8:479–488 (1964).

30. S. Katz and D. G. Comb, A new method for the determination of the base composition of ribonucleic acid, *J. Biol. Chem.* 238:3065–3067 (1963).

31. M. Jacob, J. Samec, J. Stevenin, J. P. Garel, and P. Mandel, Polysomes and polysomal RNA of rat brain, *J. Neurochem.* 14:169–178 (1967).

32. C. Vesco and A. Giuditta, Pattern of RNA synthesis in rabbit brain, *Biochim. Biophys. Acta* 142:385–402 (1967).

33. D. Elson, Ribosomal enzymes, in *Enzyme Cytology* (D. B. Roodyn, ed.) pp. 407–473, Academic Press, New York (1967).

34. J.-P. Waller and J. I. Harris, Studies on the composition of the protein from *Escherichia coli* ribosomes, *Proc. Nat. Acad. Sci. U.S.* 47:18–23 (1961).

35. S. Roberts, Regulation of cerebral metabolism of amino acids—II. Influence of phenylalanine deficiency on free and protein-bound amino acids in rat cerebral cortex: Relationship to plasma levels, *J. Neurochem.* 10:931–940 (1963).

36. S. Moore, On the determination of cystine as cysteic acid, *J. Biol. Chem.* **238**:235–237 (1963).

37. S. Yamagami, M. Masui, and Y. Kawakita, Preparation and properties of ribosomes from guinea pig brains, *J. Neurochem.* **10**:849–850 (1963).

38. K. Tsukada and I. Lieberman, Protein synthesis by liver polyribosomes after partial hepatectomy, *Biochem. Biophys. Res. Comm.* **19**:702–707 (1965).

39. P. Mandel, T. Borkowski, S. Harth, and R. Mardell, Incorporation of ^{32}P in ribonucleic acid of subcellular fractions of various regions of the rat central nervous system, *J. Neurochem.* **8**:126–138 (1961).

40. D. H. Adams, Some observations on the incorporation of precursors into ribonucleic acid of rat brain, *J. Neurochem.* **12**:783–790 (1965).

41. D. H. Adams, The relationship between cellular nucleic acids in the developing rat cerebral cortex, *Biochem. J.* **98**:636–640 (1966).

42. P. Mandel, H. Rein, S. Harth-Edel, and R. Mardell, in *Comparative Neurochemistry* (D. Richter, ed.) pp. 149–163, Pergamon Press, Oxford (1964).

43. R. Landolt, H. H. Hess, and C. Thalheimer, Regional distribution of some chemical structural components of the human nervous system—I. DNA, RNA and ganglioside sialic acid, *J. Neurochem.* **13**:1441–1452 (1966).

44. E. G. Gray, in *Electron Microscopic Anatomy* (S. M. Kurtz, ed.) pp. 369–417, Academic Press, New York (1964).

45. H. Hydén, in *The Neuron* (H. Hydén, ed.) pp. 179–219, Elsevier Publishing Co., New York (1967).

46. H. Hydén and A. Pigon, A cytophysiological study of the functional relationship between oligodendroglial cells and nerve cells of Deiters' nucleus, *J. Neurochem.* **6**:57–72 (1960).

47. B. Daneholt and S.-O. Brattgård, A comparison between RNA metabolism of nerve cells and glia in the hypoglossal nucleus of the rabbit, *J. Neurochem.* **13**:913–921 (1966).

48. E. Egyházi, Microchemical fractionation of neuronal and glial RNA, *Biochim. Biophys. Acta* **114**:516–526 (1966).

49. A. Edström, The ribonucleic acid in the Mauthner neuron of the goldfish, *J. Neurochem.* **11**:309–314 (1964).

50. E. Koenig, Synthetic mechanisms in the axon—II. RNA in myelin-free axons of the cat, *J. Neurochem.* **12**:357–361 (1965).

51. N. Miani, A. Di Girolamo, and M. Di Girolamo, Sedimentation characteristics of axonal RNA in rabbit, *J. Neurochem.* **13**:755–759 (1966).

52. L. Austin, J. J. Bray, and R. J. Young, Transport of proteins and ribonucleic acid along nerve axons, *J. Neurochem.* **13**:1267–1269 (1966).

53. K. R. Brizzee, J. Vogt, and X. Kharetchko, Postnatal changes in glia/neuron index with a comparison of methods of cell enumeration in the white rat, *Prog. Brain Res.* **4**:136–149 (1964).

54. A. Hughes and L. B. Flexner, A study of the development of the cerebral cortex of the foetal guinea-pig by means of the ultra-violet microscope, *J. Anat.* **90**:386–394 (1956).

55. U. Ringborg, Composition and content of RNA in neurons of rat hippocampus at different ages, *Brain Res.* **2**:296–298 (1966).

56. S. Roberts, C. E. Zomzely, and S. C. Bondy, Developmental alterations in cerebral ribonucleic acid and protein synthesis, in *Cellular Aspects of Growth and Differentiation in Nervous Tissue* (D. C. Pease, ed.) Univ. of Calif. Press (1971).

57. S. C. Bondy and S. Roberts, Hybridizable ribonucleic acid of rat brain, *Biochem. J.* **109**:533–541 (1968).

58. M. Wender and M. Heerowski, Activation of amino acids by "pH 5 enzymes" in the developing nervous system of guinea pigs, *Acta Neurologica Scandinavica* **38**:Suppl. 1, 24–25 (1962).

59. A. Edström, Effect of spinal cord transection on the base composition and content of RNA in the Mauthner nerve fibre of the goldfish, *J. Neurochem.* **11**:557–559 (1964).

60. G. Toschi, E. Dore, P. U. Angeletti, R. Levi-Montalcini, and Ch. de Haën, Characteristics of labeled RNA from spinal ganglia of chick embryo and the action of a specific growth factor (NGF), *J. Neurochem.* **13**:539–544 (1966).

61. J. A. Burdman, Early effects of a nerve growth factor on the RNA content and base ratios of isolated chick embryo sensory ganglia neuroblasts in tissue culture, *J. Neurochem.* **14**:367–371 (1967).

62. M. R. V. Murthy and D. A. Rappoport, Biochemistry of the developing rat brain. V. Cell-free incorporation of L-[1-^{14}C]leucine into microsomal protein, *Biochim. Biophys. Acta* **95**:121–131 (1965).

63. M. V. Simpson and J. R. McLean, The incorporation of labeled amino acids into the cytoplasmic particles of rat muscle, *Biochim. Biophys. Acta* **18**:573–575 (1955).

64. S. S. Oja, Studies on protein metabolism in developing rat brain, *Ann. Acad. Sci. Fennicae* **131**:1–81 (1967).

65. L. Austin and I. G. Morgan, Incorporation of ^{14}C-labeled leucine into synaptosomes from rat cerebral cortex *in vitro*, *J. Neurochem.* **14**:377–387 (1967).

66. R. Balázs and W. A. Cocks, RNA metabolism in subcellular fractions of brain tissue, *J. Neurochem.* **14**:1035–1055 (1967).

67. J. H. Sinclair, B. J. Stevens, N. Gross, and M. Rabinowitz, The constant size of circular mitochondrial DNA in several organisms and different organs, *Biochim. Biophys. Acta* **145**:528–531 (1967).

68. M. Rabinowitz, J. Sinclair, L. DeSalle, R. Haselkorn, and H. H. Swift, Isolation of deoxyribonucleic acid from mitochondria of chick embryo heart and liver, *Proc. Nat. Acad. Sci. U.S.* **53**:1126–1133 (1965).

69. H. G. Du Buy, C. F. T. Mattern, and F. L. Riley, Comparison of the DNA's obtained from brain nuclei and mitochondria of mice and from the nuclei and kinetoplasts of *Leishmania enriettii, Biochim. Biophys. Acta* **123**:298–305 (1966).

70. D. G. Humm and J. H. Humm, Hybridization of mitochondrial RNA with mitochondrial and nuclear DNA in agar, *Proc. Nat. Acad. Sci. U.S.* **55**:114–119 (1966).

71. M. E. Pullman and G. Schatz, Mitochondrial oxidations and energy coupling, *Ann. Rev. Biochem.* **36**:539–610 (1967).

72. T. W. O'Brien and G. F. Kalf, Ribosomes from rat liver mitochondria. II. Partial characterization, *J. Biol. Chem.* **242**:2180–2185 (1967).

73. D. T. Dubin and R. E. Brown, A novel ribosomal RNA in hamster cell mitochondria, *Biochim. Biophys. Acta* **145**:538–540 (1967).

74. S. Yamagami, Y. Kawakita, and S. Naka, Base composition of RNA of the subcellular fractions from guinea pig brains, *J. Neurochem.* **11**:899–900 (1964).

75. F. E. Samson, Jr., W. M. Balfour, and R. J. Jacobs, Mitochondrial changes in developing rat brain, *Am. J. Physiol.* **199**:693–696 (1960).

76. A. T. Miller, Jr., D. M. Conoly, M. Gabriel, and M. S. Handy, Mitochondrial regulation of tissue respiration, *Am. J. Physiol.* **197**:653–656 (1959).

77. D. M. Prescott, in *Progress in Nucleic Acid Research and Molecular Biology* (J. N. Davidson and W. E. Cohn, eds.) Vol. 3, pp. 33–57, Academic Press, New York (1964).

78. G. P. Georgiev, in *Progress in Nucleic Acid Research and Molecular Biology* (J. N. Davidson and W. E. Cohn, eds.) Vol. 6, pp. 259–351, Academic Press, New York (1967).

79. K. Matsuda and A. Siegel, Hybridization of plant ribosomal RNA to DNA: The isolation of a DNA component rich in ribosomal RNA cistrons, *Proc. Nat. Acad. Sci. U.S.* **58**:673–680 (1967).

80. R. P. Perry, in *Progress in Nucleic Acid Research and Molecular Biology* (J. N. Davidson and W. E. Cohn, eds.) Vol. 6, pp. 219–257, Academic Press, New York (1967).

81. S. H. Barondes, Studies with an RNA polymerase from brain, *J. Neurochem.* **11**:663–669 (1964).
82. S. C. Bondy and H. Waelsch, RNA polymerase in the central nervous system, *Life Sci.* **3**:633–636 (1964).
83. S. C. Bondy and H. Waelsch, Nuclear RNA polymerase in brain and liver, *J. Neurochem.* **12**:751–756 (1965).
84. I. D. Herriman and G. D. Hunter, Cytoplasmic protein synthesis in mouse brain, *J. Neurochem.* **12**:937–947 (1965).
85. S. Yamagami, R. R. Fritz, and D. A. Rappoport, Biochemistry of the developing rat brain. VII. Changes in the ribosomal system and nuclear RNA'S, *Biochim. Biophys. Acta* **129**:532–547 (1966).
86. S. C. Bondy and S. Roberts, Messenger ribonucleic acid of cerebral nuclei, *Biochem. J.* **105**:1111–1118 (1967).
87. M. H. Samli and S. Roberts, Properties of RNA fractions from nuclei of brain cells which stimulate incorporation of amino acids by brain ribosomes, *J. Neurochem.* **16**:1565–1580 (1969).
88. J. R. Warner, P. M. Knopf, and A. Rich, A multiple ribosomal structure in protein synthesis, *Proc. Nat. Acad. Sci. U.S.* **49**:122–129 (1963).
89. P. Volpe and A. Giuditta, Biosynthesis of RNA in neuron- and glia-enriched fractions, *Brain Res.* **6**:228–240 (1967).
90. H. Koenig, An autoradiographic study of nucleic acid and protein turnover in the mammalian neuraxis, *J. Biophys. Biochem. Cytol.* **4**:785–792 (1958).
91. H. Hydén and E. Egyházi, Glial RNA changes during a learning experiment in rats, *Proc. Nat. Acad. Sci. U.S.* **49**:618–624 (1963).
92. D. R. Dahl and F. E. Samson, Jr., Metabolism of rat brain mitochondria during postnatal development, *Am. J. Physiol.* **196**:470–472 (1959).
93. D. I. Kurtz and F. M. Sinex, Age related differences in the association of brain DNA and nuclear protein, *Biochim. Biophys. Acta* **145**:840–842 (1967).
94. S. Roberts and C. E. Zomzely, in *Protides of the Biological Fluids, Proceedings of the Thirteenth Colloquium, Bruges, 1965* (H. Peeters, ed.) Vol. 13, pp. 91–102, Elsevier Publishing Co., Amsterdam (1966).
95. L. Lim and D. H. Adams, Microsomal components in relation to amino acid incorporation by preparations from the developing rat brain, *Biochem. J.* **104**:229–238 (1967).
96. M. Satake, Y. Takahashi, K. Mase, and K. Ogata, Protein biosynthesis in a cell-free system of the guinea pig brain. I. Incorporation *in vitro* of C^{14}-leucine into protein by brain microsomes, *J. Biochem.* **56**:504–511 (1964).
97. D. H. Adams and L. Lim, Amino acid incorporation by preparations from the developing rat brain, *Biochem. J.* **99**:261–265 (1966).
98. Y. Takahashi, K. Mase, and S. Abe, The protein biosynthesis in a ribosomal system of the brain, *J. Biochem.* **60**:363–371 (1966).
99. G. Acs, A. Neidle, and N. Schneiderman, The effect of fatty acids on amino acid incorporation, *Biochim. Biophys. Acta* **56**:373–374 (1962).
100. A. L. Rubin and K. H. Stenzel, *In vitro* synthesis of brain protein, *Proc. Nat. Acad. Sci. U.S.* **53**:963–968 (1965).
101. G. Acs, A. Neidle, and H. Waelsch, Brain ribosomes and amino acid incorporation, *Biochim. Biophys. Acta* **50**:403–404 (1961).
102. A. Lajtha, in *Chemical Pathology of the Nervous System* (J. Folch-Pi, ed.) pp. 268–275, Pergamon Press, New York (1961).
103. M. K. Campbell, H. R. Mahler, W. J. Moore, and S. Tewari, Protein synthesis systems from rat brain, *Biochemistry* **5**:1174–1184 (1966).

104. D. H. Clouet, M. Ratner, and N. Williams, [^{14}C]leucine incorporation into brain ribosomes, *Biochim. Biophys. Acta* **123**:142–150 (1966).

105. K. H. Stenzel, R. F. Aronson, and A. L. Rubin, *In vitro* synthesis of brain protein. II. Properties of the complete system, *Biochemistry* **5**:930–936 (1966).

106. M. R. V. Murthy, Protein synthesis in growing rat tissues. I. Effect of various metabolites and inhibitors on phenylalanine incorporation by brain and liver ribosomes, *Biochim. Biophys. Acta* **119**:586–598 (1966).

107. S. H. Appel, Inhibition of brain protein synthesis: An approach to the biochemical basis of neurological dysfunction in the amino-acidurias, *Trans. N.Y. Acad. Sci.* **29**:63–70 (1966).

108. Y. Takahashi and Y. Akabane, Protein metabolism of rat brain slices, *Can. J. Biochem. Physiol.* **38**:1149–1157 (1960).

109. J. Folbergrová, Incorporation of S^{35}-methionine into proteins of brain cortex slices, *Physiol. Bohemosloven.* **10**:122–129 (1961).

110. J. Folbergrová, The effect of potassium and some other factors upon the incorporation of S^{35}-methionine into proteins of brain cortex slices, *Physiol. Bohemosloven.* **10**:130–138 (1961).

111. J. Folbergrová, Incorporation of labeled amino acids into the proteins of brain cortex slices *in vitro* in the presence of other non-radioactive amino acids, *J. Neurochem.* **13**:553–562 (1966).

112. K. Mase, Y. Takahashi, and K. Ogata, The incorporation of [^{14}C]glycine into the protein of guinea pig brain cortex slices, *J. Neurochem.* **9**:281–288 (1962).

113. S. Roberts and B. S. Morelos, Regulation of cerebral metabolism of amino acids—IV. Influence of amino acid levels on leucine uptake, utilization and incorporation into protein *in vivo*, *J. Neurochem.* **12**:373–387 (1965).

114. F. Orrego and F. Lipmann, Protein synthesis in brain slices. Effects of electrical stimulation and acidic amino acids, *J. Biol. Chem.* **242**:665–671 (1967).

115. S. C. Bondy and S. V. Perry, Incorporation of labeled amino acids in the soluble protein fraction of rabbit brain, *J. Neurochem.* **10**:603–609 (1963).

116. M. Revel and H. H. Hiatt, Magnesium requirement for the formation of an active messenger RNA-ribosome-S-RNA complex, *J. Mol. Biol.* **11**:467–475 (1965).

117. A. Gierer, Function of aggregated reticulocyte ribosomes in protein synthesis, *J. Mol. Biol.* **6**:148–157 (1963).

118. J. O. Bishop, Reticulocyte ribosome fraction with an exceptional capacity for polyphenylalanine synthesis, *Nature* **208**:361–365 (1965).

119. H. Waelsch and A. Lajtha, Protein metabolism in the nervous system, *Physiol. Rev.* **41**:709–736 (1961).

120. P. Weiss and H. B. Hiscoe, Experiments on the mechanism of nerve growth, *J. Exp. Zool.* **107**:315–396 (1948).

121. B. Droz and C. P. LeBlond, Axonal migration of proteins in the central nervous system and peripheral nerves as shown by radioautography, *J. Comp. Neurol.* **121**:325–337 (1963).

122. S. H. Barondes, Delayed appearance of labeled protein in isolated nerve endings and axoplasmic flow, *Science* **146**:779–781 (1964).

123. A. Edström, Amino acid incorporation in isolated Mauthner nerve fibre components, *J. Neurochem.* **13**:315–321 (1966).

124. B. Droz, Synthèse et transfert des protéines cellulaires dan les neurones ganglionnaires étude radioautographique quantitative en microscopie électronique, *J. Microscopie* **6**:201–228 (1967).

125. L. Z. Pevzner, Topochemical aspects of nucleic acid and protein metabolism within the neuron—neuroglia unit of the superior cervical ganglion, *J. Neurochem.* **12**:993–1002 (1965).

126. D. H. Clouet and H. Waelsch, Amino acid and protein metabolism of the brain—VIII. The recovery of cholinesterase in the nervous system of the frog after inhibition, *J. Neurochem.* **8**:201–215 (1961).

127. E. Koenig and G. B. Koelle, Mode of regeneration of acetylcholinesterase in cholinergic neurons following irreversible inactivation, *J. Neurochem.* **8**:169–188 (1961).

128. E. Koenig, Synthetic mechanisms in the axon—I. Local axonal synthesis of acetylcholinesterase, *J. Neurochem.* **12**:343–355 (1965).

129. E. Koenig, Synthetic mechanisms in the axon—IV. *In vitro* incorporation of [^{3}H]precursors into axonal protein and RNA, *J. Neurochem.* **14**:437–446 (1967).

130. I. G. Morgan and L. Austin, Synaptosomal protein synthesis in a cell-free system, *J. Neurochem.* **15**:41–51 (1968).

131. M. K. Gordon, K. G. Bench, G. G. Deanin, and M. W. Gordon, Histochemical and biochemical study of synaptic lysosomes, *Nature* **217**:523–527 (1968).

132. Y. Takahashi, M. Nomura, and S. Furusawa, *In vitro* incorporation of [^{14}C]-amino acids into proteins of peripheral nerve during Wallerian degeneration, *J. Neurochem.* **7**:97–102 (1961).

133. A. Edström, Inhibition of protein synthesis in Mauthner nerve fibre components by actinomycin-D, *J. Neurochem.* **14**:239–243 (1967).

134. F. T. Mérei and F. Gallyas, Quantitative determination of the uptake of [^{35}S]methionine in different regions of the normal rat brain, *J. Neurochem.* **11**:257–264 (1964).

135. J. Altman, Differences in the utilization of tritiated leucine by single neurones in normal and exercised rats: An autoradiographic investigation with microdensitometry, *Nature* **199**:777–780 (1963).

136. G. P. Talwar, S. P. Chopra, B. K. Goel, and B. D'Monte, Correlation of the functional activity of the brain with metabolic parameters—III. Protein metabolism of the occipital cortex in relation to light stimulus, *J. Neurochem.* **13**:109–116 (1966).

137. H. Hydén and P. W. Lange, Protein synthesis in the hippocampal pyramidal cells of rats during a behavioral test, *Science* **159**:1370–1373 (1968).

138. F. T. Mérei and F. Gallyas, Protein metabolism of the CNS at high plasma sodium level studied with [^{35}S]methionine, *J. Neurochem.* **11**:265–270 (1964).

139. T. C. Johnson and M. W. Luttges, The effects of maturation on *in vitro* protein synthesis by mouse brain cells, *J. Neurochem.* **13**:545–552 (1966).

140. K. Suzuki, S. R. Korey, and R. D. Terry, Studies on protein synthesis in brain microsomal system, *J. Neurochem.* **11**:403–412 (1964).

141. Y. Takahashi and S. Abe, Distribution of amino acid activating enzymes in rabbit's brain, *Experientia* **19**:186–187 (1963).

142. S. H. Appel, Turnover of brain messenger RNA, *Nature* **213**:1253–1254 (1967).

143. B. M. Lippe and C. M. Szego, Participation of adrenocortical hyperactivity in the suppressive effect of systemic actinomycin D on uterine stimulation by oestrogen, *Nature* **207**:272–274 (1965).

144. R. Soeiro and H. Amos, mRNA half-life measured by use of actinomycin D in animal cells—a caution, *Biochim. Biophys. Acta* **129**:406–409 (1966).

145. M. K. Gaitonde and D. Richter, The metabolic activity of the proteins of the brain, *Proc. Roy. Soc.* (*London*) *B* **145**:83–99 (1956).

146. A. Lajtha, S. Furst, A. Gerstein, and H. Waelsch, Amino acid and protein metabolism of the brain—I. Turnover of free and protein bound lysine in brain and other organs, *J. Neurochem.* **1**:289–300 (1957).

147. R. J. Schain, M. J. Carver, J. H. Copenhaver, and N. R. Underdahl, Protein metabolism in the developing brain: Influence of birth and gestational age, *Science* **156**:984–986 (1967).

148. A. A. Abdel-Latif and L. G. Abood, *In vivo* incorporation of L-(^{14}C) serine into phospholipids and proteins of the subcellular fractions of developing rat brain, *J. Neurochem.* **13**:1189–1196 (1966).

149. L. C. Mokrasch and P. Manner, Incorporation of ^{14}C-amino acids and [^{14}C]-palmitate into proteolipids of rat brains *in vitro*, *J. Neurochem.* **10**:541–547 (1963).

150. H. C. Agrawal, J. M. Davis, and W. A. Himwich, Postnatal changes in free amino acid pool of rat brain, *J. Neurochem.* **13**:607–615 (1966).

151. B. M. Hanking and S. Roberts, Stimulation of protein synthesis *in vitro* by elevated levels of amino acids, *Biochim. Biophys. Acta* **104**:427–438 (1965).

152. B. M. Hanking and S. Roberts, Influence of alterations in intracellular levels of amino-acids on protein-synthesizing activity of isolated ribosomes, *Nature* **207**:862–864 (1965).

153. D. S. Beattie, R. E. Basford, and S. B. Koritz, The turnover of the protein components of mitochondria from rat liver, kidney, and brain, *J. Biol. Chem.* **242**:4584–4586 (1967).

154. I. A. Michaelson, The subcellular distribution of acetylcholine, choline acetyltransferase and acetyl cholinesterase in nerve tissue, *Ann. N.Y. Acad. Sci.* **144**:387–410 (1967).

155. C. B. Klee and L. Sokoloff, Mitochondrial differences in mature and immature brain. Influence on rate of amino acid incorporation into protein and responses to thyroxine, *J. Neurochem.* **11**:709–716 (1964).

156. C. B. Klee and L. Sokoloff, Amino acid incorporation into proteolipid of myelin *in vitro*, *Proc. Nat. Acad. Sci. U.S.* **53**:1014–1021 (1965).

157. H. S. Bachelard, Amino acid incorporation into the protein of mitochondrial preparations from cerebral cortex and spinal cord, *Biochem. J.* **100**:131–137 (1966).

158. L. C. Mokrasch, Incorporation of [^{14}C]amino acids into the proteolipid of subcellular preparations of rat brain *in vitro*, *J. Neurochem.* **13**:49–58 (1966).

159. D. Neubert and H. Helge, Studies on nucleotide incorporation into mitochondrial RNA, *Biochem. Biophys. Res. Comm.* **18**:600–605 (1965).

160. S. H. Barondes, On the site of synthesis of the mitochondrial protein of nerve endings, *J. Neurochem.* **13**:721–727 (1966).

161. L. B. Flexner, in *Biochemistry of the Developing Nervous System* (H. Waelsch, ed.) pp. 281–300, Academic Press, New York (1955).

162. R. E. Kuhlman and O. H. Lowry, Quantitative histochemical changes during the development of the rat cerebral cortex, *J. Neurochem.* **1**:173–180 (1956).

163. A. von der Decken, in *Techniques in Protein Biosynthesis* (P. N. Campbell and J. R. Sargent, ed.) Vol. 1, pp. 65–131, Academic Press, New York (1967).

164. K. S. McCarty, J. T. Parsons, W. A. Carter, and J. Laszlo, Protein-synthetic capacities of liver nuclear subfractions, *J. Biol. Chem.* **241**:5489–5499 (1966).

165. L. S. Hnilica, in *Progress in Nucleic Acid Research and Molecular Biology* (J. N. Davidson and W. E. Cohn, eds.) Vol. 7, pp. 25–106, Academic Press, New York (1967).

166. J. Bonner, M. E. Dahmus, D. Fambrough, R-C. C. Huang, K. Marushige, and D. Y. H. Tuan, The biology of isolated chromatin, *Science* **159**:47–56 (1968).

167. R. S. Piha, M. Cuénod, and H. Waelsch, Metabolism of histones of brain and liver, *J. Biol. Chem.* **241**:2397–2404 (1966).

168. B. W. Moore and D. McGregor, Chromatographic and electrophoretic fractionation of soluble proteins of brain and liver, *J. Biol. Chem.* **240**:1647–1653 (1965).

169. B. W. Moore, A soluble protein characteristic of the nervous system, *Biochem. Biophys. Res. Comm.* **19**:739–744 (1965).

170. B. W. Moore and V. J. Perez, in *Physiological and Biochemical Aspects of Nervous Integration* (F. D. Carlson, ed.) pp. 343–359, Prentice-Hall, Englewood Cliffs, N.J. (1968).

171. G. Gombos, G. Vincendon, J. Tardy, and P. Mandel, Hétérogénéité électrophorétique et préparation rapide de la fraction protéique S 100, *C. R. Acad. Sci. Paris* **263**:1533–1535 (1966).

172. B. S. McEwen and H. Hydén, A study of specific brain proteins on the semi-micro scale, *J. Neurochem.* **13**:823–833 (1966).

173. B. S. McEwen, in *Physiological and Biochemical Aspects of Nervous Integration* (F. D. Carlson, ed.) pp. 361–381, Prentice-Hall, Englewood Cliffs, N.J. (1968).

174. H. Hydén and B. McEwen, A glial protein specific for the nervous system, *Proc. Nat. Acad. Sci. U.S.* **55**:354–358 (1966).

175. A. R. Dravid and J. A. Burdman, Acidic proteins in rat brain nuclei: Disc electrophoresis, *J. Neurochem.* **15**:25–30 (1968).

176. L. G. Tomasi and S. E. Kornguth, Characterization and cellular localization of a basic protein purified from pig brain, *J. Cell Biol.* **35**, 133A (1967).

177. K. Warecka and H. Bauer, Studies on "brain-specific" proteins in aqueous extracts of brain tissue, *J. Neurochem.* **14**:783–787 (1967).

178. R. H. Laatsch, M. W. Kies, S. Gordon, and E. C. Alvord, Jr., The encephalomyelitic activity of myelin isolated by ultracentrifugation, *J. Exp. Med.* **115**:777–788 (1962).

179. R. F. Kibler and R. Shapira, Isolation and properties of an encephalitogenic protein from bovine, rabbit, and human central nervous system tissue, *J. Biol. Chem.* **234**:281–286 (1968).

180. J. A. Burdman and A. R. Dravid, Isolation and separation of proteins from chick embryo sensory ganglia, *Brain Res.* **6**:355–358 (1967).

181. D. Gandini-Attardi, P. Calissano, and P. Angeletti, Protein synthesis in embryonic sensory ganglia: Effect of the nerve growth factor on soluble proteins, *Brain Res.* **6**:367–370 (1967).

182. M. Wender and Z. Waligóra, The content of amino acids in the proteins of the developing nervous system of the guinea pig—I. Cerebral white matter, *J. Neurochem.* **7**:259–263 (1961).

183. M. Wender and Z. Waligóra, The content of amino acids in the proteins of the developing nervous system of the guinea pig—II. Cerebral gray matter, *J. Neurochem.* **9**:115–118 (1962).

184. K. F. Swaiman and C. E. Nelson, Soluble protein nitrogen and total protein nitrogen in developing rabbit brain, *J. Neurochem.* **14**:905–910 (1967).

185. J. Folch-Pi, in *Protides of the Biological Fluids, Proceedings of the Thirteenth Colloquium, Bruges, 1965* (H. Peeters, ed.) Vol. 13, pp. 21–34, Elsevier Publishing Co., Amsterdam (1966).

186. F. Wolfgram, The amino acid compositions of some non-neural proteolipid proteins, *Biochim. Biophys. Acta* **147**:383–385 (1967).

187. D. S. Beattie, R. E. Basford, and S. B. Koritz, The inner membrane as the site of the *in vitro* incorporation of L-[^{14}C]leucine into mitochondrial protein, *Biochemistry* **6**:3099–3106 (1967).

188. W. Neupert, D. Brdiczka, and Th. Bücher, Incorporation of amino acids into the outer and inner membrane of isolated rat liver mitochondria, *Biochem. Biophys. Res. Comm.* **27**:488–493 (1967).

189. V. P. Whittaker, I. A. Michaelson, and R. J. A. Kirkland, The separation of synaptic vesicles from nerve-ending particles ("synaptosomes"), *Biochem. J.* **90**:293–303 (1964).

190. P. R. Lewis and C. C. D. Shute, The distribution of cholinesterase in cholinergic neurons demonstrated with the electron microscope, *J. Cell. Sci.* **1**:381–390 (1966).

191. E. Koenig, Synthetic mechanisms in the axon—III. Stimulation of acetylcholinesterase synthesis by actinomycin-D in the hypoglossal nerve, *J. Neurochem.* **14**:429–435 (1967).

192. B. C. Goodwin and I. W. Sizer, Histone regulation of lactic dehydrogenase in embryonic chick brain tissue, *Science* **148**:242–244 (1965).

193. Y. Takabatake and H. Sachs, Vasopressin biosynthesis. III. *In vitro* studies, *Endocrinology* **75**:934–942 (1964).

194. H. Borsook and C. L. Deasy, The metabolism of proteins and amino acids, *Ann. Rev. Biochem.* **20**:209–226 (1951).

195. C. P. LeBlond and B. E. Walker, Renewal of cell populations, *Physiol. Rev.* **36**:255–276 (1956).

196. A. Lajtha, in *Neurochemistry* (K. A. C. Elliott, I. H. Page, and J. H. Quastel, eds.) 2nd ed., pp. 399–430, Charles C. Thomas, Springfield, Ill. (1962).

197. C. E. Zomzely, S. Roberts, S. Peache, and D. M. Brown, Cerebral protein synthesis. III. Developmental alterations in the stability of cerebral messenger ribonucleic acid-ribosome complexes, *J. Biol. Chem.* **246**:in press (1971).

198. E. Roberts, Models for correlative thinking about brain, behavior, and biochemistry, *Brain Res.* **2**:109–144 (1966).

199. C. E. Zomzely, S. Roberts, and S. Peache, *Proc. Natl. Acad. Sci. US* **67**:644–651.

Chapter 2

PROTEIN AND POLYPEPTIDE BREAKDOWN

N. Marks and A. Lajtha

*New York State Research Institute
for Neurochemistry and Drug Addiction,
Ward's Island, New York*

I. INTRODUCTION

The dynamic state, or continuous turnover of proteins, in the brain as well as in other tissues, is an extensive but markedly heterogeneous process.[1p,4] Protein turnover includes not only synthesis but also breakdown of peptide macromolecules according to the physiological and metabolic requirements of the tissue. Knowledge of the events of protein synthesis and their control is well advanced, particularly in microorganisms[26]; in contrast, knowledge concerning protein breakdown, which forms an integral part of the regulatory mechanism affecting protein composition in any organ, is rudimentary. A full understanding of degradative processes is required if we are to interpret the many factors that influence protein turnover in animals; these include the effect of hormones and drugs,[4,20a,d] pathological states,[18r] hypothermia,[27] states of excitation and exhaustion,[23] and the nutritional status of the animal, especially in relation to the availability of essential amino acids.[19,20c] Proteinases in association with other tissue components appear to play a role in a number of important physiological processes. Aspects pertinent to nerve function include the role of proteolytic enzymes in embryogenesis and morphogenesis,[2a,18q] inflammatory and allergic conditions,[18c] metabolism of connective and arterial tissue especially as related to permeability properties,[18m] and formation of physiologically active peptides (kinins, angiotensin, substance P, pituitary hormones, releasing factors, etc.).[15,16,22]

Like other tissues, brain contains a diversity of proteins, with variable half-lives; the average is 15 days, but minor components show a range extending from less than 1 hr to several months.[4] Any value for protein turnover at present must be regarded as an estimate, not only because the true specific activity of the precursor amino acid is uncertain, but also because in a population of heterogeneous proteins precise information is required on the rates of protein breakdown in addition to rates of synthesis based on amino acid incorporation. Rates of degradation, in turn, cannot be computed easily from the rates of amino acid incorporation, or release, since it cannot always be assumed that the rates of synthesis and degradation are in balance: additional complicating factors are the local reutilization of newly liberated amino acids and the heterogeneity of the amino acid pools.[1v,29]

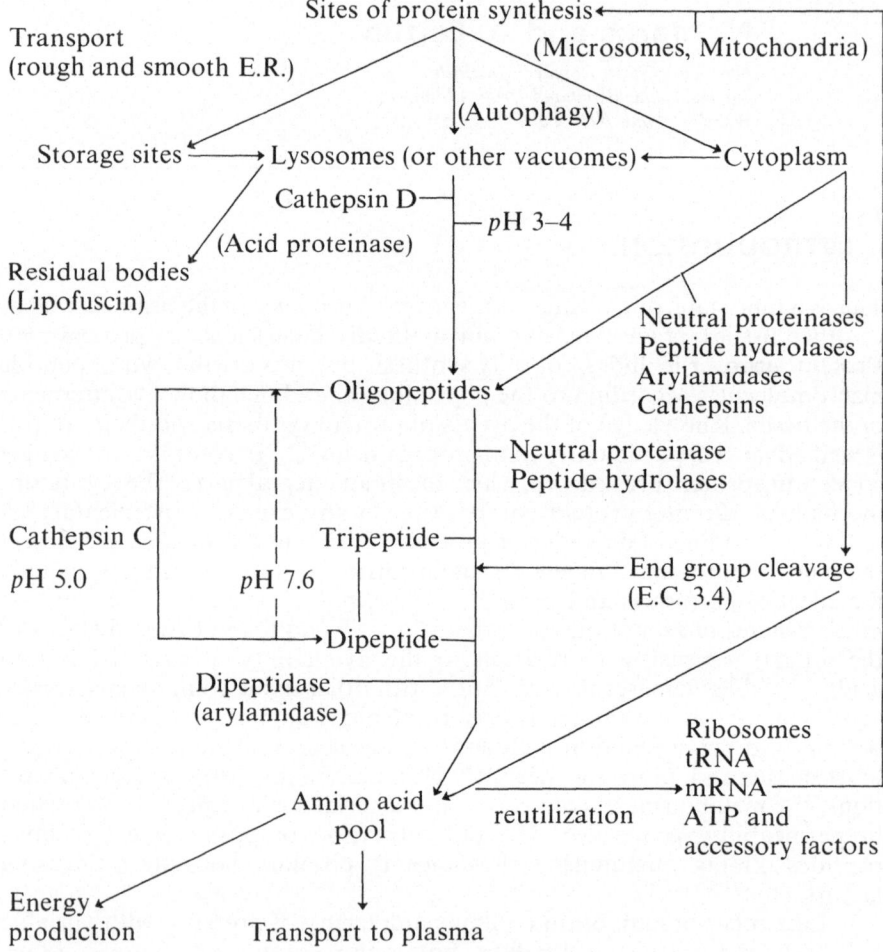

Fig. 1. Schematic diagram of postulated mechanisms in peptide breakdown and turnover in the brain. Proteins are degraded by sequential action of proteinases at specific intracellular sites (lysosomes, etc.) to yield polypeptides and amino acids. Polypeptides are exported to the cytoplasm by unknown mechanisms and further degraded by a variety of peptide hydrolases.

Studies on protein degradation in animal tissues have often relied on direct measurement of the agents chiefly involved in the hydrolysis of proteins, namely peptide or peptidyl-peptide hydrolases (E.C. 3.4 and 3.4.4). The main purpose of this review is to focus attention on the intracellular proteinases (peptidyl-peptide hydrolases), since peptide hydrolases are reviewed elsewhere in the Handbook.[1j] The major pathway of protein breakdown is considered to involve the initial cleavage of proteins by peptidyl-peptide hydrolases followed by sequential attack by the multitude of peptide hydrolases present in all tissues, including the brain[2a] (Fig. 1). Exact schemes for protein degradation can be formulated only when knowledge is available for isolating the individual enzymes and their native substrates, together with knowledge on their localization, mechanism of activation, and peptide bond specificity. In the case of brain, and also for most tissues, none of these criteria have been satisfied; intracellular proteinases are not available in a high state of purity, very little is known concerning their mechanism of action, and less information is available concerning their physiological substrates. Because of the limited information available, analogies to similar enzymes in other tissues, or microorganisms, are included in this review to help shed light on mechanisms of protein breakdown. Despite the relative abundance of degradative enzymes, a number of questions arise about the factors that determine *when* and *why* a particular protein species or structural element is broken down. These questions form part of the problem of total regulation of all enzyme activities within cells and their relationship to disease, and to the higher aspects of brain function.

A. Classification

The enzymes discussed in this review are, briefly, those cleaving *internal* peptide bonds in addition to possible end-group hydrolysis: the Enzyme Commission (E.C.)[13] assigned the name peptidyl-peptide hydrolases (E.C. 3.4.4) to this subgroup, although other convenient terms exist (endopeptidases, proteases, proteinases) (Table IA). The generic term, proteolytic enzymes, refers to all hydrolases cleaving peptide bonds (E.C. 3.4 and 3.4.4), and can lead to some confusion unless such enzymes are specified. Owing to inadequate characterization, especially of the intracellular enzymes of protein turnover, few systematic names are possible within this subgroup and many enzymes were excluded from the last (1964) classification. The trend to compile bibliographies based on the 1964 list therefore frequently results in a lack of references on many of the important enzymes involved in protein breakdown in tissues.[1c,13,344]

In one method of classification, three broad categories can be distinguished on the basis of the probable active centers: (a) serine hydrolases (trypsin and chymotrypsin), (b) sulfhydryl hydrolases (papain, and many intracellular proteinases), and (c) acid proteinases (pepsin, cathepsin D).[30] The alternative method, and certainly the one most convenient for present descriptive purposes, is to group proteinases according to their tissue of

TABLE IA

Proteolytic Enzymes Cited in the Present Review
According to Their E.C. Listing[a]

E.C. listing	Nomenclature	Substrate	Specificity (Section)
3.4.4.	Peptidyl-peptide hydrolases (proteases, endopeptidases, etc.)	—	—
3.4.4.1.	Pepsin (pepsinogen)	Proteins, peptides	Bonds adjacent to aromatic or dicarboxylic L-amino acids, pH2.0.
3.4.4.4.	Trypsin (trypsinogen)	Amides, esters, peptides	Bonds with —COOH of Arg or Lys.
3.4.4.5.6.	Chymotrypsin A or B (chymotrypsinogen)	Peptides, amides	Bonds with —COOH of aromatic amino acids.
3.4.4.7.	Pancreatopeptidase or elastase	Elastin, proteins	Bonds adjacent to neutral amino acids.
3.4.4.9.	Cathepsin C (dipeptidyl transferase or aminopeptidase I)	Dipeptides, amides	α-amino groups, Cl^- and —SH dependent.
3.4.4.13.	Thrombin (prothrombin)	Peptides, amides, esters	Fibrinogen to fibrin.
3.4.4.14.	Plasmin (fibrolysin)	Peptides, amides, esters of Arg and Lys	Fibrin into soluble products.
3.4.4.15.	Rennin	Angiotensin I	Formation of angiotensin II (octapeptide).
3.4.4.21.	Kallikrein	Kininogens	Formation of kinins (kinin 9–11) (bradykinin).
3.4.4.23.	Cathepsin D	Proteins (pH 3–4) peptides	Similar in specificity to pepsin with insulin B chain.

[a] Partial listing from Enzyme Classification of 1964.[13] Precursor forms or alternative nomenclature is noted in parentheses. An alternative classification encompassing enzymes excluded from the E.C. list is shown in Table IB.

origin, subcellular or other localization, and possible physiological role (Table IB). Several groups have been proposed[2a,279]: (a) digestive enzymes (pancreatic enzymes excreted extracellularly, etc.), (b) intracellular proteinases (cathepsins, natural proteinases, etc.), (c) circulatory enzymes (thrombin, plasmin, kininases, etc.), (d) connective tissue enzymes (collagenase, elastase, etc.). It must be emphasized that with the complex level of organization within tissues there is an inevitable overlap in the classification based on biological properties. As noted in a previous review,[2a] protein breakdown can be considered to be the complex end result of sequential breakdown of

TABLE IB

Tentative Classification Based on Biological Properties[a]

Group	Enzymes	Substrates
A. Extracellular enzymes	Trypsin, chymotrypsin, pepsin, etc.	Proteins, peptides, esters
B. Intracellular enzymes	Cathepsin A (carboxypeptidase) (pH 5–6, —SH dependent)	Z-Glu-Tyr (Section II,A)
	Cathepsin B (trypsinlike)	BANA (Section II,B)
	Cathepsin B'	BAA (Section II,B)
	Cathepsin C (dipeptidyl transferase) (pH 5–6, —SH dependent) Cl^- dependent	Gly-Phy · NH_2, Arylamides (Section II,C), polypeptides
	Cathepsin D (acid proteinase) pH 3–4, cathepsin E	Proteins, peptides, insulin B, or selected chain sequences (Section II, D)
	Peptide hydrolases (E.C. 3.4) (metalloenzymes-carboxypeptidase)	
C. Enzymes present in circulation	Thrombin, plasmin, kininases (inactivation of angiotensin, bradykinin, oxytocin, vasopressin, insulin, glucagon, substance P, etc.)	Kininogens, angiotensin I, etc. (Section V)
D. Enzymes associated with the vasculature and connective tissue	Collagenase, elastase	Native collagen or elastin (Section VIII)

[a] Scheme based on that proposed in earlier reviews.[2a,30] Description of relevant enzymes is included in the text as noted in the table. Abbr: BAA, benzoyl arginine amide; BANA, benzoyl arginine naphthylamide.

large proteins or polypeptides by different batteries of intracellular peptide hydrolases (E.C. 3.4) and proteinases (3.4.4). Peptide hydrolases were reviewed in this handbook (Vol. 3) and are discussed here only where their action cannot be clearly differentiated from true endopeptidases (see Sections V, A and H). In the absence of a rational classification, the term cathepsins is adopted for all intracellular proteinases active at acidic pH. Other enzymes will be identified, where possible, according to the nature of the substrate they attack, but nomenclature that implies specificity will be avoided (e.g., insulinase, protaminase, angiotensinase, oxytocinase).

Neurath et al.[30] have drawn attention to the immense variety of biological functions subserved by enzymes utilizing common catalytic mechanisms (digestion of dietary proteins, activation of zymogens, blood clotting, clot lysis, sensing pain, etc.). Therefore some attempt has been made to further subdivide groups according to the structure of the active centers or sites.

For example, serine hydrolases can be classified according to the tetrapeptide sequence adjacent to serine as microbial serine hydrolases, Thr-Ser-Met-Ala, and animal serine hydrolases, Gly-Asp-Ser-Gly. Technical difficulties have precluded precise studies concerned with the active centers of sulfhydryl proteinase including papain or the neutral proteinase derived from group A hemolytic streptococci. As in the case of peptide hydrolases, the question of isoenzymes in tissue, native or artifactual, has impeded structural studies.[18a,24] Although pepsin or its precursor forms was one of the first enzymes ever crystallized, recent studies show the presence of five different forms in human and porcine tissue with differing specificities when tested against insulin B chain.[31] Trypsin and collagenases from different tissues show different immunological properties, with possible significance for breakdown of native proteins.[14,32]

B. General Comment on Intracellular Enzymes

Insufficient information is available to assign intracellular proteinases to a specific category. Methods are based on the pH optima observed during autolysis of tissue, or properties of the partially purified enzyme in relation to substrate specificity and the effect of added cofactors. During autolysis, brain extracts and most other crude tissue preparations display two pH optima, one in the acid region, pH 2–6, and the other in the neutral or alkaline range (Fig. 2).[2a,33] Autolysis in tissue extracts at acid pH is a well-described phenomenon extending back to the earlier part of the century; Willstater and Baumann in 1928[34] proposed the term *Kathepsin* (from the Greek verb "to digest") for the enzymes involved. The term was adopted by Bergmann and Fruton[35,36] to describe a series of proteolytic enzymes active at acid pH. These studies were facilitated by the synthetic substrates introduced by Bergmann[35] for the characterization of trypsin and similar enzymes, and led to the reported presence in tissue of enzymes homospecific with pepsin (cathepsin A), with trypsin (cathepsin B), and with chymotrypsin (cathepsin C)[36] (Tables IA, B). In addition to the ability to degrade substrates such as denatured hemoglobin at pH 3–4, these enzymes also split synthetic substrates at pH 5–6 in the presence or absence of added metals or other cofactors. The use of synthetic substrates as the only criterion for characterizing such enzymes has been misleading since their studies failed to recognize the significance of an acid proteinase described earlier by Anson[37] and present in all tissues, which failed to hydrolyze simpler substrates. On the basis of studies in spleen and liver, Press et al.[38] suggested the term cathepsin D for this important intracellular proteinase that accounted for over two-thirds of the hemoglobin-splitting activity of tissues. Later the existence of an unstable form of cathepsin known currently as cathepsin E was reported.[39] Metrione et al.[40] suggested that the major function of cathepsin C is a transpeptidation reaction at physiological pH, and the name "dipeptidyl transferase" was proposed. Most cathepsins are associated with lysosomelike particles and are believed to participate in mechanisms of

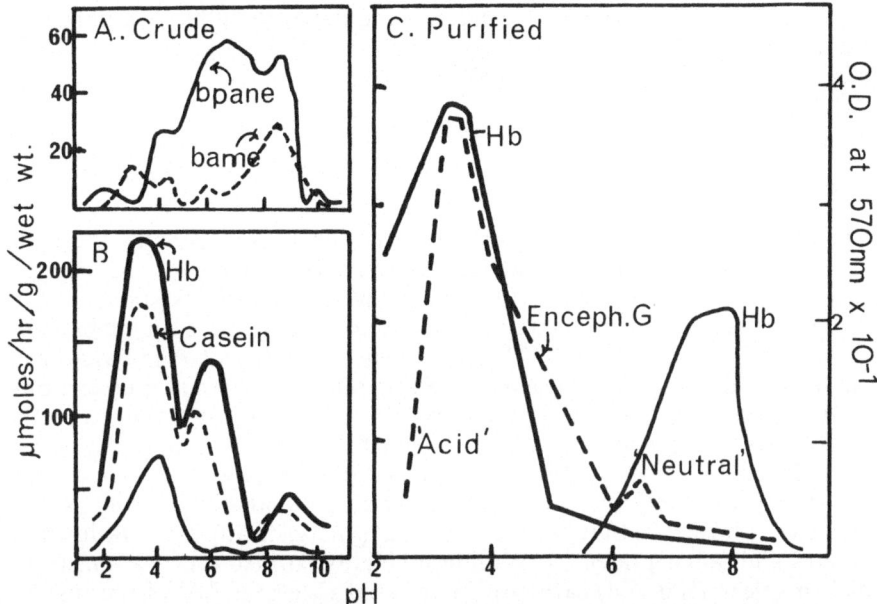

Fig. 2. Comparison of pH optima for crude (A and B) or purified (C) enzymes with a variety of protein or ester (tryptic or chymotryptic) substrates as noted on the figure. The unmarked line in B represents endogenous activity. Data for A and B are from Oja and Oja,[126] and for C on enzymes from rat brain purified according to our procedures.[2a,42] Homogenates were prepared from adult (150-day) rats with hypotonic buffer and 25 mg incubated with 5 μmoles of N^{α}-benzoyl- DL -Arg methyl ester (BAME) in a volume of 0.5 ml, or with 1 μmole N^{α}-benzoyl- DL -Phe 2-naphthyl ester (BPANE) in 2 ml for 2 hr at 37° (C). For proteinases, 10 mg homogenate was incubated with 4 mg denatured protein substrates for 1 hr (A, B).

protein breakdown during the processes of endocytosis or autophagy, or are released into the cytoplasm to participate in the turnover of cellular constituents. Catheptic enzymes are involved in Wallerian degeneration or chromatolysis, and in the turnover of connective tissues (bone resorption and tissue involution) (Sections IV and VIII).

Autolysis at neutral or slightly alkaline pH can be demonstrated readily in animal tissue, but purification of the proteinases involved, other than peptide hydrolases, has proved to be difficult.[33] Such enzymes occur in a variety of microorganisms and in many instances are now available in highly purified and crystalline form. In contrast, neutral proteinases of tissues are subject to considerable lability, which has prevented definitive studies concerning their identity. There is some debate concerning the physiological role of proteinases active at very low pH, although there is a good case for regional variations of cellular pH, especially within specific organelles. Although optimum pH for activity is undoubtedly dependent on cellular location and substrate, activity at the extremes of its pH range could make a significant contribution to breakdown. The pH range of neutral proteinases suggests that such enzymes are likely candidates to fulfill a number of physiological roles.

II. CATHEPSINS (ACID PROTEINASES)

A. Cathepsin A

The sulfhydryl-dependent hydrolase, cathepsin A, is defined by the hydrolysis of substrates homospecific with pepsin and carboxypeptidase A (e.g., Z-Glu-Tyr)[17,36,41] (Table IB). Only trace levels of carboxypeptidases have been detected in brain and it is difficult to detect this enzyme on the basis of similar synthetic substrates.[42,43] The presence of cathepsin A in tissues has been disputed[38] although it is reported present is lysosomal preparations of rat kidney, liver, and chicken spleen.[18a] Cathepsin A activity is greatly elevated in dystrophic muscles and was purified from this source[18]; at pH 5.0–5.4 it hydrolyzes glucagon and insulin at significant rates, but hemoglobin only slowly. Probably such preparations were contaminated by cathepsin D since there is evidence that both enzymes act synergistically in breakdown of proteins and polypeptides.[18r] With more highly purified preparations from chicken spleen (500-fold) cathepsin A released C-terminal Ala from insulin B chain, and a series of C-terminal amino acids from glucagon (hydrolysis was stopped by the appearance of Lys as the COOH-terminal). Since purified preparations were virtually without activity on protein substrates, such as hemoglobin, these findings probably indicate that cathepsin A is a variety of carboxypeptidase (E.C. 3.4.2.1). Once again, such studies indicate the need for careful evaluation of specificity with purified preparations before any nomenclature is applied. Matsunga et al.[44] in a survey of angiotensin activity in several tissues including brain (Table II) reported that the enzyme active at pH 5.0 (DFP sensitive) could not be easily distinguished from cathepsin A (Section V).

B. Cathepsins B and B′

The group of sulfhydryl-dependent hydrolases, including cathepsins B and B′ is defined by hydrolysis of substrates homospecific with trypsin (cathepsin B with BANA; B′ with BAA) (Table IB).[45,322] It is differentiated by the lack of effect of trypsin inhibitors including DFP, its pH optima (5–6), and the peptide bond specificity with insulin B chain. Purified preparations available from spleen (200-fold) are activated by —SH compounds (GSH, cysteine, etc.), but inhibited by PCMB and iodoacetamide, pointing to the involvement of one or more free —SH groups at the active center.[45–47] Catheptic B or tryptic-like activities are measurable in most tissue; lowest levels are in brain (5 % or less of that in spleen) (Table II).[42,43] In cases of low tissue activity it is possible that the tryptic-like activity of plasma, if present as a contamination, may make a significant contribution. As noted, brain shows barely detectable activity when BNA is employed, but at pH 7.6 it

TABLE II

Intracellular Hydrolases in Crude Homogenates of Different Tissues as Compared to Brain[a](75,62,338)

Tissue	Cathepsin B	Cathepsin C	Cathepsin D	Neutral proteinase	Acid phos-phatase	Asp^1-Val^5-Angiotensin II$^+$ pH 5.0	Asp^1-Val^5-Angiotensin II$^+$ pH 7.5
Spleen	*1.0*	*1.0*	*1.0*	*1.0*	*1.0*	*1*	*1*
Thymus	0.2	0.3	0.6	—	0.2	—	—
Blood	*	*	*	—	*	—	—
Heart	*	0.1	0.1	—	0.1	*	0.2
Lung	0.2	0.3	0.4	—	0.2	0.3	0.8
Liver	0.2	2	0.6	1.0	0.9	0.9	1.8
Kidney	0.2	0.5	0.7	2.5	0.8	1.4	2.0
Brain	*	*	0.2	0.7	0.1	*	0.2
Muscle	0.1	*	0.1	0.2	*	*	—
Adrenal	0.1	0.1	0.9	—	0.4	*	—

[a] Enzyme activities for cathepsins B and C were measured with BAA and Gly · Phe · NH$_2$ (ammonia release); for proteinases with denatured hemoglobin (ninhydrin); for acid phosphatase with β-glycerophosphate as substrates. Data were obtained with rat[42,68] or bovine tissues.[68,44] The asterisk(*) denotes activity of 0.05 or less compared to spleen. Cathepsin A with the substrate Z-Glu-Tyr (pH 5.0) is undetectable in brain tissue.[42,68] For comparison, an enzyme-inactivating angiotensin II is included since this shows some similarity to cathepsin A at pH 5.0 in other tissues.[44]

readily hydrolyzes Arg-, Arg-Arg-, Lys-Lys-, Lys-Ala-BNA.[2a] A second enzyme (B′) with a marked specificity toward BAA compared to BANA[45] is distinct in its properties from cathepsins, aminopeptidases, and arylamidases. Its molecular weight (24,400) is lower than that of cathepsin B but is similar to trypsin and chymotrypsin B. This enzyme (cathepsin B′) is inhibited by heavy metals PMSF and TPCK, but is activated by EDTA and mercaptoethanol[45,47a] and is unaffected by DFP; it can be differentiated by its peptide specificity with insulin B chain (Fig. 3). Cathepsin B (like cathepsin C) can catalyze a transpeptidation reaction by transfer of an acyl group (from an amide or ester) to the α-COOH of a peptide, NH$_2$OH, or even water.[2a] BAA reacts with Gly-Leu to form the tripeptide benzoyl-Arg-Gly-Leu with liberation of ammonia. It will be noted that the mechanism involves the carbonyl group and is similar to transpeptidation by papain.[344]

C. Cathepsin C

Cathepsin C is a thiol-activated intracellular proteinase[48,327] hydrolyzing specific dipeptidyl amides or esters at pH 5.0 or denatured hemoglobin at pH 3.5. Polymerization of some peptidyl amides by cathepsin C at neutral

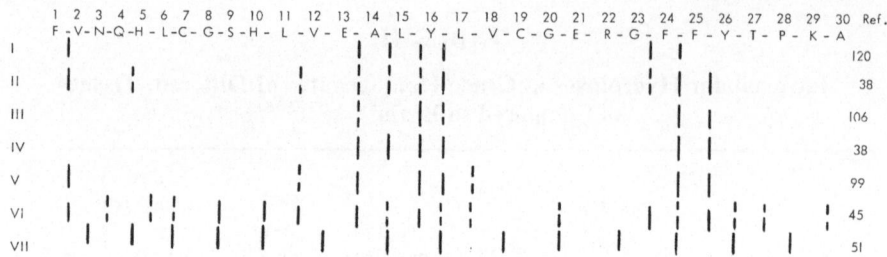

Fig. 3. Specificity of purified brain and proteinase (I and V) compared to pepsin (II), purified hog thyroid proteinase (III), spleen cathepsin D (IV), cathepsin B (VI), cathepsin C (VII). Dotted lines represent bonds less rapidly hydrolyzed. Oxidized beef insulin B chain. I.U.P.A.C. abbreviations (the first letter denotes the amino acid except Q = Glu·NH$_2$, E = Glu, Y = Tyr, F = Phe, R = Arg, K = Lys).

pH to form oligopeptides led Metrione et al.[40] to rename his enzyme "dipeptidyl transferase." In view of its dual catalytic functions, the term "cathepsin C" is retained for the present account; this enzyme has had a fascinating history during which its ability to hydrolyze proteins is emphasized less than its hydrolytic or transferase activity with peptide substrates. A significant development is the recent demonstration that some cathepsin C preparations show an absolute requirement for certain anions, notably halides, with a wider specificity than that originally proposed by Fruton.[49,50,58] Liver and spleen preparations appear to cleave polypeptides at pH 5.0 by removal of N-terminal dipeptide units; McDonald et al.[49] have suggested the alternative name dipeptidyl aminopeptidase I.

Brain and other excitable tissues (skeletal and cardiac muscle) contain 5% or less of the levels reported for spleen[42,43,62] (Table II). Early studies were performed without halide addition, but since most buffer systems fortuitously contain Cl$^-$ or other anions, the levels reported for crude preparations probably need little modification. The low levels in brain, as in the pituitary, may be due to the presence of endogenous inhibitors.[42,50,62]

1. Peptide-bond Specificity

Most preparations of purified enzyme show a very limited hydrolysis of proteins. Metrione et al.[40] reported only a 0.5% hydrolysis of albumin, 0.25% for ribonuclease, and 1.0% for the native insulin chain at pH 4.0. In contrast, recent studies with Cl$^-$ dependent preparations demonstrated a 50–90% hydrolysis of glucagon and insulin B chains (see dipeptide cleavage below).[51] It is possible that tissues contain a Cl$^-$ insensitive cathepsin with rigid specificity requirements, as proposed by Fruton,[36,40] and illustrated by the following example: A---B---X represents a dipeptide ester or amide, where B is a hydrophobic amino acid and A is variable. Substrates such as Gly-Phe-, Gly-Try-, Pro-Tyr-NH$_2$ fulfill these requirements with an N-terminal (i.e., α-amino or amino group of an L-amino acid or Gly). It has been suggested that dipeptides of this nature can aggregate to form a hexagonal conformation for interaction with the active site of cathepsin C.[52]

TABLE III

Specificity of Cathepsin C (Dipeptidyl Transferase) E.C. 3.4.4.9[a]

A. Hydrolase function (pH 5–6, —SH, Cl⁻)

   1. Polypeptides

        ACTH (N-terminal dipeptides Ser-Tyr, Ser-His · · · Lys-Pro-Val)

        Angiotensin II (N-terminal Asn-Arg, Val-Tyr)

        Insulin B (90% of all N-terminal dipeptides)

        Glucagon (50% of N-terminal dipeptides)

        Secretin (80% of N-terminal dipeptides)

        B-corticotrophin (Ser-Tyr, Ser-Met, Gly-His, Phe-Arg, Trp-Gly)

        Polyalanine (N-terminal dialanines)

   2. Peptides

        Ala_3 (and other tripeptides), Ala_4, Ala_6, Gly_4, etc.

   3. Esters

        Gly-Phe-, Ser-Met-, Ala-Ala-, etc.

   4. Amides

        Ala-Arg-, Ser-Met-, Ala-Ala-, Ser-Tyr-, Gly-Tyr-

   5. Arylamides

        Gly-Arg-,(100) Ala-Arg-,(72) Pro-Arg-,(55) Ala-Aal-,(15) Gly-His-,(11) Ser-Tyr-,

        Phe-Arg-, His-Phe-, Gly-Phe, Gly-Phe, His-Ser, Leu-Ala, Asp-Ala

B. Polymerase function (pH 7.5)

        Gly-Phe · NH_2-----Tetrapeptide: Gly-Trp-NH_2-----octapeptide:

        Gly-Arg · βNA-----(Gly-Arg)$_5$ · βNA, etc.

[a] Substrates within groups 2–5 are listed in descending order of susceptibility. For arylamides the values in parentheses indicate rates of hydrolysis relative to Gly-Arg-βNA as 100. Data for the most part represent specificity of spleen or liver enzymes[49,50] but some data represent hydrolysis of arylamides or amides by purified preparations of brain enzyme.[29]

Chloride-dependent cathepsin C preparations show a considerably wider specificity as summarized in Table III. In agreement with Fruton,[36,40] dipeptides are favored with an N-terminal Gly- or Pro- end group. There have been some unconfirmed reports that cathepsin C preparations contain metal-activated proteinases,[53,54] but closer examination has revealed these to be peptide hydrolases, capable of end group cleavage yielding free amino acids.[55] In a typical experiment, Gly-Phe-βNA or its analogues at pH 5.0 in the presence of cysteine are cleaved to release the dipeptide Gly-Phe and the free amino (measured by colorimetric or fluorimetric methods). Dipeptide cleavage has important biological implications and may play a significant role in protein breakdown.[2a,51] The precise tissue distribution of cathepsin C is unknown, although there is some evidence that it is present in lysosomes in association with other cathepsins.[18a,l] Most preparations described in the literature are based on the purification scheme of de la Habra et al.[56] with some modifications. This involves extraction from spleen, ammonium sulfate precipitation, heat treatment at 65°, acetone and ethanol–zinc precipitation. Most preparations represent a 300-fold enrichment, but there is one study reporting a 800-fold purification with crystallization of the enzyme.[57] Cathepsin C is a large molecule of 197,000 consisting of eight subunits with six N-terminal Asp or Asn and two Leu arranged in the form of two tetramers[40,52]; analysis indicates the presence of eight free —SH. The sulfhydryl groups may be involved in the formation of a free thiol-ester intermediate possibly facilitated by specific anions.[52] The most effective anion for activation of cathepsin C is Cl^-, followed by Br^-, I^-, thiocyanate, nitrate, chlorate, but not F^-.[49,51,58] Cathepsin C is activated by cysteine but is inhibited by PCMB and iodacetamide; this fact is consistent with the presence of free —SH essential for activity. The specificity of cathepsin has been studied in detail and deserves some consideration in relation to the possible role it may play in nerve tissue (Table V).

2. *Dipeptide Cleavage*

Planta *et al.*[59,60] were the first to clearly demonstrate the "dipeptide hydrolase" activity of cathepsin C preparations. With the substrate Gly-Phe-Phe-Try-Thr-Pro-Lys, cathepsin C from spleen released Gly-Phe, rapidly followed by the slow release of Phe-Try (Thr-Pro-Lys was resistant to hydrolysis). Substrates with a substituted N-terminal (succinylated hepta-peptide) were resistant to hydrolysis. Other peptides reported to be hydro-lyzed were Gly-Tyr-Gly, Gly-Gly, and Gly-Leu-ethyl esters, but not tetra-peptides such as Gly-Gly-Tyr-Gly. These studies were extended by McDonald *et al.*[49–51] to include a large number of biologically active peptides (some important for brain function). The list includes α-corticotrophin, insulin B chain, glucagon, secretin, angiotensin II, the carboxyl tetrapeptide sequence of gastrin, etc. In all cases the N-terminal dipeptide moieties were removed. With α-corticotrophin as substrate, a total of five dipeptides were removed, and further degradation was prevented by the N-terminal Pro-; with ACTH, five dipeptides were released with Lys-Pro-Val resistant to degradation; with glucagon, 16 dipeptides were removed (50% of the chain); with secretin 20 dipeptide residues were removed (80% of the chain). Cathepsin C is similar to a Cl^--dependent enzyme in liver inactivating glucagon by removal of N-terminal His-Leu at an optimal pH of 6.5. This observation may partly account for the inactivation of glucagon in liver homogenates with protection afforded by the addition of PCMB but not by the addition of soybean (tryptic) inhibitor.[51] McDonald *et al.*[50,51] reported that his cathepsin C preparations showed a preference for insulin A compared to the B chain but failed to hydrolyze intact native insulin. Some indication of competitive inhibition between different peptide substrates was provided by the preven-tion of glucagon inactivation by addition of ACTH or native insulin, or inhibition of His-Ser-βNA hydrolysis by addition of insulin A or B chains (Ki, 0.08 mM). Among other polypeptides hydrolyzed were Ala_5 to yield Ala_2 and Ala_3; Ala_6, Ala_4, Gly_4, and Phe_4 to yield only dipeptides; Glu_4 was slowly hydrolyzed; Lys_4 was resistant to hydrolysis; polyalanine was hydrolyzed slowly to yield dipeptide fragments; tripeptides gave dipeptides plus the free amino acid; some hydrolysis of dipeptides was reported, but this may represent dipeptidase contamination (Table III).

3. Brain Cathepsin C

The wide specificity of cathepsin C with addition of Cl^- presents an added difficulty in selecting suitable substrates for testing this enzyme in crude preparations. There are no studies with purified cathepsin C from brain or other nerve tissue, but this activity is presumed present since crude brain preparations hydrolyze the classical substrates of Fruton,[36] namely, Gly-Tyr-, Gly-Phe-NH_2 at pH 5.0, or Pro-Tyr-NH_2 at pH 7.0 with the addition of cysteine.[2a,42,43] The activity reported in our own study,[42] and that of Bouma and Gruber[43] is about 5% of that measured in spleen. As noted, cathepsin C from liver and spleen is capable of hydrolyzing a large range of substrates (Table III). Crude brain extracts hydrolyze Ser-Tyr-βNA at rates higher than with Gly-Phe-NH_2, but only 15% of the rate with monoacyl arylamides such as Leu- or Arg-βNA.[1j,62] In studies connected with the purification of aminopeptidases from pig brain, we separated a peak from DEAE-cellulose capable of hydrolyzing a number of dipeptidyl hydrolases, including Lys-Lys-, Arg-Arg-, Lys-Ala-, Ala Ala-βNA.[1j,2a] This activity is distinct from that of cathepsin C, since no hydrolysis was observed with protein or large polypeptide substrates, and no dependency was observed for Cl^- or other anions. Also, the criterion of N-terminal removal may be insufficient, since we have isolated from pig brain another aminopeptidase which was incapable of hydrolyzing arylamides, but which removed dipeptidyl moieties from alanyl oligopeptides.[17] Probably the polymerase activity at neutral pH is the most distinctive feature of cathepsin C and may prove to be the best method for specifying this enzyme and, as noted, is the basis for its proposed nomenclature (dipeptidyl transferase).[40] McDonald et al.[50] failed to demonstrate the hydrolysis of classical Fruton substrates in pituitary extracts even by sensitive fluorimetric procedures at pH 4.0 or 6.5. Extracts were able to hydrolyze in the presence of Cl^- the dipeptidyl arylamide Ser-Tyr-βNA, and in addition removed the N-terminal dipeptide sequence of ACTH. Failure to demonstrate hydrolysis of Gly-Phe-NH_2 or the para-nitroanilide was attributed to the presence of endogenous inhibitors.[49,50] This explanation could apply to the relatively low levels reported for such substrates in brain, and may account for the stability of polypeptidyl hormones on incubation at pH 5.0 at 37°.[51] Also, most tissues, with the notable exception of brain, rapidly inactivate ACTH in a few minutes on incubation. Even if such materials were able to penetrate the brain, there would appear to be little likelihood of rapid inactivation by cathepsin C and similar hydrolases.

4. Transamidation

As early as 1952, Jones et al.[63] reported that cathepsin preparations catalyzed the transfer of dipeptidyl units to suitable acceptors. Transfer to hydroxylamine, a nucleophilic amine, led to a convenient assay procedure of the corresponding hydroxamic acid.[17,36] Of special interest to a possible physiological role is the ability to act as a physiological pH. The substrate Ala-Phe-NH$_2$ is converted to a hexapeptide (Ala-Phe)$_3$·NH$_2$ and Gly-Phe- or Gly-Trp-NH$_2$ yield octapeptides[64] (Table III). Transfer to amine leads to a cessation of chain elongation with formation of the transamidation product; for example, with Arg-NH$_2$ and Gly-Phe-NH$_2$ the only product is Gly-Phe-Arg-NH$_2$.[60] A large variety of amides, di- and tri-peptides can act as acceptors except those with N-terminal Glu, Ala, Ser, and Gly[58] (Table III). Gly-NH$_2$, di- and tri-glycine are not acceptors, but under certain conditions Gly-Phe- and Gly-Tyr-NH$_2$ can polymerize to yield tetrapeptides.[58,66] The chain length of the oligopeptide is to a large extent determined by the solubility of the product. With the relatively soluble substrate Gly-Arg-βNA, McDonald et al.[50] reported the isolation on CM-cellulose columns of a family of oligopeptides (Gly-Arg)$_{2-5}\beta$NA.

Cathepsin C is a large molecule and may represent an aggregate of subunits with separate catalytic and binding sites for the different hydrolytic and polymerase functions. The available data are consistent with the view that a dipeptidyl-enzyme complex is formed as an intermediate at pH 5.0 (where hydrolase activity predominates) and that the complex is subject to nucleophilic attack by water with the release of the dipeptide. A question of some importance is the relative degree of hydrolase versus polymerase activity since these mechanisms can overlap under certain conditions.[65–67] The explanations available are quite complex and relate to the relative ionization of the α-amino group. In the pH range 4–6 most α-amino acids are fully protonated and hydrolase activity predominates, except in the special case of the prolyl derivatives.

5. Conclusions

Cathepsin C is a sulfhydryl-activated enzyme showing an absolute requirement for Cl$^-$ or similar anions. It is defined by its ability to hydrolyze dipeptidyl amides at pH 5.0–6.0, or its ability to catalyze the polymerization of certain dipeptidyl amides (transamidation at physiological pH). Arising from its limited ability to hydrolyze proteins at pH 3.5 or higher, and from its dual catalytic functions, several new names have been proposed. Since it is no longer considered a true peptidyl-peptide hydrolase, "dipeptidyl aminopeptidase I" could account for its hydrolytic function or "dipeptidyl transferase" could account for its polymerase function. Its remarkable ability to form oligopeptides of limited chain length may have important functional implications, although it cannot be considered as a major pathway for the formation of such materials. Another important function may be the inactivation of hormones and similar polypeptides by removal of N-terminal dipeptide sequences: purified preparations can inactivate β-corticotrophin, glucagon, insulin B-chain, and ACTH; this activity may be similar to the previously reported Cl$^-$ activated enzymes capable of removing dipeptides (N-terminal) from glucagon (His-Leu). The strict substrate requirements proposed by Fruton[36] no longer apply in the presence of Cl$^-$;

purified preparations from spleen and liver, or crude preparations hydrolyze a wide variety of peptide substrates, including Gly-Phe-, Ser-Tyr-, Arg-Arg-, Lys-Lys-, Lys-Ala-, Ala-Ala-βNA (or ρNO$_2$). Purified cathepsin C has not been studied in brain, but crude preparations hydrolyze the classical substrates of Fruton[36]: Gly-Phe·NH$_2$ at pH 5–6, or Pro-Tyr·NH$_2$ at pH 7.6.[42,43] In addition, brain preparations hydrolyze a number of dipeptidyl arylamides at physiological pH[2a] that could be characteristic of cathepsin C-like activity.

D. Cathepsin D (E.C. 3.4.4.23)

Cathepsin D is defined by its ability to hydrolyze denatured protein substrates at pH 3.0–4.0 but not the synthetic substrates used to define other cathepsins.[17,36] Its hydrolytic activity is characterized by the absence of any requirements for added cofactors, particularly sulfhydryl compounds and metal ions. It accounts for over two-thirds of the hemoglobin-splitting activity in most tissues, including brain, liver, spleen, and lung.[38,68,93] Endogenous proteolysis at low pH was described earlier in this century by Hedin and others.[18a,34] Liver enzyme was studied in detail by Anson in the 1930's[37] and subsequently in greater detail by others.[38,93] The peptide bond specificity of acid proteinase similar to cathepsin D was first studied with partially purified preparation from lung[326]; the term cathepsin D was introduced to differentiate liver and spleen enzyme from other cathepsins.[38] Current studies point to the absence of free sulfhydryl and seryl groupings at the active center, indicating that this enzyme cannot be classified as a serine or sulfhydryl proteinase.[2a,93,344]

1. Early Literature on Brain

Of historical interest are the early observations of Carlsson in 1902 of changes in Nissl (protein) substance as a consequence of chromatolysis, and Soula in 1912, who measured appearance of amino nitrogen in brain homogenates with time before and after electrical stimulation or administration of drugs (see reviews of Winterstein[69] and Hydén[70]). Subsequent studies observed changed levels of breakdown in frog or dog spinal cord with stimulation, or changes in breakdown in the visual cortex as a result of light stimulation.[23,70] Prolonged electrical stimulation was reported by Hydén[70] to alter the UV absorption in Deiters' cells, and the absorption of soluble extracts of brain was studied by Ungar.[23] Contributions in the Russian literature are cited by Palladin and Belik[2b] (e.g., Slotov and Geogievskaya, 1925) and a number of observations concerning autolysis are associated with Gibson et al.,[71] Krebs,[72] Edelbacher et al.,[73] and Krimsky and Racker.[74] A number of reservations pertain to early studies, which include assumptions concerning the source of ammonia nitrogen (which may not be the result of proteolysis), and the unspecific nature of protein changes measured by spectrophotometric methods. It can be inferred from studies of Geiger and Magnus[75] that reduction in protein nitrogen in isolated perfused brain of cat in the absence of liver extract (or uridine and cytidine)[75b] was the result of breakdown processes. Studies on brain with an attempt to characterize the enzymes involved were done by Kies and Schwimmer in 1942,[76] and were followed in chronological order by those of Ansell and Richter,[77] Polyakova, Belik, and Tsaryuk,[78a] Lajtha,[79] Palladin,[80] Polyakova and Lishko,[81] and Marks and Lajtha.[42,68]

2. Methodology

Methods for detecting and assaying acid proteinases play a crucial role in the evaluation of older results, and subsequent classification of the enzymes involved (see reviews for experimental details[17,82,343]). Although acid proteinase was among the first intracellular enzymes studied, its role was ignored in subsequent classification based on synthetic substrates.[35] Press et al.[38] showed that spleen and liver contained an acid proteinase that accounted for two-thirds of the hemoglobin-splitting activity but was unable to cleave simpler substrates. Since that time, hemoglobin denatured by the procedure of Anson[37] is the favored substrate for detecting cathepsin D in crude extracts. This substrate is readily available and is useful for rapid monitoring purposes, but it suffers from certain disadvantages inherent in such substrates: they are frequently insoluble at low pH; the extent of denaturation is difficult to define; they are nonspecific; they are often unstable on storage in aqueous form, with increase in background color; commercial samples available on a large scale contain impurities or exist in many different forms. Proteins frequently contain background materials that reduce the precision of assays. Reduction of background can be achieved by prior purification of the substrate or use of specific but often less sensitive colorimetric procedures (Folin, Sakaguchi, etc.). Prior gel filtration would reduce background of crude extracts, but in addition to losses, such procedures might be inconvenient for rapid monitoring purposes.[42] Proteins coupled to chromophores (dyes) or fluorophores (fluorescein) pose problems of standardization, and variation in the product leads to uncertainties as to the linkages involved.[84–87,193] Titrimetric (pH stat) and spectrophotometric procedures (absorption at 280 mμ) lack sensitivity and specificity but have application in purified systems for study of enzyme kinetics.[82] Radiochemical methods utilizing tritiated proteins (Hb treated with acetic anhydride-^{3}H) or materials tagged with radioactive iodine[88] require additional precautions but have been adapted for microassay procedures.

There are a number of qualitative methods for detecting the presence and purity of acid proteinase. Such methods include many different variants of electrophoresis, frequently in combination with immunoelectrophoretic procedures.[89,194] Uriel et al.[89,90] reported the direct detection of enzymes on agar gels after immersion in a solution of albumin, then incubation, followed by fixation and development of bands with amido black. Lapresele and Webb[39] employed this method to demonstrate the heterogeneity of cathepsin from spleen and bone marrow. An interesting modification developed by Merkel[91] used strips of algal chromoprotein in an agar matrix; proteolysis resulted in decolorization which can be scanned densitometrically.

Identification of enzymes can be achieved by immunological techniques. Such methods enable identification in the presence of contaminants, study of purity and homogeneity, and detection of isoenzymes. Weston et al.[92,97] recently reported preparation of specific antisera against cathepsin D from several different tissues. The combination of this technique with preliminary separation of isoenzymes by isoelectric focusing offers great potentiality for elucidating the different forms of cathepsin D in tissues.[93] Ideally, the substrate most suitable is a known peptide with a minimum number of susceptible peptide linkages. Insulin B and globin A and B chains are cleaved by brain cathepsin D with the appearance of a large number of peptides.[66] This peptide was used by Sanger and Tuppy[94] for determining the specificity of proteinases and subsequently by Dannenberg and Smith[326] for cathepsin D of lung, and by Press et al.[38] for cathepsin D from the liver and spleen. Such data indicate that the specificity of cathepsin D has similarities to that of pepsin but is slightly more restricted (Fig. 3). It is still of interest to compare specificity with other proteins, since insulin B does not contain the full spectrum of known peptide linkages. Based on the susceptible bonds of insulin B, Keilova et al.[95] have prepared simpler substrates for studies with liver cathepsin D. These include the heptapeptide or hexapeptide of insulin prepared by tryptic digestion, or a synthetic peptide related to the N-terminal sequence prepared by synthesis. Comparison of three proteolytic enzymes revealed both qualitative and quantitative differences in specificity with these substrates (Fig. 4). Pepsin released two N-terminal dipeptides from the heptapeptide sequence of insulin B chain, resulting in three end products: Gly-Phe-Phe, Gly-Phe, Phe. Free Phe probably arose by secondary cleavage of the tri- or pentapeptide. In contrast, free Phe was never observed in

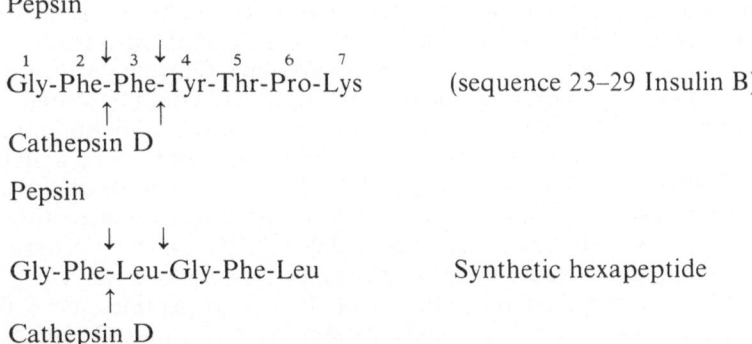

Fig. 4. Specificity of pepsin and cathepsin using the C-terminal of insulin B chain, or a synthetic hexapeptide.[95]

hydrolysis studies with cathepsin D from several different preparations. Some differences were reported in pH optima for the splitting of bond 2–3 with the two different substrates; pH 3.5–3.8 for the heptapeptide and 4.3 for the hexapeptide. Simple substrates with L-amino acids usually exist as extended and flexible structures; introduction of D-residues induces compact and folded ring molecular conformations.[95] Studies with diastereoisomers showed that a D-amino acid in the sixth position led to a 94% loss of activity as measured by the cleavage of bond 2–3.

3. Purification from Brain and Properties

Kies and Schwimmer in 1942[76] achieved a 65-fold purification by extraction of acetone powder (calf brain) with 0.2 M acetic acid at pH 3.7 followed by 0.8 M ammonium sulfate and dialysis. This preparation readily hydrolyzed denatured hemoglobin at pH 3.5 but was inactive at higher pH (e.g., pH 7.6). Ansell and Richter[77] using similar procedures isolated an enzyme from human brain stable for several months at 4°C, higher in gray than in white matter, and unaffected by the addition of CN^-, iodoacetate, cysteine, and glutathione. Palladin and his associates[81] in a number of studies reported a 2000-fold purification by a series of fractional precipitations starting with an acetone powder: (a) lead acetate at pH 5.0, (b) ammonium sulfate, (c) acetone 40–60%, (d) addition of CM-cellulose powder, (c) DEAE-cellulose chromatography and elution with 0.1 M NaCl, (e) agar-gel electrophoresis. The enzyme showed a pH optimum with denatured hemoglobin of pH 4.1 and with albumin of 3.1, was stable when stored at pH 5.5–8.6, and was unaffected by addition of divalent metal ions, ammonia, cysteine, glutathione, ascorbic acid, iodoacetic acid, EDTA, PCMB, or soybean trypsin inhibitor. In our own studies, acid proteinase was purified by extraction of acetone powder with hypotonic buffer at neutral pH followed by: (a) gel filtration on Sephadex G-100 (to separate acid proteinase from neutral proteinase and aminopeptidase) (Fig. 5), (b) ammonium sulfate precipitation at pH 5.0, (c) carrier-free electrophoresis.[42] As in the case of liver cathepsin,[38,93] we observed multiple peaks of cathepsin D on DEAE-cellulose and during the final electrophoretic steps. Enzyme purified by this method was stable when frozen for up to 1 year, and could be stored indefinitely after lyophilization. Brain enzyme could conveniently be monitored during purification by hydrolysis of denatured hemoglobin using a quantitative ninhydrin procedure (see Section II, D, 2). Evidence for its strictly endopeptidase nature was supplied by peptide maps produced on incubation with globin A and B chains, and insulin B chain[6b,28,99]; it also hydrolyzed to a lesser degree denatured albumin, globulins (γ, α, β), edestin, etc.[42] No hydrolysis was observed on addition of di- and tripeptides, or with the synthetic substrates used to specify cathepsins A, B, and C and a large number of monacyl- and dipeptidyl arylamides. The other properties of purified rat brain and proteinase are compared to neutral proteinase in Table IV. A sharp pH optima of 3.2 (Fig. 2) with a K_m for hemoglobin of

TABLE IV

Properties of Cerebral Peptidyl–Peptide Hydrolases[4]

	Acid proteinase	Neutral proteinase
pH optimum	3.2	7–8
PCMB 0.5 mM	0	$-100\%^a$
O-lodosobenzoate	0	-60%
GSH^b	0	25%
β-Mercaptoethanol	0	$+60\%$
Ca^{2+} 0.1 mM	0	$+15\%$
Co^{2+} 0.1 mM	0	-15%
Cd^{2+} 1 mM	0	-100%
Cu^{2+} 1 mM	0	-100%
K_m (Hb)	8×10^{-3} M	2×10^{-3} M
K_m (Casein)	—	3×10^{-2} M

^{a} Values for results with added materials are expressed as the percentage change compared to the controls. From the data of Marks and Lajtha.[42]

^{b} GSH was added in presence of CoA, ATP, and phosphocreatine (see text).

8×10^3 M were displayed. The enzyme was unaffected by the addition of a large number of metal ions, sulfhydryl agents, and inhibitors. Some studies have indicated the presence of sulfhydryl-dependent acid proteinases in different tissues, but such enzymes have never been isolated and studied.[54,55] It is possible that the activation seen on addition of cysteine and other sulfhydryl compounds results from an interaction with inhibitory materials or other mechanisms (see Section VI). Most preparative methods are characterized by rather low yields, and all require a final electrophoretic step associated with large losses. Three methods are currently available: (a) the agar-gel electrophoretic method of Palladin et al.,[81] (b) free-flow electrophoresis employed in our own studies,[42] and (c) isoelectric focusing as employed for purification of liver enzyme.[93] The last two methods provide evidence for the presence of at least three isoenzymes, and to some extent this is confirmed in recent studies using a series of Dialfo molecular filters.[100] About 30% of the acetone powder extract was retained by the XM-100 filter (mol. wt. cutoff 100,000); the remainder was about 80,000, consistent with earlier published studies.[42] As an alternative to an earlier method, we attempted to purify enzyme by extraction of whole brain or lysosomal-rich acetone powders with pH 3.5 formate buffer containing 20% acetone in a manner similar to the procedure described by Barrett for liver enzyme.[93,101] Elution from DEAE or DEAE-Sephadex resulted in considerable heterogeneity, with purification of some peaks in the order of 500–1000-fold. Elimination of the acetone extraction step, but extraction at pH 7.6 in hypotonic buffers, resulted in considerably greater yields. Enzyme was stable in the frozen state and could

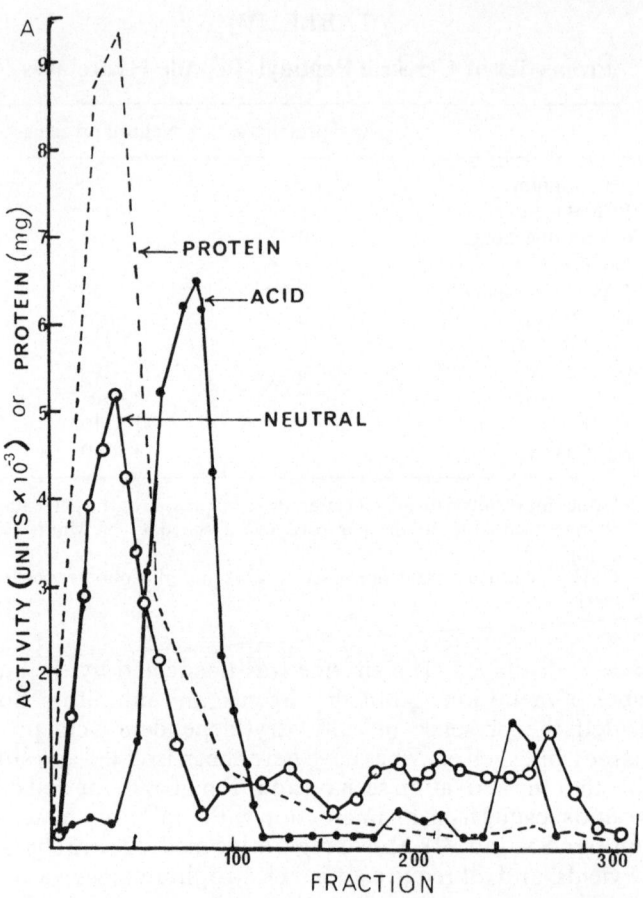

Fig. 5a. Separation of cerebral neutral and acid proteinases after passage through Sephadex G-100 (column 80 × 2.5 cm). Supernatants were prepared by extraction of acetone-dried rat brain powder and activities assayed using denatured hemoglobin at pH 7.6 or pH 3.8 in combination with the quantitative ninhydrin method.[42]

be conveniently stored as a lyophilized powder for periods of 1 year or longer. Antienzyme can be prepared by injection of purified enzyme with Freund's adjuvant into rabbits, revealing the degree of purity, and cross contamination with other species.[97] Enzyme purified from porcine brain yielded a single precipitin line against antisera (after adsorption with pig serum) without crossreactivity against bovine antisera.[100]

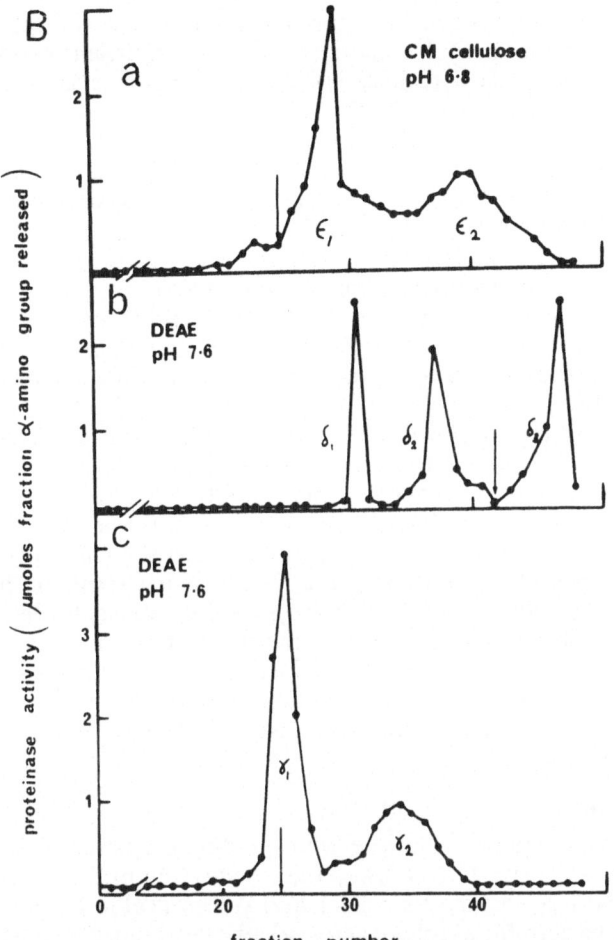

Fig. 5b. Distribution patterns of brain acid proteinase after carrier-free electrophoresis. Purified enzyme prepared by DEAE-cellulose chromatography[42] was injected at the rate of 0.25 ml/hr in 25 mM-tris-HCl buffer, pH 7.6. The different cellulose peaks were run at A, 2000 V, 120 mA at 2°; B, 2100 V, 130 mA; and C, 2000 V, 110 mA.

a. Substrate Specificity. In early studies, we demonstrated a wide specificity for purified brain enzyme using denatured protein substrates, among them hemoglobin, globin, globulins and albumin, and edestin.[68] The question of specificity with respect to native (undenatured) proteins existing in conformations inaccessible to hydrolytic agents is an important but extremely complex problem (see collagen, Section VIII). In surveying tissues for proteinases, nonspecific denatured proteins have several advantages compared to smaller peptides that may contain insufficient permutations of susceptible bonds. Indeed, cathepsin D was ignored for several years because of its inability to cleave simpler substrates used to assay other cathepsins.[38] For the study of specificity, simpler peptides (insulin B, etc.) are favored since they yield a manageable number of split products. Studies with cathepsin D from a large number of tissues, including brain, indicate similarities to pepsin with a slightly more restricted specificity (Fig. 3). It is possible that the study of breakdown with native brain proteins will reveal greater differences compared to pepsin or acid proteinases from other tissues; such proteins can be expected to contain a greater variety of peptide bonds compared to insulin B. It is of interest that basic (encephalitogenic) protein from myelin is one of the best substrates ever studied for brain acid proteinase[101] (Section IV, D).

Peptide sequences based on the susceptible bonds of insulin B chain have proved to be useful for seeking simpler substrates for the assay of purified acid proteinase. Sequences such as Gly-Ala-Leu-Tyr-Leu-Val have been used for study of uterine cathepsin D, and Gly-Phe-Phe-Tyr-Thr-Pro-Lys for liver cathepsin D.[18t,95] Probably the best sequence available is the soluble heptapeptide (insulin B chain 23–29) or the synthetic hexapeptide Gly-Phe-Leu-Gly-Phe-Leu.[95] The hexapeptide was split by brain acid proteinase to yield a large quantity of tripeptide, Gly-Phe-Leu, and smaller quantities of a di- and tetrapeptide as illustrated in Fig. 4.[100] The tripeptide itself was resistant to purified brain aminopeptidase and arylamidase, and to cathepsin D itself. These preliminary results[100] indicate some differences from liver cathepsin D.

4. Distribution

A survey of acid proteinase activities in different tissues with denatured hemoglobin as the substrate showed that the level in brain exceeded muscle but contained less enzyme than liver and spleen (Table II). In general the distribution pattern for autolytic activities (in the absence of substrate) paralleled those obtained in the presence of substrate. In different brain areas, Lajtha[79] reported a higher rate of autolysis in sheep cerebrum and cerebellum compared to hypothalamus, medulla, and white matter. In adult monkeys, the distribution was slightly different, with highest activity in the lenticular nucleus followed by cerebellum, cerebrum, mesencephalon, pons, thalamus, white matter, spinal cord, etc. The latter measurements were based on the actual percent protein hydrolyzed after incubation in a mixture containing acetate buffer, pH 3.8, in an N_2 atmosphere.

Most studies indicate that acid proteinase is particulate bound in tissues, with highest concentrations in crude mitochondrial-lysosomal fractions. Lajtha[79] in early studies reported high endogenous acid proteinase in the crude mitochondrial synaptosomal fractions of rat cortex with levels of about 10% protein hydrolyzed per hour, followed in descending order by crude nuclear (5%), debris (3%), microsome (2%), and finally supernatant fractions (1.4%). Palladin[80] reported that crude mitochondrial fractions of rabbit brain (Hb, pH 3.8) contained 82% acid proteinase compared to 10% in the microsomal soluble of 3% in the nuclei. Fractions rich in acid proteinase contain morphological structures similar to lysosomes of liver (dense osmophillic granules) but it is still premature to assign all acid hydrolases to such particles in the absence of definitive separation studies.[2a,102] Purification of crude mitochondrial-synaptosomal fractions on a variety of sucrose gradients confirms the morphological and enzymatic heterogeneity of particulates and their associated hydrolases.[102] D'Monte et al.[2c] reported one peak for acid proteinase [adjacent to the purified mitochondrial pellet (Fig. 6)] in contrast to multiple peaks for acid deoxyribonuclease and phosphatase or other "lysosomal" enzymes. Such findings support the concept of lysosomal diversity and are in agreement with the enzymatic profiles on sucrose gradients supplied by Sellinger and Hiatt.[103] Recently two groups reported intrinsic acid and neutral proteinases in purified liver mitochondrial preparations.[104,105] In one case, fragmentation with Triton X-100 led to a loss of proteinases active at pH 5.8 but without effect on the level of the pH 7.6 enzyme.[105] In our own studies, purified brain mitochondria free of morphological contamination displayed both acid and neutral proteinase activities. Fragmentation by phospholipase or hypotonic swelling procedures gave preparations with trace acid proteinase and some peptidase activities.[2c] These levels were lower than crude mitochondrial fractions but sufficient to account for the known average half-life of brain proteins.[4] Brain mitochondria are reported to contain a loosely bound proteinase active at pH 5.0 and differentiated from other proteinases by inhibition with Zn^{2+} ions.[333] Calculations show that levels of proteinase in most fractions exceed that required to maintain turnover, indicating the need for other controlling factors.[2c,102] Some consideration is given to lability of lysosomallike particulates and the role of inhibitory factors in other sections of this account. The role of proteinases in nuclear fractions is discussed in relation to nucleic acid-protein interactions (Section VI, C).

5. Acid Proteinase in Other Tissues

Acid proteinase similar in characteristics to cathepsin D is present in a very large number of tissues: thyroid,[106] liver,[38,93] spleen,[113] lung,[326] muscle,[18r] uterus,[18t] Ehrlich ascites tumor,[107] arterial tissue,[108] etc. The release of physiologically active peptides by cathepsins of other tissues may directly or indirectly affect brain function; this aspect is considered in some detail in Section V.

A good example of the sequential action of intracellular proteinases and peptide hydrolases is supplied by mechanisms for release of iodinated tyrosines in the thyroid. The enzyme purified from thyroid is similar to cathepsin D in terms of specificity on insulin B chain (similar to pepsin), and the insensitivity to metal ion addition, or sulfhydryl compounds such as cysteine and thioglycollic acid.[109–111] Also, purified thyroid proteinase is inhibited by diazoacetyl-norleucine methyl ester, a specific pepsin inhibitor also

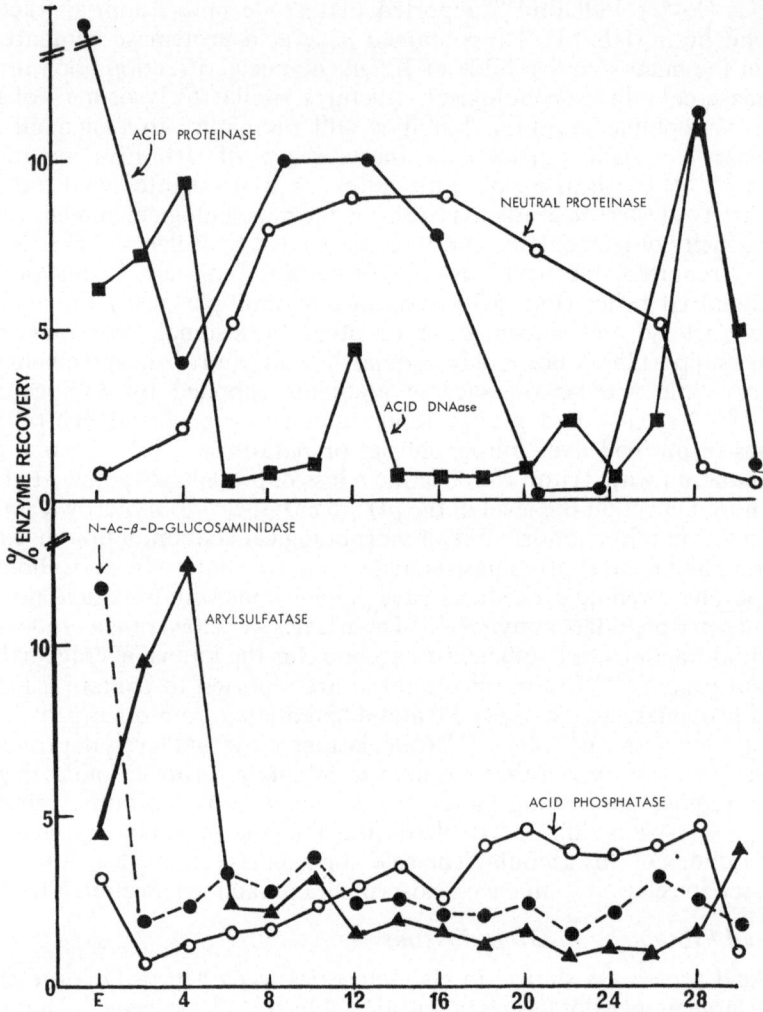

Fig. 6. Distribution of hydrolases in crude fractions containing mitochondria, lysosomes, nerve endings, myelin after separation on a continuous sucrose gradient (0.8–1.4) and centrifugation at 24,000 rpm on SW 25.1 rotor for 2 hr. Thirty fractions were collected by tube puncture. Based on the morphological appearance,[2c] these represent: fraction E, heavy mitochondria (pellet); fractions 1 and 2, mitochondria contaminated by lysosomes; fractions 3–6, lysosomal enriched; 6–20, nerve endings together with some mitochondria and lysosomes; 20–24, nerve endings and myelin figures; 24–28, myelin with some entrapped particulates; 28–30, soluble enzymes. The lysosomal region, fractions 2–6, contains high concentrations of acid proteinase, arylsulfatase, N-acetyl-β-D-glucosaminidase, and acid DNase. The nerve ending region contains a second peak of activity for acid proteinase and DNase and high activity for neutral proteinase. The myelin region contains a third peak of acid proteinase and some soluble enzymes. Data based on the results of D'Monte et al.[2c]

reported to inhibit some preparations of cathepsin D.[18a,110] The mechanism of action is related to lysosomallike granules which move under the influence of thyroid stimulating hormone from the basal to the apical region of cells, where they subsequently fuse with colloid drops before secretion into the lumen.[111] Some preparations contain enzyme with a pH optima of 6.6, and mast cells from thyroid contain neutral proteinase.[112]

Interest in the fate of protein antigens has led to study of proteolytic enzymes in liver and spleen.[38,340] The pioneer studies of Press et al.[38] led to the recognition of cathepsin D in spleen and other tissues as the most important intracellular enzyme cleaving proteins. Enzyme purified from spleen exhibited multiple chromatographic peaks on DEAE-cellulose, and starch-gel electrophoresis. Microheterogeneity was demonstrated in brain using carrier-free electrophoresis, and recently in liver with isoelectric focusing methods.[42,93] Spleen enzyme was optimal in hydrolysis of hemoglobin at pH 3.0, immunoglobulins at pH 3.5, and albumin at pH 4.2; variation of pH optima with substrate is a frequently reported phenomenon.[85] Further evidence of heterogeneity was supplied by Lapresle and Webb[39] utilizing the agar-gel method of Uriel.[89] The same group also reported the presence of an isoenzyme with a significantly higher molecular weight and different electrophoretic mobility termed cathepsin E. Press et al.[38] were the first to conclude that cathepsin D was a nonsulfhydryl enzyme, although in crude extracts cysteine caused some activation. Most probably, activation by sulfhydryl compounds represents contamination by exopeptidases that can act synergistically with cathepsin in the breakdown of proteins (see cathepsin A). Cathepsin D purified from spleen (mol. wt. 58,000; N-terminal Gly) cleaved heptapeptide of insulin at -Phe^2-Phe^3-, and -Phe^3-Tyr^4-, and the synthetic hexapeptide illustrated in Fig. 4 at -Phe^2-Leu^3-.

Barrett[93] reported a 1000-fold purification of cathepsin D from liver mitochondrial-lysosomal preparations by extraction with formate–acetone buffers prior to chromatography on DEAE–Sephadex. The enzyme displayed all the characteristics of cathepsin D: activity was unaffected by —SH agents, metal ions, PCMB, iodoacetamide, soybean inhibitor, trasylol, PMSF and TPCK; but was inhibited by 3-phenyl pyruvic acid and some reducing agents. Isoelectric focusing revealed fewer isoenzymes in liver of ox compared to man and rabbit, emphasizing possible species differences: antisera showed a close structural relationship between isoenzymes within each species.[97] Isoenzymes separated by isoelectric focusing exhibited a pH optimum of 3.0 and molecular weight of 45,000 which is approximately half previously reported values. This might indicate the presence of dimers and differentiate cathepsins D and E. Such antisera could have important therapeutic applications in dystrophy or in allergic and inflammatory conditions.

III. NEUTRAL PROTEINASES

The term neutral proteinase describes hydrolases active at physiological pH in the degradation of protein substrates. Extreme lability of enzyme from

animal tissue has prevented adequate characterization, but neutral proteinases are abundant in microorganisms and available from this source in
highly purified and often crystalline forms. The term neutral proteinase is to
some extent a misnomer, since the *p*H optima is in the *low p*H alkaline range
(*p*H 7–8) but it has come into general use and is convenient to distinguish
such enzymes from others active at considerably higher *p*H (*p*H 8–12). The
difficulties associated with characterization have led some investigators to
question the presence of neutral proteinases in animal tissues[114,115] and this
necessitates some consideration of assay and purification procedures. Some
description is accorded to localization in subcellular fractions since these
data provide information on the possible physiological roles, and are often
a prerequisite for purification studies.

There is no extensive literature on neutral proteinases in animal tissues,
and most early results fail to distinguish between autolytic activities and the
contribution of other degradative enzymes [peptide hydrolases, ribonuclease
(see next section on assay methods)]. Ansell and Richter[77] reported their
presence in aqueous extracts of brain, and distribution was subsequently
studied by Lajtha,[79] Marks and Lajtha,[68] and recently by Belik *et al.*[116]
Attempts to purify brain neutral proteinase were made by Guroff,[117] Marks
and Lajtha,[42] and recently by Riekkinnen and Clausen.[118,119] Other
aspects of neutral proteinases in tissues are reviewed by Van Heyingen and
Waley.[121]

A. Assay Procedure

As noted (Section II, D), only general principles are discussed since
specific details exist in specialized texts.[17,82,343] Because other degradative
enzymes are active at physiological *p*H (peptide hydrolases, ribonuclease),
careful consideration must be given to the end products. The most favored
methods are colorimetric[17,82] (ninhydrin, Folin, or Sakaguchi) but determination of free amino groups is not necessarily a measure of endopeptidase
activity in crude extracts. Most early experiments are difficult to evaluate for
this reason; with added protein substrates correction must be made for
background autolytic activities. Ideally, some indication of the chemical
nature of split products (i.e., peptides) is required to satisfactorily delineate
endopeptidase activity, since proteins themselves can be subject to simple
end-group cleavage by peptide hydrolases (Fig. 1).

For this purpose chromatographic procedures are suited to elucidate the
"fingerprint" patterns for complex protein substrates; column separations
determine both qualitative and quantitative aspects of peptide bond specificity and relative susceptibility.[94,107,122,339]

Other methods available for monitoring neutral proteinase include the
coupling of released tyrosine with α-nitroso-β-naphthol for fluorimetric
determination,[86,117] methylated proteins (N,N-demethylprotein) with trinitrobenzenesulfonic acid (TBS) for detection of end products,[124,277,280,281]
furylacryloyl peptide substrates (absorb in the UV)[280]; and hemoglobin

coupled with fluorescein isothiocyanate.[87,193] The use of dye-protein (chromophores of fluorophores) has been criticized on the grounds of unspecific loss of dye (bleeding),[125,278] inability to function as a substrate for Michaelis-Menten kinetics, ambiguities concerning the dye-protein linkage (examples include azocoll, congocoll, and fluorescein-treated proteins).[17,91] Spectrophotometric methods (Kunitz procedure) are criticized for lack of specificity and sensitivity; these methods cannot distinguish between aromatic amino acids and nucleic acids (liberated by ribonuclease active at physiological pH.[114] Titrimetric methods lack specificity and sensitivity compared to colorimetric procedures, but are of value in some kinetic measurements based on initial rates of reaction.[17]

Many proteolytic enzymes display esterolytic in addition to proteolytic activities; purified samples of trypsin can be assayed with BAME or TAME, and chymotrypsin with BPANE or ATEE, etc.[17] The relationship between the dual catalytic activities is unknown since there are many examples in the literature of enzymes with only one catalytic activity (esterase or protease).[17,112] In the case of carboxypeptidase B, for example, the two activities can be differentially altered by a number of diverse treatments, providing some evidence for the presence of two separate activity sites.[5] Measurement of proteases with ester substrates in crude extracts, in the presence of unspecific esterases or other hydrolases, is best avoided until the relationship between the two activities has been assessed in purified systems. Crude extracts of brain show a surprisingly high esteraselike activity with BAME or TAME, but the relationship to neutral proteinase is unknown (Figs. 2 and 16).[126]

Methods utilizing labeled precursors (isotopic methods) have been used to great advantage for estimating the rates of degradation *in vivo*[20c,79] and have theoretical implications in relation to the mechanisms of protein turnover.[4] Lajtha[79] in early studies obtained an estimate of neutral proteinase by following the decay of proteins labeled with leucine-^{14}C *in vitro*. The preparation of brain proteins with a relatively high specific activity (long half-life) is an expensive and lengthy procedure, and separation of specific proteins as substrates is dependent on the progress in purification. As yet, only the acidic (S-100) and the basic (encephalitogenic) proteins are available in sufficient quantity for such studies.[1b,1s]

As in the case of acid proteinases, it should be possible to develop qualitative methods for detecting neutral proteinase activity. Adams and Tuqan[127] reported the appearance of clear zones on contact of tissue sections with blackened photographic plates (sensitivity of a 30-μ zone equivalent to 1 pg trypsin), and several variants of this technique have appeared employing nitrocellulose or gelatin membranes.[128,129] Such histochemical techniques, however, do not premit precise localization of endopeptidases.[1h]

B. Localization

Ansell and Richter[77] observed autolysis in aqueous extracts of brains from different species (rat, rabbit, and human) higher in white compared to gray matter. Activity was labile and disappeared within 1.5 hr after incubation under anaerobic conditions without effect on addition of cysteine, CN^-, Fe^{2+}, ascorbic acid, $\alpha\alpha$-dipyridyl, but inhibition occurred with Cu^{2+} and iodoacetamide. Later studies by Lajtha[79] under aerobic conditions indicated little difference in endogenous activities in different monkey brain regions. The levels in the medulla, cortex, cerebellum, and white matter were about 3% protein hydrolyzed per hour and slightly higher compared to the pons, hypothalamus, lenticular nucleus, spinal cord, and mesencephalon. In particulate fractions obtained from sheep, activity was highest in the crude synaptosomal fraction, followed by nuclei, supernatant, and microsomes: in general, fractions from white matter exceeded levels in gray matter. With addition of hemoglobin, activity was highest in soluble cytoplasmic fractions (microsomal supernatant) and lowest in the purified particulate fractions.[79] Purification of the crude mitochondrial-synaptosomal fraction indicated that most of the activity was associated with the nerve ending particles rather than the mitochondrial pellet or myelin (Fig. 6).[68] Studies on breakdown of protamine by Belik et al.[18b,116] revealed a different distribution pattern as summarized in Table V. Some 50% of the total homogenate activity resided in the crude mitochondrial-synaptosomal fraction, 24% in the nuclear, and 14% in the microsomal supernatant fractions. Currently, there is considerable interest in the levels of neutral proteinase in spinal cord and in myelin in relation to EAE (Section IV). Belik et al.[116] reported that over 50% of the activity in crude mitochondria was associated with the myelin (Table V).

TABLE V

Distribution of Acid and Neutral Proteinase in Brain Subcellular Fractions[a]

	Acid proteinase	Neutral proteinase	
	pH 3.8 Hb	pH 7.6 Hb	pH 7.5 Protamine
Homogenate	100 (400)	100 (120)	100 (154)
Nuclei (P_1)*	30 (700)	8 (230)	24 (180)
Synaptosomal-Mt (P_2)	45 (1300)	15 (600)	50 (171)
(a) Myelin, crude	8	3	21
(b) Nerve endings, etc.	25	10	6
(c) Heavy mitochondria	12	1	5
Microsomes (P_3)	15 (600)	65 (300)	14 (180)
E.R.	13	8	
Ribosomes	2	0	
Soluble (S_4)	5 (150)	65 (300)	16 (124)
Debris	5	4	

[a] Enzyme activity is expressed relative to recovery from homogenate as 100. Values in parentheses represent specific activity (μmoles α-amino groups/g of protein/hr). Data with denatured hemoglobin as substrate are compared with rabbit fractions using protamine as substrate.[2b,116]

Consistent with other studies, protamine splitting in white matter (13 μmoles Arg/g protein/hr) was almost double that in gray (7); mesencephalon (12) was higher than the midbrain, pons, medulla, or cerebellum (11) and lowest in the spinal cord (8). We showed with hemoglobin as substrate that the levels of neutral proteinase in cat, rabbit, and rat cortex greatly exceeded the levels (expressed as total activity or specific activity) present in spinal cord (obtained from the lumbar region[130]). Purification of myelin on sucrose or cesium chloride gradients led to a complete loss of neutral proteinases.[131,132] In detailed studies by Riekkinnen and Clausen[118,119] (Section IV), the neutral proteinase of myelin from cerebral cortex was shown as an inherent component of the sheath.

In comparison to different tissues, the rate of hemoglobin hydrolysis in brain exceeded the level in muscle, but was only two-thirds the activity of liver and considerably less than that of spleen[68] (Table II). Some macrophagic cells (PMN-granulocytes) are reported to contain neutral proteinase[133] which may be involved in degenerative changes in connective tissue and during Wallerian degeneration (Section IV).

1. Purification

There are only a limited number of studies on purification because of the extreme lability of neutral proteinase and assay difficulties.[42,117–119] Partially purified preparations can be obtained from soluble supernatant (extraction with sucrose or hypotonic buffers) by $(NH_4)_2SO_4$ precipitation, gel filtration on Sephadex G-100, and finally chromatography on DEAE-cellulose. Although activity is stable in crude extracts when stored at 0°C, activity decays rapidly during purification and subsequent storage, and on lyophilization.[42] Storage is facilitated by the presence of —SH compounds (mercaptoethanol, dithiothreitol). Partially purified preparations are activated by the addition of cysteine or gluthathione; the pH optimum is 7.6 (range 6.8–8.0); activity can be assayed with denatured hemoglobin or casein. Chromatography on DEAE-cellulose revealed two major peaks (α and β) that may represent isoenzymes.[42] These observations were confirmed by Riekinnen and Rinne,[134] who also showed that the second peak displayed a higher pH optimum (pH 8.0). Purified enzyme prepared by our procedures exhibited no pronounced requirement for metal ions, although in some cases slight activation was observed (Table IV). Activity was inhibited by Cu^{2+}, o-iodosobenzoate, and PCMB, indicating a requirement of free SH group(s) for catalytic activity.

In an alternative procedure, Guroff[117] reported the presence of a Ca^{2+}-dependent neutral proteinase in soluble supernatant of rat brain. Extracts were prepared in the presence of mercaptoethanol and EDTA, subjected to ammonium sulfate treatment at pH 5.5, dialysis (against buffer containing mercaptoethanol and EDTA), and DEAE-cellulose chromatography. The enzyme showed complete dependence on Ca^{2+} for hydrolysis of casein and was activated by additional mercaptoethanol and low concentrations of chelating reagent. In the presence of Ca^{2+}, the enzyme was extremely labile, but in its absence it was stable for 3 weeks at $0°C$. Ca^{2+} could not be replaced by Mg^{2+} or other divalent metal ions. Activity was dependent on free —SH groups, as demonstrated by activation with cysteine, BAL, reduced glutathione, KCN, and ascorbic acid; the absence of inhibition by DFP would suggest that serine is not involved at the active center of the enzyme. Incubation with insulin B-chain liberated a number of peptides with the major split bond Tyr^{16}-Leu^{17}-. The enzyme was free of contaminating peptide hydrolases as judged by incubation with a large number of representative aminopeptide and carboxypeptide substrates. Ca^{2+} is known to promote the activation of trypsin from its zymogen precursor, and then subsequently stabilize trypsin against autolysis.[135] Stabilization is attributed to specific Ca^{2+} binding sites inducing conformational changes: Ca^{2+} in high concentration binds to the four N-terminal Asp on trypsinogen.[136] Calcium is not essential for trypsin activity, but it favors formation of a complex, with temperature-dependent activation of greater resistance to autolysis.[135]

C. Proteinases in Other Tissues

Van Heyingen and Waley[121] studied the levels of endogenous activity in different tissues and noted a relatively high concentration in lens tissue (Table II). Attempts to purify lens proteinases are handicapped by difficulties in separating the enzyme (present in α-crystallin) from its native substrate (α_2-crystallin). A 300-fold purification was reported from ox lens material by precipitation of soluble extracts at pH 5.0 followed by chromatography on hydroxyapaptite and DEAE-cellulose. The interrelationships of lens-soluble proteins (α, β, γ crystallins) are somewhat complex and beyond the scope of this review. The best substrate is obtained from the pH 5.0 supernatant by ethanol precipitation; untreated supernatant acts as an inhibitory substance. Another difficulty in purification is the persistence of aminopeptidase (Leu-Gly) in most fractions. Properties of lens proteinase are similar to those described for brain, namely, a requirement for —SH groups, and inhibition by PCMB, Cu^{2+}, and N-ethylmaleimide. Application of the same methods by this group to spleen and liver failed to yield enriched preparations, although others report its presence in granule (peroxisome) fraction.[336,340] Similar findings made by Marrink and Gruber[114] attribute neutral proteinase of tissues entirely to aminopeptidase as determined by the degradation of ACTH.

Stable neutral proteinases have been purified and studied extensively in B. subtilis, S. griseus, A. oryzae, group A streptococci, etc.[124,137-140] Morihua et al.[137] classified neutral proteinases according to the effects of DFP (serine proteinases) or EDTA (metal-dependent proteinases). Such classifications are dependent on purity of the preparations; pronase, for example, is now shown to contain potent endopeptidases active at pH 7.6.[17] DFP-sensitive proteinases show remarkable similarity in the splitting of insulin B chain with preference shown for peptide bonds in which Tyr, Phe, and Leu donate the —COOH group. In contrast, EDTA-sensitive enzymes split bonds in which the same amino acids donate the $-NH_2$. One such EDTA-sensitive enzyme from *B. subtilis* with a mol. wt. of 450,000 (416 residues, N-terminal Gly, and C-terminal Leu) contains Zn^{2+} at its active center and can hydrolyze casein some 5–16-fold faster than papain, chymotrypsin, or nagarase.[138] In contrast, the DFP-sensitive enzyme from *B. subtilis* with a mol. wt. of 29,000 (N-terminal Ala, C-terminal Leu or Ser) is distinctly different in all its properties.[139] An enzyme of unusual interest is the neutral proteinase from group A hemolytic streptococci that exist as a zymogen precursor activated by reduction of a disulfide bridge composed of $\frac{1}{2}$ Cys and an unknown volatile thiol.[122,140] The enzyme contains His at its active center, and its specificity has been mapped in great detail by Gerwin *et al.*[122] using insulin B chain.

D. Energy Dependence

The question of energy dependence of protein breakdown has engaged the attention of many investigators.[142–147] Energy is more likely to be a limiting factor at physiological pH (i.e., neutral proteinases) than at very acid pH (cathepsins). Schoenheimer[142] first enunciated the concept that breakdown may have an energetic dependence, and some evidence for such processes was supplied first by Simpson[143] and later by Steinberg and Vaughan,[144] Korner and Tarver,[145] and more recently by Brostrom and Jeffay.[148] Conclusions were based on observed changes during anaerobiosis, or reduced breakdown in the presence of respiratory inhibitors (these include CN^-, DNP) when added to brain or liver slices, or to their subcellular fractions.[143–145] Steinberg and Vaughan[144] also reported an inhibition by amino acid analogues (fluorophenylalanine and β-thienylalanine) on the release of phenylalanine from slices by the processes of protein catabolism. Penn[146] reported increased breakdown of albumin by liver and brain mitochondrial fractions on the addition of ATP and CoA. We observed some increase in breakdown, using soluble brain extracts at pH 7.6, on the addition of CoA, ATP, and reduced glutathione[42,68]; this was recently confirmed in the studies of Umana.[147] Explanation for these effects in soluble systems is somewhat complex and may be related to interactions with the substrate or suppression of a native inhibitor.[149,192] For example, Libensen and Yena[149] reported an increased hydrolysis of albumin in the presence of glutathione because of the formation of an intermediate complex with increased susceptibility for cleavage. In the case of rat liver, Bromstrom and Jeffay[148] consider three possible concepts in relation to earlier work: (a) protein catabolism requires respiration and thus may require energy, (b) catabolism in homogenates is different from that in slices, (c) protein catabolism may be linked to synthetic mechanisms. There is some evidence that structural integrity is a requirement for respiratory processes linked to breakdown, since considerably higher levels of inhibitors are required to reduce proteolysis in homogenates or soluble systems. Although early studies by Bergmann and Fruton[35,36] suggested that some aspects of breakdown and synthesis shared common pathways, as later elaborated by Walter,[150] the connection between the two processes is unknown. Studies on tryptophan pyrrolase induction in liver suggest that inhibition by protein synthesis inhibitors, or some respiratory inhibitors, leads to an indirect inhibition of protein catabolism by affecting synthesis of the actual catabolism enzymes[20c,29] and levels of ATP.[341]

In summary, there is no evidence at present to suggest that splitting of peptide bonds *per se* requires energy. This does not exclude the possibility that energy is required within the cellular milieu for the transport or supply of the necessary components to the actual sites of destruction. In this sense there may be a structural requirement for breakdown ultimately dependent on an energy supply for its integrity. These studies point to the paucity of information in this area and are further discussed in the companion review related to protein turnover in this volume.[1p]

IV. PROTEINASES IN SPECIFIC NEURAL ELEMENTS

The turnover of peptidyl hormones (synthesis, storage, degradation) in specific brain areas is of considerable importance to the control of many bodily activities. As emphasized earlier, difficulties associated with histo-chemical localization or assay in crude extracts have prevented precise studies on subcellular localization and the finer distinctions that may exist between different "functional" areas in the brain. Interestingly, there is some diurnal variation in pineal tyrosine possibly linked to the formation of catecholamines and melatonin.[1m,151] Brain is not characterized by a large pool of free peptides, except in the postpituitary (4% of dry weight) and pineal (0.02%) which may reflect the efficiency of endogenous peptide hydrolases.[5] The description below relates mainly to the pituitary, peripheral nerve, and spinal cord; less information is available concerning the characteristics of endopeptidases in specific areas of the central nervous system.

A. Pituitary Endopeptidases

Little consideration has been paid to possible differences in endopepti-dases in different anatomical regions of the pituitary (anterior and posterior) despite major differences in the content and spectrum of peptidyl hormones. There is some evidence that the two lobes contain different levels of enzymes hydrolyzing arylamide substrates.[5] Peptidyl hormones range in molecular size from about 1000 (oxytocin and vasopressin) to 45,000 (growth hormone): LH, FSH, and TSH are considered to represent glycoproteins, but there are still questions raised concerning their purity and structure.[1j] The pituitary contains trace quantities of nonactive peptides that may result from extrac-tion processes, or represent intermediates in synthetic or breakdown mechanisms. The role of proteolytic enzymes, particularly peptide hydro-lases, (E.C.3.4.1), assumes greater importance with the discovery of peptidyl releasing factors, one of which is a tripeptide.[152] Many possibilities exist for hormonal regulation; in addition to releasing factors, there is the question of storage in granules, transport by combination with specific proteins (neurophysin-oxytocin; thyroxine binding protein-thyroxine, etc), localization and release by degradative enzymes.[22a–c] Alteration of releasing factors, or transport proteins (neurophysin) by proteinases could represent an important control mechanism for hormonal action.[6d]

The molecular diversity of pituitary hormones and associated factors suggests that breakdown is mediated by the sequential action of both exo- and endopeptidases. Very few studies are available concerning proteinases in pituitary; the greatest emphasis has been placed on arylamidases, some of which are substrates for cathepsin C (dipeptidyl transferase) and cathepsin A (carboxypeptidase). Since these latter enzymes were previously described and are not strictly endopeptidases, they do not receive comment in this section.[1j] Most studies refer to the terminology introduced by Adams and Smith,[153] namely proteinase I and II. Proteinase I is similar in most respects to acid proteinase or cathepsin D; currently there is some question on the existence of proteinase II (active at pH 8.3) based on the exclusive use of changes in absorption at 280 mμ as a measure of proteolysis[115] (Section III, A).

Adams and Smith[153] in 1951 made the first systematic study of protein breakdown in the pituitary. They reported the hydrolysis of a large range of peptides by lyophilized porcine and beef pituitaries (anterior and posterior) and in addition the cleavage of urea-denatured hemoglobin over a wide pH range with two distinct peaks at pH 3.8 and 8.3 (proteinases I and II). There are doubts concerning the identity of proteinase II, since the assay procedure, as noted, is subject to alternative interpretations. In a detailed study, Liskowski[115] was unable to demonstrate breakdown of hemoglobin using alternative procedures (Folin, biuret, and ninhydrin methods) although neutral proteinase is reported as a contaminant in some growth hormone preparations.[154] Adams and Smith[153] reported that proteinase II was not cysteine dependent, but was inhibited by phosphate and some heavy metals, and slightly activated by Fe^{2+}. Proteinase I and II could be separated by precipitation at pH 5.5 and partially purified by ammonium sulfate treatment to yield preparations of about 10-fold enrichment. Both enzymes hydrolyzed other protein substrates (as determined by the Kunitz procedure), edestin, albumin, and globulin, and slowly hydrolyzed synthetic peptides used for the identification of serine hydrolases, pepsin, papain, carboxypeptidases, etc. Such results indicate that contamination of hormonal preparations could result in considerable endogenous breakdown, as indeed is the case in many early preparations of ACTH, and in recent samples of growth hormone and prolactin.[155]

Some success was reported by Ellis in the purification of proteinase I (pH 3.8).[155] A 50-fold enrichment was obtained by extraction of sheep pituitaries with ammonium sulfate, acid precipitation, followed by chromatography on IRC-50 and DEAE-cellulose. In many of its properties proteinase I resembles cathepsin D of brain, and cathepsins from other tissues.[156] Separation on DEAE-cellulose revealed considerable heterogeneity; the enzyme was unaffected by cysteine, KCN, and ETDA. Preincubation at different pH (optimum pH 4.0) resulted in activation, which may indicate the presence of a precursor form, or inactivation of endogenous inhibitor. Incubation with prolactin resulted in complete inactivation of biological activity and almost complete loss of growth hormone activity; the specificity with these hormones and insulin B chain suggested some similarity to pepsin.[156] The residual activity of growth hormone after enzymatic treatment raises some interesting points concerning the minimal structure required for hormonal activity. Li and Liu[157] in a series of elegant studies reported that removal of the C-terminal amino acid of growth hormone did not result in loss of activity: even limited digestion of chymotrypsin (25% for bovine, 20% for monkey, and 10% for human hormone) produced no alteration in activity despite the distinct physicochemical and immunological properties of the core material. Studies on the chemical synthesis of ACTH show that the octadecapeptide sequence Ser^1-$Tyr \cdot \cdot Lys^{15}$-Lys-Arg-Arg^{18} has steroidogenic potency in vivo comparable to the natural hormone.[157] The Kunitz assay procedure may have contributed to the difficulties by Reichert[158] and others[159] in the extraction, and attempted purification of proteinase II. It appears that efforts to purify this enzyme have been abandoned. Fragmentation of growth hormone preparations (as judged by disc-gel electrophoresis on acrylamide gels) has been attributed to neutral proteinases.[154] Breakdown occurred at pH 7.4 and 9.6 but not at pH 5.0 and was inhibited by DFP, PCMB, and EDTA. Although such preparations degraded urea-denatured hemoglobin at the same pH, the Kunitz procedure may be subject to error as earlier described (Section III, A). Additional criticisms of this method are the known presence of subunits which may give rise to multiple disc-gel patterns independent of breakdown, and the appearance of free α-amino groups not consistent with the known action of proteolytic enzymes. In other studies, Lewis and associates[160] reported the appearance of an inhibitor in sera that prevented fragmentation (as judged by disc-gel electrophoresis). The inhibitor was heat labile, nondialyzable, associated with the γ-globulin fraction, and increased in hypophysectomized or hypothyroidectomized rats. In our own studies, we were unable to detect any change in the levels of acid or neutral proteinase in crude extracts of hypophysectomized rats.[2a]

There is considerable interest in the relationships between hormonal factors and protein turnover. For example, hypophysectomy is reported to drastically reduce salivary gland proteases, which appear to be under direct hormonal control and to increase during development.[161] A number of steroid hormones have pronounced anabolic effects on protein turnover in the body during growth and development, although the mechanisms for accelerated protein synthesis have never been defined.[162] Estradiol administration is reported to alter the levels of aminopeptidases and arylamidases in the pituitary and hypothalamic regions.[163] In addition to an increase in enzymes hydrolyzing arylamide substrates (Leu-Arg, Ser-Tyr, Lys-Ala, Arg-Arg-βNA), Vanha-Perttula[163] reported a rise in acid proteinase and phosphatase following the administration of estradiol to female rats. Similar changes were observed following castration in male rats, a procedure known to result in hypertrophy and hyperplasia.[159] Meyer and Clifton[159] reported an increased level of proteinase II with estrogen administration following castration, but this will need confirmation since the Kunitz assay was employed (Section III, A). Administration of estrogen is reported to increase the secretion of prolactin but reduce the levels of LH and FSH synthesis; the higher levels of prolactin in females have been attributed to lower proteolytic activities in female pituitaries.[159]

Hypothalamic extracts are capable of inactivating substance P, bradykinin, and vasopressin, which are substrates for both exo- and endopeptidases (see Section V). In a survey by Hooper[329] the enzymes inactivating vasopressin were highest in the hypothalamus, compared to the cerebellum, thalamus, cortex, caudate nucleus, and white matter. No specific concentrations of enzymes inactivating bradykinin were observed in different brain areas. As a general comment, it should be emphasized that it is premature to relate hormone levels to the activity of a particular proteolytic enzyme without additional studies. The situation can be quite complex and may be dependent on the level of specific inhibitors, on the rates of synthesis compared to breakdown, the role of pituitary releasing factors, and their relationship to peptide hydrolases.

B. Peripheral Nerve

Some two decades ago Weiss[24a] suggested that one of the major purposes of axoplasmic flow was to replace protein or other compounds removed by breakdown processes. Although peripheral nerve (myelinated and nonmyelinated) contains a spectrum of peptide hydrolases (E.C. 3.4.1) and proteinases similar to that of other tissues,[1f,166] the relationship of these enzymes to function is unknown. It is possible that some proteolytic enzymes may be indirectly involved in the electrophysiological process since proteinases applied intraaxonally, but not topically, can affect ion permeability and resting potential[167–169]; such changes may be secondary to damage of axonal apparatus (neurotubules or fibrils). Ungar[23] reported some changes in the configuration of peripheral nerve proteins with stimulation accompanied by alteration in proteinases, using the Kunitz procedure of assay. In a study conducted with the cooperation of Derek Richter we observed a small (10%) increase in ninhydrin-positive materials with stimulation of frog sciatic nerve in situ (severed from the spinal cord but retaining normal circulation); these effects appeared to be related to changes in the amino acid pool rather than to changes in neutral proteinases.[175]

The isolated peripheral nerve preparation is an excellent system for studying changes in metabolism concomitant to function that has been overlooked in terms of protein breakdown.[166,170] This is particularly true in characterization and localization of proteinases, except in relation to extreme pathological changes associated with nerve transsection (Wallerian degeneration—WD). Interpretation of altered changes in WD is complicated by the infiltration of hydrolytic enzymes from exogenous sources, notably macrophages.[2f,171] WD has attracted interest, since it represents an excellent example of a phased breakdown followed by resynthesis and regeneration.[171-173] In myelinated nerves, WD is accompanied by loss of myelin in addition to degeneration of the axon; this condition is differentiated from typical demyelinating conditions where only loss of the myelin sheath is involved.[172] The complex phenomena associated with degeneration and regeneration are inextricably linked with the control processes of protein turnover, particularly the release and subsequent decay of proteolytic enzymes. Several excellent texts exist describing the changes induced by nerve section, tract compression, and other forms of injury[21,2f]; detailed discussion is to be found elsewhere in the Handbook.[1f,r,u] In general, ultrastructural changes precede invasion by phagocytic cells, and three distinct phases have been distinguished: (a) disintegration of the axon (proliferation of Schwann cells occurs at about 4 days), (b) chemical destruction of myelin lipids (8–32 days), (c) regenerative processes. The role of proteases (or other hydrolytic enzymes) in the initiation and progress of this sequence of events is still a question of further research. The disposition of protein and lipids within the myelin sheath is an important consideration in relation to the susceptibility to breakdown. Myelin is largely composed of proteolipid complexes which have been classified according to their solubility in chloroform–methanol, and susceptibility to proteolytic enzymes.[6a] The choice of terminology is operational and is dependent on the source of myelin (PNS or CNS, etc.) and treatments prior to extraction.[195] In early studies Folch et al.[6a] described two materials resistant to trypsin treatment; namely, a trypsin-resistant protein (TRPR) and "neurokeratin."[21d] There is now evidence that neurokeratin is artifactual, dependent on the fixation technique. Adams and Bayliss[173] consider that the myelin sheath contains both trypsin-resistant and sensitive proteins that determine the course of events during WD and in some experimentally induced (pathological) states (see below). Currently there is little information concerning the relative susceptibility of the known proteolipid components (Folch-Lees, and Wolfgram) to native brain enzymes. Basic protein is probably present as a proteolipid complex and resistant to brain cathepsin D, but when isolated is highly susceptible to breakdown.[1s,174]

Another component that deserves consideration in relation to protein breakdown is the relatively large concentration of collagen associated with the endoneural and peripheral elements.[21b] Little is known concerning the role of this element or its enzymatic activities. The same comments apply to phosphoprotein known to be distributed in a proximo-distal direction.[1f] Nerve axons also contain fibrous-like proteins that readily polymerize and exist in the form of tubules or fibrils; the integrity of this network may be critical to axonal function.[24a] Some 3 % of PN-soluble proteins is reported to consist of the acidic protein S-100, but little is known concerning its turnover.[1b,p]

Another aspect related to breakdown is the source and role of the free amino acid and peptide pools in PN preparations. Such pools could be derived by local proteolysis, and in addition be primed by axoplasmic flow, or by active transport processes, into the nerve fibers. Crustacean nerve contains levels of amino acid some 100-fold higher than rat sciatic nerve[166,175] on a wet weight basis, indicating multiple roles in arthropod tissue: (a) in osmoregulation, (b) on possible neurotransmitters, (c) in normal metabolic processes supplying energy. In relation to axoplasmic flow, Kerkut[1i,176] showed that Glu applied to the brain of a snail moved in a bound form in the axon connecting it to muscle, at which site it was released. Such findings suggest that axonal proteinases may play a role in the release of materials after transport from the nerve cell body to the appropriate target sites (synapses or effector organs). Ochs[2d] reported that the fast component of axoplasmic flow in cat spinal nerve preparations was associated with the free amino acids (Leu was used as the tracer) and accounted for most of the radioactivity present in effluents from Sephadex columns; the remainder consisted of low molecular weight polypeptides. The fate of materials transported down the axon (sometimes at relatively high rates) is rarely discussed but is very pertinent to neuronal function, particularly in relation to axoplasmic flow in undamaged nerves.[166,179]

We recently surveyed crustacean and vertebrate nerve for proteolytic activities, and in addition to exceptionally high activity with aminopeptides (Leu-Gly-Gly) and arylamides (Leu-βNA, and Arg-βNA) all nerves contained detectable quantities of acid and neutral proteins assayed with denatured protein substrates.[166] In an attempt to determine the relationships of such enzymes to the processes of axoplasmic flow, we measured the different enzymes at different locations along the nerve. In all cases acid proteinase was higher in the distal regions (both on a protein and wet weight basis), but other enzymes were more variable in the direction of the gradient when all species were examined (Fig. 7).

1. *Proteinases During Wallerian Degeneration*

Early studies were concerned primarily with altered levels of endogenous or autolytic activities consequent to nerve section.[2f,6c] In general, the time sequences for rise and decay of peptide and other hydrolases paralleled acid phosphatase. In recent studies these results have been reassessed utilizing a number of different protein substrates combined with biochemical and histochemical techniques.[2f,177] In one such study, Hallpike, Adams and Bayliss[172] observed an increased hydrolysis of casein at pH 7.6 (Kunitz assay, Section III, A), and of gelatin at acid pH (ninhydrin procedure) within 12 hr of sectioning rat sciatic nerve trunks (Fig. 8, A). Histochemical changes detected by the method of Adams and Tuqan[127] appeared only on the third day. With Leu-βNA as the substrate, changes were delayed until the fourth day (coincident with Schwann cell proliferation). The authors concluded that the proteinases may have a lysosomal source since the changes were parallel

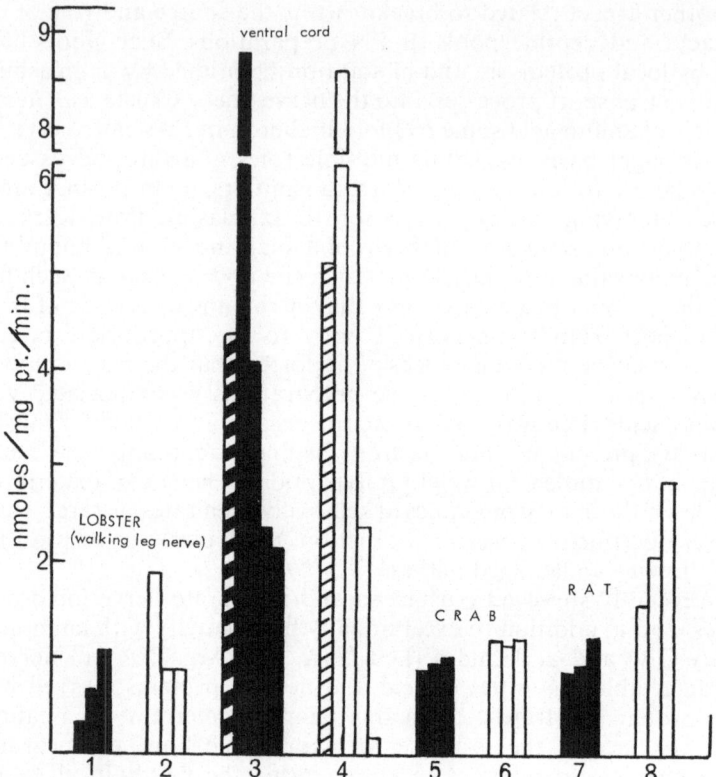

Fig. 7. Proteinase activity in different regions of the walking leg nerve preparations of lobster (1, 2), crab (5, 6), rat sciatic (7, 8) as compared to the ventral
cord and brain regions of lobster (3, 4). Acid proteinase is indicated by the
solid blocks and neutral proteinase by the open blocks. In the case of the
cord material, the first block (hatched) represents the brain regions (esophageal
ganglia) as compared to the adjacent cephalothorax, abdomen, and tail
regions. For other preparations the first block represents the proximal region,
the second the middle, and the third the distal regions. Activity is expressed as
nanomoles α-amino group released per milligram protein per minute with
denatured hemoglobin as the substrate (ninhydrin assay). Other conditions
are described elsewhere by Marks, Datta, and Lajtha,[62] and Marks.[5]

to those of acid phosphatase; in contrast, the arylamidase may be the result
of damage to the myelin sheath. These observations do not supply sufficient
information to determine the source of enzyme; it is possible that the immediate rise in activity of proteinase is due to enzyme present in the axon or
sheath rather than macrophages (which infiltrate at a later stage). In recent
studies with hemoglobin as the substrate (ninhydrin assay), Porcellati[2f]
observed a similar rise in acid proteinase following nerve section but a

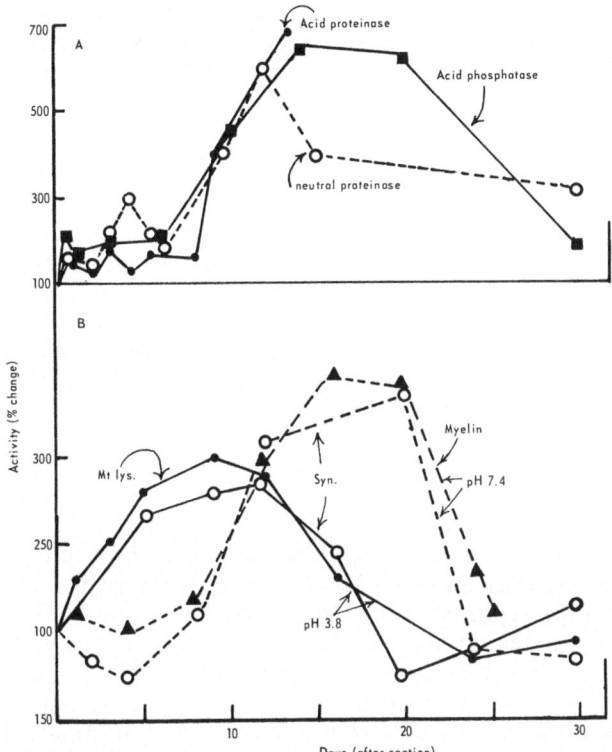

Fig. 8. Enzyme changes during Wallerian degeneration in rat sciatic A and in cat sciatic B. Data for both diagrams were based on the results obtained with aqueous homogenates (A) by Hallpike, Adams, and Bayliss,[177] or the sucrose subcellular fractions (B) by Porcellati.[2f] In A, acid proteinase was measured with gelatin (ninhydrin procedure), neutral proteinase with casein (absorption at 280 nm), and acid phosphatase with α-glycerophosphate. In B, denatured hemoglobin (ninhydrin procedure) was used for all fractions. The different pH and fractions are identified on the figure. Days on the abscissa represent the period after nerve section.

delayed rise (5–8 days) in neutral proteinase (Fig. 8,B). In subcellular fractionation studies, the localization of acid and neutral proteinases was almost identical to those we reported in earlier studies for whole brain.[68] Acid proteinase appeared in highest concentration in crude mitochondrial fractions, followed in descending order of activity by nuclear fractions, microsomes, and finally myelin; neutral proteinase was highest in myelin and synaptosomes. It is of some interest that in earlier studies, Porcellati *et al.*[165,178] reported inhibition of "endogenous" activities by a number of anticholinesterases including DFP, Parathion, and tetraethyl pyrophosphate.

This may be evidence that the "neutral hydrolases" involved contain serine at the active site. By way of contrast to the PNS, transsection of white matter tracts results in chromatolysis, which is slower and more protracted in development than WD, with the implication that different control mechanisms for myelin turnover operate in the CNS.

C. Spinal Cord

As noted in the timely review of Levi[1d] in Volume 2 of this Handbook and elsewhere,[335] the spinal cord has been neglected in terms of protein turnover and breakdown. This is somewhat surprising since spinal cord from larger animals is the most favored material for preparation of basic (encephalitogenic) proteins. Myelin obtained from whole brain of rats and pig is characterized by relatively low levels of proteolytic enzymes, compared to brain. The cord myelin purified on sucrose gradients, followed by water shock treatment to remove soluble contaminants, and further purification on CsCl gradients, still exhibited some aminopeptidase and arylamidase activities, and trace activities of acid proteinase, though recovery of activity was less than 10% of that in freshly prepared crude myelin.[131,132] Neutral proteinase, as in the case of myelin from whole brain, was totally absent in the CsCl purified material. Lajtha[79] in a comparative survey of different areas of monkey brain, reported some "endogenous" neutral acid proteinase in the spinal cord material but lower than that of the cerebrum and cerebellum, and the basal stem areas (pons, medulla, and white matter). Millo and Porcellati[164] reported an increased level of endogenous neutral proteinase in the spinal cord of hens following intoxication with organophosphorous compounds (DFP and TOCP). Adams and Tuqan[10,127] demonstrated pronounced histochemical changes in rat spinal cord following cordectomy that were consistent with a measureable decrease in the histochemical level of proteolipid protein, and TRPR components, although the concomitant proteolytic activities were not measured.

In lobster ventral cord we observed[166,175] a marked decrease in activity of all proteolytic enzymes from a proximal (adjacent to the supraesophageal ganglia) to distal regions (Fig. 7). Enzymatic activities were considerably higher than those present in the peripheral walking leg nerves. In some cases the gradient of activities was accompanied by a similar gradient for selected amino acids.

D. Degradation of Myelin

Breakdown of myelin has attracted interest in relation to clinical or experimentally induced diseases involving demyelination.[324] It has been the hope that experimental allergic encephalomyelitis (EAE) would serve as a model disease state for studying the pathogenesis of multiple sclerosis (MS).[177] Our early studies demonstrated the presence of acid and neutral proteinase in crude myelin preparations,[68] but we concluded at the time that a considerable proportion represents contamination from other sources.

Currently there is some debate concerning the presence of proteinases in myelin, and the source of enzyme during degenerative changes. In our own studies, purification of crude myelin by water shock treatment, centrifugation on sucrose, and cesium chloride gradients was accompanied by a total loss of neutral proteinase and most of the acid proteinase; the only hydrolase with unchanged or with an increased specific activity was an aminopeptidase (Leu-Gly-Gly) and a 2,3'-cyclic phosphohydrolase (Table VI).[131,132] The

TABLE VI

Distribution of Proteinases and Other Hydrolases in Myelin Proteolipid and Protein Fractions[a,b]

Enzyme	(M$_1$) Crude myelin (total activity)	(M$_2$) Water shocked (percentage recovered from crude myelin)	(M$_3$) CsCl
Acid proteinase	1.9 (1.4)*	20 (11)*	7 (0)*
Neutral proteinase	0.57 (0.62)*	43 (15)*	0 (0)*
Aminopeptidase (Leu-Gly-Gly)	4.3	51 (0.24)	20 (0.32)
2',3'-Cyclic diesterase	3.7	55 (0.24)	20 (0.28)
Arg-βNA	10.8	12 (0.14)	5 (0.20)
Leu-βNA	9.7	27 (0.4)	10 (0.5)
Ser-Tyr-βNA (Cathepsin C)	0.9	18	5
Lys-Ala-βNA	1.8	26 (0.1)	5
Arg-Arg-βNA	3.3	12	4

[a] Proteinases were determined at pH 3.8 and pH 7.6 as previously described.[42,68] The first column represents the total activity expressed as micromoles per gram of fresh tissue for the enzymes listed; the second and third columns represent the percentage recovery. The values in parentheses for the first two enzymes represent the endogenous activity in absence of added hemoglobin; for all other enzymes the values in parentheses represent the specific activities expressed as micromoles of product formed per milligram protein per hour. 2',3'-Cyclic nucleotide-3'-phosphohydrase was determined with 2',3'-cyclic AMP as substrate.
[b] Rat tissue.

presence of these two hydrolases has been reported by Benik and Davison[180] in a recent survey of enzyme activities in purified myelin. In contrast to our findings, Riekkinen and Clausen[118] reported considerable neutral proteinase (20% of that in the homogenate) and an absence of acid proteinase in myelin purified on a succession of sucrose gradients. In some cases, the neutral proteinase was demonstrated only after prolonged incubation (24 hr); activity could be extracted with 0.1% Triton X-100, suggesting the existence of a proteolipid complex.

1. *Experimental Disease States*

Hallpike *et al.*[177] have emphasized the essential differences between Wallerian degeneration (WD), EAE, and multiple sclerosis (MS). WD does not appear to be relevant to the pathogenesis of a primary demyelinating disorder (e.g., MS), since axonal damage is characteristically absent or slight. There is doubt concerning the relationship between EAE and MS although EAE has some similarities to an immune-type response involving sensitized lymphocytes. It is of interest that increased proteolytic activity in brain has been reported as associated in experimental animals with EAE.[182,183] Myelin itself is not a good substrate for purified acid proteinase, but the iso-

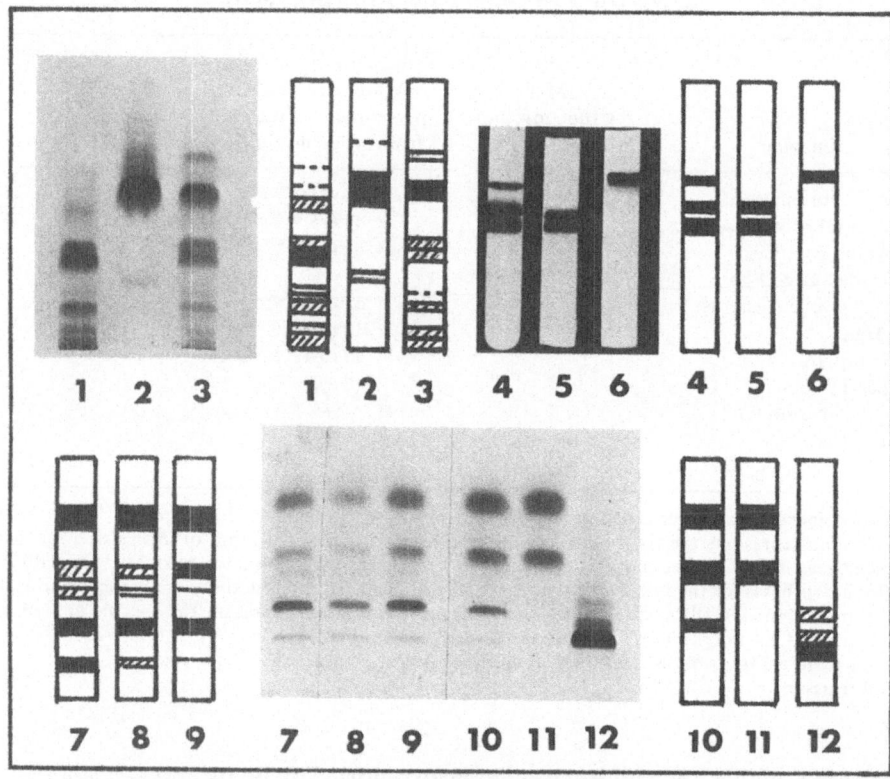

Fig. 9. Electrophoretic patterns of (1) proteolipids, (2) basic (encephalitogenic) protein, (3) myelin and the effect of purified acid proteinase on breakdown of basic proteins (4–12). Patterns 1–6 were obtained on acrylamide according to conditions used for disc-gel electrophoresis.[174] Patterns 7–12 were obtained on flat acrylamide gels according to the conditions of Einstein *et al.*[101] Incubation of the basic protein with enzyme was accompanied by the appearance of several distinct bands dependent on time of contact: 8, 2 hr; 9, 3 hr; 4 and 10, 5 hr; 5 and 11, 24 hr; gels 6 and 12 represent the basic protein starting material incubated for 24 hr in the absence of enzyme. The bands obtained after 5 hr retained potent encephalitogenic activity.

lated basic protein components are highly susceptible to breakdown[101,174] (Fig. 9). The split products show similarities to other encephalitogens, implying some breakdown during preparative procedures. Another implication of these results is that lipolysis may actually precede breakdown of the basic proteins (i.e., the reverse of WD). Thus the slow release of basic peptides from encephalitogenic proteins may be involved in the etiology of EAE; we have suggested (in collaboration with Dr. Elizabeth Roboz-Einstein[1s]) that small peptides can readily penetrate the relevant membranes and interact with lymph nodes.[174] In support of such a hypothesis, Hallpike *et al.*[184] recently reported loss of trypsin-digestible basic protein (detailed histochemically) in the plaques obtained from MS patients. Increased proteolysis was observed at the edges of plaque lesions together with an increased level of oxidative enzymes. The changes in enzymes are associated with an increase in cells at the periphery of the lesion which appear to be largely oligodendrocytes.[174,184]

Recently, important advances were made in the chemical characterization of basic proteins. Most studies indicate a mol. wt. of 18,000 (30% basic residues). In aqueous solution, basic protein has an open conformation as revealed by its susceptibility to proteolytic enzymes (cathepsin D, trypsin, etc.) and resistance to denaturation by heat and urea.[189] Eylar *et al.*[185,186] recently isolated an encephalitogenic peptide by pepsin digestion with 14 amino acid residues containing one tryptophan essential for activity. Chymotrypsin digestion of this peptide liberated the N-terminal peptides Ser-Arg-Phe and Ser-Trp (Fig. 10). Tryptic digestion of basic protein gave a peptide with 9 residues containing Trp which has been recently synthesized and shown to be active (Fig. 10). Eylar and his group[185–187] consider this peptide fragment as the major encephalitogenic determinant; modification of Trp leads to a loss of activity.[282] However, Kibler *et al.*[188] have isolated from bovine spinal cord an encephalitogenic peptide containing 45 amino acids but no Trp. Thus the possibility exists of more than one active site with considerable species differences (from bovine cord the 9-sequence is encephalitogenic in guinea pigs at about 1 μg per animal; the 45 sequence is active in rabbits at 50 μg per animal). Such peptides are highly suited for studies of peptide bond specificity as related to this important model disease state.

Studies by Chao and Einstein[189] utilizing brain acid proteinase prepared by our procedures gave peptides with N-acetyl-Ala as the N-terminal, the Ala-Arg-Arg- as the C-terminal. The existence of a blocked N-terminal would prevent breakdown by classical aminopeptidases, but not carboxypeptidases. Since brain tissue is not rich in carboxypeptidases,[5] breakdown of basic (encephalitogenic) proteins is mediated presumably by intracellular proteinases. Once larger peptides are formed by internal cleavage, then smaller fragments can be liberated by other peptide hydrolases by mechanisms of sequential breakdown as described earlier. A discussion of the extraction and partial structures of basic proteins is provided elsewhere in the Handbook.[1s] In relation to development, it is of interest that basic protein cannot

BREAKDOWN OF BASIC PROTEINS FROM MYELIN

Brain acid proteinase[101,174]

Several peptides. Major encephalitogenic component is encephalitogen G
Mol. wt. approx. 5000
Susceptible linkages of basic protein similar to pepsin
(-Phe-Phe- at bonds 44–45, and 88–89 in structure
according to Carnegie[334])

Pepsin[186]

Major encephalitogen in guinea pigs is peptide E

Ser-Arg-Phe-Ser-Trp-Gly-Ala-Glu-Gly-Gln-Lys-Pro-Gly-Phy

chymotrypsin

trypsin

Trypsin[186,189,334,342]

Peptide T-27   Phe-Ser-Trp-Gly-Ala-Glu-Gly-Gln-Lys

T-8 Thr-Thr-His-Tyr-Gly-Ser-Leu-Pro-Gln-Lys

N-terminal   N-acetyl-Ala-Ser-Gln-Lys-Arg-Pro-

trypsin

C-terminal   -Arg-Ser-Gly-Ser-Pro-Met-Ala-Arg-Arg

cyanogen bromide or
thermolysin

Fig. 10. Basic protein of myelin contains about 169 residues[334] and may exhibit more than
one encephalitogenic site. Kibler and Shapira have isolated a 45 peptide sequence from bovine
spinal cord devoid of any Trp but containing Tyr at approximately the same position that is
active when administered to rabbits but not in guinea pigs (see text and ref. 1s).[188,342] In peptide
E or T-27 the residues underlined are considered essential for activity. It is postulated that
brain acid proteinase is involved in the production of potent encephalitogenic peptides and
that these are involved in the etiology of experimental allergic encephalomyelitis.[101,174]

be easily detected in the fetus or in newborns, but increased rapidly during
myelination coincident with an increased acid proteinase.[190,191]

2. Conclusion

Acid proteinase rapidly degrades basic proteins extracted from myelin
to form potent encephalitogenic peptides. Tryptic digestion can release an
encephalitogenic peptide containing 9 amino acids, including Trp. Basic
proteins may contain multiple encephalitogenic sites active in different

species, as illustrated by larger peptides (45 amino acids), which are active in rabbits and do not contain Trp. Acid proteinase attacks myelin slowly, suggesting that lipolysis may precede protein degradation in the etiology of EAE. The liberation of toxic peptides from other (nonencephalitogenic) materials may also be an important contributory factor. The source of proteinases during WD is unknown; protein breakdown appears to precede lipolysis and its rapid increase in early phases suggests an endogenous (axon, sheath, or Schwann cell) rather than an exogenous source. In later phases, the picture is more complicated owing to the invasion of macrophages. The time sequences of WD compared to other clinical and experimental states suggest that it is not a good model for the pathogenesis of disease states.

V. BREAKDOWN OF SOME PHYSIOLOGICALLY ACTIVE PEPTIDES

During the complex sequence of events in allergic and anaphylactic states, hydrolytic enzymes are frequently activated, with the appearance of vasoactive polypeptides and biogenic amines; the mechanism of activation and/or release in most cases is unknown (Fig. 11). Sudden release of vasoactive peptides can result in cerebral apoplexy or shock with pronounced changes in capillary permeability, fluid and protein loss, hypotension, and migration of phagocytic white blood cells.[15,16] Proteinase activation during anaphylactic shock was first observed over 50 years ago (see Ref. 12b); Ungar and Hayashi[196] ascribed primary importance to this mechanism in the pathogenesis of the allergic response. Activation of the fibrinolysin system, and release of hypotensive peptides are associated with some forms of anaphylactic shock in many species. It is beyond the scope of this review to summarize the vast amount of data existing in this area[12,15,16]; aspects relevant to the role of the CNS and PNS during the inflammatory response are reviewed by Chapman and Goodel.[197] Production of vasoactive peptides represents an excellent example of sequential breakdown of proteins by the mechanisms of peptide hydrolysis.

No general agreement currently exists pertaining to nomenclature, and loose terminology has led to considerable confusion in the naming of the degradative enzymes involved.[15] Even the term kinin has led to some difficulty; it previously referred to vasoactive peptides released from α-globulins by endogenous enzymes (namely, angiotensins and bradykinins). The term kininogenase has been reserved for those enzymes liberating "kinins" from inactive protein precursors (kininogens)[204]; such enzymes would include, by definition, kallikreins, trypsin, pepsin, snake venom, bacterial proteinases, etc. Enzymes that inactivate "kinins" could logically be called "kininases." Some of the postulated mechanisms involved in kinin formation are summarized in Fig. 11.

A question of considerable importance is the existence of specific degradative enzymes for the different types of peptide hormones. Unfortunately

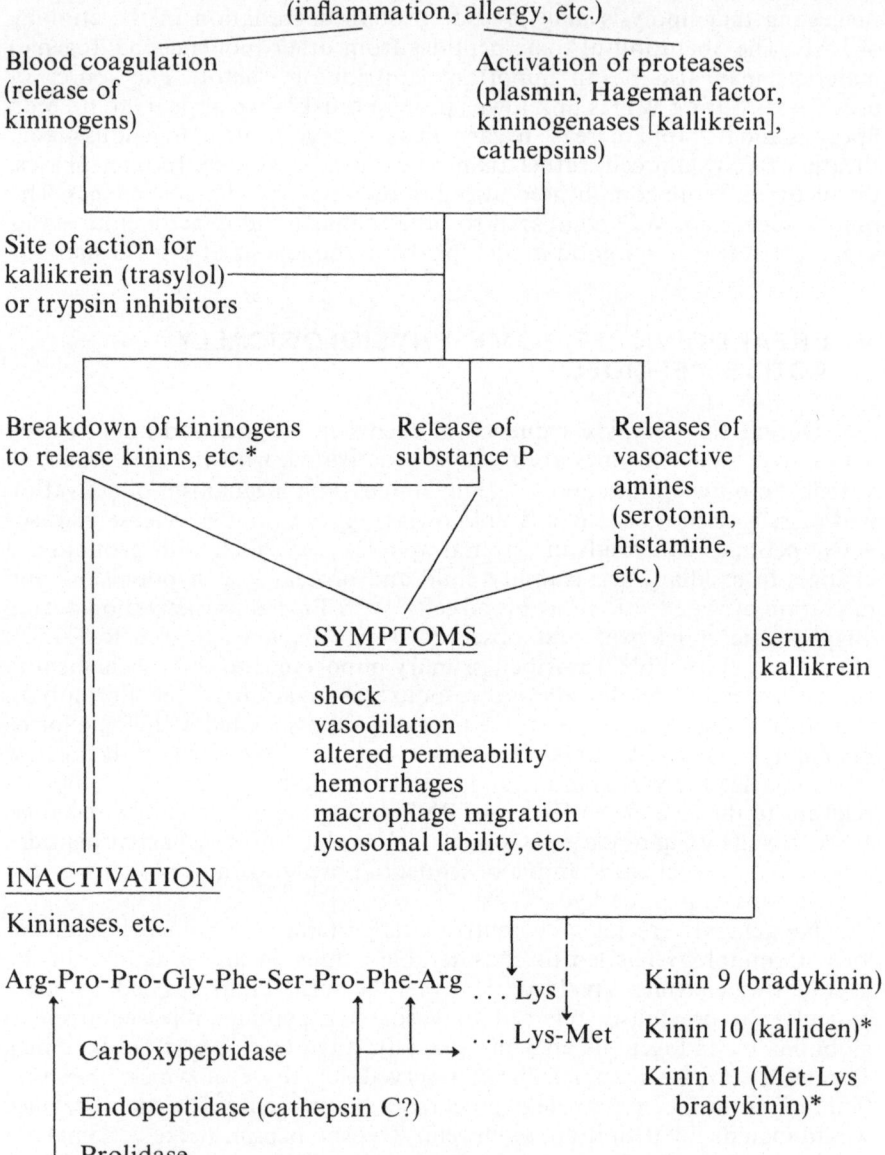

Fig. 11. Scheme for inactivation of kinins. Bradykinin and other peptides are formed by breakdown of precursor proteins or peptides (kininogens): kinins 10 and 11 can be further degraded by serum kallikrein to form kinin 9 or bradykinin. Several enzymes exist for inactivation of kinin 9 which include both exo- and endopeptidases as noted in the text.

the data available are incomplete, frequently conflicting, and difficult to evaluate. There must be considerable overlap between peptide hydrolases and endopeptidases since most (peptide) hormones contain free $-NH_2$ or $-COOH$ end groups susceptible to hydrolysis by aminopeptidases or carboxypeptidases. Peptides relevant to brain function can conveniently be described under the following headings: (a) kallikrein-bradykinins, (b) renin-angiotensins, (c) substance P. Brain itself contains exceptionally low concentrations of peptides[5]; these may resist normal methods of extraction since in some cases they are tightly bound to tissue. Often such materials are active at exceptionally low concentrations (nanograms) and represent some of the most powerful pharmacological materials known. As such, concentration may be less important than their localization and release at the target sites. Alterations in blood pressure or flow, increased capillary permeability, contraction of muscle, induction of pain can directly or indirectly affect nerve function. Physiologically active peptides exist in a variety of tissues (or in venom), particularly the skin of many amphibians, and exert similar effects as the classical kinins; nomenclature is based on their implied function or tissue of origin. Such materials include eledoisin and physalaemin (undeca-peptides, frog skin); gastrin, erythropoietin, secretin, cholecystokinin, pancreozymin, anaphylatoxin, relaxin, calcitonin, cerulein (decapeptide), etc.[16]

A. Kallikrein-Bradykinins (E.C. 3.4.423)

There are three hypotensive peptides known collectively as "kinins": bradykinin (kinin-9), kalliden (kinin-10, with addition of Lys), and Met-Lys-bradykinin (kinin-11).[15,16,22] Kinins are formed in plasma by the action of the proteolytic enzyme kallikrein on precursors known as kininogens (one such precursor, kininogen-I, has a mol. wt. of 53,000 and is present in the α-globulin of plasma: selective cleavage with cyanogen bromide has indicated that there is a kinin-11 sequence at the $-COOH$ terminus, in addition to similar sequences within the precursor.[198] Kinins can be formed also by the action of trypsin or plasmin as summarized in Fig. 11. Kallikrein (E.C. 3.4.4.21) is reported as present in brain fractions, with the highest concentration in the microsomal particulates.[199,200] Serum kallikrein has a mol. wt. of 97,000 and an electrophoretic mobility equivalent to α-globulin, and differs markedly in its properties from pancreatic enzyme (mol. wt. 29,000, mobility equivalent to albumin). Pancreatic enzyme acts on kininogen to form kinin-10, or on tryptic peptides of globulin to form kinin-11; in contrast, the enzyme from serum forms kinin-9 from all precursors, including kinin-11, to form bradykinin. Release of plasma kinin is retarded by kallikrein inhibitors which are identical in most cases to known tryptic inhibitors (Section VI, A). Kallikrein itself is formed from a precursor (prekallikrein) by several mechanisms: enzymatic, change in pH, contact of plasma with glass (activates the Hageman factor), heat, during blood coagulation[20,201,205,206] (see Fig. 11). Antiproteolytic titer is reported to rise in sera during pregnancy and a host of pathological conditions, ranging from cerebral apoplexy, arteriosclerosis, and during malignancy.[11,12] Activation of inhibitory factors may form part of a delicate bodily defense mechanism against inflammation and related allergic conditions. Some years ago, Chapman and Wolff[197,202] reported an increased formation of hypotensive peptides during disease states on incubation of cerebro-spinal fluid with globulin *in vitro*. They attributed this to an increased level of proteolytic enzyme(s), but it was pointed out in a later study by Coppen and Lewis[203] that such enzymes could result from hemorrhage due to minimal bleeding associated with the site of spinal tap. The data in this area are still meager and contradictory. Sicuteri[12d] reported only slow inactivation of bradykinin on spinal injection, or incubation *in vitro*, but rapid inactivation on infusion into ventricles of the cat. Inactivation could have occurred on contact with tissue lining the ventricle or at some specialized site such as the choroid plexus. Interest in the relationships of kinins to inflammatory conditions has stimulated considerable research, as summarized in recent symposia.[22,15,16] An example of such studies was provided by Barnhardt et al.[12c] who reported a rise in proteases in synovial fluids during arthritis; proteases were partially inhibited by a polyvalent inhibitor of trypsin (trasylol) and by an antiserum raised against cathepsin purified from human spleen. Kallikrein inhibitors are reported to alleviate symptoms accompanying cerebral stroke and subarachnoid hemorrhage,[12d] or postoperative hemorrhage, or experimental and clinical pancreatitis.[11]

The concept of specific kininases is an unresolved question since it implies enzymes capable of recognizing large polypeptide sequences. In actual fact, a host of enzymes have been described as capable of inactivating the nonapeptide sequence of bradykinin. This particular sequence would suggest susceptibility to sequential cleavage by carboxypeptidase B (removal of Arg and Phe) as indeed reported for kidney and spleen homogenates.[15,321] The possibility also exists for removal of dipeptide sequences from the C-terminal (end) by analogy to the degradation of angiotensin I (Section II, C). Degradation by cathepsin C-like enzymes (N-terminal dipeptides) has not been described but appears to be a distinct possibility. An example of the variety of enzymes that exists in tissues is illustrated by microsomes from swine kidney that are rich in carboxypeptidases, a prolidase (iminopeptidase), and endopeptidases (cathepsin).[15] The carboxypeptidase in plasma inactivating bradykinin is distinct in properties from pancreatin and kidney enzymes and has been named carboxypeptidase N, but is not available currently in highly purified form for absolute identification.[22g] Carboxypeptidases from spleen also differ from pancreatic enzymes since they are active at acid pH and require sulfhydryl compounds.[321] The term catheptic carboxypeptidase[321] (to emphasize their intracellular nature) is somewhat confusing since it implies overlap between a true endopeptidase (cathepsin) and a peptide hydrolase. Most studies point to a simple carboxypeptidase with similarities to cathepsin B (Section II, B).

Hori et al.[207] in a survey of different tissue reported a higher concentration of kallikrein (and its precursor form) in brain compared to spleen, liver, and lung, but less than pancreas and kidney as judged by the inhibition of hypotensive effects by trasylol. In a recent study, Shikimi and his group[199,200] obtained a 23-fold purification of an inactivating enzyme from brain supernatant-microsomal fractions (ammonium sulfate, Sephadex G-100, DEAE, and CM-cellulose chromatography). The enzyme displayed a pH optimum of 7.6, and was inhibited by metal ions, GSH and mercaptoethanol, trasylol, and partially by PCMB. This enzyme differed in properties from substance-P inactivating enzyme (also shown to inactivate bradykinin, see below) since it failed to act on any protein substrates. Since this enzyme liberated Arg- or Phe- end groups, its properties suggest the presence of one or more exopeptidases (amino- or carboxypeptidase). The CNS and PNS is reported to contain granules that yield kinins or substance-P on treatment with trypsin (which may be equivalent to the kininogens of other tissues.[207–209] It is of interest that a tryptic peptide obtained from globulin (kalliden-11) is itself a substrate for aminopeptidase B (or arylamidase B) of liver, yielding bradykinin.[210] This mechanism thus provides a further example of the sequential action of exo- and endopeptidases yielding active peptides.

In conclusion, a number of tissues (including brain) contain precursor materials yielding kinins or their inactivating enzymes. Inadequate characterization has prevented absolute identification of specific inactivating enzymes (kininase) in brain or other tissues. Most kinins are susceptible to inactivation by a variety of different proteolytic enzymes, including amino- and carboxypeptidases, intracellular and extracellular proteinases. Enzymes purified from a variety of tissues, or serum, display different specificities and molecular weights; in most cases properties of partially purified preparations are similar to those of peptide hydrolases (particularly carboxypeptidases). Kinins and their precursors are reported to exist in brain, with the highest concentration in the microsomal-supernatant fractions; enzyme purified from this source appeared similar to that of a peptide hydrolase and differed in its properties from other known inactivating enzymes.

B. Renin-Angiotensins (E.C. 3.4.4.17)

Formation of angiotensin represents an example of sequential breakdown of precursor glycoprotein (hypertensinogen, mol. wt. 57,000) by the proteinase renin, followed by an exopeptidase (converting enzyme) as summarized in Fig. 12. Renin is reported present in kidney, uterus, and plasma but not in nerve tissue; it is active at acid pH with different specificity compared to pepsin or cathepsin D.[211–213] Unlike the latter, plasma enzyme is inhibited by EDTA, reactivated by divalent metals. The decapeptide, angiotensin I, formed by renin is converted into the octapeptide Asp-angiotensin II, by a unique Cl^--dependent converting enzyme (carboxypeptidaselike)[211,214] by removal of the C-terminal dipeptide His-Leu. It is distinct from other Cl^--dependent hydrolases removing N-terminal dipeptides, e.g., cathepsin C and glucagon inactivating enzyme (Section II, C).

Angiotensin, like bradykinin, is inactivated by a variety of enzymes (Fig. 12)—a Ca^{2+}-activated aminopeptidase active at neutral pH, carboxy-

Angiotensin I

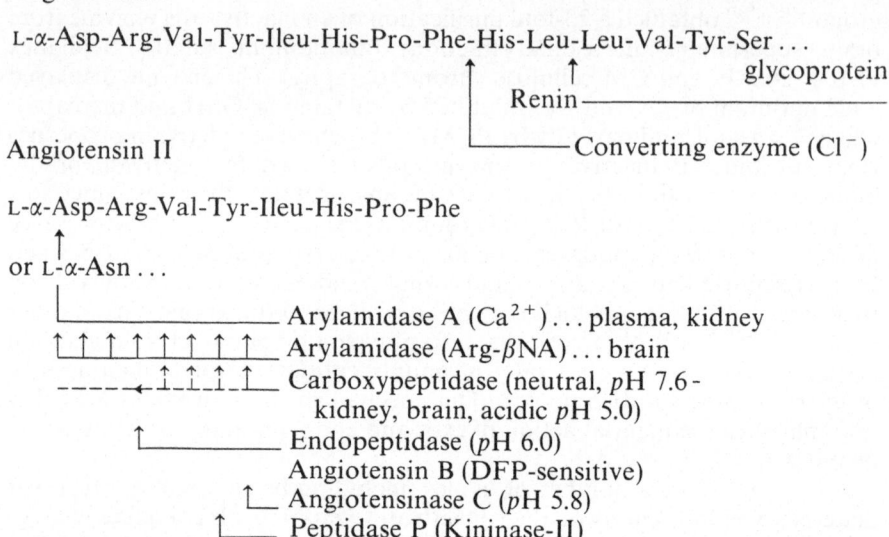

L-α-Asp-Arg-Val-Tyr-Ileu-His-Pro-Phe-His-Leu-Leu-Val-Tyr-Ser . . .
                                                            glycoprotein
                                        Renin ————————————

Angiotensin II   └————Converting enzyme (Cl⁻)

L-α-Asp-Arg-Val-Tyr-Ileu-His-Pro-Phe

or L-α-Asn . . .

└——————————— Arylamidase A (Ca²⁺) . . . plasma, kidney
↑↑↑↑↑↑↑↑↑ Arylamidase (Arg-βNA) . . . brain
————↑↑↑↑ Carboxypeptidase (neutral, pH 7.6 -
                         kidney, brain, acidic pH 5.0)
↑——————— Endopeptidase (pH 6.0)
              Angiotensin B (DFP-sensitive)
↑— Angiotensinase C (pH 5.8)
↑— Peptidase P (Kininase-II)

Fig. 12. Enzymes involved in angiotensin breakdown. It is probable that more than one enzyme exists for breakdown of angiotensin II, some of which may be examples or aminopeptidases (E.C. 3.4) or endopeptidases (E.C. 3.4.4). For description of individual enzymes see text or other sources.[234]

peptidases active at acid and neutral pH, and an endopeptidase with a specificity similar to chymotrypsin.[234] Matsunga et al.[215] recently surveyed a number of tissues, including brain, for inactivating enzymes. At pH 7.5 activity was present in soluble fractions, higher in kidney than in liver, spleen, and lung and lower in brain, heart, and skeletal muscle (Table II). At acid pH, enzyme was present in the lysosomal-rich fractions with highest activity in kidney, spleen, and liver. The neutral enzyme was Ca²⁺ dependent, was inhibited by DFP, and showed some similarities to "neutral proteinases" described earlier (Section III). The acid enzyme was optimum at 5.6 (mol. wt. 80,000) and was separated from peptide hydrolases (Leu-βNA as a substrate) and from endopeptidases (hemoglobin) by chromatography on Sephadex and DEAE-cellulose; extracts, however, were active with a carboxypeptidase A or cathepsin A substrate (Z-Glu-Tyr). In a related study, Kobuku et al.[216] reported an 8000-fold purification of an inactivating enzyme from rabbit red cells with a pH optimum near neutrality (pH 6.8). This enzyme was inactive with Asp- or Glu-βNA as substrates (i.e., arylamidase A[5]); with dipeptides Leu-Gly, Glu-Gly, and Arg-Ala (dipeptide hydrolases); with carbobenzoxy peptides, Z-Gly-Phe and Z-Gly-Arg (carboxypeptidases); or with casein and TAME (endopeptidases and esterases). The enzyme showed some similarities to aminopeptidase splitting Leu-Gly-Gly (tripeptidase) and Leu-βNA (arylamidase N). Angiotensin was cleaved at the -Tyr-Ile-bond. Incubation of angiotensin II with a purified brain arylamidase (based on Ary-βNA hydrolysis) led to the liberation of the five N-terminal amino acids while the C-terminal tripeptide was more slowly digested. Inactivation was confirmed by bioassay. Angiotensin II was unaffected by a purified aminopeptidase (Leu-Gly-Gly).[83,96]

Ischemia, anoxia, and shock are associated with labilization of lysosome-type membranes and could thus account for the release of inactivating enzymes noted under these conditions.[217,218] Currently there is considerable speculation that some vasomotor changes are mediated indirectly by the release of biogenic amines by unknown mechanisms involving angiotensin.[217] Angiotensin is reported to release catecholamines from storage sites at nerve endings; angiotensin may also affect the transport and flux of Na in arterial tissue.[15,16]

C. Substance P

The physiological role and identity of substance P is still in doubt, but its potent effects on blood pressure and smooth muscle, together with its localization in specific brain areas (generally in synaptosomal fractions) suggests a role in nerve function (see Vol. 4 of this Handbook). Substance P is present at only 0.01 mg/kg in brain [mol. wt. 1650 ($\pm$250)] and appears to be a group of peptides rather than one component.[1k,207] Such peptides require vigorous methods for extraction and are exceedingly difficult to purify. Very little is known currently concerning their activation or degradation in tissues. Substance P is reported to be released into perfusates on stimulation of cat somatosensory cortex or frog spinal cord,[331] or on incubation of microsomes with brain homogenates.[240] A distinctive feature is the resistance to breakdown *in vivo*, suggesting that release from the tissue complex is required before activation by native enzymes. Substance P is inactivated by chymotrypsin or by trypsin on prolonged incubation.[15] Claybrook and Pfiffner[219] recently obtained a 40-fold purification of inactivating enzyme from aqueous extracts of bovine brain (ammonium sulfate, CM-Sephadex, and protamine sulfate precipitation); it had a pH optimum near 6.4, and it also degraded hemoglobin at pH 4.0. Its stability at $-15°C$ or upon lyophilization suggests similarities to brain cathepsin D. In addition to substance P, this enzyme inactivated bradykinin and kalliden but not eledoisin.

The results with peptide hydrolases are conflicting, since most enzymes studied (carboxypeptidases or aminopeptidases) were contaminated with endopeptidases. Activity was unaffected by carboxypeptidase A (DFP treated to remove serine hydrolases) despite liberation of C-terminal Phe, Leu, and Tyr in that order.[16] The absence of breakdown with leucine aminopeptidase suggests the presence of a protected N-terminal amino acid.[16] Prior treatment with trypsin gave peptides susceptible to aminopeptidase, which could form an important route for inactivation in the gut. No inactivation occurred with prolidase, eliminating the possibility of N-terminal prolyl groupings.

Substance P shows some analogies to the phosphopeptide "nervside" since the peptide known as "Fc" is antagonized in its actions by atropine and morphine, and is rapidly inactivated by chymotrypsin compared to trypsin.

D. Parathyroid Hormone

Inactivation can occur in slices of liver, spleen, and skeletal muscle (studies with brain are unavailable); inactivation is prevented by prior heat treatment, extreme pH, and proteinase inhibitors.[220] More recently an enzyme was isolated from rat kidney, purified 700-fold from microsomal fractions, and shown to hydrolyze parathyroid hormone at pH 8.0–9.5, but its exact nature cannot be defined since it hydrolyzed a pepsin substrate (Z-α-Glu-Tyr), arylamidase substrate (α-Glue-βNA), a chymotryptic substrate (glutarylphenylalanyl-βNA), and a tryptic substrate (BANA). The enzyme bears some analogy to an aminopeptidase specific for acid analogues of β-naphthylamide.[5] This peptide is considered to affect indirectly CNS function.

E. Nerve Growth (NGF) and Other Factors

Materials promoting growth of specific nerve fibers (e.g., NGF), or growth of selected tissues (epidermal growth factor, thymolytic factor, spinal tubule factor, etc.) display immunological and other properties akin to proteins, but the precise structure (protein or low molecular weight haptan) is still undecided (see Schenkein, Chapter 16).[10] Such factors are associated in many cases with esterase or proteolytic activities[221] [e.g., NGF, thrombin, factor prompting growth of embryonic muscle tissue[222]]. Proteinases may be involved in their formation, or alternatively, such factors may have intrinsic catalytic properties and release active peptides (or other components) at the appropriate target sites. Answers to such questions could provide insight into control mechanisms of turnover in nerve tissue. NGF is present in spinal and sympathetic ganglia of chick or mammalian embryos, in the spinal axial region of tadpoles and spiny fishes but, surprisingly, is present in higher yield in snake venoms, and mouse submaxillary gland.[10] These sources contain endogenous esterase and proteolytic activities that could persist in purified factor preparations. According to Shooter et al.[221] the NGF in submaxillary gland (7S, 140,000 daltons) can be split into three subunits (α, β, γ) at different pH (of about 30,000); only the gamma subunit was associated with esterase activity (BAEE or TAME as substrates). Disaggregation of NGF into subunits led to loss of factor activity, and recombination led to restoration but with loss of esterase activity. Only trace activities with casein were observed (Kunitz method). Angeletti,[223] however, reported the absence of hydrolases (esterase, protease, peptide hydrolase, etc.) in highly purified preparations of snake venom NGF. Thus, the presence or absence of hydrolases may be dependent on the source and method of isolation. Submaxillary gland itself is known to contain several different hydrolytic enzymes (without NGF activity) active at pH 8.0[10]; one such enzyme hydrolyzed

basic substrates (polyarginine, histones, protamine, etc.) in addition to esters
(TAME, BAEE). In some of its properties it resembled "neutral proteinase"
(Section III). The role of hydrolytic enzymes in factor activity is unknown;
in some cases inhibition of endogenous hydrolases can result in inactivation
(embryonic muscle tissue factor).[10]

F. Hypothalamic Releasing Factors

Little information is available concerning the identity, origin, and degradation of hypothalamic releasing factors. Most factors appear to be polypeptides, present in trace quantities in hypothalamus, that affect release of
anterior pituitary hormones.[152] Only the structure of *bovine*-TSH releasing
factor (TRF) is known with any certainty, pyroglutamyl-His-Pro · NH_2;
that from pigs may be similar since its activity is reproduced by synthetic
derivatives.[225] Since such peptides contain protected end groups, their
resistance to proteolytic enzymes casts doubt on early studies on their polypeptide nature. Plasma and serum are reported to contain an inactivating
enzyme, possibly a pyroglutamyl peptidase.[152] The nature of other factors
is still speculative; those releasing luteinizing, or corticotrophin hormones are
inactivated by proteolytic enzymes, indicating a polypeptide nature.[337]
Burgus and Guillemin[152] concluded that such factors may be formed by
degradation of precursor hormones, a mechanism that bears some analogy
to release of kinins.

G. Insulin Breakdown

Insulin peptides are used as model substrates for specificity studies but
the breakdown of insulin *per se* in tissues could be important to processes
affecting brain function.[226] Perfusion studies in liver with insulin show
inhibition of amino acid release and urea production[227]; mechanisms involved could include turnover, either increased synthesis or decreased breakdowns, or alterations in transport. Mortimore and Morton[227] reported that
release of valine-^{14}C from prelabeled protein in perfused livers of normal
rats was inhibited by addition of insulin; since incorporation of added extracellular amino acids was unaffected, it was concluded that the release during
perfusion resulted from breakdown. The effects of insulin are complex; it
has been linked to hormonal changes, K^+ loss from tissues, and altered levels
of cyclic AMP.[227] Also, insulin *in vitro* can affect transport of sugars in spinal
cord or the phrenic nerve muscle preparation; *in vivo*, insulin hypoglycemia
is associated with elevated serum levels of amino acids.[226] Other effects of
insulin include an increased rate of free fatty acid esterification, to be contrasted with an increased rate of hydrolysis of triglycerides by glucagon,
ACTH, MSH, and vasopressin.[228] Turnover of such peptides can be considered critical to a number of important functional processes and accounts
for the great interest in specific degrading enzymes, particularly "insulinase".[229,230] As noted earlier, most tissues contain cathepsins that can
degrade insulin B, and these may be identical to the "insulinase" described

in earlier studies. As in the case of other polypeptides containing a disulfide ring structure (oxytocin or vasopressin), cleavage by a specific transdehydrogenase is frequently a prerequisite for complete degradation.[226,229,231] In crude extracts the degradation of insulin may be preceded by such mechanisms; the liberated A and B chains are then readily susceptible to cathepsin D [or alternatively to dipeptide cleavage by cathepsin C (Section II, C)]. A crude insulin-degrading enzyme from adipose tissues showed a similar specificity to cathepsin D when tested with insulin B chain.[228]

Insulin itself is reported to act as a weak proteolytic enzyme, an activity that can be destroyed by disruption of the disulfide bridges (iodination or photooxidation).[226] In connection with this phenomenon, a number of synthetic peptides can act in this capacity,[232,233] indicating that only a portion of enzyme protein is required for actual catalysis. It is possible that peptides with active catalytic properties are formed from inactive proteins, rather in the manner suggested for zymogen activation. Insulin itself or its constituent chains are resistant to digestion when bound to albumin, suggesting a possible control mechanism for peptide transport and turnover.[229]

H. Oxytocin and Vasopressin Breakdown

Oxytocin and vasopressin are rapidly inactivated by a variety of body fluids, tissue homogenates, and extracts, and by a large number of exo- and endopeptidases.[22,235] Oxytocin is a unique peptide with a disulfide ring and a C-terminal glycyl amide (Fig. 13); as in the case of insulin, disulfide cleavage by transdehydrogenases yields an open chain structure with greater susceptibility to hydrolytic cleavage. This factor complicates interpretation in crude tissue extracts that contain a mixture of enzymes. The cleavage of different proteolytic enzymes is indicated in Fig. 12; for convenience the chromogenic substrate cystine-di-βNA was introduced for assay based on analogies to the disulfide ring structure. Tuppy[22a] at one time considered that enzymes hydrolyzing this particular synthetic substrate or its congeners were related to a specific oxytocin-inactivating enzyme present in pregnancy serum and termed in an earlier study by Werle[22a] "oxytocinase." The fact that purified preparations split other arylamide substrates (Leu-, Ala-βNA) at even higher rates, in addition to a variety of polypeptides (angiotensin, oligopeptides prepared by tryptic digestion of cytochrome c), does not support the contention of a specific enzyme solely recognizing the nonapeptide sequence of oxytocin. An interesting corollary in such studies is the polypeptide specificity of enzymes assayed on the basis of arylamide substrates; it would seem on this basis that arylamidase activity *per se* does not reveal the scope or function of this category of enzyme. The term cystine aminopeptidase (CAP) has been employed to describe enzymes purified by means of an assay using the synthetic substrate cystine di-βNA. Preparations of 9000-fold enrichment can be obtained by ammonium sulfate precipitation of sera of pregnant animals, followed by Rivanol treatment, and DEAE-cellulose chromatography.[235] Two forms of the enzyme have been described, CAP_1 and CAP_2, based on the susceptibility to inhibitors and electrophoretic

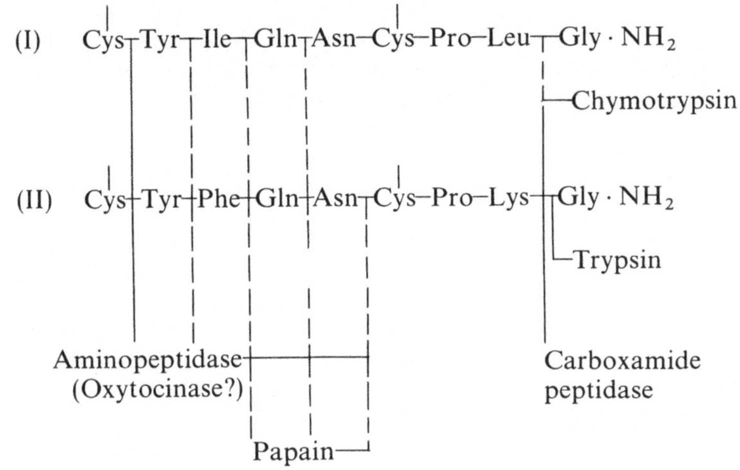

(I)   Cys–Tyr–Ile–Gln–Asn–Cys–Pro–Leu–Gly · NH$_2$

—Chymotrypsin

(II)   Cys–Tyr–Phe–Gln–Asn–Cys–Pro–Lys–Gly · NH$_2$

—Trypsin

Aminopeptidase
(Oxytocinase?)

Carboxamide
peptidase

Papain

Fig. 13. Cleavage of octapeptide hormones. Data based on the review of Tuppy and the results of Walter et al.[237] The Cys residues are joined by a disulfide bridge that prevents hydrolysis by pepsin and other hydrolytic enzymes (see text). Release of Gly amide by chymotrypsin is considerably slower than release of the C-terminal group by trypsin. Aminopeptidase and "oxytocinase" split the bond between Cys and Tyr, yielding a linear polymer with subsequent release of other N-terminal amino acids. Oxytocinase is similar to properties of monoacyl arylamides as described in the text.

mobility. The dominant form in tissue is CAP$_2$, although some interconversion between the forms has been described.[22a] The purified enzyme is a glycoprotein of 300,000 mol. wt. with an N-terminal neuraminic acid; removal of sialic acid by neuraminadase is accompanied by the appearance of yet another electrophoretic form termed CAP$_n$, but without loss of biological activity. The pH optimum is about 7.0–7.5 (slightly lower for synthetic substrates); K_m is 10^{-5} M, but lower for the arylamide synthetic substrates. Sjoholm and Yman[236] reported cleavage of the disulfide bridge with liberation of Tyr, Ile, Gln, and Asn, presumably by N-terminal cleavage, leaving the pentapeptide core Cys^1-Cys^6-Pro-Leu-Gly^9 · NH$_2$ (or in the case of vasopressin Lys- instead of Leu- at position B). In relation to arylamidase described for other tissues, inactivating enzyme differs in being only slightly inhibited by puromycin; the enzyme is slowly inactivated by dialysis against EDTA (as in the case of brain enzymes) with reactivation by divalent metal ions. Cystine aminopeptidase is unaffected by DFP or trasylol but is competitively inhibited by substrate analogues such as cystinyl-di-Tyr · NH$_2$, or oxytocin itself. There appears to be no requirement for a free —SH essential for activity.

In recent studies, Walter and his group[237–239] reported the presence of enzymes inactivating oxytocin or/and vasopressin in a variety of tissues. All enzymes inactivated the neurohypophyseal hormone by removal of the glycinamide moiety; however, the peptide was recognized as a substrate on the basis of different structural features. An enzyme extracted from toad urinary bladder releases glycinamide from both oxytocin and lysine-vasopressin.[239] This enzyme also cleaves glycinamide from several hormone analogues possessing in position 8, instead of leucine (in oxytocin) or lysine (in vasopressin), an arginine, serine, glutamine, phenylalanine, valine, or alanine residue, respectively; however, analogues in which the C-terminal glycinamide moiety is replaced by a glycine, glycine methylamine, or glycine dimethylamide are not cleaved between the 8 and 9 position. All these findings militate against the enzyme being a mixture of trypsin and chymotrypsin. This enzyme termed carboxamidopeptidase may play a role in the inactivation of other physiologically active peptides containing the C-terminal amide grouping, namely, gastrin, secretin, cerulein, phyllocerulein, thyrocalcitonin, etc.

An enzyme capable of inactivating oxytocin by the release of glycinamide is also present in rat uterus. This enzyme exhibits chymotrypsinlike activities and is similar to enzyme obtained from rat kidney.[237,238]

In summary, at least three mechanisms exist for the inactivation of pituitary peptides such as oxytocin[237,242,276]: (a) removal of the C-terminal amino acid amide moiety, (b) reductive scission of the disulfide bridge followed by peptide- or peptidyl-peptide hydrolases, (c) specific cleavage of the peptide bond associated with the N-terminal cysteine bridge. There is now evidence that the first mechanism[237] is the major route for oxytocin inactivation in most tissues. As a final comment, there has been much discussion of the relationship between oxytocinase levels and the events associated with pregnancy and parturition[22a,c,275]; as yet, the role of hydrolytic enzymes in these processes is unknown. It is even possible that inactivation of angiotensin by the same enzyme(s) is more pertinent, since time events of oxytocinase rise and decay in serum do not coincide with the events of pregnancy, and are not consistent with the known half-life of oxytocin in sera.[22b]

VI. INHIBITORS OF PROTEIN BREAKDOWN

Information gained on isolated enzymes with inhibitors, largely synthetic, has shed light on active sites and enzyme mechanisms. The question of proteinase inhibition *in vivo* is more complex than generally assumed; although tissues are rich in inhibitory materials, the question of their role within the cell, or even accessibility of enzyme to the inhibitor is rarely posed. The demonstration of an *in vitro* inhibition may have little bearing on the effects in a multienzyme system, or the role in tissues with their implied structural complexities. Inhibition of an enzyme by the large macromolecule itself poses formidable technical problems, as noted in several treatises.[11,85,243] J. L. Webb[243] has offered some excellent criteria for seeking and testing inhibitory materials, some of which are emphasized in this account.

In addition to the theoretical interest of inhibitors in enzymatic mechanisms, such materials have attracted interest in a relation to a number of clinical applications, to their role in secretory mechanisms and protein turn-over, and in food preservation.[11,12] Such studies include their role in blood-clotting mechanisms, in collagen and connective tissue turnover, and in tissue permeability and inflammatory mechanisms (release of kinins) etc.[11,12] Many of these studies have implications in relation to CNS function, as cited below. The available data are described under three headings: (a) naturally occurring factors, (b) synthetic compounds, (c) nucleic acid–protein inter-actions.

A. Naturally Occurring Factors

Tissues and body fluids contain an enormous range of inhibitory mate-rials, as summarized in the excellent monograph of Vogel, Trautschold, and Werle.[11] Inhibitors can be classified according to mol. wt. (small, 6000–14,000, heat-resistant, and nondialyzable; large, 20,000–60,000), or tissue or origin (plant or animal) or specificity (e.g., antitrypsin, or in the event of wider activity as polyvalent).[11,244] In some cases smaller polypeptides exhibit the same potency as larger proteins. Tschesche et al.[245] recently isolated from a modified porcine trypsin inhibitor the pentapeptide Des-Thr-Ser-Pro-Gln-Arg with potent antitrypsin activity. Interestingly, the remaining fragment as well as the pentapeptide retained activity illustrating their allosteric nature. Most tissue inhibitors are poorly characterized and are frequently named after the original investigators; for example, "Kunitz" type inhibitors exist in plants and bovine tissues; bovine tissues contain materials known as "Kazal," "Werle," "Kalser–Grossman."[11,12,245] The anatomical localization of the Kunitz (bovine organs) and the Kazal (pan-creatic juice), both of about mol. wt. 6100, suggests that the first is related to a secretory role *within* cells in contrast to a possible protective role of the second material for the newly secreted proteins.[246,247] Bovine factors from different organs show great similarity in structure and function, and this has led to the commercial introduction of the material known as trasylol with possible therapeutic applications.[12,15] Serum is also characterized by a large number of inhibitory factors: for example, chymotryptic inhibitor associated with postalbumins; polyvalent inhibitors associated with α, and α_2-globulins; a specific inhibitor of plasmin, etc.[11,248] Inhibitors occur also in a variety of other body fluids including cerebrospinal fluid, urine, and semen.[11,249] In some cases inhibitors are genetically linked, as demonstrated by the absence of an antitrypsin α-globulin in some families.[11]

1. Brain Extracts

Our early studies based on a net increase in enzyme activity during purification suggested the presence of inhibitory material in crude homog-enates.[42,68] To a limited extent, addition of α-globulin or albumin resulted in some inhibition of endogenous activities of acid and neutral protein-ases.[68] Despite many reports that tissues contain inhibitors of cathepsin B (trypsinlike hydrolases), this material has resisted attempts at isolation and purification.[250,251] Brain displays only trace activity with BANA (cathepsin B substrate), and it is possible that an inhibitor-enzyme complex prevents hydrolysis, in contrast to Arg-, Arg-Arg-βNA, BAME, and TAME (substrates for trypsin).[1j,126] Laskowski[12a] postulated the existence of enzyme-inhibitor complexes (named temporary inhibitors) in tissues with activation dependent on the stability of the complex by other proteolytic enzymes. Studies by Brecher and his group[181,252–254] indicate that brain extracts contain materials that interfere with the hydrolysis of BANA by trypsin or of glutaryl-βNA by chymotrypsin. Such factors are present in high concentration in brain microsomal extracts but are generally absent in the supernatant (except possibly in human astrocytoma).[252] It must be emphasized that this experimental system is different from attempts to isolate factors inhibiting intracellular brain proteinases (rather than exogenous extracellular enzymes).

Since inhibitors could play an important role in turnover, we surveyed different brain fractions for inhibitory factors using purified brain enzymes (proteinases, aminopeptidases, monoacryl- and dipeptidyl arylamidases) in comparison with known inhibitors of trypsin or kallikrein.[255,256] Purified acid proteinase was inhibited by crude homogenate, with highest activity in the nuclear (70%) inhibition of hemoglobin hydrolysis and in mitochondrial (80%) fractions (Table VII). Brain particulates potently inhibited the tryptic hydrolysis of BANA, as earlier reported by Brecher and Quinn,[252] with 33% of the homogenate antitrypsin units (assessed by the method of Vogel et al.[11]) present in the nuclear fraction. No inhibitory activity was present in the soluble supernatant (Fig. 14). The nuclear inhibitor was particulate-bound, nondialyzable, heat-resistant, and not reproduced by protamine or histone, or lipid extracts of nuclei, or by the level of amino acids expected

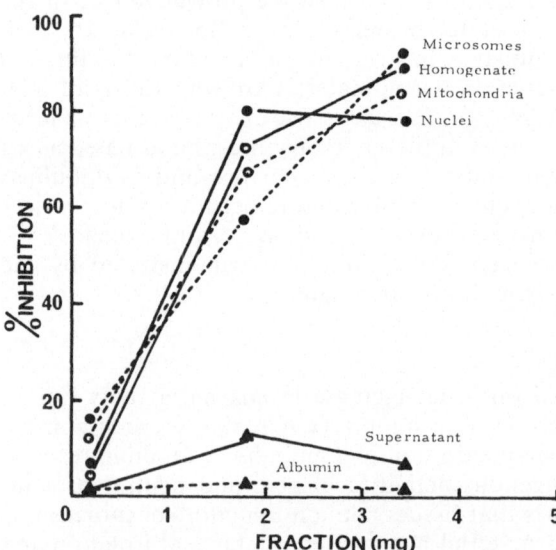

Fig. 14. Particulate fractions of rat brain were exhaustively dialyzed and then added to an assay system containing 10 μg trypsin at pH 7.6 [tris-HCl, 40 mM and 0.25 mM BANA (benzoyl arginyl naphthylamide)] and incubated for 30 min at 37°C. Albumin at similar concentrations was added as a control to determine nonspecific protein–protein interactions.

TABLE VII

Effect of Brain Particulate Fractions on Brain Proteinases, Pepsin, and Trypsin[a]

*Fraction or inhibitor	Conc. ratio or mM	% Inhibition			
		Acid proteinase (Hb, pH 3.2)	Neutral proteinase (Hb, pH 7.6)	Pepsin Hb (pH 3.2)	Trypsin BANA (pH 7.2)
Homogenate	1:35	93	—	—*	82
Nuclei	1:35	70	—	—*	80
Crude Mt.	1:35	80	—	—*	80
Microsomal Supt.	1:35	10	—	10	10
Albumin	1:35	0	—	0	5
Soybean	1:4	0	80+	0	100
Pancreatic	1:4	0	14+	0	100
Lima bean	1:4	0	41+	0	100
TPCK	0.2	0	46	0	0
PMSF	1.0	10	—	0	—
β-Phenyl pyruvate	0.2	36	—	100	0
Benzethonium Cl⁻	2.5	36	100	0	0
Indole-3-pyruvic	12.5	0	0	0	100

[a] Particulate fractions were exhaustively dialyzed before being added to purified enzymes at concentrations indicated by the ratio of protein to enzyme protein. The assay systems contained 50 μg purified acid proteinase,[100,101] 1 μg pepsin or 10 μg trypsin and were incubated with substrates at the pH indicated in the table.[255,256] Neutral proteinase was not tested with particulate fractions because of high endogenous activity: inhibition was not observed with some naturally occurring materials added at 1:40 concentration ratios (+). In the case of pepsin, activation was observed with particulate fractions (*). Abbr: TPCK, L-1-tosylamide-2-phenylethyl chlormethyl ketone: PMSF, phenylmethyl sulfonyl chloride.

to be present in the free pool. Known tryptic inhibitors from soybean, ovomucoid, bovine pancreas, and lima bean inhibited brain aminopeptidase (Leu-Gly-Gly) and crude neutral proteinase (hemoglobin) 50–80% but were inactive against acid proteinase.

In a brief report, the diurnal variation reported for hydrolysis of the tryptic substrate Z-Gly-Gly-Arg-βNA by brain extracts was attributed to variation in the level of endogenous inhibitor.[257]

In summary, results suggest that tissues contain endogenous inhibitors active against intracellular proteinases (or other hydrolases); however, there is no instance where such factors have been isolated and characterized. Data concerning brain are open to several interpretations if assessed according to the criteria of J. L. Webb[243]: (a) use of a low concentration of enzyme to determine the effect of inhibitors on initial rates, (b) determination whether linear with enzyme concentration (nonlinearity may imply limitation by diffusion, relative depletion of coenzymes or cofactors, impurities), (c) use of several different analytical plots to determine the inhibition mechanism (competitive versus noncompetitive), (d) use of low substrate concentration (arylamidases, for example, are inhibited at high substrate concentration), (e) examination of effects of pH and nonspecific effects in the medium (coprecipitation),[258] (f) determination of the reversibility of the inhibition (may serve to demonstrate secondary inactivation or denaturation), (g) use of purified inhibitor preparations, (h) determination of protective effects of other macromolecules (permitting enzymatic expression but preventing combination with the inhibitor).

To illustrate the complex protein–protein interaction possible in crude mixtures, Haverback *et al.*[260] identified a serum glycoprotein that protected proteolytic enzymes in the presence of inhibitors but which did not prevent normal hydrolytic activity. In contrast, some hydrolytic enzymes bind to α-macroglobulin so that only synthetic rather than protein substrates are hydrolyzed; this complex is resistant to the effect of high molecular weight tryptic inhibitors.[261,262]

B. Synthetic Inhibitors

There are only a few studies on the effect of synthetic inhibitors on brain proteinase and related intracellular enzymes.[255] Compounds selected on the basis of their known role in inhibition of pepsin (indole-3-pyruvic acid),[263] trypsin (TPCK, etc.),[259] carboxypeptidase (β-phenyl propionic acid),[264] or aminopeptidases (benzonium chloride)[265] caused some inhibition of acid proteinase but were less effective compared to the inhibition of peptide hydrolases (Table VII). At relatively high concentrations, neutral proteinase was inhibited, suggesting some similarity to the active sites of peptide hydrolases. Evidence for actual covalent binding by known site-specific reagents requires corroboration by isolation of the residues affected.[259] Frequently, an inhibitor or metal ion will affect enzymatic activity by alteration of an allosteric site, precluding a simple interpretation.[266] In an earlier report we concluded that TPCK inhibited brain acid proteinase, but subsequently could not show an effect with acid proteinase purified by an alternative procedure.[101] A number of inhibitors known to inactivate pepsin were reported to inhibit cathepsin D of liver, spleen, and thyroid; these include diazoacetylnorleucine methyl ester,[267] and azoacetyl-N^8-(2,4-dinitrophenyl) ethylenediamine.[18a] Barrett[93] reported that cathepsin D of rabbit liver is inhibited partially by α-oxo-amino acid and related aromatic amino acids; the inhibition by 3-phenylpyruvic was competitive in nature. The lack of effect of PCMB, iodoacetamide, trasylol, PMSF, and TPCK suggests that the active center does not involve seryl and sulfhydryl groupings. As noted in a review by Shaw,[259] the specificity of site-specific reagents is affected by adjacent residues and the conformation of the peptide chain. A variety of enzymes, including chymotrypsin, trypsin, thrombin, plasmin, and kallikrein, inactivated by DFP are classed as serine proteinases. However, TPCK showed unexpected selectivity since it inhibited chymotrypsin but not trypsin. The active center of chymotrypsin contains a His residue promoting ionization of Ser by hydrogen bonding. Three-dimensional studies have shown that one His and one Ser can form a "charge relay system" capable of reacting with TPCK (alkylation at His[57]). The chloromethyl ketone of a typical trypsin substrate tosyl-L-lysine (TLPK) inactivated trypsin at His[46] but was ineffective against chymotrypsin. TLPK also inactivated thrombin and plasmin (but no kallikrein), suggesting active centers akin to trypsin. The inaccessibility of cathepsin D to these reagents does not preclude Ser at the active center but must indicate an unusual conformation compared to other known proteases.

C. Nucleic Acid–Protein Interactions

The possible roles of histones in gene regulation by suppression of genetic expression has focused attention of the stability and turnover of deoxyribonucleic acid–protein complexes. Several years ago Harris and Wiseman[268] concluded that peptidyl-nucleotidate esters could represent an intermediate in the recyclization of proteins. Ribose esters accumulate during the stationary phase of yeast culture and their level appears to parallel the growth cycle and formation of adaptive enzymes. The extent of nuclear protein degradation is variable and is dependent on the source and purity of the nuclei concerned. As early as 1949, Maver and Greco[269] reported digestion of nucleoproteins by acid proteinases prepared from calf thymus nuclei. The presence of proteinases was confirmed by the finding of proteinases in histone preparations, and the degradation of nuclear proteins at acid and neutral pH.[270,271] It has been suggested that some minor components of histones reported in the literature are the result of degradation during preparation,[272,273] a feature which is similar to breakdown to myelin proteins (Section IV, D). There is still some question whether the DNA-histone complex is resistant to degradation. Bartley and Chalkey[274] observed a tightly bound "neutral" proteinase in nuclear histone isolated from tissues subject to rapid cell division such as thymus and intestinal cell, and virtually absent in lung and brain nucleohistone. Comparison of the susceptibility of five histone preparations was shown to be dependent upon whether the histones in question form a complex with DNA. In most cases intact nucleohistone is totally resistant to proteolytic cleavage, but in two cases (lysine-rich and a sulfhydryl-containing histone) it is rapidly attacked, as observed by disc-gel electrophoresis studies and determination of end groups by the dansylation method. The authors posed the interesting question of the role of proteases in cell or protein turnover and concluded that it mainly is related to tissue subject to extensive cell renewal. Generally the level of histone relative to DNA is constant in a nondividing tissue (brain), and one could conclude that the only function of proteases is degradation, with little advantage to the tissue with respect to turnover. There are no reports that nucleic acids *per se* inhibit proteinases, and this was confirmed in our recent studies in relation to purified brain proteinases and peptidases.[255]

It is possible that nucleic acids may interfere with degradation of prot-amine or other proteins which according to Palladin *et al.*[2b] serve as good substrates for acid proteinase. Most studies report some association of acid proteinases with brain nuclear fractions, although this may be a question of contamination, as indicated in recent studies of Furlan and Jericijo[272] for calf thymus nuclei. They reported the presence of both acid and neutral proteinases (*p*H optima 4.4 and 7.8) capable of degrading histones, but purification led to a loss of acid proteinase; neutral proteinase was largely associated with the chromatin fraction. Polyanions are known to inhibit pepsin; materials such as RNA, heparin, and chondroitin-4-sulfate promote the activation of zymogens (pepsinogen) at *p*H 4.0.[283]

In summary, there is some evidence to suggest that proteinases play a role in breakdown of the DNA-histone complex. The critical factors may be the integrity of the complex and the accessibility to proteinases. Evidence in calf thymus indicates a close association with neutral, and to a lesser extent with acid proteinases, which may represent contamination. Nucleic acids themselves do not appear to affect proteinases, so it must be assumed that the primary effect is on the histone components. The similarity between histones and myelin basic proteins suggests that such materials would form good substrates for brain acid proteinase.

VII. LYSOSOMES AND BREAKDOWN

Central to any concept of protein turnover in the nervous system is some consideration of the role that "lysosomes and associated particulates" play in the intracellular digestive process. The activation or release of enzymes bound in a "latent" form can result in profound perturbations within the cell, with pathological consequences and often demise of the cell. It is con-sidered that a selective and patterned cell death is essential to the remodeling of many tissues during morphogenesis and during tissue involution. The ramifications of the "lysosomal" concept now touch upon many areas critical to brain function, and only aspects associated with proteolytic enzymes are discussed in the present account. Other areas are covered by Koenig else-where in the Handbook[1e] and are reviewed by Holtzman[18g] and Novi-koff.[325] Areas involving proteolytic enzymes and covered in a recent com-prehensive treatise[18] include allergic and inflammatory conditions, wasting diseases, deposition diseases linked to genetic defects, vitamin deficiency states, teratogenesis, tissue involution, connective tissue metabolism. Relatively few biochemical studies have been performed in brain owing to difficulties of isolation *in vitro* as described below.

A. Definition and Preparation

DeDuve[18d] now regards the term lysosome as an operational entity to describe the complete digestive system within cells, and has proposed in a timely article the alternative term vacuome to define the appropriate membrane-bounded spaces within the cytoplasm. This term encompasses the vesicular entities associated with the formation of lysosomes from the Golgi apparatus (GERL or Golgi endoplasmic reticulum lysosomes; primary lysosomes, etc.) or vacuoles created by the processes of endocytosis or autophagy (see below). Previously, morphologists defined a lysosome as an organelle with a single limiting membrane containing dense osmophilic components, but these criteria are unsatisfactory to describe the structural diversity of the vacuome system. DeDuve[18d] has postulated the existence of permanent or transient channels within the cytoplasm linking different parts of the vacuome system to form a complex but sophisticated digestive system within cells.[284] Such structures may be involved in the transport of newly formed polypeptides from surfaces of ribosomes (attached to rough and smooth membranes) into the cavity of the endoplasmic reticulum.[284,285] The membranes of the Golgi complex may play a role in "condensing" newly formed secretory proteins, which are subsequently stored in the form of zymogen granules to prevent breakdown or autodigestion. Nascent polypeptide chains before release are protected from proteolysis, probably by the juxtaposition of the microsomal membranes.[286]

The assignment of proteolytic activities to a particulate is dependent on methods of isolation, and purity of the final preparation. An excellent review of current methods is provided by Beaufay[18b] who has emphasized the problems encountered in the preparation of liver and kidney lysosomal components. The median equilibrium density of liver lysosomes is 1.22 in sucrose gradients (compared to 1.19 for mitochondria and 1.23–1.25 for peroxisomes[25]); however, the range of lysosomal particulates gives rise to contamination in all areas of the gradient. We reported in early studies the enrichment of some particulate fractions with respect to cathepsins in purified mitochondria suspensions after centrifugation in sucrose 0.8–1.4 M.[6b,68] Other groups have also reported the enrichment of similar fractions with respect to other lysosomal hydrolases, which is good presumptive evidence for the presence of these particulates in the relevant fractions.[102,103] In subsequent studies (utilizing continuous sucrose gradients and isopycnic centrifugation), fractions enriched with lysosomelike organelles were present in sucrose layers with a density of about 1.3–1.15,[2c,102] although other fractions, particularly those in the so-called synaptosomal regions, contained considerable contamination (Fig. 6). Similar heterogeneity was also reported by Sellinger and his group[103] using continuous sucrose gradients of 0.8–1.6 M. The ability of lysosomes to engulf foreign material (endocytosis has been exploited by Wattiaux et al.[18b] to alter lysosomal density and facilitate isopycnic separation on sucrose gradients. Material used for this purpose includes Triton WR-1339, dextran, and Thorotrast, but their application to brain has not been as successful. Sellinger et al.[287] in a recent study reported that Triton WR-1339 given intrathecally over a period of several days could alter lysosomal density by making the particle lighter, as judged by the migration of typical lysosomal enzymes, and these could be further purified by recentrifugation on additional sucrose gradients. Lysosomal-enriched fractions prepared in this manner exhibited only trace proteolytic and aryl-amidase activities but were enriched with respect to arylsulfatase.[287a] Intra-cisternal or intravenous injection of Triton WR-1339 in our own studies gave some alteration in the different peaks of hydrolase enzymes (Fig. 6). The presence of neutral proteinases within lysosomes or other particu-lates[336] is of considerable importance, since most hydrolases are regarded as active only at acid pH (pH 3–6). Beaufay[18b] has criticized the methodology used by several groups claiming the presence of hydrolases in lysosomes active at neutral pH. Baudhuin[288] also concluded that as much as 80% of the protein in so-called "purified lysosomes" prepared by other groups[289] was from a nonlysosomal source, throwing some doubt on the reputed presence of arylamidases active at neutral pH. Our studies indicate acid proteinases as present in particulates rich in lysosomes and arylamidases, or proteinases active at physiological pH as present in soluble cytoplasmic components.[68,287] The heterogeneity of liver lysosomes is illustrated in recent attempts to purify such preparations by carrier-free electrophore-sis.[290]

B. Autophagy and Proteolysis

Turnover of intracellular components is critical to tissues, such as brain, which are incapable of cellular renewal or division. Autophagic vacuoles formed by sequestration of cytoplasmic materials are less common in nerve compared to other tissues except in experimental or pathological conditions (e.g., Wallerian degeneration, mechanical injury, nerve section, radiation, glucose starvation of organ-cultured ganglia, injection of colchicine).[18g] Some autophagic vacuoles are seen to contain myelin figures, remnants of mitochondria and endoplasmic reticulum, or other inclusion bodies, notably lipofuscin, which increases with age, or in vitamin D deficiency and degen-erative diseases.[291] Lipofuscin constitutes a type of inclusion body of unknown origin and composition; its relative resistance to degradation could account for its accumulation. Lipofuscin or similar inclusions could rep-resent a failure of the cell to remove residual bodies by a reversed pinocytosis; the cell is said then to suffer from indigestion and lacks the mechanism for "defecation."[9,18i] Daems[18e] postulated a "hypothetical loading capacity" for lysosomes (telolysosomes) which later become inactive (postlysosomes). Histochemical studies show that lipofuscin granules can show some cath-epsin and hydrolase activities and it is believed that the pigment is derived from the same groups of degraded mitochondria.[181,292] The association with the aging process of pigments resistant to degradation is an interesting but unresolved question. Inclusion bodies may be related to gene-linked defects, leading to a deficiency in a specific lysosomal hydrolase. In many cases the defective enzyme is a lipase, leading to the well-defined group of lipidoses.[18k]

Autophagy may play a role in the secretion of important physiological products, as reported for the release of thyroxine from thyroglobulin within thyroid lysosomes. The transport mechanisms for materials from lysosomes have never been defined, although most breakdown products, amino acids and sugars, can escape into the cytoplasm or into artificial media (see below). In most tissues, autophagic processes are considered to account for bulk degradation of intracellular components, which evidently plays a crucial role in the remodeling of tissues in growth or in tissue involution.[18t,293] Fletcher and Sanadi[323] reported a half-life of 10 days for liver mitochondria, which if extrapolated to the number of cells would indicate the degradation of at least one mitochondrion per cell every 15 min. In conditions of starvation, autophagy could represent a source of nutrients necessary for survival.

C. Endocytosis and Proteolysis

The manner in which nerve and other tissue cells acquire extracellular macromolecules is unknown, but it involves the following considerations: (a) the brain barrier systems, (b) mechanisms for formation of vacuoles, (c) whether vacuoles once formed can coalesce with so-called "primary" lysosomes, (d) the accumulation of materials resistant to digestion, (c) the degree of digestion within the vacuole and the subsequent transport of breakdown products to the cytoplasm. The ability of cells to occlude material is known now as "endocytosis," although other terms are used, as described in an excellent review by Jacques[18h] (pinocytosis, phagocytosis, and heterophagy). Dependent on the experimental conditions, the plasma membrane of nerve cells is able to occlude a variety of materials, including fluorescein-labeled proteins, horseradish peroxidase, ferritin, serum albumin and globulin, dextran, Triton WR-1339, heavy metals, cationic dyes, colloidal carbon, etc.[18h,338] Klatzo and Miguel[291] reported that microglia, neuroglia, and choroid epithelium acquire inclusions when incubated in the presence of fluorescin-labeled proteins. *In vivo* these materials do not readily penetrate the brain parenchyma except in the areas with poorly developed barriers (hypothalamus, epiphysis, area postrema), although some penetration into astrocytes, oligodendrocytes, and macrophages can occur in brain edema. Studies with spinal neurons of toad or mouse, neurons grown in culture, or with myelinated and unmyelinated axons of the adrenal medulla indicate protein uptake along the neuronal surface in addition to uptake within the perikarya.[18g] Protein is reported to enter cells as small coated vesicles and then merge with multivesicular bodies, which show histochemical evidence for acid phosphatase. Evidently the route of injection of foreign material is important with respect to brain, as demonstrated by entry into cells when given subarachnoidally or intracethecally.[1e,291] In myelinated fibers the primary site of uptake may be at the nerve cell body, with subsequent transport down the axon as indicated by electron microscopy.[175] It is well known that Schwann and glial cells play a prominent role during Wallerian degeneration, presumably by the process of endocytosis, in removal and digestion of myelin and other nerve components (Section IV, B).

There is good evidence that some proteins occluded by the processes of autophagy of endocytosis are digested within lysosomes with subsequent release of breakdown products. Maunsbach[18k] demonstrated the progressive release of monoiodotyrosine by liver and kidney lysosomes incubated with iodinated globulin at pH 5.0; activity was increased with cysteine and inhibited by iodoacetate, suggesting the involvement of a cathepsin B-like enzyme. Coffey and deDuve[294] observed a 70% degradation of acid-denatured globin by liver lysosomes *in vitro*, with release of most amino acids (except cysteine and tryptophan) and a mixture of dipeptides. Such dipeptides were hydrolyzed by cytoplasmic peptidases when incubated at pH 8.0, indicating that protein breakdown can occur in more than one compartment. In a similar study Beck[295] observed an increased protein content of liver lysosomes incubated with albumin at 37°C at pH 5.7 with subsequent liberation of free amino acids and smaller peptides (presumed to be dipeptides since higher peptides were not detected by gel filtration). Amino acid concentration within liver lysosomes was five- to tenfold higher than in the homogenate and was reduced by dialysis; subsequent incubation led to the appearance of additional amino acids, presumably by hydrolysis by enzymes within the organelle. Tappel[181] suggested that these amino acids were derived from digestion of structural protein of mitochondria. In other studies, Fowler and deDuve[296] reported degradation by liver lysosomes of glycoproteins (fetuin, orosomucoid, lipoproteins) and proteins derived from mitochondrial and microsomal sources. DeDuve[18d] suggested that removal of sialyl groups from sugars and denaturation of the protein moiety are necessary prerequisites for digestion. Lysosomes contain available hydrolases for this purpose (glucuronidase, N-acetylglucosamindiase, sialidase, etc.), and the low intralysosomal pH could denature the protein.

D. Structural Lability and Proteolysis

The lysosome is bounded by a membrane that exhibits structural latency with respect to its enzymatic activities. Latency phenomena are of current interest in relation to a number of degenerative diseases affecting the nervous system. In vitamin A deficiency states, degeneration of myelin is accompanied by lesions within the axonal sheath attributed to the release of lysosomal enzymes. Vitamin excess also results in bone malformation and inflammatory conditions probably linked to detergent-like effects on the lysosomal membrane. It is reported that some lysosomal membranes are "stabilized" by steroids (cortisol), or phenothiazines at low concentrations.[18s] Some steroids can antagonize the harmful effects of vitamin A excess *in vivo* and *in vitro*. Woessner[18t] has suggested that steroids play a role in the control mechanisms responsible for the involution of the uterus *postpartum*.

Very little information is available concerning the permeability and composition of lysosomal membranes. Lysosomal vacuoles have diverse origins, ranging from the endoplasmic reticulum, Golgi apparatus, and associated membranes, to the plasma membrane. Several groups have reported the presence of phospholipids in lysosomes characteristic of such membranes, but few data are available for similar brain preparations.[1e]

H. Koenig[1e] has attributed latency of lysosomal enzymes to glyco-proteins present in the membrane, and to the internal milieu of lysosomes. The osmotic properties of lysosomes, according to deDuve,[18d] suggest that enzymes are not bound but are present in a soluble form. Lucy[18j] has attempted to explain coalescence of some vacuomes to form lysosomes by "interdigitation" of proteins and lipids forming a membrane in accord with the Dannieli–Davson hypothesis. Based on the effects of inhibitors, the transport of materials has been linked to the presence of sulfhydryl groups and ATPase in lysosomal membranes.[297] In connection with transport of protein to the organelle itself, Malawesta and Bodel[298] reported some inhibition of intralysosomal digestion by colchicine, possibly indicating movement of material destined for degradation along microtubules. The ability of lysosomes to occlude pharmacological agents could serve an important therapeutic purpose in delivery of drugs to their sites of action.[18d]

VIII. BREAKDOWN OF CONNECTIVE TISSUE

Brain tissue, unlike peripheral nerve, contains only trace levels of components considered to be characteristic of connective tissue. Collagen and similar components are likely to be associated with the cerebral vasculature, and their breakdown and turnover is relevant to some degenerative diseases, with possible implications for cerebral circulation and permeability. Collagen is a unique polymer capable of forming tensile fibers, and its metabolism is the subject of much current research, as attested by recent reviews.[14] Nerve tissue is rich in an acidic protein unrelated to collagen that is capable of polymerization to form fibrils and tubules related to the structure and transport properties of axons.[24a] Collagen imparts rigidity to tissues and its destruction followed by resynthesis is required for growth or remodeling of tissues.[18p] Experimental evidence supports the concept that specific "collagenases" together with lysosomal hydrolases are involved in muscular dystrophy and denervation atrophy, bone resorption, metamorphosis, and tissue involution.[18p–s] The mechanisms controlling the equilibria between synthesis and catabolism are unknown; recent views concerning the control mechanisms for cartilage breakdown are summarized by Reynolds[18m] and are given in a comprehensive treatise edited by Balazs.[14]

A. Proteolysis in Cerebral Vascular Epithelium

Surprisingly few biochemical studies pertain to cerebral blood vessels despite many observations that degenerative cerebrovascular diseases can be related to mental deterioration and frequently are a cause of death.[10,306] Disease states include those resulting from vitamin deficiency, Paget's disease, Cushing's "syndrome," experimentally induced lathryism.[14] Arteriosclerosis results from accumulation and infiltration of lipids in connective tissue and calcification.[10] The cerebral vasculature is less susceptible than systemic arteries, with a milder and later onset of symptoms. This may be part of the remarkable ability of brain to resist changes produced by starvation, disease, and to some extent the aging process. Occlusion of cerebral capillaries, even though slight, could have serious repercussions resulting in cerebral ischemia.[10] The pathological appearance of lesions adjacent to arteriosclerotic vessels may be the result of impaired nourishment resulting from permeability changes. Calcification adjacent to arteries in Fahrs' disease (nonarteriosclerotic) characterized by the accumulation of proteins and polysaccharide may be related to the faulty transport of materials out of glial cells and to the blood vessels.[9] Little information is available concerning protein turnover in blood vessels or its associated proteolytic enzymes. In terms of collagen content, brain itself (free of blood vessels) and vitreous humor contain the lowest levels (0.05 %), followed by skeletal muscle (2 %), internal organs (2–8 %), blood vessel walls (5–10 %), bone and cartilage (10–20 %). Peripheral nerve is unusual in the relatively high levels associated with the sheath (40–45 % of dry weight).[1f] Collagen forms as much as 25 % of total body protein and is subject to considerable heterogeneity in turnover states, with half-lives for "labile" fractions of 1–5 days, compared to 50–100 days for most fractions except that of adult aorta (300 days).[14] The extremely low levels associated with brain or its vasculature preclude accurate estimates of turnover. In terms of breakdown, several studies including the histochemical (gelatin-film) methods of Adams and Tuqan, suggest the presence of elastase in cerebral blood vessels,[127,305] and an antiplasmin factor.[299] Systemic arteries contain cathepsin D, which has been purified some 1000-fold and shown to be capable of hydrolyzing polyamino acid peptides, serum albumin, insulin B, and chrondroitin-4-sulfate.[108] Antisera raised against spleen cathepsin is reported to inhibit release of chrondroitin sulfate in limb bone rudiments.[14] It has been suggested that the increased levels of cathepsin D during arteriosclerosis act synergistically with hyaluronidase in production of polysaccharide peptides.[18r]

B. Collagenases

Enzymes degrading native (undenatured) collagen at neutral pH and physiological temperatures occur in bacteria, but their existence in higher animals has been questioned.[300,301] The breakdown of cartilage first demonstrated by Mellanby[18m,n] during vitamin A deficiency states is believed to involve labilization of lysosomal enzymes. Vaes[18p] has proposed an acid secretion at some limited target areas to facilitate bone resorption followed by the release of lysosomal enzymes and digestion. Since collagenase is an example of a specific neutral proteinase, its presence in lysosomes was considered unlikely; however, there are recent reports that some polymorphonuclear leukocytes present in rabbit peritoneal exudate contain such enzymes.[133,302] The granule fractions of these cells (considered rich in lysosomes) hydrolyzed histones, collagen, and elastin, but not hemoglobin at neutral pH, and were devoid of arylamidase (Leu-βNA) and cathepsin B (BAA, Section II, B) activities.[303] Injection of granulocyte protein results in vascular injury, with damage to the basement membrane and an impairment in vascular permeability.[133] Neutral proteinases may play a role in remodeling of tissues, but Woessner[18t] failed to detect such changes during involution of the uterus despite a significant rise in cathepsin D. In muscle-wasting diseases linked to vitamin A deficiency, there are reports of increased neutral and acid proteinases.[304]

True collagenases from animal sources are reported as present in explants of tadpole tail, and mammalian bone, uterus, and skin grown in culture media.[14,300] Enzyme from tadpole purified 300-fold gave a pH optimum of 8–9 with native collagen, but was inactive with cathepsin substrates at lower pH. Enzymes purified on the basis of synthetic substrates (Pro- -Gly-Pro-Y, sensitive bond indicated) prepared by analogy to polymeric sequences of collagen generally display wider specificity (similar to that of bacterial enzyme) and may even represent a class of iminopeptidase. The use of an *undenatured* protein substrate to assay a neutral proteinase (collagenase) suggests some important criteria in seeking similar enzymes in nerve tissue.

IX. PROTEINASE CHANGES IN DEVELOPMENT

The complex growth pattern of brain affords a valuable situation for studying the relationship of protein breakdown to morphogenesis and other aspects of development. Our early studies indicated an increase of most proteinases with age,[2a] and some aspects of this work were confirmed in recent studies by Oja,[8,126] Palladin and Belik,[2b] and Einstein et al.[190] During development the rates of protein turnover are considerably higher than in the adult, and a relatively small change in the differential rate of synthesis compared to breakdown can materially affect the rate of protein deposition and ultimately the maturational process. Aspects affecting growth can be summarized as: (a) cathepsin changes during embryogenesis, (b) pathological states, (c) states of excitation and exhaustion, (d) level of hormones, (e) nutritional status, (f) availability of essential amino acids.[2a,3,162] There appears to be a complex relationship between the level of circulating amino acids and that of the tissue, partly as a consequence of interactions in membrane transport and variations in intracellular metabolism. The increased demands during development may require considerable reutilization of liberated amino acids, and this in turn is related to changes in the extracellular space and the amino acid pool. Estimates based on degradative rates must take into account the possibility of local reutilization of liberated amino acids, and the heterogeneity of the amino acid pool. In addition to alterations in protein breakdown, there is an increase in total brain protein accompanied by a change in solubility and extractability; some components appear only during the active phases of myelination. Most enzymes follow characteristic patterns during growth; the majority increase, but some are known to decrease, and there is also a change in the relative proportions of proteins with fast and slow turnover.[2a,4]

Maturation is accompanied by an increased (intracellular) population of organelles, accounting in part for increased protein levels in brain subcellular fractions (Fig. 15). Lysosomes contain a family of hydrolases, which might be expected to accompany in a quantitative fashion the brain growth cycle. Measurement of an enzyme in a whole brain extract does not reflect regional differences, or the rates of development in specific organelles. The medulla oblongata, and thalamus are reported to show the most advanced degree of myelination, and caudal-rostral gradients are reported for many proteins, phospholipids, cholineacetylase, and cholinesterase in developing rabbits.[2a,9] Ontogenetic changes can be measured in different ways: (a) absolute quantitative changes based on total brain weight or some estimate of cell number based on DNA, (b) differential growth patterns based on the rate of change of a specific component in relation to the levels of protein or lipids.[3] Brain extracts represent a mixture of heterogeneous proteins, with the possibility of a complex relationship between hydrolases and competing substrates. Different methods of extraction can be expected to yield different patterns. In sucrose homogenates, acid proteinase reached maximum levels at 10–15 days, and in contrast to the aqueous-Triton extracts, neutral proteinase fell shortly after birth and continued to fall during later phases of development[2a] (Fig. 15). A possible explanation for such findings is the variable presence of endogenous inhibitors dependent on the extraction method. Lewis and Vanderlaan[307] reported an inhibitory factor of hypophyseal proteinases in hypophysectomized rats; hypophysectomy is reported also to reduce salivary gland proteases which are affected by hormones and increase during development.[161]

Ontogenetic changes resulting in the appearance of new protein substrates may be accompanied by altered kinetic properties of the related enzymes (pH, K_m, V_{max}, etc.).[2a] Partially purified enzymes from 5-day-old rats exhibited different bond specificities based on peptide maps with insulin B chain, or goblin A and B chains.[6b,28] End products of degradation could form a control mechanism by feedback inhibition. The ninhydrin-positive pools increase following birth, with maximum levels at 10 days in sucrose homogenates, at 24 days in crude mitochondrial-synaptosomal fractions, at 30 days for nuclear fractions (Fig. 15). Although such pools contain some 60 components, the preponderance of reacting material can be attributed to acidic amino acids such as glutamate.[2a,3] In the present results the increase in total pool size after birth is consistent with the known increase in glutamate and aspartate; finer analysis of individual amino acids has shown considerable variation of levels with development in some species.[1a] During development three groups can be distinguished: (a) acidic components, which increase with age, notably Glu, Asp, GABA, and the tripeptide GSH; (b) neutral amino acids, which decrease (Ala, Gly, Tyr, Tau, etc.); and (c) other amino acids, which do not change significantly, or alternatively, are present in levels that prevent accurate measurements (Fig. 15). As noted, conclusions based on whole brain extracts do not preclude resolution of finer temporal changes parallel to development for the pools associated with subcellular or

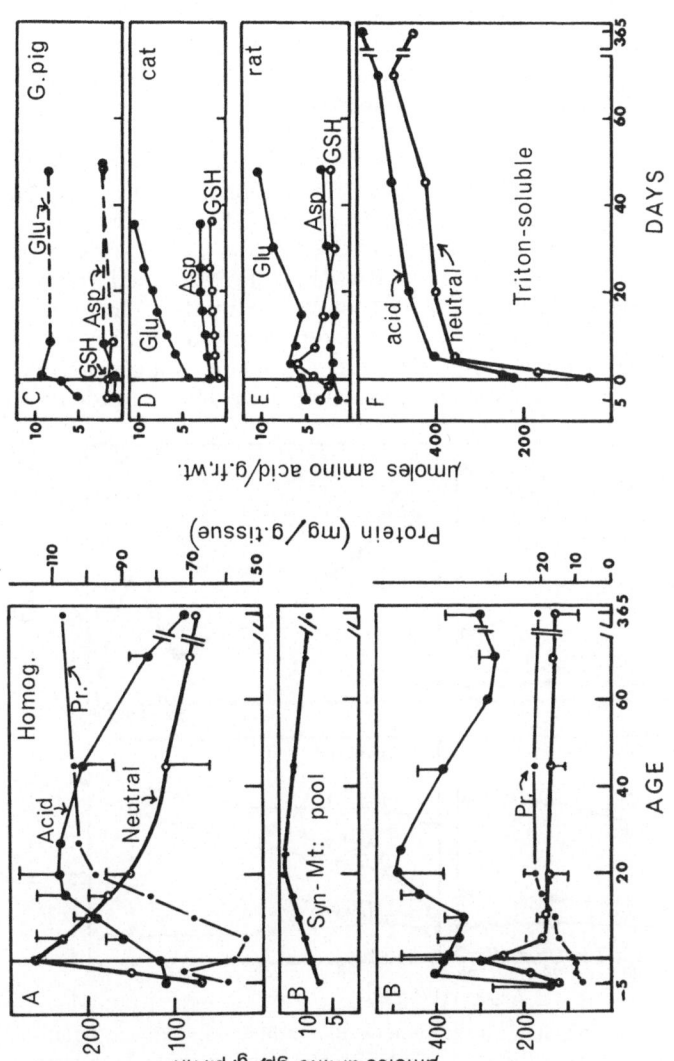

Fig. 15. Alteration in proteinase levels and pool size of amino acids and peptides with maturation in Cobbs River rat brain fractions. A, sucrose homogenate; B, crude synaptososomalmitochondrial fraction; F, Triton X-100 0.2% in tris-HCl 40 mM pH 7.6.[2a] As noted in the text, the change in proteinases varied according to the method of extraction. For comparison, the change in protein and amino acids with age is included in the figure. Data for pool size in guinea pig and rat (C, E) are from Oja,[8] and for the cat from Berl.[332] In A and B, the different components analyzed were acid proteinase with hemoglobin (darkened circles), neutral proteinase with hemoglobin (open circles); protein and pool size as shown in the figure. Brain fractions were prepared and analyzed by previously published procedures.[42,68]

regional fractions, or in relation to possible caudal-rostral changes. Arylamidases as an index of some proteolytic activities in crude extracts increased
with age, with a maximum at 15–20 days in the rat,[2a] and 9–12 days in
rabbit.[190] As noted previously, such enzymes may be involved in the maturation of physiologically active peptides, or are involved in the inactivation of
hormonal peptides, and in some instances may release physiologically active
materials from polypeptide or protein precursors.

Increased proteolytic activity during development in brain homogenates
has been confirmed in a number of studies.[2b,126,190,308] Oja and Oja[126]
reported a three-fold increase in breakdown up to 300 days of age in the
presence or absence of hemoglobin or casein at pH 3.0, with a smaller increase at pH 6.5 or 8.5 (Fig. 16). Hydrolysis of BANA was significantly lower
than with other tryptic substrates such as BAME or TAME; the latter and
chymotryptic substrates BPANE and ATEE were maximally hydrolyzed in
extracts from 2-week-old animals. The authors reported some inconsistency
with synthetic substrates, which might be evidence for the variable presence
of tissue inhibitors. The use of esters as a measure of protein breakdown in

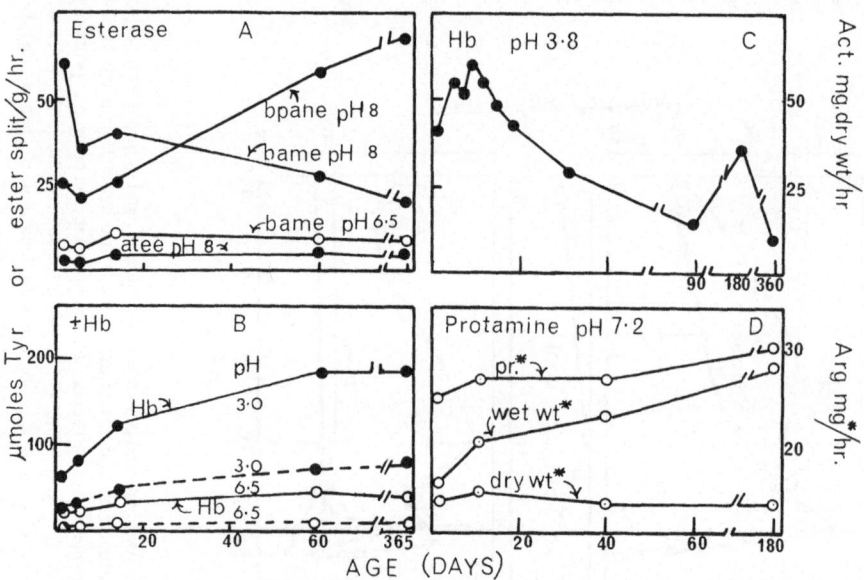

Fig. 16. Change in hydrolase activity with age in rat (A, B) or rabbit brain extracts (C. D) at
different pH and in the presence of protein or ester (tryptic or chymotryptic) substrates.
Data quoted in the figure are based on the results of Oja and Oja,[126] Einstein et al.,[190]
Belik et al.[2a,116] Experimental conditions for BPANE and ATEE, and for BAME are
identical to those of Fig. 2. Acid proteinase (C) was determined on extracts prepared from
acetone powders and then submitted to gel filtration to remove background ninhydrin-
positive materials.[190] In D, 3 mg brain homogenized in saline were incubated with 2 mg
protamine for 40 min (results are quoted for Arg released per hour based on protein, dry
or wet weight[116]).

the presence of unspecific esterases has yet to be established (Section II, A). In rabbits, Belik et al.[116] reported maximum acid proteinase in homogenates at 11 days coincident with eye opening, and Einstein et al.[190] in hypotonic extracts at 7–9 days coincident with the rapid onset of myelinization. Palladin and Belik[2b] and Belik et al.[116] also studied protamine degradation at pH 7.2 in rabbit brain and reported maximum values at about the same time period. The subsequent change with age varied with the manner of assessment; in terms of milligrams of protein or dry weight, activity increased with age, but decreased in terms of milligrams of dry weight.

X. PROTEIN BREAKDOWN AND THE INTRACELLULAR POOL OF FREE AMINO ACIDS

In the postulated scheme representing protein turnover (Fig. 1) the intracellular concentration of free amino acids was shown as the equilibrium established between (a) rate of amino synthesis, (b) transport of amino acids into the pool, (c) rate of utilization, (d) formation by the breakdowns of proteins and oligopeptides. Free amino acids in adult brains from different species, although differing quantitatively, show a surprising similarity when ranked in order of their concentrations.[28] Little is known about the remarkable control mechanisms that maintain the steady-state concentrations as well as the individual amino acid ratios; presumably elevated plasma levels lead to a decreased uptake (by complex interactions on amino acid carriers described elsewhere[1v]), or alternatively to increased rates of utilization or catabolism. Conversely, low plasma levels would produce the opposite effect, a lowered utilization accompanied by an increased rate of formation by normal metabolic processes or by protein breakdown. Catabolism can therefore be involved as a component of the basic control mechanism maintaining the dynamic equilibrium within tissues. Alteration of the body steady-state conditions by the infusion of labeled protein or amino acids, or starvation is reported to affect intracellular degradative rates.[309–311] The increased rate of proteolysis of albumin in mice after injection of ^{131}I-labeled albumin was considered by Friedberg[309] to directly correlate with the pool of serum albumin. Serum albumin normally remains within relatively narrow limits despite a rapid protein turnover; normal catabolic rates for albumin are 2.4% and 0.16% in the human.[309]

Infusion of labeled amino acids results in an increase in the radioactivity of the liver and muscle pool of amino acids, but interestingly their levels never reach that of the plasma.[310,311] The failure of plasma and tissue levels to fully equilibrate has been attributed, in part, to the dilution resulting from unlabeled amino acids formed by the breakdown of protein. As noted, increased plasma levels may affect rates of transport and utilization; thus such results are difficult to interpret entirely from the aspect of catabolic mechanisms.

Interpretation of such experiments is improved by the use of essential amino acids not readily metabolized, or synthesized in tissues (e.g., lysine or valine in rat tissue).[227] In adults on a normal diet it was estimated that catabolism contributed as much as 50% of the liver and 30% of the muscle pool of free amino acids.[310] The source and identity of rapidly metabolized proteins is unknown. Munro[20a] suggested that the presence in tissues of a "labile" or "functional" protein subject to rapid fluctuation dependent on the nutritional status of the animal. Liver, pancreas, and the mucosal lining of the intestine are very sensitive to dietary change, in contrast to brain, which undergoes little change in the adult with altered intake of protein. Munro[20a] suggested that starvation resulted in a rapid diminution of a "labile" or "functional" protein from liver, pancreas, or kidneys for the purpose of priming the plasma levels and thus maintaining the supply for brain and muscle. During starvation in the rat, 90% of the liver amino acid was reported by Gan and Jeffay[310] to be derived from intracellular protein. After depletion of the reserve protein liver, muscle protein then was degraded to supply 65% of its own intracellular pool of amino acids. Several years ago, Geiger and Magnus[75] demonstrated a large loss of protein in the isolated brain of cats perfused with salines containing blood cells, but without including liver extract (or uridine and cytidine). The loss of protein might suggest a similar labile reserve in addition to the increase in catabolism. Starvation in rats led to some alteration in the levels of proteolytic enzymes.[312] Nutritional deficiency is known to affect brain weight and development in young animals, which may be the result of lowered plasma levels of amino acids on transport, utilization, and catabolism previously described. Normally, during development, amino acids in brain show wider fluctuation than the adult and this is presumably related to ontogenic mechanisms.[1a] An abnormally high level of Phe, as in phenylketonuria, during development has been linked to a decreased synthesis of myelin protein and incomplete myelination in the CNS.[313] High doses of Phe in rabbits led to an inhibition of cerebroside synthesis in rabbits[314] and up to a 70% reduced incorporation of labeled Leu or Met into protein from rat brain.[315] Dietary deficiency in Phe in rats led to higher plasma and cerebral levels of threonine and urea that may reflect enhanced protein metabolism (Thr is not rapidly utilized in brain).[316]

A. *In Vitro*

There are many studies showing an enhanced incorporation of radioactivity into proteins of different tissues *in vitro* with changes in the level of extracellular amino acids in the incubation media. Increased incorporation with increased extracellular concentration of specific amino acids is reported for the protein or rat liver slices, diaphragm, Ehrlich's ascites tumor cells, blood cells, etc.[224,317] Even growth in tissue culture is proportional to the levels of individual amino acids in the extracellular fluid.[318] Few of these studies attempted to evaluate the contribution of proteolysis to alteration of the intracellular pool of amino acids.[2a,144] Hanking and Roberts[319] showed that an elevated level of Phe in the incubation media led to an increased release of amino acids from hepatic protein in liver slices; the level of amino acids was correlated with the content of amino acids in the hepatic protein. Levels of Phe lower than the physiological range in plasma (50 mμ) resulted in a net release of Phe into the medium. The precise effect of altered amino levels on protein breakdown is difficult to evaluate in experiments of this nature owing to utilization of amino acids and their transport into and out of the cell. Hanking and Roberts[319] estimated that the overall increase in amino acids, especially in response to elevated Phe levels, was the result of protein breakdown estimated to be in excess of 1 mg protein/g wet wt. liver/hr (or 8 μmoles of glutamic acid equivalents). Calculated *in vivo* rates for brain based on a half-life of 15 days indicate a rate of 0.15 mg protein/g wet wt. (or 1 μmole glutamic equivalent).

Recent studies with brain slices have demonstrated reactions favoring breakdown with the appearance of amino acids in the medium or in perfusates. From our laboratory, Neidle *et al.*[320,328] reported that slices with a 1 : 20 media ratio established an equilibrium within 30 min with respect to all amino acids except Gln (which continued to leak from slices). Transfer of the slice to fresh medium led to an additional outflow with the establishment of a new equilibrium. Comparison to levels of amino acids native to the slice pool at zero time clearly indicated that the major source for the efflux was derived from tissue breakdown. Calculations based on the appearance of Leu indicate that breakdown actually occurred at a rate higher than that required for protein turnover *in vivo*.

There was some indication that the equilibrium was established by the effect of the extracellular amino acid on transport processes: interference with such processes with CN$^-$ addition led to a continued efflux.

XI. REFERENCES

1. A. Lajtha (ed.), *Handbook of Neurochemistry*, Plenum Press, New York (1969–1972).
   Vol. 1, Chemical Architecture of the Nervous System (a) Chap. 3, W. A. Himwich and H. C. Agrawal (amino acids). (b) Chap. 6, B. W. Moore (acidic proteins). (c) Chap. 16, N. Seiler (enzymes).
   Vol. 2, Structural Neurochemistry (d) Chap. 5, G. Levi (spinal cord). (e) Chap. 12, H. Koenig (lysosomes). (f) Chap. 16, G. Porcellati (peripheral nerve). (g) Chap. 21, B. Droz (radioautography). (h) Chap. 22, C. W. M. Adams (enzyme histochemistry: applications and pitfalls). (i) Kerkut (arthropods).
   Vol. 3, Metabolic Reactions of the Nervous System (j) Chap. 5, N. Marks (peptide hydrolases).
   Vol. 4, Control Mechanisms of the Nervous System (k) Chap. 6, G. Zetler (substance P). (l) Chap. 17, H. Sachs (neurosecretion). (m) Chap. 19, R. J. Wurtman (pineal hormones).
   Vol. 5, Metabolic Turnover in the Nervous System (n) Chap. 1, S. Roberts (synthesis). (o) Chap. 16, Schenkein (NGF). (p) Chap. 18, N. Marks and A. Lajtha (turnover).
   Vol. 6, Alterations of Chemical Equilibrium in the Nervous System (q) Chap. 9, J. Dobbing (undernutrition and the developing brain). (r) Chap. 19, G. Porcellati (demyelination).
   Vol. 7, Pathological Chemistry of the Nervous System (s) Chap. 6, E. R. Einstein (basic protein. (t) Chap. 11, G. E. Gaull (amino acids and brain dysfunction). (u) Chap. 12, G. Porcellati (Wallerian degeneration). (v) S. R. Cohen and A. Lajtha (transport).

2. A. Lajtha (ed.), *Protein Metabolism of the Nervous System*, Plenum Press, New York (1970). (a) Chap. 2, N. Marks and A. Lajtha (developmental changes in peptide-bond hydrolases). (b) Chap. 3, A. V. Palladin and Ya. V. Belik (breakdown of proteins: protamine-splitting enzymes). (c) Chap. 8, Brian D'Monte, N. Marks, R. K. Datta, and A. Lajtha (protein turnover in membranous fractions). (d) Chap. 13, S. Ochs (fast axoplasmic flow of proteins and polypeptides in mammalion nerve fibers). (e) E. M. Shooter and S. Varon (macromolecular aspects of the nerve growth factor proteins). (f) Chap. 31, G. Porcellati (studies on proteinase enzymes during Wallerian degeneration). (g) Chap. 34, E. R. Einstein and Li-Pen Chao (problems relating to the protein-eliciting experimental allergic encephalomyelitis).

3. W. A. Himwich, *Developmental Neurobiology*, Charles C. Thomas, Springfield, Ill. (1970). Chap. 7, J. Dobbing (nutrition). Chap. 9, H. C. Agrawal and W. A. Himwich (amino acids).

4. A. Lajtha, *Protein metabolism of the nervous system*, Int. Rev. Neurobiol. 6:1–98 (1964).

5. N. Marks, *Exopeptidase of the nervous system*, Int. Rev. Neurobiol. 11:57–97 (1968).

6. H. Peeters (ed.), *Protides of the Biological Fluids*. (a) J. Folch (proteolipids). (b) A. Lajtha and N. Marks (cerebral protein breakdown). (c) G. Porcellati (degeneration). (d) H. Sachs (neurosecretion).

7. Z. Lodin and S. P. R. Rose (eds.), *Macromolecules and Function of the Neuron*, Excerpta Medica Foundation, Amsterdam (1968). (a) A. G. E. Pearse and K. Wächtler (histochemistry). (b) C. W. M. Adams (histochemistry) p. 111. (c) N. Marks, R. K. Datta, and A. Lajtha (arylamidases and breakdown). (d) B. Jakoubek and E. Gutmann (turnover).

8. S. S. Oja, Postnatal changes in the concentration of nucleic acids, nucleotides and amino acids in the rat brain, *Ann. Acad. Sci. Fenn.* **125** (5A) 1–69 (1966); and Studies on protein metabolism in developing rat brain, *Ann. Acad. Sci. Fenn.* **131** (5A) 7–78 (1967).

9. R. Friede, *Topographic Brain Chemistry*, Academic Press, New York (1966).

10. C. W. M. Adams (ed.), *Neurohistochemistry*, Elsevier, Amsterdam (1965).

11. R. Vogel, I. Trantschold, and E. Werle, *Natural Proteinase Inhibitors*, Academic Press, New York (1968).

12. *Structure and Function of Biologically Active Peptides*, Ann. N.Y. Acad. Sci. **104**:1–464 (1963), and, *Chemistry, Pharmacology and Clinical Application of Proteinase Inhibitors*, **146**:361–787 (1968). (a) M. Laskowski (inhibitor complexes), pp. 374–380. (b) N. Back, H. Wilkens, and R. Steger (shock). (c) M. I. Barnhardt, C. Quintana, H. L. Larson, C. B. Bhelan and J. M. Riddle (inflammation). (d) F. Sicuteri (intracranial hemorrhages).

13. (a) *Enzyme Nomenclature*, Elsevier, Amsterdam (1965). (b) *A Bibliographic Guide to Neuroenzyme Literature* (D. J. Hoijer, ed.) IFI-Plenum Press, New York (1969).

14. E. A. Balazs (ed.), *Chemistry and Molecular Biology of the Intracellular Matrix*, Academic Press, New York, Vols. 1–3 (1970–1971).

15. E. G. Erdos, N. Back, and F. Sicuteri (eds.), *Hypotensive Peptides*, Springer-Verlag, Berlin (1966).

16. *Bradykinin and Related Kinins. Cardiovascular, Biochemical and Neural Actions* (F. Sicuteri, M. Rocha, E. Silva, and N. Back, eds.), Advances in Experimental Medicine and Biology, Vol. **8**, pp. 1–636, Plenum Press, New York (1970).

17. G. Perlmann and L. Lorand (eds.), *Proteolytic Enzymes*, Methods in Enzymology, Vol. 19, Academic Press, New York (1970). (a) M. Mycek (cathepsins). (b) N. Marks and A. Lajtha (aminopeptidase), pp. 534–543.

18. J. T. Dingle and H. B. Fell (eds.), *Lysosomes in Biology and Pathology, Vols. I–II* (pp. 1–567 and 1–689). (a) A. J. Barrett (enzymology). (b) H. Beaufay (preparation). (c) R. R. Combs and H. B. Fell (allergy). (d) C. De Duve (concepts). (e) W. Zh. Daems (lysosome formation). (f) H. B. Fell and H. G. Hers (genetics). (g) E. Holtzman (brain lysosomes). (h) P. J. Jacques (endocytosis). (i) H. Koenig (brain lysosomes). (j) J. A. Lucy (membranes).

(k) A. B. Mannsbach (kidney). (l) A. L. Tappel (enzymology). (m) J. J. Reynolds (connective tissue). (n) O. A. Roels (vitamins). (o) A. L. Tappel (preparation and enzymology), (p) P. G. Vaes (bone resorption). (q) R. Weber (tissue involution). (r) I. M. Weinstock and A. A. Iodice (dystrophy). (s) S. Weissman (steroids). (t) J. F. Woessner (uterus).

19. *Relationship of Nutrition to CNS Development and Function, Fed. Proc.* **26**:134–138 (1967), and *Nutrition and Cell Development, Fed. Proc.* **29**:1489–1516 (1970).

20. *Mammalian Protein Metabolism* (H. N. Munro and J. B. Allison, eds.), Vols. I–II; (H. N. Munroe, ed., Vols. III–IV) Academic Press, New York (1969–1970).
   Vol. I (a) Chap. 10, H. N. Munro (hormones-regulation).
   Vol. II (b) Chap. 12, J. B. Allison (nutrition).
   Vol. IV (c) Chap. 32, R. Z. Schimke (degradation). (d) Chap. 33, K. L. Manchester (hormones).

21. G. H. Bourne (ed.), *The Structure and Function of the Nervous System*, Academic Press, New York, Vols. 1–111 (1969).
   Vol. I (a) P. W. Lambert (myelin). (b) T. R. Shantna and G. H. Bourne (perineural epithelium). (c) M. Cole (retrograde degeneration).
   Vol. II (d) M. Wolman (neurokeratin).
   Vol. III (e) C. M. M. Adams and S. Leibowitz (demyelination).

22. O. Eichler, A. Farah, H. Herken, and A. D. Welch (eds.), *Handbook of Experimental Pharmacology*, Springer-Verlag, Berlin, Vol. I, *Oxytocin* (1968). (a) H. Tuppy (enzymology) pp. 67–130. (b) M. Ginsburg (metabolism) pp. 286–371. (c) H. Sachs (synthesis and transport). Vol. II, *Bradykinin, Kalliden and Kallikrein.* (d) E. Werle (Kallikrein and inhibitors). (e) J. L. Prado (proteolytic enzymes). (f) E. Habermann (Kininogens). (g) E. G. Erdos and H. Y. T. Young (Kiminases).

23. G. Ungar, *Excitation*, Charles C. Thomas, Springfield (1963), pp. 127, 143.

24. E. S. Vesell (ed.), *Multiple Molecular Forms of Enzymes*, *Ann. N.Y. Acad. Sci.* **151**:pp. 1–689 (1968).

24a. Axoplasmic transport and Neuronal fibrous proteins, *Neurosciences Res. Bull.* **5**:309–416 (1967) and **6**:155–216 (1968).

25. *The nature and function of perioxisomes* (microbodies), *Ann. N.Y. Acad. Sci.* **166**:209–381 (1970).

26. F. Lipmann, Polypeptide chain elongation in protein biosynthesis, *Science* **164**:1024–1031 (1969).

27. Gr. Benato and E. Gabrielescu, Influence de l'hypothermie sur l'histochemie fonctionelle de la phosphatase acide et de proteases du cerveau de rat, *Rev. Roum Physiol.* **5**:257–266 (1968).

28. A. Lajtha and N. Marks, Dynamics of protein metabolism of the nervous system, in *The Future of the Brain Sciences* (S. Bogoch, ed.), Plenum Press, New York (1969).

29. R. Z. Schimke and D. Doyle, Control of enzyme levels in animal tissues, *Ann. Rev. Biochem.* **39**:929–976 (1970).

30. H. Neurath, K. A. Walsch, and W. P. Winter, Evolution of structure and function of proteases, *Science* **158**:1638–1644 (1967).

31. D. J. Etherington and W. H. Taylor, The pepsins of normal human gastric juice, *Biochem. J.* **113**:663–668 (1969) and The sites of action of human and swine pepsins on B chain of oxidized insulin, *Biochem. J.* **113**:29P (1969).

32. E. A. Bauer, A. Z. Eisen, and J. J. Jeffrey, Immunological relationship of a purified human skin collagenase to other human and animal collagens, *Biochim. Biophys. Acta* **206**:152–160 (1970).

33. H. Hanson, Die Antolyse als biochemisches Phänomen in ihrer Verknüpfung mit der Pathophysiologie der Organe und des Blutes, *Wiss. Z. Univ. Halle* **25**:225–235 (1968).

34. R. Willstätter and E. Baumann, Über die Proteasen der Magenschleimhaut, *Hoppe-Seylers Z. Physiol. Chem.* **180**:124–143 (1929).

35. M. Bergmann, A classification of proteolytic enzymes, *Adv. Enzymol.* **2**:49–68 (1942).

36. J. S. Fruton, Cathepsins, in *The Enzymes* (P. D. Boyer, H. Lardy, and K. Myrbäck, eds.), 2nd ed. Vol. 4, pp. 233–241, Academic Press, New York (1960).

37. M. L. Anson, The estimation of pepsin, trypsin, papain and cathepsin with hemoglobin, *J. Gen. Physiol.* **22**:79–89 (1939), and Purification of cathepsin, *J. Gen. Physiol.* **23**: 695–704 (1939).

38. E. M. Press, R. R. Porter, and J. Cebra, The isolation and properties of a proteolytic enzyme. Cathepsin D from bovine spleen, *Biochem. J.* **74**:501–514 (1960).

39. C. Lapresle and T. Webb, The purification and properties of a proteolytic enzyme, rabbit cathepsin E, and further studies on rabbit cathepsin D, *Biochem. J.* **84**:455–462 (1962).

40. R. M. Metrione, A. G. Neves, and J. S. Fruton, Purification and properties of dipeptidyl transferase (cathepsin C), *Biochemistry* **5**:1597–1604 (1966).

41. N. Lichtenstein and J. S. Fruton, Studies on beef spleen cathepsin A, *Proc. Nat. Acad. Sci. (U.S.),* **46**:787 (1960).

42. N. Marks and A. Lajtha, Separation of acidic and neutral proteinase of brain, *Biochem. J.* **97**:74–83 (1965).

43. J. M. W. Bouma and M. Graber, The distribution of cathepsin B and C in rat tissues, *Biochim. Biophys. Acta* **89**:545–547 (1964), and Intracellular distribution of cathepsin B and cathepsin C in rat liver, *Biochim. Biophys. Acta* **113**:350–358 (1966).

44. M. Matsunga, N. Saito, J. Kira, K. Ogino, and M. Takagasu, Acid angiotensinase as a lysosomal enzyme, *Jap. Circul. J.* **32**:137–150 (1968); **33**:545–551 (1969).

45. K. Otto and S. Bhahdi, Zur Kenntnis des Kathepsins B′:Spezifizität und Eigenschaften, *Z. Physiol. Chem.* **348**:1449–1462 (1967) and **350**:1577–1588 (1969).

46. L. M. Greenbaum and J. S. Fruton, Purification and properties of cathepsin B, *J. Biol. Chem.* **174**:851–871 (1957).

47. H. Keilova and B. Keil, Isolation and specificity of cathepsin B, *FEBS Letters* **4**:295–298 (1969).

47a. J. Ken McDonald, B. B. Zeitman, and S. Ellis, Leucinenaphthyamide: an appropriate substrate for the histochemical detection of cathepsins B and B′, *Nature* **225**:1048–1049 (1970).

48. H. R. Gutmann and J. S. Fruton, On the proteolytic enzymes of animal tissues. VIII. Intracellular enzyme related to chymotrypsin, *J. Biol. Chem.* **174**:851–858 (1948).

49. J. K. McDonald, S. Ellis, and T. J. Reilly, Properties of dipeptidyl arylamidase I of the pituitary. Chloride and sulfhydryl activation of seryltyrosyl β-naphthylamide, *J. Biol. Chem.* **241**:1494–1501 (1966).

50. J. K. McDonald, B. B. Zeitman, T. J. Reilly, and S. Ellis, New observations on the substrate specificity of cathepsin C (dipeptidyl aminopeptidase I), *J. Biol. Chem.* **244**:2693–2709 (1969).

51. J. Ken McDonald, P. X. Callahan, B. B. Zeitman, and S. Ellis, Inactivation and degradation of glucagon by dipeptidyl aminopeptidase I (cathepsin C) of rat liver including a comparative study of secretin degradation, *J. Biol. Chem.* **244**:6199–6208 (1969).

52. I. M. Voynick and J. S. Fruton, The specificity of dipeptidyl transferase, *Biochemistry* **1**:40–44 (1968).

53. B. Sylven and O. Snellman, A new metal-dependent enzyme present in cathepsin C preparations, *Biochim. Biophys. Acta* **65**:350–353 (1962).

54. B. Sylven and I. Bois, Studies on the histochemical "leucine aminopeptidase" reaction. I. Identity of the enzymes possibly involved, *Histochemie* **3**:65–78 (1962); **3**:484–486 (1964).

55. T. Vanha-Pertulla, V. K. Hopsu, V. Sonninen, and G. G. Glenner, Cathepsin C activity as related to some histochemical substrates, *Histochemie* 5:170–181 (1965).

56. G. de la Habra, P. S. Cammarata, and S. N. Timasheff, The partial purification and some physical properties of cathepsin C from beef spleen, *J. Biol. Chem.* 234:316–319 (1959).

57. S. C. Dhar and S. M. Bose, Purification, crystallization and properties of cathepsin C from beef spleen, *Leather Sci.* 11:309–320 (1964).

58. J. Gorter and M. Gruber, Cathepsin C: an allosteric enzyme, *Biochim. Biophys. Acta* 198: 546–555 (1970).

59. R. J. Planta and M. Gruber, Specificity of cathepsin C, *Biochim. Biophys. Acta* 53:443–444 (1961); and Chromatographic purification of the thiol enzyme cathepsin C, *Biochim. Biophys. Acta* 89:503–510 (1964).

60. R. J. Planta, J. Gorter, and M. Gruber, The catalytic properties of cathepsin C, *Biochim. Biophys. Acta* 89:511–519 (1964).

61. S. Kakiuki and H. H. Tomizawa, Properties of a glucagon-degrading enzyme of beef liver, *J. Biol. Chem.* 239:2160–2164 (1964).

62. N. Marks, R. K. Datta, and A. Lajtha, Partial resolution of brain arylamidases and aminopeptidases, *J. Biol. Chem.* 243:2882–2889 (1968).

63. M. E. Jones, W. R. Hearn, M. Fried, and J. S. Fruton, Transamidation reactions catalyzed by cathepsin C, *J. Biol. Chem.* 195:645–656 (1952).

64. C. P. Heinrich and J. S. Fruton, The action of dipeptidyl transferase as a polymerase, *Biochemistry* 7:3556–3565 (1968).

65. K. K. Nilsson and J. S. Fruton, Polymerization reactions catalyzed by intracellular proteinases. IV. Factors influencing the polymerization of dipeptide amides by cathepsin C, Biochemistry 3:1220–1224 (1964).

66. H. Würz, A. Tanaka, and J. S. Fruton, Polymerization of dipeptide amides by cathepsin C, *Biochemistry* 1:19–29 (1962).

67. I. M. Voynick and J. S. Fruton, The specificity of dipeptidyl transferase, *Biochemistry* 7: 40–44 (1968).

68. N. Marks and A. Lajtha, Protein breakdown in the brain: Subcellular distribution of neutral and acid proteinases, *Biochem. J.* 89:438–477 (1963).

69. H. Winterstein, *Handbuch der Normalen und Pathologisch Physiologie* Vol. 9, p. 515 (1929).

70. H. Hydén, Protein metabolism in the nerve cell during growth and function, *Acta. Physiol. Scand.* 6, Suppl. 17:5–136 (1943).

71. C. A. Gibson, F. Umbreit, and H. C. Bradley, VII. Autolysis of brain, *J. Biol. Chem.* 47: 333–339 (1921).

72. H. A. Krebs, Über die Proteolyse der Tumoren, *Biochem. Z.* 238:174–196 (1931).

73. S. Edelbacher, E. Goldschmidt, and V. Schläppi, Über die Enzyme des Gehirns, *Hoppe-Seylers Z. Physiol. Chem.* 227:118–123 (1934).

74. I. Krimsky and E. Racker, Proteins of glycolysis in mouse brain homogenates by amides and esters of amino acids, *J. Biol. Chem.* 179:903–914 (1949).

75. (a) A. Geiger and J. Magnus, Perfusion of brain, *Amer. J. Physiol.* 149:517–526 (1947) and (b) A. Geiger and S. Yamasaki, Cylidine and uridine requirement of the brain, *J. Neurochem.* 1:93–100 (1956).

76. M. W. Kies and S. Schwimer, Observations on proteinase in brain, *J. Biol. Chem.* 145:685–691 (1942).

77. G. B. Ansell and D. Richter, The proteolytic activity of brain tissue, *Biochim. Biophys. Acta* 13:87–91 (1954); Evidence for neutral proteinase in brain tissue, *Biochim. Biophys. Acta* 13:92–97 (1954).

78. N. M. Polyakova and V. K. Lishko, Isolation and purification of brain proteinase, *Ukr. Biokhim. Zh.* 34:208–216 (1962).

78a. L. A. Tsaryuk, Assay of brain proteinases, *Ukr. Biokhim. Zh.* **34**:815–823 (1962); **36**:334–342 (1964) (*Chem. Abstr.* **61**:8697).

79. A. Lajtha, Observations of protein catabolism in brain, in *Regional Neurochemistry* (S. S. Kety and J. Elkes, eds.), pp. 25–36, Pergamon Press, London (1961).

80. A. V. Palladin, Distribution of enzymes among various subcellular units of the brain, in *Regional Neurochemistry* (S. S. Kety and J. Elkes, eds.), pp. 8–18, Pergamon Press, London (1961).

81. A. V. Palladin, N. M. Polyakova, and V. K. Lishko, Purification of brain proteinase, *J. Neurochem.* **10**:187–194 (1963).

82. H. U. Bergmeyer, *Methods of Enzymatic Analysis*, 2nd ed., Academic Press, New York (1965).

83. M. Pirotta and N. Marks, Purfication of arylamidase. Cleavage of dipeptidyl arylamides and angiotensin (unpublished observations).

84. G. L. Moore, Use of azo-dye-bound collagen to measure reaction velocities of proteolytic enzymes, *Anal. Biochem.* **32**:122–127 (1969).

85. M. Dixon and E. C. Webb, *Enzymes*, Academic Press, New York (1964).

86. S. Udenfriend, *Fluorescence Assay in Biology and Medicine*, Vols. I, II, Academic Press, New York (1962, 1969).

87. B. O. DeLumen and A. L. Tappel, Fluorescein-hemoglobin as a substrate for cathepsin D and other proteases, *Anal. Biochem.* **36**:22–29 (1970).

88. M. B. Hille, A. J. Barrett, J. T. Dingle, and H. B. Fell, Microassay for cathepsin D shows an unexpected effect of cycloheximide on limb bone rudiments in organ culture, *Exp. Cell. Res.* **61**:470–472 (1970).

89. J. Uriel, Characterization of enzymes in specific immune-precipitates, *Ann. N.Y. Acad. Sci.* **103**:956–979 (1963); A method for direct detection of proteolytic enzymes after electrophoresis in agar gel, *Nature* **188**:853 (1960).

90. J. Uriel and S. Avraneas, Systematic fractionation of swine pancreatic hydrolases, *Biochemistry* **4**:1740–1757 (1965).

91. J. R. Merkel, Direct detection of proteolytic enzymes that have been separated by zone electrophoresis, *Anal. Biochem.* **17**:84–92 (1966).

92. P. D. Weston, A. J. Barrett, and J. T. Dingle, Specific inhibition of cartilage breakdown, *Nature* **222**:285–286 (1969).

93. A. J. Barrett, Cathepsin D. Purification of isoenzymes from human and chicken liver, *Biochem. J.* **117**:601–607 (1970).

94. F. Sanger and F. Tuppy, The amino-acid sequence in the phenylalanyl chain of insulin. The specificity of pepsin, chymotrypsin and trypsin, *Biochem. J.* **49**:481–500 (1951).

95. H. Keilova, K. Blaha, and B. Keil, Effect of steric factors on digestibility of peptides containing aromatic amino acids by cathepsin D and pepsin, *Eur. J. Biochem.* **4**:442–447 (1968) and *Coll. Czech. Chem. Commun.* **33**:131 (1968).

96. L. Abrash, N. Marks, M. Pirotta, and R. Walter, Inactivation of peptidyl hormones by brain enzymes, in *Proc. Amer. Neurochem. Soc.* (2nd meeting), Hershey, Pa. (1971).

97. P. D. Weston, Specific antiserum to lysosomal cathepsin D, *Immunology* **17**:421–428 (1969).

98. J. H. Hageman and B. C. Carlton, An enzymatic and immunological comparison of two proteases from transformable *Bacillus subtilis* with "subtilisin," *Arch. Biochim. Biophys.* **139**:67–79 (1970).

99. N. Marks and A. Lajtha, Specificity of brain acid proteinase, *Fed. Meeting* **25**:793 (1966).

100. N. Marks and B. D'Monte, Purification and specificity of cathepsin with synthetic substrates (in preparation).

101. E. R. Einstein, J. Csejtey, and N. Marks, Degradation of the encephalitogen by purified brain acid proteinase, *FEBS Letters* **1**:191–195 (1968).

102. N. Marks, B. D'Monte, C. Bellman, and A. Lajtha, Protein metabolism in cerebral mitochondria. I. Hydrolytic enzymes and amino acid incorporated into mitochondrial membranes, Brain Res. 18:304–324 (1970).

103. O. Z. Sellinger and R. A. Hiatt, Cerebral lysosomes. IV. The regional and intracellular distribution of arylsulfatase and evidence for two populations of lysosomes in rat brain, Brain Res. 7:191–200 (1968).

104. K. Fitzpatrick and R. J. Pennington, Evidence for proteolytic activity in liver mitochondria, Biochem. J. 111:29P–30P (1969).

105. K. G. Alberti and W. Bartley, The localization of proteolytic activity in rat liver mitochondria and its relation to mitochondrial swelling, Biochem. J. 111:763–776 (1969).

106. T. A. A. Dopheide and P. E. E. Todd, Hydrolysis of the A and B chains of oxidized insulin by adrenal acid proteinase, Biochim. Biophys. Acta 86:130–135 (1964) and 181:105–115 (1969).

107. H. Keilova, B. Keil, and F. Sorm, Proteases of Ehrlich ascites tumor I–VI, Coll. Czech. Chem. Commun. 27:2193–2201; 27:2186–2192 (1962); 29:2216–2222 (1964); 2272–2276: 33:131 (1968).

108. G. Reich and E. Buddecke, Darstellung und Eigenschaften eines Kalthepsin aus Rinderarteriengewebe, Z. Physiol. Chem. 348:1616–1628 (1967).

109. L. F. Kress, R. J. Peanasky, and H. M. Klitgaard, Purification, properties, and specificity of hog thyroid proteinase, Biochim. Biophys. Acta 113:375–389 (1965).

110. G. D. Smith, M. A. Murray, L. W. Nichol, and V. M. Trikojus, Thyroid acid proteinase properties and inactivation by diazoacetyl-norleucine methyl ester, Biochim. Biophys. Acta 171:288–298 (1969).

111. P. Jablonski and M. T. McQuillan, The distribution of proteolytic enzymes in the thyroid gland, Biochim. Biophys. Acta 132:454–471 (1967).

112. I. Pastan and S. Almqvist, Nonidentity of esterase and neutral protease in rat thyroid, Endocrinology 77:415–416 (1965).

113. H. Keilova, O. Marcovic, and B. Keil, Characterization and chemical modification of cathepsin D from bovine spleen, Coll. Czech. Chem. Commun. 34:2154–2157 (1969).

114. J. Marrink and M. Gruber, Proteolytic activity at neutral pH in bovine spleen, FEBS Letters 1:68–73 (1968) and Biochim. Biophys. 118:438 (1966).

115. J. Lisowski, Pituitary enzymes II. The problem of existence of proteinase II, Arch. Immunol. et. Ther. Exper. 12:645–665 (1964).

116. Ya V. Belik, L. S. Smerchinska, Y. T. Terletska, and O. G. Grineko, Proteolytic activity in nerve tissue, Ukr. J. Biochm. 40:543–548 (1968); (with O. H. Hrinenko) Ukr. J. Biochm. 40:332 (1968).

117. G. Guroff, A neutral, calcium activated proteinase from the soluble fraction of rat brain, J. Biol. Chem. 239:149–155 (1964).

118. P. J. Riekkinen and J. Clausen, Proteinase activity of myelin, Brain Res. 15:413–430 (1969).

119. P. J. Riekkinen, J. Clausen, and U. Arstila, Further studies on neutral proteinase activity of CNS myelin, Brain Res. 19:213–227 (1970); and with M. G. Rumsby, ibid. 24:495–516 (1970).

120. V. K. Lishko, Studies of the properties of brain cathepsin, Ukr. Biokhim. Zh. 35:874–880 (1963); 36:565–573 (1964); 37:163–168 (1965). (See Chem. Abstr. 57, 5008g, 60:12292a; 61:14962.

121. S. G. Waley and R. van Heyingen, Neutral proteinases in the lens, Biochem. J. 83:274–283 (1962) and Partial purification and properties, Biochem. J. 86:92–101 (1963).

122. B. I. Gerwin, W. H. Stein, and S. Moore, On the specificity of streptococcal proteinase, J. Biol. Chem. 241:3331–3339 (1966).

123. N. Marks and A. Lajtha, Levels of cerebral acid and neutral proteinase during development, in *Variations in Chemical Composition of the Nervous System* (G. B. Ansell, ed.), p. 74, Pergamon Press, London (1966).

124. J. Feder and J. M. Schuck, Studies on the *B. subtilis* neutral proteinase and proteolytic themolysin-catalyzed hydrolysis of dipeptide substrates, *Biochemistry* 9:2784–2791 (1970).

125. G. L. Moore, Use of azo-dye bound collagen to measure reaction velocities of proteolytic enzymes, *Anal. Biochem.* 32:122–127 (1969).

126. S. S. Oja and H. Oja, Cerebral proteinases in the growing rat, *J. Neurochem.* 17:901–912 (1970).

127. C. W. M. Adams and N. A. Tuqan, The histochemical demonstration of protease by a gelatin-silver substrate, *J. Histochem. Cytochem.* 9:469–472 (1961). Histochemistry of myelin, II. Proteins, lipid-protein dissociation and proteinase activity in Wallerian degeneration, *J. Neurochem.* 6:334–341 (1961).

128. P. Gaddum and R. J. Blandu, Proteolytic reaction of mammalian spermatozoa on gelatin membranes, *Science* 170:749–750 (1970).

129. T. I. Pristoupil, Microchromatography and microelectrophoresis on intracellulose membranes, *Chromatographic Rev.* 12:109–125 (1970).

130. S. Serra, A. Grynbaum, A. Lajtha, and N. Marks, Enzymes of protein breakdown in spinal cord, *Neurology* 21:415 (1971).

131. B. D'Monte, P. Mela, and N. Marks, Protein metabolism in rat brain myelin fractions, *Fed. Proc.* 29:472 (1970).

132. B. D'Monte, P. Mela, and N. Marks, Turnover of myelin *in vivo*, *Eur. J. Biochem.* (submitted, 1971).

133. A. Janoff and J. O. Zeligs, Vascular injury and lysis of basement membrane *in vitro* by neutral protease of human leucocytes, *Science* 166:702–704 (1968).

134. P. J. Riekkinen and U. K. Rinne, A new neutral proteinase from the rat brain, *Brain Res.* 9:126–135 (1968).

135. T. Sipos and J. R. Merkel, An effect of calcium ion on the activity, heat stability, and structure of trypsin, *Biochemistry* 9:2766–2775 (1970).

136. J. P. Abita, M. Delaage, M. Lazdunski, and J. Savrda, The mechanisms of activation of trypsinogen. The role of the four-N-terminal aspartyl residues, *Eur. J. Biochem.* 8:314–324 (1969).

137. K. Morihara and H. Tsuzuki, Comparison of the specificities of various serine proteinases from microorganisms, *Arch. Biochim. Biophys.* 129:620–634 (1969).

138. J. D. McConn, D. Tsura, and K. T. Yasunobu, *B. subtilis* neutral proteinase. I. A zinc enzyme of high specific activity, *J. Biol. Chem.* 239:3706–3715 (1964).

139. H. P. Rappaport, W. S. Riggsby, and D. A. Holden, A *B. subtilis* proteinase. I. Production, purification and characterization of a proteinase from a transformable strain of *B. subtilis*, *J. Biol. Chem.* 240:78–86 (1965).

140. T. Y. Liu, N. P. Neumann, S. D. Elliott, S. Moore, and W. H. Stein, Chemical properties of streptococcal proteinase and its zymogen, *J. Biol. Chem.* 238:251–256 (1963).

141. H. Iwata, T. Shikimi, and T. Oka, Pharmacological significance of peptidase and proteinase in the brain. Enzymatic inactivation of bradykinin in rat brain, 18:119–128 (1969) and 19:1399–1407 (1970).

142. R. Schoenheimer, *The Dynamic State of Body Constituents*, Harvard University Press, Cambridge, Mass. (1941), p. 38.

143. M. B. Simpson, Release of labeled amino acids from the proteins of rat liver slices, *J. Biol. Chem.* 201:143–154 (1953).

144. P. Steinberg, M. Vaughan, and C. B. Anfinsen, Kinetic aspects of assembly and degradation of proteins, *Science* 124:389–395 (1956).

145. A. Korner and H. Tarver, Studies on protein synthesis *in vitro*. II. Incorporation and release of amino acids in particulate preparations from livers of rats, *J. Gen. Physiol.* **41**:219–231 (1958).

146. N. W. Penn, The requirements for serum albumin on subcellular fractionation of liver and brain, *Biochim. Biophys. Acta* **37**:55–63 (1960) and **53**:190 (1961).

147. R. Umana, Re-evaluation of the method of Kunitz for the assay of proteolytic activities in liver and brain homogenates, *Anal. Biochem.* **26**:430–438 (1968).

148. C. O. Bostrom and H. Jeffay, Protein catabolism in rat liver homogenates. A re-evaluation of the energy requirement for protein catabolism, *J. Biol. Chem.* **245**:4001–4008 (1970).

149. L. Libensen and M. Yena, Role of glutathione in the catheptic hydrolysis of plasma albumin, *Arch. Biochim. Biophys.* **104**:292–296 (1964).

150. H. Walter, Protein catabolism, *Nature* **188**:643–645 (1960).

151. R. J. Wurtman, T. Axelrod, and D. E. Kelly, *The Pineal Gland*, Academic Press, New York (1968).

152. R. Burgus, G. R. Guilemin, Hypothalamic releasing factors, *Ann. Rev. Biochem.* **39**:499–526 (1970).

153. E. Adams and E. L. Smith, Proteolytic activity of pituitary extracts, *J. Biol. Chem.* **191**:651–664 (1951).

154. U. J. Lewis, Factors that affect the fragmentation of growth hormone and prolactin of hypophyseal proteinases, *J. Biol. Chem.* **238**:3330–3335 (1963).

155. S. Ellis, Pituitary proteinase. I. Purification and action on growth hormone and prolactin, *J. Biol. Chem.* **235**:1694–1699 (1960).

156. P. X. Callahan and S. Ellis, Action of pituitary proteinase I on the insulin box chain, *Biochem. Biophys. Acta* **122**:413–416 (1970).

157. C. H. Li and W. K. Liu, Human pituitary hormone, *Experientia* **20**:169–240 (1964). (Reviewed: *Perspective Biol. Med. II:*498–521 and *Ann. N.Y. Acad. Sci.* **148**:284–571 (1968).

158. L. E. Reichert, Effect of urea on the suitability of bovine pituitary proteinase. II. *Biochim. Biophys. Acta* **50**:191–193 (1961).

159. R. K. Meyer and K. H. Clifton, Effect of diethylstilbestrol on the quantity and intracellular distribution of pituitary proteinase activity, *Arch. Biochim. Biophys.* **62**:198–209 (1956).

160. U. J. Lewis and W. P. Vanderlaan, A relationship between thyroid function and a naturally occurring inhibitor of hypophyseal proteinases, *J. Biol. Chem.* **238**:3336–3340 (1963).

161. L. M. Sreebny, Studies of salivary gland proteinases, *Ann. N.Y. Acad. Sci.* **85**:182–188 (1960).

162. F. Cross (ed.), *Protein Metabolism Influence of Growth Hormone, Anabolic Steroids and Nutrition in Health and Disease*, Springer-Verlag, Berlin (1962).

163. T. Vanha-Perttula, Aminoacyl dipeptidyl arylamidases of the pituitary gland as related to function, *Endocrinology* **85**:1062–1069 (1970).

164. A. Milo and G. Porcellati, The action of some organio-phosphorous compounds on neutral proteinase *in vitro*, *J. Biochem.* **5**:333–343 (1961).

165. G. Porcellati, A. Milo, and I. Manocchio, Proteinase activity of nervous tissue in organophosphorous compound poisoning, *J. Neurochem.* **7**:317–320 (1961).

166. N. Marks, R. K. Datta, and A. Lajtha, Distribution of amino acids of exo- and endopeptidases along vertebrate and invertebrate nerves, *J. Neurochem.* **17**:53–63 (1970).

167. E. Rojas, Membrane potentials, resistance and ion permeability in squid giant axons injected or perfused with proteinases, *Proc. Nat. Acad. Sci. (U.S.)* **53**:306–311 (1965).

168. I. Tasaki and T. Takenaka, Effects of various potassium salts and proteinases upon suitability of intracellularly perfused squid giant axons, *Proc. Nat. Acad. Sci. (U.S.)* **52**:804–810 (1964).

169. T. Narahashi and J. M. Tobias, Properties of axon membrane as affected by cobra venom, digitonin, and proteinases, *Amer. J. Physio.* **207**:1441–1446 (1964).

170. A. Lajtha and N. Marks, Dynamics and significance of cerebral protein metabolism, *Dis. Nerv. Sys.* **30**:36–43 (1969).

171. Y. Olsson and J. Sjostrand, Origin of macrophages in Wallerian degenerations of peripheral nerves demonstrated autoradiographically, *Exper. Neurol.* **23**:102–112 (1969).

172. J. F. Hallpike, C. W. M. Adams, and O. B. Bayliss, Histochemistry of myelin. IX. Neutral and acid proteinase in early Wallerian degeneration, *Histochem. J.* **2**:209–218 (1970).

173. C. W. M. Adams and O. B. Bayliss, Histochemistry of myelin. III. Peripheral nerve cathepsin, *J. Histochem. Cytochem.* **9**:473–476 (1961); V. Trypsin-digestible and trypsin-resistant nerve cathepsin, *J. Histochem. Cytochem.* **16**:110–114 (1968).

174. E. R. Einstein, J. Csejtey, W. J. Davis, A. Lajtha, and N. Marks, Enzyme degradation of the encephalitogenic protein, *Pathogenesis and Etiology of Demyelinating Diseases*, *Int. Arch. Allergy.* **361**:(Suppl.) 336–375, Karger, Basel (1969).

175. N. Marks, Some correlates of axoplasmic flow, *Dis. Nerv. System* **31**:5–27 (1970) (Suppl.).

176. G. A. Kerkut, A. Shapiro, and R. J. Walker, The transport of labelled materials from CNS-muscle along a nerve trunk, *Comp. Biochem. Physiol.* **23**:729 (1968).

177. J. F. Hallpike and C. W. M. Adams, Proteolysis and myelin breakdown: A review of recent histochemical and biochemical studies, *Histochemistry J.* **1**:559–578 (1969).

178. G. Porcellati and B. Curti, Proteinase activity of peripheral nerve during Wallerian degeneration, *J. Neurochem.* **5**:277–282 (1960).

179. F. Orrego, Protein degradation in squid giant axons, *Proc. Int. Neurochem. Soc.* 310 (2nd meeting, Milan) Tamburini Editore, Milan (1969) and *J. Neurochem* (in press, 1971).

180. N. L. Banik and A. N. Davison, Enzyme activity of myelin and subcellular fractions in the developing rat brain, *Biochem. J.* **115**:1051–1062 (1969).

181. A. S. Brecher, B. B. Oliphant, and R. E. Sobel, The proteolytic activity in brain, *Arch. Int. Physiol. Biochim.* **74**:429–434 (1966).

182. G. Benato and E. Gabrielescu, Histochemistry of proteinases in allergic reactions, *Ann. Histochem.* **9**:295–304 (1964); (with J. Boros). *Rev. Rouman Physiol.* **2**:379–384 (1965).

183. M. F. Kerekes, T. Feszt, and A. Kovacs, Cathepsin activity in the cerebral tissue of the rabbit during EAE, *Experientia*, **21**:42–43 (1965).

184. J. F. Hallpike, C. W. M. Adams, and O. B. Bayliss, Histochemistry of myelin. VIII. Proteolytic activity around multiple sclerosis plaques, *Histochem. J.* **2**:209–218; XI. Loss of brain protein in early myelin; Breakdown and multiple sclerosis plaques, *Histochem. J.* **2**:323–328 (1970).

185. E. H. Eylar and G. A. Hashim, Allergic encephalomyelitis: The structure of the encephalitogeic determinant, *Proc. Nat. Acad. Sci. (U.S.)* **61**:644–650 (1968).

186. E. H. Eylar, J. Caccam, J. J. Jackson, F. C. Westall, and A. B. Robinson, Experimental allergic encephalomyelitis: Synthesis of disease-inducing site of the basic protein, *Science* **168**:1220–1223 (1970); and *Nature* **229**:22–24 (1971).

187. E. H. Eylar, J. Salk, G. C. Beveridge, and L. V. Brown, EAE and encephalitogenic brain protein from bovine myelin, *Arch. Biochem.* **132**:34–48 (1969).

188. R. F. Kibler and R. Shapira, Isolation and properties of an encephalitogenic protein from bovine, rabbit, and human CNS-tissue, *J. Biol. Chem.* **243**:281–286 (1968).

189. L. P. Chao and E. R. Einstein, Isolation and characterization of an active fragment from enzymatic degradation of encephalitogenic protein, *J. Biol. Chem.* **243**:6050–5 (1968), Physical properties of the bovine encephalitogenic protein. M. and conformation, *J. Neurochem.* **17**:1121–1132 (1970).

190. E. R. Einstein, K. B. Dalal, and J. Csejtey, Biochemical maturations of the CNS, II. Protein and proteolytic enzyme changes, *Brain Res.* **18**:35–49 (1970).

191. M. K. Gaitonde and R. E. Martenson, Metabolism of highly basic protein of rat brain during postnatal development, *J. Neurochem.* **17**:551–563 (1970).

192. R. L. Peake, K. Balasubramaniam, and W. P. Deiss, Effect of reduced glutathione on the proteolysis of intraparticulate and native globulin, *Biochim. Biophys. Acta.* **148**:689–702 (1967).

193. F. A. Green, Proteolysis of fluoresceinated protein, *Immunochemistry* **2**:287–292 (1965).

194. E. Kaminski and W. Bushuk, Detection of multiple forms of proleolytic enzymes by Starch-gel electrophoresis, **46**:1317–1320 (1968).

195. B. Norton, Myelin, in *Cellular and Molecular Basis of Neurology and Disease* (E. Goldenson and S. Appel, eds.), Lee and Febiger, Philadelphia (in press).

196. G. Ungar and H. Hayashi, Enzymatic mechanisms in allergy, *Ann. Allergy.* **16**:542–562 (1958).

197. L. F. Chapman and H. Goodell, The participation of the nervous system in the inflammatory response, *Ann. N.Y. Acad. Sci.* **116**:990–1017.

198. M. Yano, S. Nagasawa, and T. Suzuki, Evidence for location of Met-Lys-bradykinin moieties at the carboxyl terminus and inside of bovine kininogen, *Biochem. Biophys. Res. Comm.* **40**:914–918 (1970).

199. T. Shimkimi and H. Iwata, Pharmacological significance of peptidase and proteinase in the brain. II. Purification and properties of a bradykinin inactivating enzyme from rat brain, *Biochem. Pharmacol.* **19**:1399–1407 (1970) with T. Oka **18**:119 (1969).

200. T. Shikimi, H. Shigemitsu, and I. Heitanoh, Substrate specificity and amino acid composition of partially purified enzyme inactivating bradykinin in rat brain, *Jap. J. Pharmacol.* **20**:169–170 (1970).

201. F. Fieldler, C. Hirshauer, and E. Werle, Anreicherung von Präkallikrein B aus Schweinepankreas und Eigenschaften verschiedener Formen des Pancreas Kallikreins, *Hoppe-Seyler's Z. Physiol.* **351**:225–238 (1970).

202. L. F. Chapmann and H. G. Wolff, Studies of proteolytic enzymes in the c.s.f., *Arch. Int. Med.* **103**:86–96 (1959).

203. A. J. Copen and G. P. Lewis, An enzyme forming vasodilator polypeptides in human cerebrospinal fluid, *J. Neurochem.* **7**:155–160 (1961).

204. K. Nustad, Localization of kininogenase in the rat kidney, *Brit. J. Pharmacol.* **39**:87–98 (1970).

205. E. Habermann, Reinigung und einige Eigenschaften eines Kallikreins aus Schweineserum, *Biochem. Z.* **337**:440–448 (1967) and **346**:133–158 (1966).

206. D. J. McConnell and B. Mason, Isolation of human plasma prekallikrein, *Brit. J. Pharmacol.* **38**:490–502 (1970).

207. S. Hori, The presence of bradykinin-like materials, kinin releasing and destroying action in brain, *Jap. J. Physiol.* **18**:772–787 (1968); Zetlers sattelite polypeptides of substance P in subcellular particles of bovine peripheral nerves, *Jap. J. Physiol.* **18**:746–771 (1968).

208. A. Inouye, K. Kataoka, and T. Tsujioka, On a kinin-like substance in the nervous tissue elements treated with trypsin, *Jap. J. Physiol.* **11**:319–334 (1961).

209. U.S. von Euler, Substance P in subcellular particles in peripheral nerve, *Ann. N.Y. Acad. Sci.* **104**:449–463 (1963).

210. V. K. Hopsu-Havu, K. K. Makinen, and G. G. Glenner, Formation of bradykinin from kalliden-10 by aminopeptidase B, *Nature* **212**:1271–1272 (1966).

211. L. T. Skeggs, K. E. Lentz, J. R. Kahn, and H. Hochstrasser, Kinetics of the reaction of renin with nine synthetic peptide substrates, *J. Exper. Med.* **128**:13–34 (1968).

212. B. Foltman, Studies on renin IX, *Comp. Rend. Tran. Lab. Carlsberg* **34**:319–325 (1964).

213. J. W. Ryan and D. C. Johnson, The renin-like enzyme of rabbit uterus, *Biochim. Biophys. Acta* **191**:386–396 (1969).

214. Y. Piguillard, A. Reinharz, and M. Roth, Studies on the angiotensin converting enzyme with different substrates, *Biochim. Biophys. Acta* **206**:136–142 (1970).

215. M. Matsunga, N. Saito, J. Kira, K. Ogino, and M. Takayasu, Acid angiotensinase as a lysosomal enzyme, *Jap. Circul. J.* **32**:137–157 (1968); **33**:545–551 (1969).

216. T. Kokubu, H. Akutsu, S. Fujimoto, E. Ueda, K. Hiwada, and Y. Yamamura, Purification and properties of endopeptidases from rabbit red cells and its process of degradation of angiotensin, *Biochim. Biophys. Acta* **191**:668–676 (1969).

217. H. D. Itskovitz, L. Miller, W. Ural, J. Zapp, and R. White, Inactivation of angiotensin in shock, *Amer. J. Physiol.* **216**:5–10 (1969).

218. W. Feldberg and C. P. Lewis, The action of peptides on the adrenal medulla. Release of adrenaline by bradykinin and angiotensin, *J. Physiol.* **171**:98–108 (1964).

219. D. L. Claybrook and J. J. Pfiffner, Purification and properties of a substance P-inactivating enzyme from bovine brain, *Biochem. Pharmacol.* **17**:281–293 (1968).

220. H. Orimo, T. Fujita, H. Morii, and K. Nakao, Inactivation *in vitro* of parathyroid hormone activity by kidney slices, *Endocrinology* **76**:255–258 (1965) and M. Yoshikawa, Mechanism of inactivation, *Endocrinology* **77**:428–432 (1965).

221. L. A. Greene, E. M. Shooter, and S. Varon, Enzymatic activity of mouse NGF and its subunits, *Proc. Nat. Acad. Sci. (U.S.)*, 1383–1388 (1968); *Biochemistry* **8**:3735–3741 (1969).

222. D. G. Attardi, J. M. Schlesinger, and S. Schlesinger, Submaxillary gland of mouse: properties of a purified protein affecting muscle tissue *in vitro*, *Science* **156**:1253–1255 (1967).

223. R. H. Angelleti, Nerve growth factor from cobra venom, *Proc. Nat. Acad. Sci.* **65**:668–674 (1970).

224. D. Dunlop and A. Lajtha, Turnover of protein in brain slices (in preparation).

225. C. Y. Bowers, A. Weil, J. K. Chang, H. Sievertson, F. Enzmann, and K. Folkers, Activity-structure relationships and the thyrotropic releasing hormone, *Biochim. Biophys. Res. Comm.* **40**:683–691 (1970).

226. P. Rieser, *Insulin Membranes and Metabolism*, Williams and Wilkins, Baltimore (1967).

227. G. E. Mortimore and C. E. Mondon, Inhibition by insulin of valine turnover in liver. Evidence for general control of proteolysis, *J. Biol. Chem.* **245**:2375–2383 (1970).

228. D. Rudman, L. A. Garcia, M.Ch Girolamo, and P. W. Shank, Cleavage of bovine insulin by rat adipose tissue, *Endocrinology* **78**:169–171 (1966) and *Biochemistry* **7**:1864–1875 (1968).

229. H. Grimmer and E. Kallee, Die Insellhormone: Der Abbau von Insulin, in *Diabetes Mellitus*, Vol. 1, pp. 257–280, J. F. Lehmans Verlag (1969).

230. I. A. Mirksy, G. Perisutti, and F. J. Dixon, Destruction of I^{131}-labelled insulin by liver slices, *J. Biol. Chem.* **214**:397–408 (1955).

231. H. M. Klatzen, F. Tietze, and D. Stellan, Further studies on the properties of hepatin glutathione-insulin transdehydrogenase, *J. Biol. Chem.* **238**:1006–1011 (1963).

232. H. Yammoto and J. Noguchi, Studies on the catalytic activity of polyamino-α-amino acids, *J. Biochem. (Japan)* **67**:103–121 (1970).

233. A. Ohnishi, S. Tokura, and J. Noguchi, Studies on the catalytic action of poly-α-amino acids, *J. Biochem. (Japan)* **68**:97–108 (1970).

234. H. Y. Yang, E. G. Erdos, and T. S. Chiang, New enzymatic route for the inactivation of angiotensin, *Nature* **218**:1224–1225 (1968).

235. L. Yman, Human serum aminopeptidases. Properties of oxytocinase, human serum aminopeptidase A and leucine aminopeptidase and their purification from retroplacental serum, *Acta Pharm. Sueccia* **7**:75–86 (1970).

236. J. Sjoholm and L. Yman, Preparation of highly purified oxytocinase (cystine aminopeptidase) from retroplacental serum, *Acta Pharm. Sueccia* **3**:377–388; and 389–396 (1966).

237. J. D. Glass, I. L. Schwartz, and R. Walter, Enzymatic inactivation of peptide hormones possessing a C-terminal amide group, *Proc. Nat. Acad. Sci. (U.S.)* **63**:1426–1430 (1969).

238. J. D. Glass, B. M. Dubois, F. L. Schwartz, and R. Walter, Inactivation of neurohypophyseal peptides by a uterine enzyme. Characterization of the hydrolytic cleavage, *Endocrinology* **87**:730–737 (1970).

239. M. Koida, J. D. Glass, F. L. Schwartz, and R. Walter, Mechanisms of inactivation of oxytocin by rat kidney enzymes, *Endocrinology* **88**:633 (1971).

240. K. Kataoka, The subcellular distribution of substance P in nervous tissue, *Jap. J. Physiol.* **12**:81–96 (1962).

241. R. M. G. Nair, J. F. Barrett, C. V. Bowers, and A. V. Schally, Structure of porcine thyrotropin releasing hormone, *Biochemistry* **9**:1103–1106 (1970).

242. T. Barth, H. J. Hütter, V. Piska, and F. Sorm, Inactivation of deamino-oxytocin deaminocarba′-oxytocine (D-Lys^8)- and (D-Arg^8) vasopressins by kidney cell particles, *Experientia* **25**:646–647 (1969).

243. J. L. Webb, *Enzyme and Metabolic Inhibitors*, Chap. 17, Vol. 1, pp. 879–888, Academic Press (1963).

244. H. Fritz, B. Greif, W. Schramm, K. Hochstrasser, and E. Werle, Zur Identität des polyvalenten Protease Inhibitor aus Schafslunge mit dem Trypsin Kallikrein Inhibitor aus Rinderorganen, *Hoppe-Seyler's Z. Physiol. Chem.* **351**:139–144 (1970).

245. H. Tschesche, E. Wachter, and G. Kallup, Isolierung eins Des-Thr-Ser-Pro-Gln-Arg-Trypsin Inhibitors vom Schwein mit voller biologischer Aktivität, *Hoppe-Seyler's Z. Physiol. Chem.* **350**:1662–1668 (1969); **351**:1449–1459 (1970).

246. R. E. Feeney, G. E. Means, and J. C. Bigler, Inhibition of human trypsin plasmin and thrombin by naturally occurring inhibitors of proteolytic enzymes, *J. Biol. Chem.* **244**:1957–1960 (1969).

247. L. J. Greene, M. Rigbi, D. S. Fackre, and J. R. Broich, Trypsin inhibitor from bovine pancreatic juice, *J. Biol. Chem.* **241**:5610–5618 (1966).

248. A. Rimon, Y. Shamash, and B. Shapiro, The plasmin inhibitor—human plasma, *J. Biol. Chem.* **241**:5102–5107 (1966).

249. J. Bieth, F. Miesch, and P. Metais, Enzymic and immunologic inhibitors of proteases in the CSF, *Clin. Chim. Acta* **24**:203–209 (1969).

250. C. Kaye and D. Dabich, An inhibitor of trypsin-like activity in rat liver, *Proc. Soc. Exp. Biol. Med.* **131**:1366–1368 (1969).

251. C. Blackwood and I. Mandl, An improved test for the quantitative determination of trypsin-like enzymes and enzyme inhibitors, *Anal. Biochem.* **2**:370–379 (1961).

252. A. S. Brecher and N. M. Quinn, The occurrence of trypsin inhibitor in brain, *Biochem. J.* **102**:120–121 (1967) and The presence of distinct soluble and particulate trypsin inhibitors in human astrocytoma tissue, *J. Neurochem.* **14**:203–206 (1967).

253. A. S. Brecher, B. B. Oliphant, and V. P. Wasilanskas, The binding and inhibition of chymotrypsin by mammalian brain cell fractions, *Arch. Int. Physiol. Biochim.* **76**:287–297 (1968).

254. B. B. Oliphant, J. B. Suszkin, and A. S. Brecher, The distribution of a particulate trypsin inhibitor in brain, *Proc. Soc. Exp. Biol. Med.* (in press).

255. B. D'Monte, N. Marks, A. Lajtha, and P. Mela, Inhibitors of cerebral exo- and endoproteinases, *Trans. Amer. Soc. Neurochem.* **1**:37 (1970).

256. B. D'Monte, A. Lajtha, and N. Marks, Inhibitors of cerebral exo- and endopeptidases, *J. Neurochem.* (in preparation).

257. G. F. Buletza and W. B. Quay, Tryptic activity and its inhibition by brain tissue according to brain region and time of day, *Trans. Amer. Soc. Neurochem.* **1**:29 (1970).

258. C. H. Johnson and J. R. Knowles, The binding of inhibitors to α-chymotrypsin, *Biochem. J.* **101**:56–62 (1966).

259. E. Shaw, Selective chemical modification of protein, *Physiol. Rev.* **50**:244–296 (1970).

260. B. J. Haverback, B. Dyce, H. F. Bundy, and S. K. Wirtshafler, Protein binding of pancreatic proteolytic enzymes, *J. Chem. Invest.* **41**:972–980 (1962).

261. J. Bieth, M. Pichoir, and P. Metais, The influence of α_2-macroglobin on the elastolytic and esterolytic activities of elastase, *FEBS Letters* **8**:319–321 (1970).

262. S. Nagasawa, H. Suqihara, B. H. Han, and T. Suzuki, Studies on α_2-macroglobulin in bovine plasma, *J. Biochem. (Japan)* **67**:809–819, and 821–832 (1970).

263. J. D. Geratz, α-Keto analogues of amino acids as inhibitors of α-chymotrypsin carboxy-peptidase A, and pepsin, *Arch. Biochim. Biophys.* **111**:134–141 (1965).

264. P. H. Petra and H. Neurath, The heterogeneity of bovine carboxypeptidase A, *Biochemistry* **8**:2466–2475 (1969).

265. K. K. Mäkinen, Inhibition of arylamidase and proline iminopeptidase of human whole saliva by benzonium chloride, *FEBS Letters* **2**:101–104 (1968).

266. B. L. Vallee and J. F. Riodan, Chemical approaches to the properties of active sites of enzymes, *Ann. Rev. Biochem.* **38**:733–794 (1969).

267. H. Keilova, Inhibition of cathepsin D by diazoacetyl-norleucine methyl ester, *FEBS Letters* **6**:312–314 (1970).

268. G. Harris and A. Wiseman, Changes in peptidyl-nucleotidates during development of yeast cultures, *Biochim. Biophys. Acta* **61**:395–399 (1962).

269. M. E. Marver and A. E. Greco, The hydrolysis of nucleoproteins by cathepsins from calf thymus, *J. Biol. Chem.* **181**:853–860 (1969).

270. R. H. Stellwagen, B. R. Reid, and R. D. Cole, Degradation of histones during the manipulation of isolated nuclei and desoxyribonucleoprotein, *Biochim. Biophys Acta* **155**:581–592; and Further studies on the biosynthesis of very lysine-rich histones in isolated nuclei, *Biochem. Biophys. Acta* **155**:593–602 (1968).

271. A. L. Dounce and R. Umana, The proteases of isolated nuclei, *Biochemistry* **1**:811–819 (1962).

272. M. Furlan and M. Jencijo, Protein catabolism in thymus nuclei I and II, *Biochim. Biophys. Acta* **147**:135–144, and 145–153 (1967).

273. R. Vendrely, J. Bonner, and P. Ts'o, *The Nucleotides*, Holden-Day, San Francisco (1964), p. 307.

274. J. Bartley and R. Chalkley, Further studies of a thymus nucleohistone-associated protease, *J. Biol. Chem.* **245**: 4286–4292 (1970).

275. I. Sjohölm, Oxytocinase and its possible significance in the degradation of oxytocin during pregnancy, *FEBS Letters* **4**:135–139 (1969).

276. V. Pliska, J. Rudinger, T. Dousa, and J. H. Cort, Oxytocin activity and the integrity of the disulfide bridge, *Am. J. Physiol.* **215**:916–920 (1968).

277. C. Wilson, R. F. Kibler, and R. Shapira, Analysis with trinitrobenzenesulfonic acid of peptides separated on paper, *Anal. Biochem.* **35**:371–376 (1970).

278. H. Rinderknect, P. Silverman, M. C. Geokas, and B. J. Haverback, Determination of trypsin and chymotrypsin with remazobrilliant blue-hide, *Clin. Chim. Acta* **28**:239–246 (1970).

279. B. S. Hartley, Proteolytic enzymes, *Ann. Rev. Biochem.* **29**:43–72 (1960).

280. J. Feder, A spectrophotometric assay for neutral proteinase, *Biochem. Biophys. Res. Comm.* **32**:326–332 (1968).

281. Y. L. Lin, G. E. Means, and R. E. Feeney, The action of proteolytic enzymes on N_1N-dimethyl proteins, *J. Biol. Chem.* **244**:789–793 (1969) and An assay for carboxypeptidases A and B on polypeptides from protein, *Anal. Biochem.* **32**:436–445 (1969).

282. L. P. Chao and E. R. Einstein, Localization of the active site through chemical modification of the encephalitogenic protein, *J. Biol. Chem.* **245**: 6397–6403 (1970).

283. M. I. Horowitz, T. Palmer, and G. B. Jerzy-Glass, Promotion of activation of pepsinogen by polyanions including RNA and sulfated mucosubstances, *Proc. Soc. Exp. Med. Biol.* **133**: 853–857 (1970).

284. D. Jamieson and G. E. Palade, Intracellular transport of secretory proteins in the pancreatic exocrine cell. I. Role of the peripheral elements of the Golgi complex, *J. Cell. Biol.* **34**: 577–596 (1967). II. Transport to condensing vacuoles and zymogen granules, *J. Cell. Biol.* **34**: 597–615 (1967).

285. P. N. Campbell, Functions of polyribosomes attached to membranes of animal cells, *FEBS Letters* **7**: 1–7 (1970).

286. G. Blobel and D. D. Sabatini, Controlled proteolysis of nascent polypeptides in rat liver cell fractions I and II, *J. Cell Biol.* **45**: 130–145 and 146–157 (1970).

287. O. Z. Sellinger and L. M. Nordrum, The cellular physiology of brain lysosomes, *J. Cell Biol.* **128a**: 43 (1969).

287a. N. Marks and O. Z. Sellinger, Arylamidase and acid proteinase levels in purified lysosomal preparations (unpublished observations).

288. P. Baudhuin, L'analyse morphologique quantitative des fractions subcellulaires, Thesis, University of Louvain, Belgium, 1968.

289. S. Mahadevan and A. L. Tappel, Arylamidases of rat liver and kidney, *J. Biol. Chem.* **242**: 2369–2374 (1967), and β-aspartyl glucosylamine amido hydrolase of liver and kidney, *J. Biol. Chem.* **242**: 4568–4576 (1967).

290. R. Stahn, K.-P. Maier, and K. Hannig, A new method for the preparation of rat liver lysosomes—Separation of cell organelles of rat liver by carrier-free continuous electrophoresis, *J. Cell. Biol.* **46**: 576–591 (1970).

291. I. Klatzo and J. Miguel, Observations on pinocytosis in nervous tissue, *J. Neuropathol. Exp. Neurol.* **19**: 475–487 (1960).

292. H. E. Hirsch, Enzyme levels of individual neurons in relation to lipofucsin content, *J. Histochem. Cytochem.* **18**: 268–270 (1970).

293. H. J. Helurinan and J. L. E. Fricsson, Quantitation of lysosomal enzyme changes during enforced mammary gland involution, *Exper. Cell. Res.* **60**: 419–426 (1970).

294. C. deDuve and J. W. Coffey, Digestive activity of lysosomes I. The digestion of proteins by extracts of rat liver lysosomes, *J. Biol. Chem.* **243**: 3255–3263 (1968).

295. C. I. Beck, Thesis, University of California (cited by A. L. Tappel, ref. 18).

296. S. Fowler and C. deDuve, Digestive activity of lysosomes, III. Digestion of lipids by extracts of rat liver lysosomes, *J. Biol. Chem.* **244**: 471–481 (1969).

297. M. A. Verity and A. Reith, Effect of mercurial compounds on structure-linked latency of lysosomal hydrolases, *Biochem. J.* **105**: 685–690 (1967).

298. S. F. Malewista and P. T. Bodel, The dissociation of colchicine of phagocytosis from increased consumption in human leukocytes, *J. Clin. Invest.* **46**: 786–796 (1967).

299. C. Citterio and A. Cunego, Determinazione del principio elastico nelle arterie cerebrali comuni, *G. Gerontol.* **13**: 353–356 (1965).

300. E. Harper and J. Gross, Separation of collagenase and peptidase activities of tadpole tissues in culture, *Biochim. Biophys. Acta* **198**: 286–292 (1970).

301. J. Gross, C. M. Lapiere, and M. L. Tanzer, in *Cytodifferentiation and Macromolecular Synthesis* (M. Locke, ed.), p. 175, Academic Press, New York (1963).

302. P. Davies, K. Krakauer, and G. Weissman, Subcellular distribution of neutral protease and peptidases in rabbit polymorphonuclear leucocytes, *Nature* **228**: 761–762 (1970).

303. R. H. Mitchell, M. J. Karnovsky, and M. L. Karnovsky, The distribution of some granule-associated enzymes in guinea pig polymorphonuclear leucocytes, *Biochem. J.* **116**: 207–216 (1970).

304. R. R. Kohn, Mechanism of protein loss in denervation muscle atrophy, *Amer. J. Pathol.* **45**:435–447 (1964).

305. S. G. Sereni Nevi, P. L. Rossi Ferrini, P. Paoletti, and G. Mascolti, Attivitá della parete arteriosa sul processo della coagulazione, VI. Attivitá fibrinolitica e antifibrinolitica della parete vasale, *G. Gerontol.* **13**:53–64 (1965).

306. J. E. Kirk, *Enzymes of the Arterial Wall*, Academic Press, New York (1969), pp. 311–353.

307. U. J. Lewis and W. P. Vanderlaan, A relationship between thyroid function and a naturally occurring inhibitor of hypophysial proteinases, *J. Biol. Chem.* **238**:3336–3340 (1963).

308. E. Albert, Cathepsin and phosphoprotein phosphatase in soluble fractions of mouse brain and rat, *Z. Klin. Chem. Biochem.* **6**:38–41 (1968).

309. W. Friedberg, Relation between pool size and catabolism of albumin in the mouse, *Amer. J. Physiol.* **204**:501–504 (1963).

310. J. G. Gan and H. Jeffay, Origins and metabolism of the intracellular amino acid pools in rat liver and muscle, *Biochim. Biophys. Acta* **148**:448–459 (1967).

311. R. B. Loftfield and A. Harris, Participation of free amino acids in protein synthesis, *J. Biol. Chem.* **219**:151–159 (1956).

312. R. V. Datta and N. Marks, Relationship of brain cathepsins and nutritional deficiency (unpublished observations).

313. N. Malamud, Neuropathology of phenylketonuria, *J. Neuropath. Exp. Neurol.* **25**:254–268 (1966).

314. L. Barbato, I. W. M. Barbato, and A. Hamanaha, The *in vivo* effect of high levels of phenylalanine on lipids and RNA of the developing rabbit brain, *Brain Res.* **7**:399–406 (1968).

315. H. C. Agrawal, A. H. Bone, and A. N. Davison, Effect of phenylalanine on protein synthesis in the developing rat brain, *Biochem. J.* **117**:325–331 (1970).

316. S. Roberts, Regulation of cerebral metabolism of amino acids—II. *J. Neurochem.* **10**:931–940 (1963).

317. M. O. Huttunen, Protein and ribonucleic acid metabolism in rat brain cortex slices, Dissertation 1969, Medical Faculty (University of Helsinki).

318. H. Eagle, K. A. Piez, and M. Levy, The intracellular amino acid concentrations required for protein synthesis in cultured human cells, *J. Biol. Chem.* **236**:2039–2042 (1961).

319. B. M. Hanking and S. Roberts, Stimulation of protein synthesis *in vitro* by elevated levels of amino acids, *Biochim. Biophys. Acta* **104**:427–438 (1965).

320. A. Neidle, J. Kandera, and M. Chedekel, Protein breakdown in mouse brain slices, *Trans. Amer. Soc. Neurochem.* **1**:59 (1970).

321. L. M. Greenbaum and K. Yamafuji, The *in vitro* inactivation and formation of plasma kinins by spleen cathepsins, *Brit. J. Pharmacol.* **27**:230–238 (1966), and Catheptic carboxypeptidase, *J. Biol. Chem.* **237**:1082–1085 (1962).

322. L. M. Greenbaum and J. S. Fruton, Purification and properties of beef spleen cathepsin B, *J. Biol. Chem.* **226**:173–180 (1957).

323. M. J. Fletcher and D. R. Sanadi, Turnover of rat liver mitochondria, *Biochim. Biophys. Acta* **51**:356 (1961).

324. R. J. Rossiter, The biochemistry of demyelination, in *Neurochemistry* (K. A. C. Elliott, I. H. Page, and J. H. Quastel, eds.), pp. 696–714, Charles C. Thomas, Springfield, Ill. (1955).

325. A. B. Novikoff, Lysosomes in nerve cells, in *The Neuron* (H. Hydén, ed.), pp. 319–366, Elsevier, Amsterdam (1967).

326. A. M. Dannenberg and E. L. Smith, Proteolytic enzymes of lung, *J. Biol. Chem.* **215**:45–53 (1955), and Action of proteinase I of bovine lung. Hydrolysis of the oxidized B chain of insulin, polymer formation from amino acid esters, *J. Biol. Chem.* **215**:55–65 (1955).

327. H. H. Tallan, M. E. Jones, and J. S. Fruton, On the proteolytic enzymes of animal tissues X Beef spleen cathepsin C, *J. Biol. Chem.* **194**:793–805 (1952).

328. A. Neidle, J. Kandera, and M. Chedekel, Amino acid effect and protein turnover in mouse brain slices, *Fed. Proc.* **29**:911 (1970).

329. K. C. Hooper, The distribution of hypothalamic peptidases in pregnant and non-pregnant dogs, *Biochem. J.* **90**:584–587 (1964).

330. A. C. M. Camargo and F. G. Graeff, Subcellular distributions and properties of the bradykinin system in rabbit brain homogenates, *Biochem. Pharmacol.* **18**:548–549 (1969).

331. J. F. Shaw and P. W. Ramwell, Release of substance P polypeptide from cerebral cortex, *Amer. J. Physiol.* **215**:262–269 (1968).

332. S. Berl, in *Progress in Brain Research*, Vol. 9, pp. 178–182, Elsevier, Amsterdam (1964).

333. D. W. Pumplin and W. O. McClure, A loosely bound mitochondrial protease maximally active at *p*H 5, *Trans. Amer. Soc. Neurochem.* **1**:94 (1970).

334. P. R. Carnegie, Properties, structure, and possible neuroreceptor role of the encephalitogenic protein of human brain, *Nature* **229**:25–28 (1970).

335. L. L. Ross, V. M. Andreoli, and R. M. Marchbanks, A morphological and biochemical study of subcellular fractions of the guinea pig spinal cord, *Brain Research* **25**:103–119 (1971).

336. R. W. Gray, C. Arsenis, and H. Jeffay, Neutral proteinase activity associated with the rat liver peroxismal fraction, *Biochim. Biophys. Acta* **222**:626–636 (1970).

337. A. V. Shally, Y. Baba, A. Arimura, and T. W. Redding, Evidence for peptide nature of LH- and FSH-releasing hormone, *Biochim. Biophys. Res. Comm.* **42**:50–56 (1971).

338. A. Brun and U. Brunk, Histochemical indications for lysosomal localization of heavy metals in normal rat brain and liver, *J. Histochem. Cytochem.* **18**:820–830 (1970).

339. M. L. Kornguth, L. G. Tomasi, D. L. Keyes, and S. Kornguth, The proteolytic degradation of a purified pig brain protein, *Biochim. Biophys. Acta* **229**:167–175 (1971).

340. J. L. Lospallo, K. Fehr, and M. Ziff, Degradation of immunoglobulins by intracellular proteases in the range of neutral *p*H, *Immunology* **105**:886–897 (1970); and acid proteinases, *ibid.* **105**:973–983 (1970).

341. A. Hershko and G. M. Tomkins, Studies on the degradation of tyrosine aminotransferase in hepatoma cells in culture. Influence of the composition of the medium and ATP dependence, *J. Biol. Chem.* **246**:710–714 (1971).

342. R. Kibler, S. McKneally, E. Urban, and R. Shapira, A basic protein encephalitogenic locus for rabbit and monkey, *Trans. Amer. Soc. Neurochem.* **2**:88 and 104 (1971).

343. M. L. Laskowski, B. Kassel, R. J. Peansky, and M. Laskowski, "Endopeptidases," in *Handbuch der Physiologish und pathologisch-chemischen analyse*, **VI** C, 229–304 (1966).

344. A. J. Barrett and J. T. Dingle, *Tissue Proteinases*, North-Holland, Amsterdam (1971).

Chapter 3

CEREBRAL PHOSPHOPROTEINS

R. Rodnight

Department of Biochemistry
Institute of Psychiatry
(British Postgraduate Medical Federation
University of London)
London, England

I. INTRODUCTION

Phosphorus has been found bound covalently to cerebral proteins in two ways: in a relatively stable ester linkage to the hydroxyl groups of serine or threonine; and in a labile acyl linkage involving the γ-carboxyl groups of acidic amino acids. So far the proteins that have been found in liver[1] containing N-phosphorylhistidines and N-phosphoryllysine have not been reported in brain.

This account will be mainly concerned with phosphoproteins in which the phosphorus is esterified to serine or threonine. The presence in brain and other tissues of protein-bound phosphorus in this form has been known for years, although we still know virtually nothing of its function. The acyl-phosphate bond in proteins is a recent discovery; it occurs in protein fractions rich in the Na^+ and K^+ stimulated adenosinetriphosphatase (Na, K, Mg-ATPase) enzyme complex of the cell membrane and may be intimately involved in the mechanism of ATP hydrolysis and cation transport.[2-6] Only a brief account of its properties will be given here.

For descriptive purposes it is convenient to consider these two categories of protein-bound phosphorus separately. It is important to realize, however, that we have no information on the extent to which both types of bond may be present in the same protein. That this situation may occur is suggested by observations which indicate that phosphorylserine as well as acyl phosphate groups may be involved in the action of the Na, K, Mg-ATPase enzyme complex.[7]

II. PROTEIN-BOUND PHOSPHORUS IN ESTER LINKAGE

Compared with the levels of phosphorylserine, only traces of phosphorylthreonine are found in cerebral proteins; again, there is no evidence

141

to indicate whether the two phosphoamino acids occur in the same or different protein molecules. Metabolically, however, phosphorylthreonine appears to behave as phosphorylserine and for present purposes it will be ignored.

Since no individual phosphoproteins in this category have been isolated and characterized from brain, it is necessary to consider them as a group or fraction. There is very little direct evidence as to the nature of the proteins comprising this fraction, but there is little doubt that it contains proteins of an enzymatic nature and others which lack enzymatic function. The enzymatic components have been calculated[8] to comprise only 1–2% of the total, but this may well prove an underestimate as more enzymes containing phosphorylserine residues, either at the active center or elsewhere in the protein molecule, are discovered. At the time of writing, unequivocal demonstration of a phosphorylserine residue in an enzyme of cerebral origin does not appear to have been made, though in the case of glycogen phosphorylase from brain, the kinase and phosphatase concerned with the phosphorylation and dephosphorylation of the enzyme have been described.[9]

The nonenzymatic proteins containing phosphorylserine are believed to be analogous to the classical phosphoproteins, such as the milk caseins and egg phosvitins. This is suggested by two kinds of indirect evidence. First, brain contains enzymes phosphorylating and dephosphorylating casein and phosvitin[10–12] as well as native phosphoprotein fractions lacking enzymatic activity[13,14]; these enzymes do not utilize as substrates phosphorylated enzymes, either of the phosphoglucomutase or phosphorylase types,[10,11] even in their dephospho forms. Second, the partial acid hydrolysis of cerebral proteins yields polyserylphosphate peptides[15,16] first demonstrated in acid hydrolysates of casein and phosvitin.[17] We have found that a "structural protein" fraction[18] isolated from cerebral microsomes contains phosphorylserine residues (M. Weller and R. Rodnight, unpublished observations).

A. Methods of Analysis

Quantitative analysis depends on the lability of protein-bound phosphorylserine to treatment with mild alkali at 37°C. This lability was first observed many years ago in casein and phosvitin.[19] When later it was found by Schmidt and Thannhauser[20] and Schneider[21] that the same treatment of denatured proteins prepared from mammalian tissues yielded a small amount of Pi, the procedure became accepted as giving a measure of the phosphoprotein fraction. The denatured protein has first to be washed with acid to remove soluble phosphates and then with organic solvents to remove lipids. If desired, nucleic acids can be extracted from the residue with hot TCA, but this step is not necessary and does not affect the amount of alkali-labile P in the final product.[22]

The specificity of this procedure for protein P bound to serine is now reasonably established, although some doubt will remain until a more direct method is devised for determining intact phosphorylserine residues in protein. The doubt (see also Ref. 23) arises not so much from the possible presence in the washed residue of other types of phosphoamino acid (N—P bonds, for instance, would not survive the acid washing procedure and the acyl—P bond only appears to be present after incubation with ATP under specific conditions), as from P bound to nonprotein constituents. These are particularly significant in brain where the alkali-labile P only constitutes 8–10% of the total organically bound P in the residue before, and 11–23% after, extraction of the nucleic acids; the two figures quoted are for white and gray matter respectively.[24] Most of the alkali-stable P remaining after removal of the nucleic acids can, however, be extracted with a chloroform–methanol mixture acidified with HCl; the proportion of the total P labile to alkali then approaches 80% (Table I). The alkali-stable P occurs in the phosphatidyl-peptide fraction[25] which is known to be extracted by the acidified solvent.[26]

TABLE I

Distribution of Phosphorus in Residue after Extraction with Neutral and Acidified Solvents[a]

	mμmoles/mg protein	
Solvent treatment	Total P	Alkali labile P
Chloroform–methanol (2:1 v/v)	25.0	8.1
Chloroform–methanol– conc. HCl (200:100:1 v/v)	8.5	6.3

[a] Ox brain microsomes were washed with cold and hot TCA to remove soluble phosphates and nucleic acids before treatment with solvents.

Partial acid hydrolysis of the Schmidt and Thannhauser residue yields several acidic phosphates with electrophoretic mobilities of the inositol and glycerol phosphates that arise from the hydrolysis of the phosphoinositides.[27,28] These extra phosphates are almost completely absent after extraction with acidified solvent (Fig. 1). Similar conclusions have been reached by electrophoretic analysis of the alkaline digest.[26]

It may be noted that some alkali-labile P is extracted by acidified solvents. This apparently consists of the phosphorylserine contained in the peptide moiety of the phosphatidylpeptide fraction.[26,29,30]

Due to the lability of the serine-phosphate bond in hydrolysis by acid or alkali, it has proved difficult to obtain a direct estimate of the level of phosphorylserine in proteins. The complete enzymatic hydrolysis of casein has been reported but without details of the total yield of phosphorylserine.[31]

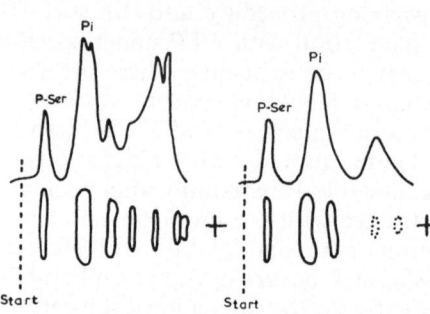

Fig. 1. Distribution of phosphate and radio-activity in protein hydrolysates of guinea pig brain before and after extraction with acidified solvent. Slices of cerebral cortex were incubated with $[^{32}P]$ H_3PO_4, precipitated, and washed with TCA, treated with solvent and the residue hydrolyzed with 2M-HCl for 6 hr at 100°C. The hydrolysates were freeze-dried and analyzed by paper electrophoresis at 2 kV, pH 1.9 for 90 min. (a) After treatment with neutral chloroform–methanol. (b) After acidified chloroform–methanol (see Table I). In each case the upper trace gives the distribution of radioactivity determined by scanning the paper.

The usual procedure is to rely on partial acid hydrolysis to release a fraction of the phosphorylserine. The optimum conditions may well vary with the source of the protein; Fig. 2 shows the time course of the release from cerebral cortex proteins on hydrolysis with 2 N HCl at 100°C. A maximum yield of 2.5 mµmoles of phosphorylserine/mg of protein was obtained after 8 hr, or 25 % of the Pi released by hydrolysis with 1 N NaOH for 18 hr at 37°C. Variations in temperature, acid strength, or hydrolysis time have not been found to improve yields.

In the experiment of Fig. 2 the phosphorylserine was isolated from the freeze-dried hydrolysates by high voltage electrophoresis and the maximum yield is considerably lower than that of 8 mµmoles/mg of protein reported earlier.[32] In the earlier work the hydrolysates were fractionated by an ion-exchange procedure[33] which has since been shown to yield a phos-

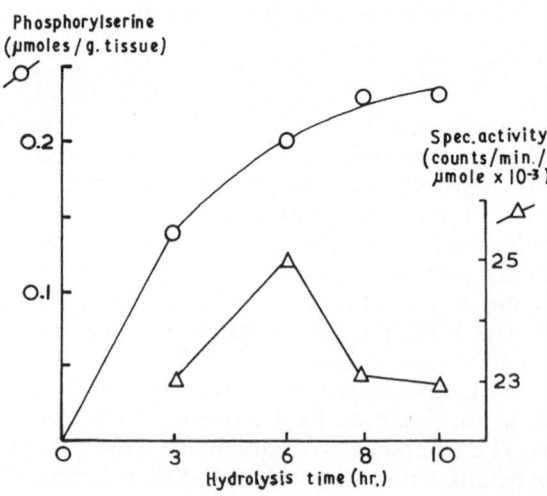

Fig. 2. Time course of release of labeled phosphorylserine during partial acid hydrolysis at 100°C with 2 N HCl of protein residue from slices of guinea pig cerebral cortex. See Fig. 1. for further details of preparation; the protein was extracted with neutral solvents.

phorylserine fraction contaminated with other phosphates (D. A. Jones and R. Rodnight, unpublished observations).

Analysis of phosphorylserine in partial acid hydrolysates from tissues labeled with ^{32}P is a particularly valuable method for studying the phosphorus metabolism of this class of proteins. In the electrophoretic procedure the phosphorylserine area can be eluted from the paper and its specific radioactivity determined by conventional methods. With adequately sensitive methods for Pi analysis, about 200 mg wet weight of gray matter are required for one determination. For some purposes a measure of total counts incorporated into the phosphorylserine fraction suffices; smaller quantities of tissue can then be used and carrier phosphorylserine added to assist location.

B. Regional and Subcellular Distributions

The levels of alkali-labile P are similar in white and gray matter from different areas of the brain and in peripheral nerve.[24] In fresh tissue this amounts to about 1 μmole P/g of tissue.

Proteins containing phosphorylserine are found in all parts of the cell.[23,43] Expressed per unit weight of protein, membrane-enriched fractions contain the highest levels and mitochondria the lowest, but the differences are not striking (Table II). In brain tissue incubated *in vitro* with H_3PO_4-^{32}P

TABLE II

Subcellular Distribution of Phosphoproteins
(as Alkali-Labile P) in Guinea Pig
Cerebral Cortex[a]

Fraction (g-min)	Phosphoprotein P	
	mμmoles/ mg tissue	mμmoles/ mg protein[b]
Original dispersion	1010	9.7 $\pm$ 0.2
Nuclear (8×10^3)	151	10.7 $\pm$ 0.3
Mitochondrial (1.5×10^5)	295	6.6 $\pm$ 1.4
Microsomal (7×10^6)	235	12.3 $\pm$ 2.2
Supernatant	266	10.1 $\pm$ 1.4

[a] From Trevor, Rodnight, and Schwartz.[23]
[b] Means in the second column are followed by the standard deviation.

labeled phosphorylserine can be isolated from proteins in all parts of the cell[32,35]; metabolic activity appears to be a general feature of the cellular phosphoproteins, suggesting that they possess a variety of functions.

C. Stability: General Features

Despite the rapid turnover of protein-bound phosphorylserine described below, the gross level of alkali-labile P in the brain is relatively stable *post-mortem*, hardly changing over a period of 6 hr or more at 0°C. After disruption of the cell structure, longer periods of storage at low temperatures result in some loss, but even after several months at $-20°$ the level remains more than 50% of the fresh value. Fixation of the brain *in vivo* by immersion in liquid N_2 does not result in a higher value for the alkali-labile P than found in brain removed after killing by conventional methods.

Few observations have been made of phosphoprotein stability in pathological states, but in Wallerian degeneration of both peripheral and central tracts the level does not change appreciably.[36,37]

D. Metabolism *in vivo*

Incorporation of P into the alkali-labile fraction is very rapid *in vivo*. In cats injected intracisternally with H_3PO_4-^{32}P the specific radioactivity of the alkali-labile P relative to that of the tissue Pi approached 1.0 after 72 hr; incorporation into the nucleic acids and phospholipids, on the other hand, was much slower.[38] Similar results have been obtained using smaller animals. In the guinea pig specific radioactivities of 0.6–0.7 relative to the acid-soluble fraction were reached 4 hr after injection of the isotope intracisternally. After intraperitoneal injection into rats, equilibration with Pi was reached after 48 hr, but the time course of the labeling suggested the presence of a fraction turning over very much faster (Fig. 3); here also the phospholipid fraction was labeled at a considerably slower rate. No differences in the rate of labeling of particulate and soluble proteins were observed nor did injection of chlorpromazine 2 hr before the isotope have any effect

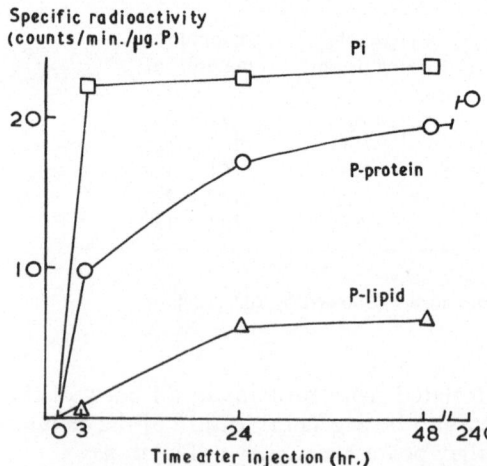

Fig. 3. Time course of labeling of phosphates in rat cerebral hemispheres after intraperitoneal injection of H_3PO_4-^{32}P. The dose was 3 mC/kg to 20-day-old rats.

TABLE III

Labeling of Cerebral Phosphoproteins and Phospholopids after Intraperitoneal Injections of $^{32}P^a$

	Chlorpromazine	Phosphoprotein P		Phospholipid P	
		μmoles/g tissue	RSA[b]	μmoles/g	RSA
Original	−	1.1, 1.1	0.7, 0.9	48, 49	0.12, 0.14
dispersion	+	0.8, 1.1	0.8, 0.8	58, 56	0.16, 0.12
Particulate	−	0.58, 0.55	0.6, 0.6		
fraction					
(3×10^5 g-min)	+	0.66, 0.66	0.7, 0.7		
Supernatant	−	0.39, 0.39	0.5, 0.7		
fraction	+	0.36, 0.39	0.7, 0.8		

[a] Rats (90 g) were injected intraperitoneally with 2.75 mC/kg of H_3PO_4-^{32}P and killed 15 hr later. Where indicated, chlorpromazine (1.1 mg/kg) was given by the same route 2 hr before the isotope. Phosphoproteins were determined as alkali-labile P.

[b] $RSA = \dfrac{\text{specific radioactivity in counts/min/}\mu\text{mole of protein or lipid P}}{\text{specific radioactivity of total acid soluble P}}$

(Table III). Convulsions induced by drugs or electrically 4 or 15 hr after intraperitoneal injection of isotope did not alter the specific radioactivity of protein-bound phosphorylserine in rat brain (unpublished observations of the author).

E. Turnover *in vitro*

1. *Cell-Containing Systems.*

Slices and other cell-containing preparations of cerebral cortex respiring in the presence of substrate and O_2 rapidly incorporate PO_4^{3-}-^{32}P into the alkali-labile fraction[39-41] at a rate several times faster than that given by the phospholipids or nucleic acids. Incorporation is energy dependent and is favored by substrates or other factors (such as the ionic composition of the medium) promoting synthesis and maintenance of ATP in the cell.[39,42] Anaerobiosis in slices of cerebral cortex for more than a few minutes results in an irreversible loss of alkali-labile P[41].

Labeling of protein-bound phosphorylserine in an *in vitro* system proceeded linearly for 2–3 hr after addition of isotope at a rate approximately one third of that of the tissue Pi and the γ-P of ATP (Fig. 4). In this experiment the ^{32}P was introduced into the system after a preliminary period of incubation in the presence of glucose and O_2 to allow time for the establishment of steady levels of energy-rich phosphates.

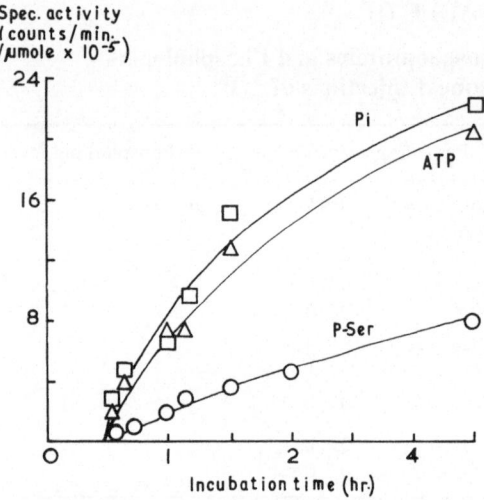

Fig. 4. Time course of uptake of H_3PO_4-^{32}P into Pi, ATP (γ-P) and phosphorylserine of respiring slices of guinea pig cerebral cortex. The slices were preincubated for 30 min before addition of isotope to allow time for resynthesis of creatine phosphate and ATP. Phosphorylserine was isolated by electrophoresis of partial and hydrolysates of the washed and extracted slice protein. Neutral solvents were used (see Fig. 1).

2. Electrically Stimulated Cell-Containing Systems

Much of the current interest in cerebral phosphoproteins originated in Heald's observation[43] that the apparent rate of labeling of the alkali-labile P in cortical slices can be increased by the brief application of electrical pulses to the tissue. Later work[32,33] confirmed that changes in the radioactivity of protein-bound phosphorylserine groups were responsible for the effect. The increase in radioactivity appears to reflect an increase in the actual turnover of protein-P rather than changes in the activity of the γ-P of ATP, the immediate precursor of protein phosphorylation. This interpretation is supported by the observation that in the first few minutes after introduction of isotope to the system and when the tissue phosphates are only lightly labeled, stimulation induces, along with the increase in phosphorylserine radioactivity, either no change or a fall in ATP-specific radioactivity. However, the actual sequence of events appears to be considerably more complex than in the scheme originally proposed[8,43]; study of the many variables involved suggests that interpretation of specific radioactivity changes on stimulation is vitiated by the existence within the tissue of different pools of ATP, which in the early stages of labeling vary in activity. Nevertheless, the fact that stimulation under these conditions does not induce parallel changes in the specific radioactivity of the nucleic acids or lipids, which are also labeled from ATP as precursor, supports the conclusion that the change reflects a specific role for a phosphoprotein in the events accompanying excitation.

Stimulation of respiring brain tissue by this technique (see Ref. 22) depolarizes the neuronal membrane and results in major changes in cation flux into and out of the cell[44,45]; to restore ionic balance and excitability, energy reserves are depleted and oxidative metabolism is stimulated. On the face of it, therefore, the increase in phosphoprotein turnover might be connected with (a) energy-yielding reactions in the cytoplasm or mitochondria, (b) energy utilization for transport processes at the membrane, or (c) in the changes in membrane permeability which initiate the excitation process; other hypothetical explanations for the response have been discussed elsewhere.[46]

One explanation can be ruled out immediately: the time course of the response is too slow for it to be involved in the initiation of the permeability changes in the membrane. A role in energy production is also unlikely in view of fractionation studies[32,35] showing that the major part of the increase in radioactivity is confined to phosphoproteins present in the microsomes and the synaptosome subfraction of the mitochondrial fraction. Mitochondrial and cytoplasmic phosphoproteins did not change on stimulation and it was thought unlikely that these would be less stable during the fractionation procedure than the phosphoproteins of the cell membrane.[32] Participation in some aspect of energy utilization at the membrane seems more likely for several reasons. Foremost is the fact that the microsomal and synaptosomal fractions are enriched in fragments of the outer cell membrane and in the Na,K,Mg-ATPase system, now generally believed to be involved in cation transport; intermediation in the action of this enzyme was therefore an attractive hypothesis to explain the response. In fact, there is scant evidence to support a direct involvement of phosphorylserine groups in transport. On the positive side there is the finding that the increased turnover is blocked by the addition of ouabain (an inhibitor of cation transport and the ATPase) to the incubation medium[32,47] or by the substitution of the Na^+ of the medium by Li^+ or choline.[47] High concentrations of di*iso*propylphosphofluoridate (DFP), which irreversibly phosphorylates serine residues in cerebral preparations, do in fact inhibit the Na,K,Mg-ATPase,[7] as do other agents reacting with serine hydroxyl groups such as tetraethylpyrophosphate,[7] methane sulfonyl chloride,[48] and diethyl-4-nitrophenylphosphate.[48] The specificity of the inhibition with DFP has been questioned,[49] but on the whole the evidence that a serine hydroxyl group is in some way involved is good. As against these supporting observations there is no firm evidence that Na^+ stimulates the incorporation of the γ-P of ATP into protein-bound serine in membrane fragments prepared from sub-cellular fractions, as might be expected if the hypothesis were fact. An earlier report[50] to this effect has not been confirmed in this laboratory despite repeated attempts (M. Weller and R. Rodnight, unpublished observations). The reasons for the discrepancy are not clear, but the negative result is more in accord with the now well-documented finding[51–53] already mentioned, that phosphate bound to the Na,K,Mg-ATPase during ATP hydrolysis is in acyl rather than ester linkage.

If a direct participation of protein-bound phosphorylserine in ATP hydrolysis by the transport enzyme now appears very unlikely, is it possible that in the intact cell phosphorylserine groups either have an indirect function or are labeled through a reaction occurring secondarily to the main hydrolytic path way? The latter suggestion might conceivably arise through a transphosphorylation from the acyl phosphate formed in the course of ATP hydrolysis. It is known that under suitable conditions phosphate in acyl linkage in denatured membrane protein can be transferred to hydroxyl groups of aliphatic alcohols instead of water[53]; in the intact tissue electrical stimulation may induce or accelerate transfer of acyl-bound phosphate to serine hydroxyl groups situated on the same or adjacent protein molecules. It is of interest in this connection that the phosphorylserine residues sensitive to stimulation apparently occur in peptides containing mainly aspartic and glutamic acids (D. A. Jones and R. Rodnight, unpublished observations).

In our present state of knowledge it is difficult to see what role a side reaction such as the above would play in membrane function. A more rational approach, therefore, is to postulate the presence in the transport enzyme of serine hydroxyl groups which regulate enzyme activity according to whether they are phosphorylated or not. The paradigms here are the enzymes metabolizing glycogen phosphorylase mentioned above and which possess their own specific kinases and phosphatases. An important feature of protein kinase activity associated with the glycogen enzymes is its stimulation by 3'5'-adenosinemonophosphate (cyclic AMP). Now there is evidence (Ref. 54 and see below) that the cell membrane contains an intrinsic protein kinase system phosphorylating serine hydroxyl groups in the membrane proteins which is cyclic AMP dependent. It is also known[55] that electrical stimulation of cortical slices under conditions that lead to the increased turnover of protein-bound phosphorylserine in the membrane also results in an accumulation of cyclic AMP in the tissue, presumably through an activation of the membrane-bound adenyl cyclase. As yet, very little is known of the nature of cyclic AMP-sensitive kinase system, except that it is tightly bound to the membrane and cannot be released by disrupting agents, features which distinguish it sharply from the kinases of glycogen metabolism.

A regulatory role for cyclic AMP in the membrane, activated by stimulation, is one possible way of explaining the stimulated increase in phosphorylserine turnover. At present, however, there is no evidence either way as to whether the Na,K,Mg-ATPase system, or part of it, is the acceptor for the cyclic AMP-sensitive kinase. Cyclic AMP, in concentrations that stimulate phosphorylation in membrane fragments, does not stimulate the Na,K, Mg-ATPase from the same source, but it must be remembered that such membrane preparations contain adenyl cyclase and therefore under normal control conditions of ATPase assay some cyclic AMP may be produced *pari passu* with ATP hydrolysis.

Possible roles for a phosphoprotein in the transport of glucose or other substrates, or of phosphate itself, which may well increase in rate on stimulation, have also been suggested,[46] but await experimental investigation.

3. *Intrinsic Protein Kinases in Cell-Free Systems*

Cell-free preparations from brain contain kinases transferring the γ-P of ATP to intrinsic serine hydroxyl groups of the tissue proteins. In fresh preparations incorporation is not accompanied by any appreciable increase in the quantity of alkali-labile P in the tissue; thus it would appear that as well as phosphorylation of vacant groups there is occurring either exchange or a rephosphorylation of groups made vacant by the action of the intrinsic protein phosphatase present in membrane (see below). The reaction has been studied by isotope techniques in acetone powders from whole brain,[14] and in untreated membrane fragments.[56] Transfer is Mg-dependent, but inhibited by other divalent cations; the pH optimum lies in the range 7–7.6. Sulfydryl reagents (e.g., p-chloromercuribenzoate) inhibit the reaction, but activity can be restored by cysteine. Univalent cations, in the concentration range 20–200 mM, are also inhibitory (Fig. 5). The mechanism of this effect is obscure: Na^+ and K^+ are equally effective and the inhibition can be demonstrated in suspensions of membrane fragments or in "solubilized" protein from the membrane. It seems possible that in some way the univalent cations modify the physical state of the protein acceptor molecule(s) or their association with enzyme. An effect on substrate rather than enzyme seems more likely in view of the finding noted below that univalent cations stimulate rather than inhibit the activity of a purified kinase enzyme using extrinsic soluble phosphoproteins as substrates.

Intrinsic protein kinase activity occurs in all parts of the cell with the highest rates in the mitochondrial and microsomal fractions. In cerebral preparations it has not been separated from ATPase activity; the time course of the transfer depends therefore on the level of ATPase activity in the preparation and the concentration of ATP in the medium. In a microsomal system containing 1 mM ATP and 0.4 mg of protein/ml, transfer at 20°C to phosphorylserine followed a curvilinear course at an initial rate of about 0.1 mμmoles P/min to a maximum at 1 hr, when most of the ATP had been

Fig. 5. Time course at 20°C of labeling of protein-bound phosphorylserine in a cerebral microsomal preparation. The reaction mixture contained 30 mM tris HCl (pH 7.4), 1 mM MgATP (labeled in the γ-position with ^{32}P), 100 mM NaCl (where indicated) and microsomal protein at a concentration of 0.4 mg/ml. The values given on the ordinate are derived from the total counts incorporated into the free phosphorylserine fraction isolated from partial acid hydrolysates of the protein residue by the procedure outlined in Fig. 1; the results therefore do not represent the total incorporation since some phosphorylserine is lost during hydrolysis.

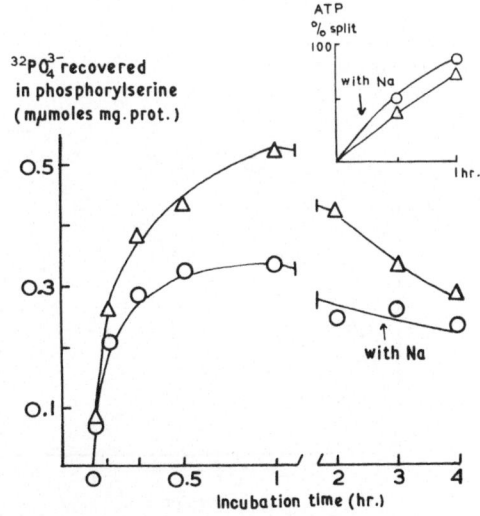

hydrolyzed; further incubation for 4 hr then resulted in a slow loss of protein-bound P (Fig. 5). The actual rate of transfer to phosphorylserine is approximately five times higher because of losses occurring during partial acid hydrolysis; thus in an earlier study the total protein-bound P transferred in 1 min equaled 0.5 mμmole/mg.[56] The reaction has not been studied in an ATP-generating system. As already indicated, there is no evidence to link protein kinase activity with the Na,K,Mg-ATPase. In fact, the observed transfer is probably the result of more than one kinase system, though little progress has yet been made in isolating proteins with acceptor function.

As was mentioned earlier, at least one of the membrane-bound intrinsic kinases is stimulated by cyclic AMP.[54] A typical time course with 10 μM cyclic AMP is shown in Fig. 6. Maximum stimulation occurs over the first minute of incubation, but subsequently the stimulated time course falls off faster than the control so that at 10 min there is little difference. This falloff is not due to hydrolysis of the cyclic AMP by the phosphodiesterase present in the membrane, because it is not affected by theophylline, an inhibitor of this enzyme, and also occurs when higher concentrations of cyclic AMP are used. It suggests that the nucleotide is activating the phosphorylation of molecules with relatively few available serine hydroxyl groups so that saturation is reached in a short time. Stimulation was first evident at 0.01 μM cyclic AMP and was maximal at 10 μM; these concentrations are similar to those reported to stimulate the phosphorylation of casein by the kinase system involved in the activation of glycogen phosphorylase.[57] However, transfer of phosphate from ATP to casein and phosvitin by the extrinsic protein kinase system in membranes, described below, is not stimulated by cyclic AMP.

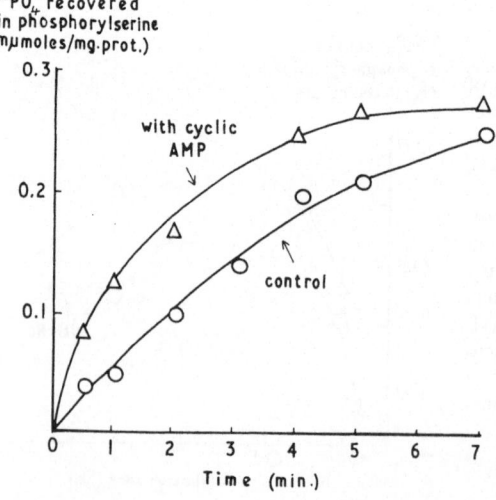

$^{32}PO_4^{3-}$ recovered in phosphorylserine (mμmoles/mg.prot.)

with cyclic AMP

control

Time (min.)

Fig. 6. Time course of the phosphorylation of a microsomal preparation in the presence or absence of 10 μM cyclic AMP at 20°C. The incubation mixture contained 30 mM Tris–HCl (pH 7.4) and 1 mM MgATP (γ-^{32}P-labeled). A similar time course was observed with a preparation of synaptosomes.

Cerebral tissues also contain a soluble protein kinase system activated by cyclic AMP.[58] This has been studied using histones, protamine, and phosvitin as acceptor. Using histone, which was the most active acceptor, 0.5 μM cyclic AMP increased phosphorylation fourfold. The relation of this enzyme to the phosphorylase kinase system of muscle described by Walsh, Perkins, and Krebs[57] is not clear, but the two have several features in common.

4. Intrinsic Protein Phosphatase Activity of Membrane Preparations

Part of the phosphate incorporated into membrane proteins by kinase action in microsome preparations is unstable. This is evident from Fig. 5. where it can be seen that after reaching a maximum at 1 hr (by which time most of the ATP had been hydrolyzed) the level of phosphorylserine ^{32}P slowly declines. Properties of the dephosphorylating reaction have been studied by labeling a membrane preparation until maximum incorporation has occurred and then separating and washing the labeled material by high-speed centrifugation at 0°C. About 75% of the counts incorporated into phosphorylserine are retained during the washing. On reincubation at 37°C in fresh medium, about one half of the remaining counts are released by enzymatic action as Pi. The dephosphorylation is accelerated by Mg^{2+} and inhibited by EDTA (Fig. 7). Acid denaturation of the membrane protein

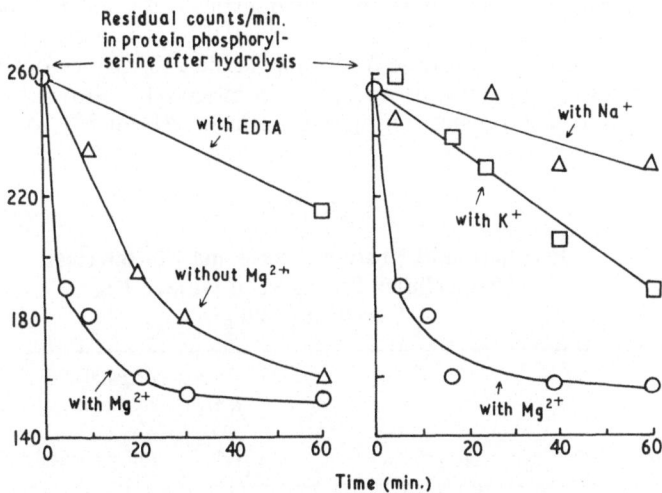

Fig. 7. Dephosphorylation of protein-bound phosphorylserine in a microsomal preparation. The microsomes were first labeled with ATP-^{32}P and then washed centrifugally. Residual protein-bound phosphorylserine was determined after partial acid hydrolysis and electrophoresis. (a) Effect of 10 mM EDTA and 1 mM $MgCl_2$; (b) Effect of 100 mM NaCl and 100 mM KCl.

stabilizes the phosphorylserine label to incubation under these conditions. The pH optimum for the dephosphorylation reaction is 7.4, a feature which distinguishes the enzyme responsible from the protein phosphatase in brain acting on casein, which has a pH optimum of 5.9.[11] The reaction is also inhibited by Na^+ and K^+ (Fig. 7b); this is an interesting feature in view of the inhibition of the intrinsic kinase system by these univalent cations already mentioned. The dephosphorylation reaction is potently inhibited by Cu^{2+} and Zn^{2+} and to some extent by SH reagents.

5. Enzymes in Brain Phosphorylating and Dephosphorylating Casein and Phosvitin

The milk caseins, egg yolk phosvitins, and polypeptides derived from the partial enzymatic hydrolysis of these proteins, can act as acceptors for a protein kinase system in brain.[10,12,14,59] The enzyme activity can be extracted in a soluble form from acetone powders and further subfractionated with ammonium sulfate.[10] The reaction is Mg^2-dependent, but inhibited by Ca^{2+}. In contrast to the situation with the intrinsic kinase system, the activity of the partially purified enzyme transferring to phosvitin is markedly stimulated by the chlorides of certain univalent cations in the range 100–200 mM.[12,14] Na^+, K^+, and Rb^+ are most effective; Li^+ on the other hand, has little action or may inhibit slightly. Choline, NH_4^+, and tris salts also stimulate to some extent.

Protein kinase activity with these properties is found in all parts of the cell. Approximately half the activity is found in the cytoplasm; the remainder occurs mainly in the membrane fractions (Table IV). Under optimum conditions and using phosvitin as acceptor, homogenates show an activity of about 0.1 μmole of P transferred/mg of protein/hr. About 75% of the activity

TABLE IV

Distribution of Protein Kinase and Phosphatase in Subcellular Fractions of Guinea Pig Cerebral Cortex

Fraction (g-min)	Percent original activity recovered	
	Kinase[a]	Phosphatase[b]
Nuclear (8–10 $\times$ 10^3)	6	12
Mitochondrial (2 $\times$ 10^5)	25	26
Microsomal (6 $\times$ 10^6)	20	3
Supernatant	47	56

[a] Data from Rodnight and Lavin.[12]
[b] Data from Rose.[60]

can be readily extracted from membrane fractions by brief exposure to high pH,[13,59] or from acetone powders by neutral salts.[14] However, even after repeated extractions and treatment with detergents, some activity remains tightly bound to the insoluble residue. Kinase activity displayed toward extrinsic phosphoproteins differs from the intrinsic system not only in respect of the effect of univalent cations, but also in that it is much less sensitive to inhibition by SH reagents.

There is no clear evidence as to the extent to which the extractable kinase activity corresponds to the activity observed in the intrinsic system, but some acceptor function for native proteins using a soluble kinase enzyme has been demonstrated. Added to a soluble protein fraction with low intrinsic phosphorylating activity[13] or to heat-denatured proteins lacking intrinsic activity,[14] the partially purified enzyme from whole gray matter stimulated incorporation into protein-bound phosphorylserine in the acceptor preparation. It seems likely that a fraction of the intrinsic activity is due to this extractable kinase system.

Rather less is known about cerebral phosphatases acting on extrinsic phosphoproteins. A Mg-dependent enzyme acting on casein and phosvitin with a pH optimum of 5.5–5.9 has been described by Rose and Heald.[11,60] With casein as substrate, the rate in cerebral dispersions was around 0.2 μmole of P hydrolyzed/mg of protein/hr, a value similar to the kinase rate with phosvitin as acceptor. The subcellular distribution of this acid protein phosphatase, however, is different from the phosvitin kinase: only 3% of the activity was found in the microsomes, 25% in the mitochondrial fraction and the rest in the nuclear fraction and cytoplasm (Table IV). The phosphatase also appears to be more labile than the kinase, the partially purified preparation losing activity in a few days, even at low temperatures.

The protein kinase and phosphatase enzymes acting sequentially on their native substrates constitute an ATPase system and presumably account for the observed turnover of protein-bound phosphorylserine in the intact tissue. Judging from the rates these enzymes display *in vitro*, the actual rate of ATP hydrolysis effected intracellularly by the phosphoprotein system would appear to be insignificant when compared with the rates of the major ATPases in brain. However, lack of knowledge of the precise rates of turnover of phosphoproteins in the intact cell makes it difficult to be sure on this point. Possibly the actual turnover is faster than suggested by studies of the intrinsic system in subcellular fractions or using casein and phosvitin as substrates.

F. Attempts to Prepare Purified Phosphoproteins from Brain

No significant purification of a phosphoprotein from brain has yet been achieved. In an early attempt[61] labeled particulate proteins were solubilized with an anionic detergent and then fractionated with solvents and neutral salts. A small enrichment of alkali-labile P was found in one fraction, but the yield was very small and the product heterogeneous. Kothary[13] (see also

Ref. 62) surveyed possible procedures for fractionating membrane-bound phosphoproteins in brain. A variety of mild methods were used for extracting a soluble fraction, which was then fractionated with solvents, neutral salts, or by ion-exchange chromatography on substituted celluloses. In all the procedures investigated, protein-bound phosphorus was found to be distributed in all fractions with no significant variation. It is difficult to avoid the conclusion that a considerable number of the proteins constituting cell membrane structures in brain are phosphorylated. In another study of soluble proteins in brain, at least three phosphoproteins were found with differing mobilities on starch-block electrophoresis.[63]

A characteristic of the classical phosphoproteins of the type represented by casein is the occurrence in their constituent peptide chains of sequences of adjacent phosphorylserine residues.[17] Such structures have not been found in the phosphoproteins that function as enzymes and their occurrence in the tissue proteins would point to, though not establish, the existence of tissue phosphoproteins of the classical type. Tentative evidence for the presence in brain particulate proteins of polyserylphosphate peptides has been reported[15,16,62] but only in amounts suggesting that the parent molecules constitute less than 0.1 % of the membrane protein.

III. PHOSPHORUS BOUND TO PROTEIN IN ACYL LINKAGE

When cerebral membrane preparations containing the Na,K,Mg-ATPase enzyme are incubated for a few seconds with low concentrations of MgATP and Na^+, phosphate bound in acyl linkage to the carboxyl groups of acidic amino acids can be detected in the protein of the membrane after denaturation with acid or detergent. Under optimum conditions, a plateau of labeled protein equal to about 100 $\mu\mu$moles/mg is reached within 1 sec of starting the reaction; this plateau is maintained until ATP hydrolysis is

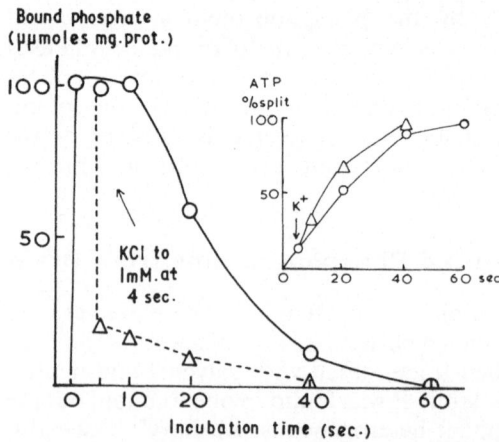

Fig. 8. Time courses at 37°C of formation of acyl-phosphate groups in a cerebral microsomal preparation. The reaction mixture contained 30 mM Tris–HCl (pH 7.4), 2.5 μM MgATP (labeled in the γ-position with ^{32}P), 50 mM NaCl, 1 mM KCl (where indicated) and microsomal protein at a concentration of 200 mg/ml. The values given on the ordinate are derived from the total counts in the acid-washed protein residue without extraction of lipids.

nearly complete, when it rapidly declines to zero (Fig. 6). The rapid formation of acyl-bound phosphate is an Na-dependent reaction: if K^+ as well as Na^+ is present in the initial reaction mixture, either less or no acyl-bound phosphate is formed; if K^+ is added after the plateau of bound phosphate is established, the latter immediately falls to a new level (Fig. 8). These and many other features of the rapid binding reaction correlate well with the notion, now widely accepted, that the Na-dependent acyl-bound phosphate represents a covalent intermediate in the mechanism of action of the Na,K, Mg-ATPase enzyme of cell membranes.[2–6] On this interpretation, the action of K^+ is one of increasing the turnover of the phosphoprotein "intermediate" and therefore reducing its steady-state level when ATP is limiting. However, alternative views[64,65] on the relation of the phosphorylation to ATPase action and the role of K^+ in discharging the bound phosphate have been advanced. Complex mechanisms involving more than one phosphorylated intermediate have also been suggested.[6,67,68]

Using crude membrane fractions, the Na-dependent transfer of phosphate becomes progressively swamped as the ATP concentration in the medium is increased by the quantitatively much greater (though initially slower) transfer of phosphate to protein-bound serine hydroxyl groups. For example, using untreated cerebral microsomes at 400 μg of protein/ml of reaction mixture, a clear demonstration of the properties of the Na-dependent reaction requires an ATP concentration of 25 μM or less.[2] On the other hand, membrane protein treated with NaI by a procedure[69] which yields a preparation with an enhanced Na,K,Mg-ATPase activity, largely loses its ability to incorporate the γ-P of ATP into phosphorylserine.[70] In this preparation, therefore, it is possible to demonstrate the formation of Na-dependent protein-bound phosphate at higher levels of ATP in the medium.[4]

The reason for this loss of intrinsic protein kinase activity toward serine hydroxyl in the NaI preparation is unknown. The product contains phosvitin kinase activity and the same level of alkali-labile P as the starting material. Activity cannot be restored by recombination of the various fractions produced in the procedure; nor is appreciable activity found in these other fractions. Either an irreversible disruption of the enzyme-acceptor complex or an inactivation of the kinase enzyme(s) seems the most likely explanation.

The evidence for the acyl nature of the Na-dependent protein-bound phosphate is derived from observations on the stability of the phosphate bound in this way after acid denaturation of the membrane protein.[51–53] Liberation of the bound phosphate as Pi from the acid-washed protein residue suspended in buffer of pH 3.5 at 40° follows first-order kinetics, indicating that the phosphate is bound as a single species.[52] The rate of release is temperature-dependent, minimal at pH 2, and greatly accelerated by hydroxylamine.[52,53] These are properties of the acyl-phosphate bond and distinguish the Na-dependent bound phosphate from other possible types of P-linkage to protein, such as the ester, $N—P$, or thiophosphate bonds. Very similar results have been obtained with labeled membrane material denatured at neutral pH with sodium dodecylsulfate (D. R. Alexander and R. Rodnight, unpublished observations).

Further evidence has come from attempts to isolate labeled fragments of the protein after proteolytic digestion of the denatured membrane preparation. This approach has proved difficult because of the lability of the acylphosphate bond, but labeled fragments rich in glutamic and aspartic acids have been isolated. In another interesting experiment,[71] labeled peptides were treated with a tritiated hydroxylamine derivative, N-(n-propyl-2,3^{3}H) hydroxylamine, which would be expected to form a stable labeled hydroxamate with the reactive group. From further digests a tritiated compound with the properties of authentic L-glutamyl–γ–propylhydroxamate was isolated. This is reasonable evidence that at least in the denatured protein the Na-dependent phosphate is bound as L-glutamyl-γ-phosphate.

IV. CONCLUSIONS

The cerebral phosphoproteins are now seen to constitute a much wider and more complex subject than appeared to be the case a few years ago when it was considered possible that most, if not all, of the phosphorus bound to the tissue proteins would be found in combination with only a few proteins. Now we know of at least four enzymes which are phosphorylated at some stage during activity and it is reasonable to infer that more will be found. More problematical is the nature and extent of the nonenzymatic fraction because here enzymatic properties are unavailable to assist in devising isolation procedures. Notable recent contributions to this aspect of the subject in noncerebral tissues have been the isolation of proteins from liver[72] and thymus nuclei[73] with phosphorus contents comparable to that of casein. Nuclear phosphoproteins have yet to be studied in brain though it is known that a protein kinase in brain activated by cyclic AMP catalyzes the phosphorylation of a histone preparation from calf thymus.[58] Possibly the concepts now being put forward suggesting roles for protein phosphorylation in the regulation of nuclear functions in noncerebral tissues[72,74,75] may also be found applicable in some respects to brain.

Much of the present review has concerned phosphorylserine proteins occurring in cell membranes from brain. This covalent binding in a labile linkage of the metabolically active P atom with proteins constituting the neuronal membrane is obviously of great interest in view of the paramount importance of the neuronal membrane in the functioning of the brain as an organ. There is no clear experimental evidence, however, as to which particular functions of the membrane are involved in phosphoprotein turnover, although there are no lack of candidates. It is reasonable to speculate, for example, that conformational changes induced by cyclic phosphorylation and dephosphorylation of membrane proteins will alter permeability and other characteristics of the membrane; such cyclic changes may also be involved in the regulation of membrane-bound enzymes. It is also possible that the free negative charges of protein-bound phosphorylserine are involved in the binding of the positively charged transmitters such as acetylcholine and the catecholamines, or of calcium and potassium. These are examples of agents that become bound on the outside of the membrane, but there is no evidence indicating whether serine hydroxyl groups in this situation can be phosphorylated by intracellular ATP. The turnover of protein-bound phosphorylserine groups accessible to the inside of the cell presents no such difficulties, but direct evidence for the earlier suggestions[76,77] that the turnover is involved in sodium transport out of the cell has not been forthcoming. However, the molecular complexity of membrane transport is little understood at present and it would be premature to dismiss altogether the possibility that the phosphorylserine proteins have a part to play.

ACKNOWLEDGMENT

Original research described in the review was supported in part by generous grants (No. 05502) from the U.S. Public Health Service.

V. REFERENCES

1. O. Walinder, O. Zetterqvist, and L. Engstrom, Purification of a bovine liver protein rapidly phosphorylated by adenosine triphosphate, *J. Biol. Chem.* **243**:2793–2798 (1968).
2. R. Rodnight, D. A. Hems, and B. E. Lavin, Phosphate binding by cerebral microsomes in relation to adenosinetriphosphatase activity, *Biochem. J.* **101**:502–515 (1966).
3. K. Nagano, N. Mizuno, M. Fujita, Y. Tashima, T. Nakao, and M. Nakao, On the possible role of the phosphorylated intermediate in the reaction mechanism of $(Na^+ - K^+)$-ATPase, *Biochim. Biophys. Acta* **143**:239–248 (1967).
4. T. Kanazawa, M. Saito, and Y. Tonomura, Properties of a phosphorylated protein as a reaction intermediate of $Na^+ - K^+$-sensitive ATPase, *J. Biochem., Tokyo* **61**:555–566 (1967).
5. J. S. Charnock and L. J. Opit, in *The Biological Basis of Medicine* (E. E. Bittar, ed.) p. 69, Academic Press, New York (1968).
6. R. L. Post, A. K. Sen, and H. Bader, in *Protides of the Biological Fluids* (H. Peeters, ed.) Vol. 15, pp. 237–240, Elsevier, Amsterdam (1968).
7. L. E. Hokin, A. Yoda, and R. Sandhu, Irreversible inhibition of ATPases, diglyceride kinase and phosvitin kinase of brain by DFP, *Biochim. Biophys Acta* **126**:100–116 (1966).
8. P. J. Heald, *Phosphorus Metabolism of Brain*, Pergamon Press, Oxford (1960).
9. G. I. Drummond, J. Keith, and M. W. Gilgan, Brain glycogen phosphorylase, *Arch. Biochem.* **105**:106–162.
10. M. Rabinowitz and F. Lipmann, Reversible phosphate transfer between yolk phosphoprotein and adenosine triphosphate, *J. Biol. Chem.* **235**:1043–1050 (1960).
11. S. P. R. Rose and P. J. Heald, A phosphoprotein phosphatase from ox brain, *Biochem. J.* **81**:339–347 (1961).
12. R. Rodnight and B. E. Lavin, Phosvitin kinase from brain; activation by ions and subcellular distribution, *Biochem. J.* **93**:84–91 (1964).
13. K. Kothary, Studies on the Nature of Phosphoproteins in Mammalian Brain, Ph.D. Thesis, London University (1964).

14. H. Matsui, E. Orikabe, S. Ishikawa, and N. Shimazono, Studies on phosphoprotein in brain. II. Reactions catalysed by protein kinase, *J. Biochem., Tokyo* **57**:131–141 (1965).

15. P. J. Heald, Studies on the phosphoproteins of brain: phosphorylserine sequences in brain phosphoprotein, *Biochem. J.* **80**:510–514 (1961).

16. C. N. Cook and R. Rodnight, Acidic peptides derived from proteins of a cerebral membrane fraction, Abstract 307 of the Fourth Meeting of the Federation of European Biochemical Societies, Oslo (1967).

17. J. Williams and F. Sanger, The grouping of serine phosphate residues in phosvitin and casein, *Biochem. Biophys. Acta* **33**:294–296 (1959).

18. K. Got, G. M. Polya, and J. B. Polya, Structural proteins of microsomal subfractions from sheep brain, *Nature (London)*, **212**:714–715 (1966).

19. R. H. A. Plimmer and W. M. Bayliss, The separation of phosphorus from caseinogen by the action of enzymes and alkali, *J. Physiol.* **33**:439–461 (1906).

20. G. Schmidt and S. J. Thannhauser, A method for the determination of desoxyribonucleic acid, ribonucleic acid and phosphoproteins in animal tissues, *J. Biol. Chem.* **161**:83–89 (1945).

21. W. C. Schneider, Phosphorus compounds in animal tissues. I. Extraction and estimation of desoxypentose nucleic acid and pentose nucleic acid, *J. Biol. Chem.* **161**:293–303 (1945).

22. H. McIlwain and R. Rodnight, *Practical Neurochemistry*, J. and A. Churchill, London (1962).

23. A. J. Trevor, R. Rodnight, and A. Schwartz, The subcellular distribution of cerebral phosphoproteins, *Biochem. J.* **95**:883–888 (1965).

24. R. J. Rossiter, in *Neurochemistry. The Chemistry of Brain and Nerve*, 2nd ed. (K. A. C. Elliott, I. H. Page, and J. H. Quastel, eds.) pp. 10–54, C. C. Thomas, Springfield, Ill. (1964).

25. F. N. LeBaron, G. Hauser, and E. E. Ruiz, Occurrence and metabolism of protein-bound phosphoinositides in several lipid-protein complexes from brain, *Biochim. Biophys. Acta* **60**:338–349 (1962).

26. M. Ledig, H. Feigenbaum, and P. Mandel, Sur la nature des phosphopeptides qui contaminent la fraction ribonucleotidique au cours dosage d'acides nucleiques selon Schmidt and Thannhauser, *Biochim. Biophys. Acta* **72**:332–334 (1963).

27. D. M. Brown and J. C. Stewart, The structure of triphosphoinositide from beef brain, *Biochim. Biophys. Acta* **125**:413–421 (1966).

28. R. M. C. Dawson and D. C. Dittmer, Evidence for the structure of brain triphosphoinositide from hydrolytic degradation studies, *Biochem. J.* **81**:540–545 (1961).

29. P. Mandel and M. Ledig, Amidated phosphopeptides in nervous tissue, *Biochem. Biophys. Res. Comm.* **24**:275–279 (1966).

30. E. Orikabe, H. Matsui, S. Ishikawa, and N. Shimazono, Studies on phosphoprotein in brain. I. Phosphorylation of bound phosphopeptide and its isolation, *J. Biochem., Tokyo* **57**:346–354 (1965).

31. R. L. Hill and W. R. Schmidt, The complete enzymic hydrolysis of proteins, *J. Biol. Chem.* **237**:389–396 (1962).

32. A. J. Trevor and R. Rodnight, The subcellular localisation of cerebral phosphoproteins sensitive to electrical stimulation, *Biochem. J.* **95**:889–896 (1965).

33. P. J. Heald, Phosphorylserine and cerebral phosphoprotein, *Biochem. J.* **68**:580–584 (1958).

34. M. Ledig and P. Mandel, Distribution of phosphatidopeptides and phosphoproteins in subcellular fractions of rat brain, *C. R. Acad. Sci. Paris* **260**:514–543 (1965).

35. P. J. Heald, Studies on the phosphoproteins of brain: the intracellular localization of a phosphoprotein involved in the metabolic response of cortical slices to electrical stimulation, *Biochem. J.* **73**:132–141 (1959).

36. J. E. Logan, W. A. Mannell, and R. J. Rossiter, Chemical studies of peripheral nerve during Wallerian degeneration, *Biochem. J.* **51**:482–487 (1952).

37. R. E. McCaman and E. Robins, Quantitative studies of Wallerian degeneration in the peripheral and central nervous system. I. *J. Neurochem.* **5**:18–31 (1959).

38. K. P. Strickland, Nucleic acids and other protein-bound phosphorus compounds of cat brain. Incorporation of P^{32} after intracisternal injection, *Canad. J. Med. Sci.* **30**:484–493 (1952).

39. M. Findlay, K. P. Strickland, and R. J. Rossiter, Incorporation of radioactive phosphate into non-nucleotide protein–bound phosphorus fractions of respiring brain slices, *Canad. J. Biochem. Physiol.* **32**:504–514 (1954).

40. N. P. Lisovskaya and N. B. Livanova, Effects of some drugs on the respiration, adenosine triphosphate content, and phosphoprotein metabolism of slices of cat cerebral cortex, *Biokhimiya* **24**:735–745 (1959).

41. A. J. Trevor, Studies on the Distribution and Metabolism of Cerebral Phosphoproteins, Ph.D. Thesis, London University (1963).

42. M. Findlay, W. L. Magee, and R. J. Rossiter, Incorporation of inorganic phosphate into lipids and pentose-nucleic acid of cat brain slices. The effect of inorganic ions, *Biochem. J.* **58**:236–243 (1954).

43. P. J. Heald, The incorporation of phosphate into cerebral phosphoprotein promoted by electrical impulses, *Biochem. J.* **66**:659–663 (1957).

44. H. McIlwain, *Chemical Exploration of the Brain; A Study of Cerebral Excitability and Ion Movement*, Elsevier, Amsterdam (1963).

45. J. C. Keesey, H. Wallgren, and H. McIlwain, The sodium, potassium and chloride of cerebral tissues: maintenance, change on stimulation and subsequent recovery, *Biochem. J.* **95**:289–300 (1965).

46. R. Rodnight, in *The Scientific Basis of Medicine Annual Reviews*, pp. 304–320, Athlone Press, University of London (1967).

47. K. Ahmed, J. D. Judah, and H. Wallgren, Phosphoproteins and ion transport of cerebral cortex slices, *Biochim. Biophys. Acta* **69**:428–430 (1963).

48. G. Sachs, E. Z. Finley, T. Tsuji, and B. I. Hirschowitz, Effect of irreversible inhibitors on transport ATPase, *Arch. Biochim. Biophys.* **134**:497–499 (1969).

49. H. Yoshida, K. Nagai, M. Kamali, and Y. Nagakawa, Irreversible inactivation of (Na^+, K^+)-dependent ATPase and K^+-dependent phosphatase by fluoride, *Biochim. Biophys. Acta* **150**:162–164 (1968).

50. K. Ahmed and J. D. Judah, Identification of active phosphoprotein in a cation-activated adenosinetriphosphatase, *Biochim. Biophys. Acta* **104**:112–120 (1965).

51. L. E. Hokin, P. S. Sastry, P. R. Galsworthy, and A. Yoda, Evidence that a phosphorylated intermediate in a brain transport adenosinetriphosphatase is an acyl phosphate, *Proc. Nat. Acad. Sci.* **54**:177–184 (1965).

52. K. Nagano, T. Kanazawa, N. Mizuno, Y. Tashima, T. Nakao, and M. Nakao, Some acyl phosphate-like properties of ^{32}P-labeled sodium-potassium-activated adenosinetri-phosphatase, *Biochim. Biophys. Res. Comm.* **19**:759–764 (1965).

53. D. A. Hems and R. Rodnight, Properties of phosphate bound to cerebral microsomes during adenosine-triphosphatase activity, *Biochem. J.* **101**:516–523 (1966).

54. M. Weller and R. Rodnight, Stimulation by cyclic 3′5′-adenosinemonophosphate of intrinsic protein kinase activity in membrane preparations from ox brain, *Nature (London)* **225**:187–188 (1970).

55. S. Kakiuchi, T. W. Rall, and H. McIlwain, The effect of electrical stimulation upon the accumulation of adenosine 3′5′-phosphate in isolated cerebral tissue, *J. Neurochem.* **16**:485–491 (1969).

56. R. Rodnight and B. E. Lavin, Enzyme transfer of phosphate from adenosine triphosphate to protein-bound serine residues in cerebral microsomes, *Biochem. J.* **101**:495–501 (1966).

57. D. A. Walsh, J. P. Perkins, and E. G. Krebs, An adenosine 3'5'-monophosphate-dependent protein kinase from rabbit skeletal muscle, *J. Biol. Chem.* **243**:3763–3765 (1968).

58. E. Miyamoto, J. F. Kuo, and P. Greengard, Adenosine 3'5'-monophosphate-dependent protein kinase from brain, *Science* **165**:63–65 (1969).

59. L. Decsi and R. Rodnight, The phosvitin kinase enzyme from cerebral microsomes, *J. Neurochem.* **12**:791–796 (1965).

60. S. P. R. Rose, The localisation of cerebral phosphoprotein phosphatase, *Biochem. J.* **83**: 614–622 (1962).

61. P. J. Heald, Studies on the phosphoproteins of brain. Partial purification of a phosphoprotein attached to subcellular particles, *Biochem. J.* **78**:340–348 (1961).

62. R. Rodnight, in *Protides of the Biological Fluids* (H. Peeters, ed.) Vol. 13, pp. 39–48, Elsevier, Amsterdam (1966).

63. N. A. Fedorova, A. I. Komkova, and T. T. Zarabailo, The phosphoproteins of brain tissue, *Tret'ya Vses. Konf. po Biokhim. Nervnoi sistemy, Sb.Dokl.* 77–80 (1962).

64. S. J. Masiak and J. W. Green, Relationship of Na$^+$, K$^+$ and Mg-ATP binding sites for the human erythrocyte stromal (Na$^+$ + K$^+$)-dependent ATPase, *Biochim. Biophys. Acta* **159**:340–344 (1968).

65. J. C. Skou, Enzymatic basis for active transport of Na$^+$ and K$^+$ across the cell membrane, *Physiol. Revs.* **45**:596–617 (1965).

66. J. C. Skou and C. Hilberg, The effect of cations, g-strophanthin and oligomycin on the labeling from (^{32}P) ATP of the (Na$^+$ + K$^+$)-activated enzyme system and the effect of cations and g-Strophanthin on the labeling from (^{32}P)-ITP and ^{32}Pi. *Biochim. Biophys. Acta* **185**:198–219 (1969).

67. S. Fahn, G. J. Koval, and R. W. Albers, Sodium-potassium-activated adenosine triphosphatase of electrophorus electric organ. 1. An associated sodium-activated transphosphorylation, *J. Biol. Chem.* **241**:1882–1889 (1966).

68. S. Fahn, G. J. Koval, and R. W. Albers, Sodium-potassium-activated adenosine triphosphatase of electrophorus electric organ. V. Phosphorylation by adenosine triphosphate-^{32}P, *J. Biol. Chem.* **243**:1993–2002 (1968).

69. T. Nakao, Y. Tashima, K. Nagano, and M. Nakao, Highly specific sodium-potassium-activated adenosine triphosphatase from various tissues of rabbit, *Biochem. Biophys. Res. Comm.* **19**:755–758 (1965).

70. M. Weller and R. Rodnight, Protein kinase activity in the cerebral membrane fraction: Effect of treatment with sodium iodide, *Biochem. J.* **111**, 16P.

71. A. Kahlenberg, P. R. Galsworthy, and L. E. Hokin, Sodium-potassium adenosinetriphosphatase. Acyl phosphate "intermediate" shown to be L-glutamyl-γ-phosphate, *Science* **157**:434–436 (1967).

72. T. A. Langan, in *Regulation of Nucleic Acid and Protein Biosynthesis* (V. V. Koningsberger and L. Bosch, eds.) **16**:233–242, Elsevier, Amsterdam (1966).

73. L. J. Kleinsmith and V. G. Allfrey, Nuclear phosphoproteins-1. Isolation and characterization of a phosphoprotein fraction from calf thymus nuclei, *Biochim. Biophys. Acta* **175**:123–235 (1969).

74. L. J. Kleinsmith, V. G. Allfrey, and A. E. Mirsky, Phosphoprotein metabolism in isolated lymphocytic nuclei, *Proc. Nat. Acad. Science* **55**:1182–1189 (1966).

75. M. G. Ord and L. A. Stocken, Variation in the phosphate content and thiol/disulphide ratio of histones during the cell cycle. Studies with regenerating rat liver and sea urchins. *Biochem. J.* **107**:403–410 (1968).

76. P. J. Heald, Phosphoprotein metabolism and ion transport in nervous tissue: a suggested consideration, *Nature (London)* **193**:451–454 (1962).
77. J. D. Judah and K. Ahmed, The biochemistry of sodium transport, *Biol. Revs.* **39**:160–193 (1964).

Chapter 4

RNA METABOLISM

Paul Mandel and Monique Jacob

Centre de Neurochimie du
Centre National de la Recherche Scientifique
Strasbourg, France

I. INTRODUCTION

The ultimate dream of the neurochemist who is dealing today with the metabolism of RNA in brain is to describe the mechanism of the regulation of the genetic code transcription and translation which leads the nerve cells to perform their specific function. Once this very fine mechanism is known, he could start the next step of finding how to modify it.

Unfortunately the goal is far from being reached. One of the difficulties is that the bulk of our knowledge about the genetic code and the way it is expressed in cells derives mainly from work dealing with bacteria. The results concerning eukaryotic cells have not yet reached the state of clarity which is nowadays that of bacterial RNA biochemistry.

In order to survive or to grow, every cell has to maintain or synthesize its proteins, and we must assume that the main characteristics of the transcription and translation of the genetic code are similar in all cells, from bacteria to brain. However, the mechanisms become more complex when evolution proceeds. The appearance of a distinct nucleus in a cell is paralleled by the necessity of new reactions, and we can assume for instance that supplementary transport mechanisms will take place. Moreover, in the most differentiated organisms, there is a specialization of functions, as proved by the existence of various tissues, which requires the introduction of regulatory processes. Our final aim, which is to know what makes the specificity of a given tissue or cell type, is still at an embryonic stage.

Another difficulty is that, in the nervous system, each cell type or even each cell should be studied separately and an adaptation of the methods of molecular biology or the elaboration of new methods seems to be necessary.

Thus when dealing with brain cells, we are confronted with two problems. We must know the main pathways of RNA biosynthesis in animal brain considered as a complex of cells and find out the regulatory mechanisms of differentiation and those determining the specificity of the different cell types.

II. THE MAIN FEATURES OF THE TRANSCRIPTION AND THE TRANSLATION OF THE GENETIC CODE IN ANIMAL CELLS

We shall first attempt to summarize what we already know about the biosynthesis of RNA in animal cells; the data that we present have been obtained by studies with brain, when they are available, or with other animal cells.

We would like first to note that the characterization of given types of RNA can be done with the aid of various methods. It must be emphasized that none of them, however, is sufficient by itself to ascertain that one is in the presence of a specific type of RNA. Thus 18S ribosomal RNA which has been isolated as a single peak in sucrose density-gradients can still contain dRNA with a similar sedimentation coefficient. Other techniques are necessary to make sure of its purity and affirm that, for instance, the hybridization properties or the ability to stimulate protein synthesis *in vitro* is truly due to 18S rRNA. The problem is even more complex with messenger RNA which represents a heterogeneous population. If the bulk of mRNA may have a base composition "close" to that of DNA, it is far from being proved that individual species all have this average composition. Moreover, all the dRNA perhaps do not have a messenger function in the cell, although they can, like many polynucleotides, stimulate *in vitro* protein synthesis and hybridize easily with DNA.

This critical comment should not be considered as negative. Its only purpose is to emphasize the complexity of the situation and the care that must be taken in interpreting experimental results.

A. The Transcription of the Genetic Code

1. *The Fraction of the Genome Coding for RNA in Brain*

All the information required during the whole life span of an organism is coded in its DNA, but evidently it is not all needed in an adult specialized tissue like brain. How much of the DNA is really active in brain?

The hybridization technique provides a tool to answer this question. Saturation of DNA with *in vivo* synthesized RNA would furnish the expected answer. Since the specific activity (SA) of the RNA must be known to determine the amount of RNA hybridized with a given amount of DNA, each type of RNA must be dealt with separately.[1] Brain ribosomal RNA prepared from polysomes is complementary to 0.15% of the DNA and the saturation is reached for low RNA/DNA input (Table I).[1] That this is not due to the hybridization of another kind of RNA or to artifacts was checked by competition experiments and by determination of base composition of the hybridized RNA. The problem is less simple when RNA of the messenger type is studied. Indeed, it is not possible to prepare it uncontaminated by other RNA, and the specific activity cannot be determined directly. It was assumed that the specific activity of the phosphorus of this RNA was that

TABLE I

Saturation of Brain DNA with Various Brain RNA

	μg RNA/100 μg DNA at saturation	
Nature of RNA	Average	Limit values
Total	1.20	1.15–1.28
Giant-size	1.23	1.04–1.33
Microsomal	0.80	0.72–0.90
Ribosomal (18S + 28S)	0.15	0.13–0.16

of the Pα of their direct precursors, the nucleoside triphosphates, when the isotopic equilibrium of the nucleotide pool was reached.*

This hypothesis proved to be true when it was checked by a direct determination of the specific activity of the hybridized RNA. In these conditions it was possible to show that 1.2% of the DNA was complementary to DNA-like RNA (dRNA) (Table I, Fig. 1).

Thus 0.15% + 1.2% = 1.35% of the DNA active in brain. If we assume that only one of the two strands is used for transcription, we find that 2.7% of the genome is expressed in brain. The number of corresponding cistrons in a diploid cell can be evaluated.† We calculated that about 6000 cistrons were coding for the two types of ribosomal RNA and 50,000 to 500,000 for dRNA. This suggests an important redundancy of the cistrons coding for rRNA. Since rRNA is synthesized from the DNA of nucleolar organizers, one can also wonder if the existence of several organizers can be postulated, a few of them only being active in a given cell. As for the cistrons coding for dRNA,

---

* Adult brain, in contrast to liver, proved to be a good material for this kind of experiment since the specific activity of the Pα of the nucleoside triphosphates showed little variation even after a long lapse of time.

† The DNA content of a rat diploid cell is 6.7×10^{-12} g, the average molecular weight of rRNA, 10^6. We thus calculate

$$\frac{6.7 \times 10^{-12} \times 0.15 \times 10^{-2} \times 6 \times 10^{23}}{10^6} \text{ moles, or } 6 \times 10^3 \text{ molecules}$$

For dRNA we assumed that mRNA were coding for proteins containing from about 100 to 1000 amino acids and found

$$\frac{6.7 \times 10^{-12} \times 1.2 \times 10^{-2} \times 6 \times 10^{23}}{10^5} = 480,000 \text{ to}$$

$$\frac{6.7 \times 10^{-12} \times 1.2 \times 10^{-2} \times 6 \times 10^{23}}{10^6} = 48,000 \text{ molecules}$$

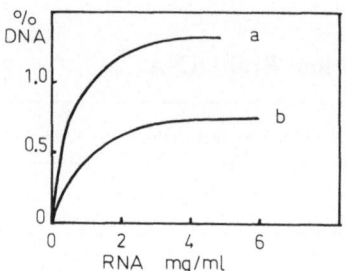

Fig. 1. Saturation of DNA by brain dRNA. Curve a: total dRNA and giant-size dRNA. Curve b: microsomal dRNA.

it must be kept in mind that all dRNA is not messenger RNA. We have demonstrated that 35 % of the dRNA species is never found in the cytoplasm. That reduces the number of cistrons coding for messenger RNA (mRNA) to 30,000–300,000. In the actual state of our knowledge it is not possible to ascertain if as many protein species are synthesized in an adult brain or if we must assume for mRNA a redundancy similar to that of rRNA. This last possibility seems highly probable[2,3] and one can even wonder whether we are able to detect the RNA species which are not synthesized from redundant cistrons. An organ specific population of RNA was detected in salmon brain[4] after a short labeling time but the interpretation of the results should take into account the above considerations.

2. *The DNA-Dependent RNA Polymerase*

RNA nucleotidyl transferase was first shown by Weiss[5] to occur in an associated form with insoluble nucleoprotein gels in rat liver nuclei. The existence of this enzyme was demonstrated later in different mammalian tissues,[6–8] in various microorganisms,[9–12] and in plants.[13]

RNA polymerase is directly responsible for the synthesis of cellular RNA and is therefore a key enzyme in the transcription of the genetic message. The sequential assembly of ribonucleoside triphosphates directed by DNA ensures the transfer of genetic information into a form that can be used for a specific protein synthesis. For this reason the mode of action of this enzyme is of considerable interest.

There has been an increasing tendency to study the development and the differentiation of animal tissue and the protein metabolism in connection with the molecular mechanism of RNA synthesis.

The purest preparations which enable studies on the specificity and mechanism of action of RNA polymerase have been isolated from bacteria. Up to now most investigations on the RNA polymerase of animal tissues have been conducted with "aggregate" enzyme preparations, that is, an insoluble gel which contains firmly bound DNA. Recent advances have been made in the purification of RNA polymerase from animal tissue.[14–17] It is most likely that purified polymerase will have properties very similar to those of enzymes isolated until now. The physical structure of the purified bacterial enzyme remains relatively unexplored. A sedimentation coefficient of 24S

$$\begin{array}{c} \text{nATP} \\ \text{nGTP} \\ \text{nCTP} \\ \text{nUTP} \end{array} \xrightarrow[\substack{\text{RNA polymerase} \\ \text{Mg}^{++} \text{ or Mn}^{++}}]{\text{DNA}} (^{+}\text{ApGpCpUp})n + 4\,\text{nPP}$$

Fig. 2.

was measured in the analytical centrifuge, and electron microscopy revealed particles apparently made of subunits. Dissociation of subunits is obtained at high ionic strength and reassociation by dialysis into buffers at low ionic strength.[18]

In brain as well as in other tissue nuclear RNA polymerase exists either in a DNA-free form, the "soluble RNA polymerase," or associated with chromatin deoxyribonucleoprotein, the "particulate" or "aggregate" RNA polymerase.

a. General Characteristics of RNA Polymerase Activity. This enzyme catalyzes the synthesis of RNA *in vitro* in the presence of the DNA template and the four nucleoside triphosphates as follows (Fig 2).

The different steps in the overall reaction of RNA synthesis are shown schematically in Fig. 3.

(*i*) *Binding.* The binding of the RNA polymerase molecules to specific sites on the DNA template is a rapid and reversible reaction. The specificity and number of binding sites depend on the DNA template. The nature of these specific bindings on certain DNA sites and their recognition by RNA polymerase are not yet known. *In vitro* RNA polymerase binds to strand breaks, free ends, or denatured single-stranded regions. The binding reaction is inhibited by competing polyanions[19] or transfer RNA[20-22] and also by proflavin which is known to intercalate in the DNA.[22,23]

There is evidence in bacteria that a genetic element besides the operator, the promotor, corresponds to the binding site, or initiation site of RNA polymerase,[24] but nothing is known in mammalian cells and brain.

(*ii*) *Initiation.* The second step is the initiation reaction in which a local strand separation of the DNA appears necessary in order to fit the nucleoside triphosphates to their complementary bases on the transcribed strand.

It has been shown that RNA chains are initiated almost exclusively by purine nucleoside triphosphates in animal cells as well as in bacteria and that the number of initiation sites per DNA molecule is limited and represents a fixed number for a given template.[25-27] The formation of the DNA enzyme complex (step 1) can be inhibited at high ionic strength, for instance

1. Enzyme + DNA $\rightleftharpoons$ Enzyme-DNA (binding)
2. Enzyme-DNA + n_1 purine NTP $\rightarrow$ Enzyme-DNA-nucleotide (initiation complex)
3. Enzyme-DNA + n NTP $\rightarrow$ Enzyme-DNA-polyribonucleotide + PP

Fig. 3.

by preincubating the complete medium with ammonium sulfate 0.3 M. When initiation occurs, a DNA–enzyme complex is found which is not as easily dissociated by high ionic strength. The exact nature of this complex is not established, but for its formation, a relatively high level of purine nucleotides is required. The effect of purine nucleotides on initiation correlates with the observations of Maitra and Hurwitz,[25] that the purine nucleoside triphosphates are the predominant 5′-terminal nucleotides found in RNA synthesized *in vitro*.[28] Sites of initiation may be determined by some specific base sequences.[29] The incorporation of nucleoside triphosphates labeled in the γ and in β-phosphate therefore provides an excellent method for determining the number of RNA chains formed and for the direct study of the processes of chain initiation.

(*iii*) *Elongation.* The elongation process proceeds owing to a mechanism which involves condensation of the α-phosphate of the incoming nucleotide with the 3′-OH of the preceding one. In this process the first nucleotide in the chain remains as a triphosphate.

(*iv*) *Termination.* Nothing is known about the termination of the RNA chains which should occur at the end of each operon; signals on the DNA probably exist.

b. *Soluble DNA-Dependent RNA Polymerase in Brain Nuclei.* The presence of a soluble RNA polymerase essentially free of DNA has been reported.[30] The characteristics are summarized in Table II. Omission of DNA, or of one of the nucleoside triphosphates, or addition of deoxyribonuclease, or ribonuclease, markedly reduces the incorporation of the labeled nucleotide in the polyribonucleotide. Omission of Mn^{2+} or Mg^{2+} results in practically no incorporation. Actinomycin D at low concentrations and inorganic pyrophosphate are very potent inhibitors.

TABLE II

Soluble RNA Polymerase Activity of Brain Nuclei

Reaction mixture	Incorporation GMP α-^{32}P $\mu\mu$moles/mg protein
Complete	49.1
Omit UTP + CTP	3.0
Omit DNA	1.0
+pyrophosphate (10 mM)	2.5
+inorganic phosphate	48.5
+ammonium sulfate 0.4 M at 0°C	2.0
+ammonium sulfate 0.4 M after 3 min pre-incubation enzyme + DNA + NTP	29.0

That the reaction product is an RNA, is suggested by phosphodiesterase stepwise degradation and by alkaline hydrolysis. After alkaline hydrolysis, all of the 2′,3′-ribonucleoside monophosphates are radioactive when one $\alpha^{32}P$ nucleoside triphosphate, ATP or GTP, is added to the incubation medium. As it was shown for the soluble DNA-dependent RNA polymerase in liver nuclei, native DNA is a more efficient primer than heat-denatured DNA. The rate at which each nucleotide is incorporated reflects the composition of the DNA used in the reaction.

The relation between this soluble enzyme and the two bound RNA polymerases separated until now[31] is not obvious. It may be suggested that the whole cellular RNA polymerase present is partially bound to DNA, partially free, until an induction of RNA biosynthesis occurs.

The transcription of DNA is asymmetric in the proper conditions *in vitro* as it is *in vivo*. Only one strand of native double-stranded DNA is transcribed.[16,32,33] Magnesium and manganese salts stimulate the reaction.

c. *Particulate RNA Polymerase in Brain Nuclei.* The activity of an RNA polymerase similar to that described by Weiss[5] in the liver, was found in brain nuclei simultaneously by Barondes,[34] Bondy and Waelsch[35] and by us.[8] This polymerase activity as usual requires the presence of the four nucleoside triphosphates, as it appears from Table III, as well as of DNA and of divalent cations: magnesium and manganese.

Goldberg[6] observed that an increase of ionic strength by addition of ammonium sulfate or other salts to the incubation medium stimulates RNA synthesis catalyzed by the animal nuclear aggregate RNA polymerase. This observation could be explained in several ways: (1) Stimulation of a latent RNA polymerase activity.[36] (2) Dissociation of histones from DNA, increasing thus the template activity.[19,37] (3) Inhibition of nucleolytic enzyme.[38,39]

TABLE III

Particulate RNA Polymerase Activity of Brain Nuclei

Reaction mixture	Incorporation GMP α-^{32}P $\mu\mu$moles/mg protein
Complete	110.0
Omit UTP + CTP	12.0
+deoxyribonuclease	13.5
+actinomycin (50 μg)	10.0
+pyrophosphate (10 mM)	11.0
+inorganic phosphate (10 mM)	112.0
+ribonuclease (5 μg)	18.0
+ammonium sulfate (0.4 M)	420.0

The first possibility can be supported by the recent finding of two types of RNA polymerases in mammalian cells and the results which showed different cation requirements and pH optima for the reactions conducted either in the presence or in the absence of ammonium sulfate.

There are many data to demonstrate that the increase of the aggregate RNA polymerase activity in a high ionic strength medium may be due to a dissociation of some of the bound histones.[19,37,40] Experiments carried out with a reconstituted DNA-RNA polymerase–histone complex strongly suggest that the effect of ammonium sulfate on the aggregate RNA polymerase enzyme is due to an increase of the template efficiency of nuclear deoxyribonucleohistone.[37]

It was also found in our laboratory that the template efficiency of rat liver chromatin increases and is equal to that of deproteinized liver DNA when the incubation is carried out in the presence of either ammonium sulfate or a strong polyanion such as polyethylene sulfonate (PES) which interacts with histones as do other polyanions.[41] The results of Pegg and Korner,[42] showing that trypsin digestion of rat liver nuclei stimulates aggregate RNA polymerase activity when assayed at low ionic strength but not at high ionic strength, also support the suggestion that the stimulating effect of ammonium sulfate is related to the removal of histones.

The question whether an RNA sensitive to ribonuclease is synthesized during the incubation of the aggregate RNA polymerase and is immediately degraded in the absence of the ammonium sulfate or PES has to be examined. Nuclear ribonuclease was described by Cunningham and Steiner[38] and preferential localization of ribonuclease in the aggregate enzyme fraction was demonstrated in brain nuclei.[43] Since the ribonucleases are strongly inhibited by ammonium sulfate or by PES, this inhibition could account for the observed stimulation. However, the stimulating effect of high ionic strength or of PES, which interacts with histones like other polyanions, still exists at 17°C when the nucleases are quite inactive. Thus, the main effect of high ionic strength is not related to an inhibitory action of nucleolytic enzymes.

One can ask what is the exact physiological meaning of the *in vitro* RNA polymerase assays. The RNA synthesized at high ionic strength either by the aggregate enzyme or by an exogenous RNA polymerase with rat liver chromatin as template has the base composition and the nearest neighbor base frequencies of DNA.[41] It was also found by Chambon and Mandel, Jr.,[44] using the hybridization method, that all RNA species which were synthesized *in vitro* and which hybridized with rat liver DNA were also present *in vivo* in rat liver nucleus. Only a fraction of this *in vitro* synthesized RNA[44] exists *in vivo* in the cytoplasm. This result suggests that the RNA synthesized *in vitro* at high ionic strength by the aggregate RNA polymerase retains most of the specificity of RNA synthesized *in vivo*.

It appears also from the results obtained by using a reconstituted DNA-RNA polymerase histone complex and from assays with chromatin that the enzymatic activities measured in the presence of ammonium sulfate are proportional to the amount of RNA polymerase molecules present in the chromatin.[19,41] Thus, it can be assumed that the activity of the nuclear aggregate enzyme measured in the presence of ammonium sulfate reflects the amount of bound enzyme. These observations may be of a great interest for the studies of organ specificity and of drug or hormone actions. Several examples can be reported in favor of this statement.[45–47] Thus, we found[8,48] as did Bondy and Waelsch[35] and Barondes,[34] that at 37°C at high ionic strength the RNA polymerase activity attached to the chromatin that is the aggregate Weiss' enzyme is higher in brain than in liver, although at a low ionic strength the activity is slightly higher (Table IV). However, the ribonuclease activity is much lower in the aggregate when brain nuclei are tested.[49]

TABLE IV

RNA Polymerase Activity of the Aggregate RNA Polymerase of Rat Liver and Brain Nuclei

	μmoles ^{32}P incorporated/mg DNA at 17°C	
	without $(NH_4)_2SO_4$	with $(NH_4)_2SO_4$
Brain	1.2	7.0
Liver	0.75	2.5

One can assume that at 17°C, when the ribonuclease activity is low, the aggregate RNA polymerase activity represents the actual template activity of the chromatin, whereas at a high ionic strength the whole potentiality of the RNA polymerase present in chromatin is determined.[50] Thus, the potential as well as the actual RNA polymerase activity in brain seems to be higher than in liver. These data are in agreement with the *in vivo* observations showing a high rate of RNA synthesis in neurons.

d. Regulation of Transcription. Different possibilities exist for the control of RNA synthesis in the living cells.

The regulation of single messenger RNA synthesis is mediated by specific repressors. Different signals on the DNA and different factors for the initiation in the cell may intervene for different types of RNA transcription. This kind of regulation may exist in differentiation processes where groups of genes must be activated and inactivated. A mechanism for regulating gene activity via transcription may also involve multiple RNA polymerases with different template specificities. Some data suggest that the enzyme responsible for ribosomal RNA synthesis is distinct from the enzyme involved in the production of the remainder of the cellular RNA. The ribosomal genes are localized in the nucleoli[51-54] whereas the extranucleolar DNA is more dispersed through the nucleoplasm.[55-57] In the presence of magnesium, at low ionic strength, nuclei synthesizes predominantly a G-C rich ribosomal like RNA[58,59]; in contrast, in the presence of manganese at high ionic strength, the nuclei synthesizes a more DNA like RNA.[58,59] At least two forms of RNA polymerase were found in rat liver[17] and thymus nuclei.[60,61] Isolated nucleoli contain predominantly polymerase 1, whereas the nucleoplasmic fraction is greatly enriched in polymerase 2. Whether these polymerases are products of separate genes or regulatory modification of a single gene product remains to be demonstrated. The existence of regulatory processes related to a higher functional activity in the central nervous system is not yet established. There are, however, some indications of an increase of RNA polymerase activity in these conditions (see Chapter III.C.1.).

A high proportion of the DNA in cerebral chromatin cannot normally support RNA synthesis because it is masked by a complex with proteins or lipoproteins. Assuming that the purity of the chromatin preparations is similar, regional differences of template activity and a decline of this activity in whole brain chromatin for RNA synthesis with increasing age were reported.[62] Conflicting results have been presented concerning the effect of Moore's S100 protein on the template activity of cerebral chromatin,[62,63] and the existence of this protein in cerebral nuclei.[64,65]

3. The Biosynthesis of the Various Types of RNA

The biosynthesis of RNA has been studied in neural tissue both *in vivo* and *in vitro*. Incorporation of a variety of precursors into RNA has been obtained with tissue slices.[66-70] The data obtained are in general in good agreement with the experiments *in vivo* showing the aptitude for uptake of different precursors. However, taking into account the swelling of brain slices and the alteration of the free nucleotide pool,[71] the results have to be examined with caution.

The experiments performed with cell suspension[72] or even with cell cultures[73,74] are more reliable.

a. The Ribosomal RNA. Ribosomal RNA is synthesized with DNA as a template in brain[1] as well as in bacteria[75-77] or animals.[53] Abundant evidence shows that the nucleolus is the site of synthesis.[52,53,78-80] In animal cells, the first molecule to be synthesized is a preribosomal RNA with a sedimentation coefficient of 45S.[81-85] This polycistronic precursor is cut down into 32S and 20S rRNA. 20S rRNA is the precursor of 18S rRNA which is directly transported to the cytoplasm. 32S rRNA is transformed into 28S rRNA and after a passage in the extranucleolar space of the nucleus goes also into the cytoplasm.

Cytoplasmic rRNA is methylated and there is evidence that the main methylation process occurs on 45S rRNA[85,86] with secondary methylation afterwards.[87] These results were not confirmed in adult rat brain,[88] but

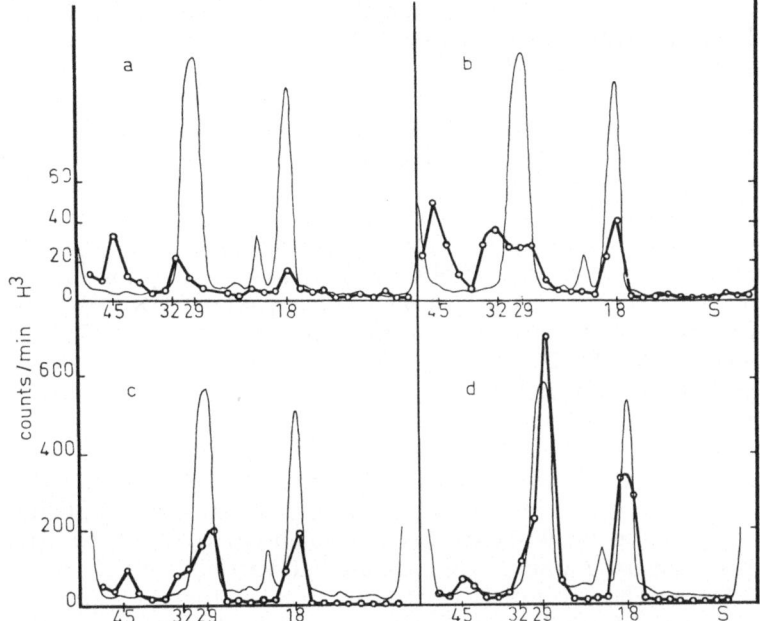

Fig. 4. Polyacrylamide gel electrophoresis of RNA from 14-day-old chick embryos. Slices were incubated *in vitro* in Eagle's medium in the presence of methyl-³H labeled methionine for (a) 30 min, (b) 1 hr, (c) 2 hr, (d) 4 hr. RNA was extracted and analyzed. ——, optical density at 260 nm; ○——○, radioactivity.

methylated 45S rRNA was found in cerebral hemispheres of chick embryo[89] (Fig. 4).

The half-life of rRNA was determined in brain; it varies between 6 and 12 days.[90,91] Values of the same order were reported for major fractions of rat brain microsomes (13–15 days).[92]

b. The DNA like RNA. DNA like RNA is often assimilated to messenger RNA. However, if the bulk of mRNA has a base composition "close" to that of DNA, it does not mean that all dRNA is messenger, functioning at the level of polysomes (for comprehensive discussion of mRNA characterization).[93,94]

The main features of dRNA biosynthesis in brain and other animal cells can be described as follows:

After injection of a radioactive precursor nuclear RNA is labeled first.[90,95–109] Most of the label is found after short pulses in RNA species sedimenting more rapidly than 28S rRNA. In brain the proportion of precursors of dRNA to those of rRNA seem to be high.[88,110–114]

The incorporation of the precursor into the cytoplasmic RNA seems to be slower. Most of the label is usually found in RNA sedimenting between 6 and 20S.[96,99,100,102,104,106,115,116]

Following these observations, several questions arise. Is there any relationship between the nuclear high molecular weight dRNA and the cytoplasmic 6–20S dRNA? Are there nuclear-restricted RNA's and if so, are they giant-sized or not?

Here again the hybridization techniques can furnish an answer. The saturation level of microsomal dRNA* was determined: 0.8% of the DNA was complementary to this dRNA as compared to 1.2% for total dRNA (Table I, Fig. 1). Therefore 0.4% of the DNA is coding for RNA species that remained inside the nucleus. This means also that only 65% of dRNA species are probable mRNA.[117,118] The existence of nuclear specific dRNA was also demonstrated in L cells[119] where cytoplasmic mRNA represented only 20% of the total dRNA species.

The nature of the nuclear-specific RNA was, however, still unknown. The giant-sized dRNA was a good candidate for this role. But it could also be a polycistronic precursor of cytoplasmic mRNA, as suggested analogically by the study of the biosynthesis of rRNA. Giant-sized dRNA was prepared by sucrose density gradient and was used to saturate the corresponding sites of DNA. The saturation level corresponded to 1.23% of the DNA (Table I, Fig. 1). Thus all the dRNA species, the nuclear as well as the cytoplasmic ones, were found in the giant-sized dRNA.[117,118] The first conclusion that can be reached from these results, and from the kinetic data, is that cytoplasmic dRNA is first found under the form of high molecular weight precursor. In order to fully explain the results, a choice must still be made between the two following possibilities: (1) giant-sized dRNA is a mixture of giant-sized nuclear-specific dRNA and of polycistronic precursors of mRNA; (2) nuclear-specific dRNA and mRNA are part of the same giant-sized molecules. Some sequences (or cistrons) would be transported to the cytoplasm, whereas others would stay in the nucleus where their role and fate would have to be determined.

The preference of the authors goes to the second possibility, but much work remains to be done to prove their point and to explain how the regulation mechanism would take place.

The kinetics of appearance of the various types of RNA in the cytoplasm can furnish some more indications concerning their metabolism; mRNA is probably synthesized and transported rapidly since it is found attached to preformed ribosomes. The 18S rRNA is released from its polycistronic precursor since it is found in cytoplasmic ribosomes before 28S rRNA.

From the above results concerning both rRNA and dRNA, a tentative outline of the synthesis of RNA in brain can be proposed. RNA is first synthesized in the nucleus under a polycistronic form (45S for rRNA, 30 to 60S for dRNA). The polycistrons are then cleaved at specific sites, with or without

---

* The specific activity was calculated according to the same principle as that of total dRNA. Labeling time was 16 hr.

intermediary steps, into their constitutive cistrons (18 and 28S rRNA, 6–20S mRNA). The site of the cleavage process is unknown yet but is probably to be found either inside the nucleus or nucleolus or at the nuclear membrane.

The presence of at least two populations of mRNA in cortex and subcortex was suggested: a major, more labile fraction with a half-life of less than 4 hr and a smaller one, stable for 10–12 hr and even for longer than 20 hr.[120] A mean half-life of 2.6 hr was estimated for the unstable template RNA in brain slices.[70,121]

B. The Transport of RNA from the Nucleus to the Cytoplasm

In organisms without any nucleus, like bacteria, the RNA synthesized on the DNA matrix are immediately taken in charge by ribosomes so that incompleted RNA chains are already part of a polysomal structure and can perhaps start protein synthesis.[122,123] Such a mechanism must probably be divided into two or more steps in animal cells where transcription and translation occur at two different cellular sites. In order to elucidate the mechanism of the transport, we must take into account that the various RNA's are usually part of subcellular structures.

1. Ribosomal RNA

45S, 18S, and 28S rRNA are not found in a free form in the cells. "Native particles" containing this RNA have been isolated from nuclear extract and from postmitochondrial supernatant of various animal cells. 45S and 32S preribosomal RNA were isolated from particles sedimenting between 60 and 110S (presomes), probably localized in the nucleolus.[124–128] The native particles carrying 18 and 28S rRNA in the cytoplasm have a sedimentation coefficient higher than that of the ribosomal subunits obtained, for instance, by EDTA action on ribosomes[115,129–135]; they are also less dense, as shown by the measurement of their buoyant density in CsCl.[132] 45S and 60S particles carry respectively 18S and 28S rRNA, with perhaps some dRNA, although it was not demonstrated yet whether this was due to a contamination by particles carrying dRNA. The existence of native ribosomal particles has been mentioned in brain.[112,130,132,136]

Although all the details of the transport mechanism are not clear at the present time, there is a high probability that the native particles are the precursors of ribosomes.

2. DNA like RNA

Particles carrying exclusively dRNA have been first isolated from sea urchin eggs and fish embryos.[137,138] They were baptized "informosomes," their authors implying thus that they were transporting information; such particles have since been isolated from the nucleus and from the cytoplasm of adult animal cells.[139,140] The nuclear particles carrying dRNA are heterogeneous in size[136,141]; in the best conditions they were found sedimenting between 30 and 120S in liver,[136,140,141] 80 and 300S in brain.[142] They are very susceptible to the action of ribonuclease, and the presence of ribonuclease inhibitor is indispensable but not always sufficient to obtain them under a high sedimentation coefficient form. When ribonuclease degradation occurs, either accidentally or on purpose, RNA is degraded to 6–10S and the particles are found at 30–50S. This suggests a polysomal-like structure made of giant-size dRNA and of proteins called "informoferes." The reality of the existence of such structures was shown by electron microscope examination.[140]

In the cytoplasm of brain, particles carrying dRNA, sedimenting between 20 and 40S with a buoyant density in CsCl of about 1.36, were found attached to the microsomes.[136] Possibly similar particles sediment also between 40 and 60S and further; it is not clear yet whether they are free or bound to the 45S native ribosomal subunit.

Particles 30–50S containing a dRNA 6–10S with a buoyant density of 1.40 were also described in the postmicrosomal supernatant of rat brain[143] and of other tissues.[134,135]

C. The Cytoplasmic RNA and the Translation of the Genetic Code

1. Polysomal RNA

The characteristics of brain polysomes or ribosomes have been described by various authors[116,144–147] and their capacity to synthesize proteins *in vivo* and *in vitro* is well known. The main features of the biosynthesis of RNA follows the same general sequence as it does in other tissues. The sedimentation profile of their RNA is also similar, suggesting that the proteins have the same average number of amino acids in brain as in other tissues. In one case, however, a high molecular weight RNA ($> 28S$) was described in brain microsomes.[112]

When animal polysomes are submitted to the action of EDTA, in conditions achieving the separation of subunits, their messenger RNA is released not under a free form but bound to proteins with a buoyant density of 1.40, close to that of nuclear and cytoplasmic particles carrying dRNA.[143,148,149] It would be of interest to know whether these "derived" particles are related to the native nuclear and cytoplasmic ones.

2. Transfer RNA

Little is known about transfer RNA in brain. The exact site of synthesis of transfer RNA in animal is not known with certainty but is probably non-nucleolar.[150] Pre-tRNA of somewhat larger size than tRNA has been described.[151–153] tRNA is probably methylated by cytoplasmic enzymes[154,155] different from those methylating RNA. Pre-tRNA is not methylated.

3. Mitochondrial RNA

The various types of RNA as well as DNA have been described in mitochondria. They are distinct from other cellular RNA and are not of nuclear origin, as shown particularly by hybridization experiments. To all appearances mitochondria possess their own transcription and translation machinery.[156] The nucleic acids of synaptosomes can be accounted for, at least in part, by mitochondrial nucleic acids.

D. Discussion

As apparent from this brief survey of our knowledge of the mechanism of transcription and translation of the genetic code, much remains to be done in order fully to understand it. Among the main problems that have to be solved and whose solution can be expected in a not too far future we shall note:

The mechanism of transformation of nuclear polycistronic RNA into cytoplasmic (monocistronic?) RNA. The general mechanism might be similar for rRNA and dRNA even if the possible sites where the cleavage occurs are different. Such study will require a more precise knowledge of the physiochemical properties of the nuclear and cytoplasmic particles carrying these RNA. If it is demonstrated that nucleases have a role to play in the cleavage, they will have to be isolated and their specificity determined.

The mechanism of selection between the polynucleotide sequences transported to the cytoplasm and those remaining within the nucleus. It would also be interesting to know the fate of the latter and if they play a specific role in the nucleus.

The mechanism of the assembly of polysomes. We ignore yet how mRNA attaches to ribosomes in animal cells, to produce polysomes. The answer to this question is fundamental if we wish to be able to realize an in vitro protein synthesis with exogenous RNA. The characterization and a study of the metabolism of the cytoplasmic particles carrying rRNA and dRNA, the knowledge of their interrelations, might bring interesting information. Moreover, we will need to know the various factors involved in the reaction.

One can imagine that the regulation of the metabolism of RNA in animal cells occurs at least at three different sites: transcription, transport, translation. Only a more precise knowledge of the reactions involved in these global processes will allow us to determine how and where they are modified in various physiological or experimental situations.

III. SOME CHARACTERISTICS OF THE REGULATION OF RNA BIOSYNTHESIS IN THE NERVOUS SYSTEM

A. The Distribution of RNA and Its Metabolism in the Various Parts of the Central Nervous System

1. Regional Distribution

The relative richness of a tissue or a cell in RNA can often be taken as an index of the cellular or functional activity. In brain the RNA concentration of a specific region depends upon the relative abundance of neurons and glial cells and can grossly reflect the importance of the protein synthesis. The comparison of the RNA concentration, of the RNA/DNA ratios, or of the rate of incorporation of labeled precursors between various regions of a brain or between the same region of the brain of various animal species leads to interesting confrontations which can be used as guides for future investigations.[157-165]

The incorporation of radioactive precursors into the RNA of subcellular fractions of several brain regions showed considerable differences between the zones examined. The results supported the existence of a metabolic autonomy of the various zones.[157-159,162,163,166]

2. The RNA of Neurons and Glial Cells

The RNA content of specific neurons and glial cells can be determined with the aid of microtechniques. The development of these techniques has permitted a fractionation of neuronal and glial RNA in several cytoplasmic and nuclear parts. The determination of their base composition and of their specific radioactivity could be carried out, and the presence of rapidly labeled RNA in neurons and glial cells could be ascertained.[167,168] A comparison between the base composition of neurons and glial cells in the hypoglossal nucleus of the rabbit showed that neurons contained RNA with a higher ratio of cytosine whereas glial RNA was richer in adenine and uracil.[169] The incorporation of precursors into adenine, guanine, and uracil is 2.5 times greater in glial than in nerve cells, and the incorporation into cytosine was similar. The suggestion was made that the synthesis of RNA is twice as active in glial RNA as in neuronal RNA at the time considered. However, the knowledge of the specific activity of the precursor pool, essential to support this assumption, is not available.

The possibilities of qualitative RNA analysis at the level of a few cells are still limited by a low amount of material; methods of fractionation of neurons and glial cells in bulk have been devised and they should offer a good material to study the regulation mechanisms of RNA metabolism with the techniques of molecular biology. An attempt in this direction showed that glial RNA was labeled more rapidly than neuronal RNA, which in turn reached a higher specific activity after longer times of labeling.[170,171] This was true for the various sizes of RNA and might be due either to variations in the labeling of the precursor pool or to a more rapid synthesis of dRNA in glial cells. On the other hand, it was shown that RNA polymerase activity was higher in neurons than in glial cells.[172] Caution in the interpretation of results is clearly necessary, particularly concerning cytoplasmic RNA. Indeed, it was shown that cytoplasmic fragments containing RNA were stripped off during the dissociation of the cells.[89]

The independent metabolism of RNA in neurons and glial cells can also be studied with isolated cells in tissue culture. Such techniques have been developed in recent years, and synthesis of RNA in isolated growing neurons has been demonstrated.[173,174] But here again, there is a limitation of the amount of material available, especially in the case of neurons which do not multiply in culture.

3. Axonal RNA

The existence of specific axonal RNA and its possible role are still controversial problems. It must be remembered that mitochondria, Schwann cells, and synaptosomes are probable contaminants of an axon preparation and that mitochondria which are present in axons and nerve endings do contain mRNA, rRNA, and tRNA. Moreover, the presence of ribosomes in dendrites close to the synaptic junction has been shown by electron microscopic examination.[175] Very low concentrations of RNA have been found in myelin-free axons of nerve cells,[176] although the total content in RNA might be quite high relative to that of the cell body. This is particularly true for very long axons.[177] The RNA of axonal origin seems to be made of ribosomal, transfer, and perhaps messenger RNA,[178–181] and in some cases it was ascertained that it did not derive from Schwann cells, perineurium, endoneurium, connective tissue, or myelin sheaths. Some of this RNA seems to be ribonuclease resistant.[182]

If specific axonal RNA exists, one can wonder whether it is synthesized *in situ* or transported from the perikaryon. Here again, the answer is not a clear-cut one.[179–181,183–185]

Does axonal RNA play a role in axonal protein synthesis? That a large part of the axonal proteins are synthesized in the nerve cell body was demonstrated by an increase of incorporation of RNA into the neurons of the hypoglossal nucleus after section of the axon[186] and by the existence of a transport of proteins along the axon.[187] On the other hand, a local synthesis of proteins and specific enzymes in the axon has been demonstrated[188,189]; an inhibition of incorporation of amino acids into proteins was obtained after actinomycin D treatment in spite of interruption of the connection with the cell body.[190]

Axonal RNA might be related to function activity. Changes in RNA concentration were observed in the Mauthner axon of the fish in relation to activity[191] and β,β-iminodipropionitrile provoked a decrease of A/U ratio of RNA from axonal balloons.[192]

B. RNA Metabolism During the Development of the Brain

The increase of size of the brain during development is caused by an increase of the number of cells and of their size and therefore of the RNA and DNA content of brain.

At the same time, morphologically undifferentiated cells are transformed into neurons and glial cells. The increase of the DNA and RNA is common to all tissues, and the specific problems of the central nervous system would be to explain how this differentiation occurs; here again we are faced with the methodological problem already stressed above.

It was reported that the RNA content of individual nerve cells increases during the development period[193,194] and the appearance of numerous ribosomes either free or membrane-bound accounts for a large part of this increase.[195] Histoautoradiographic studies show that there is an accumulation of tritiated uridine in the nuclear RNA of the differentiating chick embryo hemispheres.[196] The passage from nucleus to cytoplasm seems to be accelerated during the differentiation of neuroblasts.[196]

The variations in the metabolism of the different species of RNA, and their ability to be translated into proteins have been studied during the development of the brain. It was shown that the net synthesis of ribosomal RNA stopped at the same time as that of DNA.[197] The incorporation of radioactive precursor into nuclear RNA proceeds at a similar rate in 4-day-old and adult animals.[197] A decrease in the incorporation of precursors into RNA was observed by other authors.[72] The half-life time of the most labile component of nuclear RNA was estimated to be 20–30 min.[198] The ability of nuclear RNA to enhance protein synthesis is much greater in young animals.[199] Heavy polysomes are more abundant in newborn rats than in adults. The ability of ribosomes to incorporate phenylalanine into proteins decreases with age.[200] Modifications in the ratio of transfer RNA to ribosomal RNA were observed during the development of the rat [201] and were accompanied by qualitative changes of transfer RNA and amino acid–transfer RNA ligases.[201,202] The ligase activity slightly increases during the development.[155]

The effect of the nerve growth factor on the metabolism of RNA was studied. It seems to accelerate the synthesis of RNA in spinal ganglia without any apparent difference, either in the sedimentation profile of RNA or in the base composition of individual neuroblasts.[74,203] The RNA content of the neuroblasts increases.[203] On the other hand, corticosterone provokes a decrease of the RNA/DNA ratio of brain cells from growing mice,[204] and neonatal thyroidectomy provokes also a decrease of this ratio in rats.[205]

In an attempt to study the possible differences in the synthesis of RNA during the transformation of morphologically undifferentiated cells into neuroblasts and spongioblasts, we have examined the kinetics of incorporation of radioactive precursors into dRNA and rRNA from chick embryo brains at different ages. In spite of a considerable increase of the net synthesis of RNA and important morphological changes, the rate of incorporation of the precursors into the various types of RNA remained the same. The results suggest a very strong coupling of the synthesis of dRNA and rRNA during ontogenesis. No specific pattern could be demonstrated, although the probability that different species of messenger RNA are synthesized is high.[85]

C. Relation between RNA Biosynthesis and the Specific Function of the Nervous System

Different approaches can be used to ascertain whether or not RNA plays a role in the specific function of the nervous system. The conflicting experiments, where the consequences of an ingestion or an injection of RNA or of brain extracts supposed to contain RNA are examined, will not be considered here[206-222]; the fate of this RNA in the organism is not known and it is difficult to express the results of these controversial experiments on a molecular basis. We shall mainly refer to two types of experiments: the consequences of a functional stimulation of the nervous system on the biosynthesis of RNA; the consequences of an inhibition of the biosynthesis of RNA on the function of the nervous system.

1. Effect of the Functional Activity of the Brain on RNA Biosynthesis

The effects of a stimulation were studied with two main methods. On one hand the variations of the incorporation of a radioactive precursor into the RNA of a region or of total brain were examined upon stimulation. Thus the totality of neurons and glial cells was analyzed. On the other hand, individual cells which were known to be related to the functional stimulus were analyzed for their RNA content or base composition.

a. Sensory Stimulation. A vestibular stimulation provoked an increase of the RNA content of specifically localized Purkinje cells, of other neurons from the cerebellum, and of the Deiters' nerve cells.[223,224,225] Small variations of base composition of RNA were observed in cerebellum neurons.[223] Changes in the RNA concentration of Deiters' nerve cells were detected after gravitational changes.[226]

Different kinds of olfactory stimulation also resulted in an increase in brain nuclear RNA and a change of its base composition.[227] These are, however, differences in this respect with the various odorants. Some of them do not change the base ratio of the nuclear RNA to any extent.[228]

The role of light on the biosynthesis of RNA of visual pathways was studied during ontogenesis.[229-232] The visual stimuli were necessary for a quantitatively normal RNA biosynthesis. A retardation in the evolution of this biosynthesis was observed when animals were brought up in total darkness. A unilateral section of the optic tract in monkeys resulted in a rapid increase of the RNA content of the lateral geniculate nucleus in the first days after denervation.[223] The RNA increase was accompanied by an increase in the incorporation of amino acids into proteins. Polysomes of brain and not of liver decreased in darkness and increased in light.[234] In adult mammals and primates, it is difficult to obtain a synchronized activated or inhibited state of all neurons of the visual cortex. However, using a rhythmically flickering light, a lower incorporation of ^{32}P or uridine ^{3}H into RNA of the occipital cortex was observed in comparison with monkeys kept in darkness.[235] Acoustic stimulation increased the RNA content of rabbit cochlear ganglion cell.[236]

b. Electrical and Chemical Stimulation. Moderate stimulations for a short time may increase the concentration of RNA in nervous tissue. Prolonged or very intense stimulations may lead to a reversible decrease.[236]

Electroconvulsive treatment of rats produces a disaggregation of brain polysomes, suggesting a modification of nucleic acid metabolism.[237] In electrically stimulated brain cortex slices, the rate of RNA synthesis decreased but there was no change in the rate of breakdown.[238] On the other hand, generation and conduction of spike potentials over a period of several hours did not alter the amount of total RNA of the nerve cell body. There were, however, some changes in the A/U ratio of the cellular RNA.[239]

An increase in the cytoplasmic and nuclear content of RNA in the ganglion cells of the supraoptic nucleus after stimulation with sodium chloride was observed.[240] A variation of the K^+ ion concentration in medium where brain slices were incubated[241] or a local application of K^+ ions caused a variation in the biosynthesis of RNA[242,243]

Although an increase of the turnover of the different RNA fractions of mouse brain was observed under audiogenic seizures, metrazol convulsions caused a fall in brain RNA in rats,[244,245] and topical application of metrazol of neuronally "isolated" cerebral cortex slabs also caused a depletion of RNA. Decrease of the RNA content of isolated Purkinje cells of rats was observed after administration of large doses of insulin to the animals.[246] Amphetamine was shown to increase the incorporation of labeled precursors into free nucleotides and into RNA without any net synthesis of RNA.[247,248] It must also be noted that stress provoked an increased incorporation of precursors into brain RNA.[249] This might be of importance for the interpretation of the results of other experiments.

c. RNA in Learning Processes and in Memory. It was reported that learning experiments were accompanied in specific nerve cells by a net synthesis of RNA with a base composition different from that of the bulk cellular RNA.[250–253] An increased incorporation of precursors into brain RNA was also observed after short-term training experiences.[237,254–256] Training experiments provoked a qualitative modification of the transfer RNA pool during development.[201] A unique RNA species was detected by hybridization experiments after a shock avoidance task.[257] However, due to the difficulties in interpretation, the results of such experiments must still be taken with caution.

Magnesium pemoline, a mild stimulant of the central nervous system, was shown to stimulate RNA polymerase *in vitro* and *in vivo* and to enhance the incorporation of uridine-^{3}H into RNA.[258,259] However, completely contradictory results have been obtained by other authors.[260–263] The effect of magnesium pemoline is to facilitate the acquisition of training rather than its retention.[261]

2. Effect of an Inhibition of RNA Synthesis on the Function of the Brain

The transcription and translation of the genetic code can be blocked at several specific, more or less well-known sites. This, in turn, can impair the functioning of the cell. Since our main concern here is the inhibition of

RNA synthesis, we shall not consider the action of drugs like puromycin or cycloheximidine which are known to inhibit primarily the translation process and therefore the synthesis of proteins, although they may have a secondary action on the production of RNA. Actinomycin D, which inhibits the transcription of DNA by RNA polymerase, and antimetabolites like 8-azaguanine, which lead to the synthesis of azaguanine-containing, nonfunctional RNA, have been used. It might be relevant to remember that actinomycin D is more actively inhibitory of rRNA than of dRNA biosynthesis.

In conditions where most of the RNA synthesis was inhibited in mouse brain by actinomycin D, the animals appeared to learn and remember as well as controls in short-term experiments.[264,265] In goldfish brain, however, it was demonstrated that intact RNA synthesis was required during long-term memory formation.[266] Injury of the neurons was observed after injection of the drug.[267]

The injection of 8-azaguanine is followed by some modifications in behavior.[268,269] The interpretation of the experimental results is difficult since these drugs impair the general synthetic machinery of the cell.

The participation of RNA in reflectoral activity of the mouse is suggested by experiments where intracerebral injection of ribonuclease inhibited this activity.[270]

D. Discussion

The experiments that we reported were done according to three main approaches. We shall discuss their possibilities and limitations.

The study of total brain or of specific zones of the brain permits the use of the techniques of molecular biology. However, the results concern a mixture of cells of different types, and the interpretation of these results is therefore rendered difficult. One does not know, for instance, whether variations must be attributed to one type of cells or to all types of cells, or whether they are in the same direction for all. Moreover, important variations in a few specific cells may be masked when a large number of cells is examined.

The difficulties encountered above can be partially overcome, while the advantages of working with a large number of cells are kept, when a bulk fractionation of brain cells is used. This can be particularly useful when one expects to study modifications affecting a large group of cells, as in differentiation.

When one works with isolated cells, the limitations are of another kind. It is advantageous to be able to choose very definite cells, but the techniques do not permit yet the qualitative analysis of RNA that one wishes to obtain. Even with the micromethods now available, it is not possible to work with single cells unless they are particularly large, and when working with a high number of cells, it is difficult to draw any conclusion if the variations only take place at the level of a few cells. Indeed, looking for a variation in one cell among a million or in one cell among a hundred may furnish variations which are, in both cases, within experimental error, although changes have actually occurred.

These possible pitfalls in the interpretation of the results being kept in mind, interesting conclusions can nevertheless be drawn from the available experiments. The nature of the RNA and their metabolism vary from one cell type to the other. If this could be expected from the morphological and functional differences, it would, however, be of interest to learn more about it. For instance, the RNA of neurons which are rich in Nissl bodies, i.e., in ribosomes, can be expected to contain more RNA of the ribosomal type than the glial cells. In addition, what we need to know is the regulation mechanism which makes cells with the same genetic material become either a neuron or a glial cell. Here again we are confronted with the necessity of studying the regulation processes of the transcription and translation of the genetic code. The study of differentiation in the central nervous system can shed some light on this point of view. But few data that can be considered as specific to brain are available yet.

The stimulation of the nervous system seems to have similar consequences in all types of receptor cells when the metabolism of their RNA is studied. Indeed, a general increase of the RNA content of the receptor cells, often accompanied by a change in base composition, was observed. The incorporation of radioactive precursors into the RNA is also enhanced. The increase has been shown to be reversible when the effect of stimulation disappears. These results can be interpreted as being due to a functional temporary hypertrophy of the receptor cells. In order to synthesize the cellular components required for the transmission of the stimulus, the nerve cell must synthesize more enzymes than the resting cells. Thus it will need to increase its synthetic machinery, and what we finally observe is an increase of RNA synthesis. It seems reasonable to believe that the quantitative variations of RNA that we observe are due to species differences in the various cells. Unfortunately the available techniques even in the field of molecular biology do not yet allow distinction to be made between the different species of messenger RNA.

A temporary hypertrophy of specific cells upon the influence of specific stimuli would account for most of the reported facts. Nevertheless, care must still be taken in interpretation, since it has been demonstrated that stress can cause an increase in brain RNA metabolism[249] and it is often difficult to determine what exactly is the consequence of stress and what of specific stimuli. If our generalization is true, the detection of hypertrophied cells or of cells with an increased RNA content could help to find receptor cells of given stimuli. Many questions arise from the possibility that a stimulus provokes a cellular hypertrophy. We can, for instance, assume that a physical stimulus like electroconvulsive shock or light causes modifications of the microenvironment or enhancement of a chemical reaction or of a chain or reactions that in turn would switch on the production of new RNA molecules. The final result would be an increased protein synthesis and a hyperfunction of the cells. In fact, this is typical of a regulation mechanism that would have to be elucidated.

IV. REFERENCES

1. J. Stevenin, J. Samec, M. Jacob, and P. Mandel, Détermination de la fraction du génome codant pour les RNA ribosomiques et messagers dans le cerveau du rat adulte, *J. Mol. Biol.* **33**:777–793 (1968).
2. R. J. Britten and D. E. Kohne, Repeated sequences in DNA, *Science* **161**:529–540 (1968).
3. J. O. Bishop, The effect of genetic complexity on the time-course of ribonucleic acid-deoxyribonucleic acid hybridization, *Biochem. J.* **113**:805–811 (1969).
4. S. Mizuno, S. Tano, and S. Shirahata, Detection of an organ-specific population of RNA in the salmon brain, *J. Biochem.* **66**:119–121 (1969).
5. S. B. Weiss, Enzymatic incorporation of ribonucleoside triphosphates into the inter-polynucleotide linkages of ribonucleic acid, *Proc. Natl. Acad. Sci.* **46**:1020–1030 (1960).
6. I. H. Goldberg, Ribonucleic acid synthesis in nuclear extracts of mammalian cells grown in suspension culture; effect of ionic strength and surface-active agents, *Biochim. Biophys. Acta* **51**:201–204 (1961).
7. R. M. S. Smellie, in *Nucleic Acid Research* (J. N. Davidson and W. E. Cohn, eds.), Vol. 1, pp. 27–58, Academic Press, New York (1963).
8. N. Pete, M. Wintzerith, L. Mandel, and P. Mandel, Activité de l'acide ribonucléique (RNA) polymérase de divers tissus du rat et de deux hépatomes, C. R. Ac. Sc. Groupe 13 **258**:5283–5286 (1964).
9. J. Hurwitz, A. Bresler, and R. Diringer, The enzymic incorporation of ribonucleotides and polyribonucleotides and the effect of DNA, *Biochem. Biophys. Res. Commun.* **3**:15–19 (1960).
10. A. Stevens, Incorporation of the adenine ribonucleotide into RNA by cell fractions from *E. coli* B, *Biochem. Biophys. Res. Commun.* **3**:92–96 (1960).
11. S. Ochoa, D. P. Burma, H. Kroeger, and J. D. Weill, Deoxyribonucleic acid-dependent incorporation of nucleotides from nucleoside triphosphates into ribonucleic acid, *Proc. Natl. Acad. Sci.* **47**:670–679 (1961).
12. S. B. Weiss and T. Nakamoto, On the participation of DNA in RNA biosynthesis, *Proc. Natl. Acad. Sci.* **47**:694–697 (1961).
13. R. C. Huang, N. Maheshwari, and J. Bonner, Enzymatic synthesis of RNA, *Biochem. Biophys. Res. Commun.* **3**:689–694 (1961)
14. P. Ballard and H. G. Williams-Ashman, Isolation of a soluble RNA polymerase from rat testis, *Nature* **203**:150–151 (1964).
15. J. J. Furth and P. Ho, The enzymatic synthesis of ribonucleic acid in animal tissue. I. The deoxyribonucleic acid-directed synthesis of ribonucleic acid as catalyzed by an enzyme obtained from bovine lympho sarcoma tissue, *J. Biol. Chem.* **240**:2602–2606 (1965).
16. M. Ramuz, J. Doly, P. Mandel, and P. Chambon, A soluble DNA-dependent RNA polymerase in nuclei of nondividing animal cells, *Biochem. Biophys. Res. Commun.* **19**:114–120 (1965).
17. R. G. Roeder and W. J. Rutter, Multiple forms of DNA-dependent RNA polymerase in eukaryotic organisms, *Nature* **224**:234–237 (1969).
18. J. P. Richardson, Some physical properties of RNA polymerase, *Proc. Natl. Acad. Sci.* **55**:1616–1623 (1966).
19. P. Chambon, M. Ramuz, P. Mandel and J. Doly, The influence of ionic strength and a polyanion on transcription *in vitro*. I. Stimulation of the aggregate RNA polymerase from rat liver nuclei, *Biochim. Biophys. Acta* **157**:504–519 (1968).
20. A. Tissieres, S. Bourgeois, and F. Gros, Inhibition of RNA polymerase by RNA, *J. Mol. Biol.* **7**:100–103 (1963).
21. P. Mandel, Quelques aspects de la régulation des biosynthèses des acides ribonucléiques chez les animaux supérieurs, *Bull. Soc. Chim. Biol.* **46**:43–70 (1964).

22. J. P. Richardson, The binding of RNA polymerase to DNA, *J. Mol. Biol.* **21**:83–114 (1966).

23. O. W. Jones and P. Berg, Studies on the binding of RNA polymerase to polynucleotides, *J. Mol. Biol.* **22**:199–209 (1966).

24. J. Shapiro, L. Machattie, L. Eron, G. Ihler, K. Ippens, and J. Beckwith, Isolation of pure lac operon DNA, *Nature* **224**:768–774 (1969).

25. U. Maitra and J. Hurwitz, The role of DNA in RNA synthesis. IX. Nucleoside triphosphate termini in RNA polymerase products, *Proc. Natl. Acad. Sci.* **54**:815–822 (1965).

26. H. Bremer, M. W. Konrad, K. Gaines, and G. S. Stent, Direction of chain growth in enzymic RNA synthesis, *J. Mol. Biol.* **13**:540–553 (1965).

27. D. D. Anthony, C. W. Wu, and D. A. Goldthwait, Studies with the ribonucleic acid polymerase. II. Kinetic aspects of initiation and polymerization, *Biochemistry* (USA) **8**:246–256 (1969).

28. D. D. Anthony, E. Zeszotek, and D. A. Goldthwait, Initiation by the DNA-dependent RNA polymerase, *Proc. Natl. Acad. Sci.* **56**:1026–1033 (1966).

29. L. Hirschbein, J. M. Dubert, and C. Babinet, Affinité différentielle de la RNA polymérase pour divers polyribonucléotides synthétiques, *Europ. J. Biochem.* **1**:135–140 (1967).

30. P. Mandel, in *The Future of the Brain Sciences* (S. Bogoch, ed.) pp. 197–215, Plenum Press, New York (1969).

31. C. C. Widnell and J. R. Tata, Similarities and differences in the stimulation of nuclear ribonucleic acid synthesis by thyroid hormones, growth hormones and testosterone, *Biochem. J.* **93**:2P–3P (1964).

32. M. Hayashi, M. N. Hayashi, and S. Spiegelman, DNA circularity and the mechanism of strand selection in the generation of genetic messages, *Proc. Natl. Acad. Sci.* **51**:351–359 (1964).

33. M. H. Green, Strand selective transcription of T_4 DNA *in vitro*, *Proc. Natl. Acad. Sci.* **52**:1388–1395 (1964).

34. S. H. Barondes, Studies with an RNA polymerase from brain, *J. Neurochem.* **11**:663–669 (1964).

35. S. C. Bondy and H. Waelsch, RNA polymerase in the central nervous system, *Life Sci.* **3**:633–636 (1964).

36. C. C. Widnell and J. R. Tata, Evidence for two DNA-dependent RNA polymerase activities in isolated rat liver nuclei, *Biochim. Biophys. Acta* **87**:531–533 (1964).

37. P. Chambon, M. Ramuz, and J. Doly, Relation between soluble DNA-dependent RNA polymerase and "aggregate" RNA polymerase, *Biochem. Biophys. Res. Commun.* **21**:156–161 (1965).

38. D. D. Cunningham and D. F. Steiner, A novel ribonuclease in isolated rat liver nuclei, *Fed. Proc.* **25**:788 (1966).

39. D. F. Steiner and J. King, Insulin-stimulated ribonucleic acid synthesis and RNA polymerase activity in alloxan-diabetic rat liver, *Biochim. Biophys. Acta* **119**:510–516 (1966).

40. K. Marushige and J. Bonner, Template properties of liver chromatin, *J. Mol. Biol.* **15**:160–174 (1966).

41. P. Chambon, H. Karon, M. Ramuz, and P. Mandel, The influence of ionic strength and a polyanion on transcription *in vitro*. II. Effects on the template efficiency of rat liver chromatin for a purified bacterial RNA polymerase, *Biochim. Biophys. Acta* **157**:520–531 (1968).

42. A. E. Pegg and A. Korner, The effect of trypsin digestion and ionic strength on RNA polymerase of rat liver, *Arch. Biochem. Biophys.* **118**:362–366 (1967).

43. D. Munoz, F. Petek, and P. Mandel, Unpublished results.

44. J. L. Mandel Jr. and P. Chambon, Characterization of RNA synthesized at high ionic strength by rat liver aggregate RNA polymerase, *Biochem. Biophys. Res. Commun.* **35**:868–874 (1969).

45. For references, see *Mechanisms of Hormone Action* (P. Karlson, ed.), 275 pp., Georg Thieme Verlag, Stuttgart and Academic Press, New York (1965).

46. For references, see *Wirkungsmechanismen der Hormone*, 18. Colloquium der Gesell. für Physiol. Chemie, Mosbach, Springer-Verlag, Berlin, 257 pp. (1967).

47. For references, see *Advances in Enzyme Regulation* (G. Weber, ed.) Vol. 6, Pergamon Press, Oxford (1967).

48. D. Munoz and P. Mandel, Etude comparée de l'activité RNA polymérasique de divers tissus du Rat, *C. R. Soc. Biol.* **162**:2283–2286 (1968).

49. P. Mandel and D. Munoz, Unpublished data.

50. J. Doly, M. Ramuz, P. Mandel, and P. Chambon, Soluble DNA-dependent RNA polymerase from prostate nuclei, *Life Sci.* **4**:1961–1966 (1965).

51. R. P. Perry, The cellular sites of synthesis of ribosomal and 4 S RNA, *Proc. Natl. Acad. Sci.* **48**:2179–2186 (1962).

52. E. H. McConkey and J. W. Hopkins, The relationship of the nucleolus to the synthesis of ribosomal RNA in HeLa cells, *Biochemistry* **51**:1197–1204 (1964).

53. F. M. Ritossa and S. Spiegelman, Localization of DNA complementarity to ribosomal RNA in the nucleolus organizer region of *Drosophila melanogaster*, *Proc. Natl. Acad. Sci.* **53**:737–745 (1965).

54. H. Wallace and M. L. Birnstiel, Ribosomal cistrons and the nucleolar organizer, *Biochim. Biophys. Acta* **114**:296–310 (1966).

55. D. D. Brown and E. Littna, Variations in the synthesis of stable RNA's during oogenesis and development of *Xenopus laevis*, *J. Mol. Biol.* **8**:688–695 (1964).

56. D. D. Brown and E. Littna, Synthesis and accumulation of DNA-like RNA during embryogenesis of *Xenopus laevis*, *J. Mol. Biol.* **20**:81–94 (1966).

57. J. R. Tata, in *Progress in Nucleic Acid Research and Molecular Biology* (J. N. Davidson and W. E. Cohn, eds.) Vol. 5, pp. 191–250, Academic Press, New York (1966).

58. C. C. Widnell and J. R. Tata, Studies on the stimulation by ammonium sulphate of the DNA-dependent RNA polymerase of isolated rat-liver nuclei, *Biochim. Biophys. Acta* **123**:478–492 (1966).

59. A. O. Pogo, V. C. Littau, V. G. Allfrey, and A. E. Mirsky, Modification of ribonucleic acid synthesis in nuclei isolated from normal and regenerating liver: some effects on salt and specific divalent cations, *Proc. Natl. Acad. Sci.* **57**:743–750 (1967).

60. C. Kedinger, M. Gniazdowski, J. L. Mandel Jr., F. Gissinger, and P. Chambon, α-Amanitin: a specific inhibitor of one of two DNA-dependent RNA polymerase activities from calf thymus, *Biochem. Biophys. Res. Commun.* **38**:165–171 (1970).

61. M. Gniazdowski, J. L. Mandel Jr., F. Gissinger, C. Kedinger, and P. Chambon, Calf thymus RNA polymerases exhibit template specificity, *Biochem. Biophys. Res. Commun.* **38**:1033–1040 (1970).

62. S. C. Bondy and S. Roberts, Developmental and regional variations in ribonucleic acid synthesis on cerebral chromatin, *Biochem. J.* **115**:341–349 (1969).

63. G. R. Dutton and H. R. Mahler, *In vitro* RNA synthesis by intact rat brain nuclei, *J. Neurochem.* **15**:765–780 (1968).

64. H. Hydén and B. McEwen, A glial protein specific for the nervous system, *Proc. Natl. Acad. Sci.* **55**:354–358 (1966).

65. A. R. Dravid and J. A. Burdman, Acidic proteins in rat brain nuclei: disc electrophoresis, *J. Neurochem.* **15**:25–30 (1968).

66. H. A. Deluca, R. J. Rossiter, and K. P. Strickland, Incorporation of radioactive phosphate into the nucleic acids of brain slices, *Biochem. J.* **55**:193–200 (1953).

67. M. Findlay, W. L. Magee, and R. J. Rossiter, Incorporation of radioactive phosphate into lipids and pentosenucleic acid of cat brain slices. The effect of inorganic ions, *Biochem. J.* **58**:236–243 (1954).

68. P. J. Heald, The incorporation of phosphate into cerebral phosphoprotein promoted by electrical impulses, *Biochem. J.* **66**:659–663 (1957).

69. D. F. Cain, Biosynthesis of RNA in slices of guinea pig cerebral cortex, *J. Neurochem.* **14**:1175–1185 (1967).

70. F. Orrego and F. Lipmann, Protein synthesis in brain slices. Effects of electrical stimulation and acidic amino acids, *J. Biol. Chem.* **242**:665–671 (1967).

71. M. Ramuz, C. Judes, and P. Mandel, The distribution of free nucleotides in brain slices incubated *in vitro*, *J. Neurochem.* **11**:826–827 (1964).

72. T. C. Johnson, The effects of maturation on *in vitro* RNA synthesis by mouse brain cells, *J. Neurochem.* **14**:1075–1081 (1967).

73. G. Toschi, D. Attardi Gandini, and P. U. Angeletti, Effect of a specific neuronal growth factor on RNA metabolism by sensory ganglia from chick embryo, *Biochem. Biophys. Res. Commun.* **16**:111–115 (1964).

74. G. Toschi, E. Dore, P. U. Angeletti, R. Levi-Montalcini, and Ch. De Haën, Characteristics of labeled RNA from spinal ganglia of chick embryo and the action of a specific growth factor (NGF), *J. Neurochem.* **13**:539–544 (1966).

75. S. A. Yankofsky and S. Spiegelman, Distinct cistrons for the two ribosomal RNA components, *Proc. Natl. Acad. Sci.* **49**:538–544 (1963).

76. M. Oishi and N. Sueoka, Location of genetic loci of ribosomal RNA on *Bacillus subtilis* chromosome, *Proc. Natl. Acad. Sci.* **54**:483–491 (1965).

77. I. Smith and D. Dubnau, Chromosomal location of DNA base sequences complementary to transfer RNA and to 5 S, 16 S and 23 S ribosomal RNA in *Bacillus subtilis*, *J. Mol. Biol.* **33**:123–140 (1968).

78. J. E. Edström, Composition of ribonucleic acid from various parts of Spider oocytes, *J. Biophys. Biochem. Cytol.* **8**:47–51 (1961).

79. R. P. Perry, A. Hell, and M. Errera, The role of the nucleolus in ribonucleic acid and protein synthesis. I. Incorporation of cytidine into normal and nucleolar inactivated HeLa cells, *Biochim. Biophys. Acta* **49**:47–57 (1961).

80. D. D. Brown and J. B. Gurdon, Absence of ribosomal RNA synthesis in the anucleolate mutant of *Xenopus leavis*, *Proc. Natl. Acad. Sci.* **51**:139–146 (1964).

81. K. Scherrer, H. Latham, and J. E. Darnell, Demonstration of an unstable RNA and of a precursor to ribosomal RNA in HeLa cells, *Proc. Natl. Acad. Sci.* **49**:240–248 (1963).

82. H. Greenberg and S. Penman, Methylation and processing of ribosomal RNA in HeLa cells, *J. Mol. Biol.* **21**:527–535 (1966).

83. S. Penman, RNA metabolism in the HeLa cell nucleus, *J. Mol. Biol.* **17**:117–130 (1966).

84. R. A. Weinberg, U. Loening, M. Willems, and S. Penman, Acrylamide gel electrophoresis of HeLa cell nucleolar RNA, *Proc. Natl. Acad. Sci.* **58**:1088–1095 (1967).

85. E. F. Zimmerman and B. W. Holler, Methylation of 45 S ribosomal RNA precursor in HeLa cells, *J. Mol. Biol.* **23**:149–161 (1967).

86. E. K. Wagner, S. Penman, and V. M. Ingram, Methylation patterns of HeLa cell ribosomal RNA and its nucleolar precursors, *J. Mol. Biol.* **29**:371–387 (1967).

87. E. F. Zimmerman, Secondary methylation of ribosomal ribonucleic acid in HeLa cells, *Biochemistry* **7**:3156–3163 (1968).

88. G. R. Dutton, A. T. Campagnoni, H. R. Mahler, and W. J. Moore, Studies on labeling patterns of RNA from cerebral cortex nuclei, *J. Neurochem.* **16**:989–997 (1969).

89. C. Judes and M. Jacob, Unpublished data.

90. S. C. Bondy, The ribonucleic acid metabolism of the brain, *J. Neurochem.* **13**:955–959 (1966).

91. D. M. Dawson, Turnover of ribosomal RNA in the rat brain, *J. Neurochem.* **14**:939–946 (1967).

92. A. A. Khan and J. E. Wilson, Studies of turnover in mammalian subcellular particles: brain nuclei, mitochondria and microsomes, *J. Neurochem.* **12**:81–86 (1965).

93. M. F. Singer and P. Leder, Messenger RNA: an evaluation, *Ann. Rev. Biochem.* Part I, **35**:195–230 (1966).

94. M. Jacob, J. Stevenin, and J. Samec, in *Macromolecules and the Function of the Neuron* (Z. Lodin and S. P. R. Rose, eds.) pp. 171–181, Excerpta Medica Foundation, Amsterdam (1968).

95. R. L. Hancok, Uridine incorporation into pyramidal nuclei of the mouse brain, *Experientia* **21**:152–153 (1965).

96. W. K. Roberts, Studies on RNA synthesis in Ehrlich ascites cells extraction and properties of labeled RNA, *Biochim. Biophys. Acta* **108**:474–487 (1965).

97. K. Scherrer and L. Marcaud, Remarques sur les RNA messagers polycistroniques dans les cellules animales, *Bull. Soc. Chim. Biol.* **47**:1697–1713 (1965).

98. M. Yoshikawa-Fukada, T. Fukada, and Y. Kawade, Characterization of rapidly labeled ribonucleic acid of animal cells in culture, *Biochim. Biophys. Acta* **103**:383–398 (1965).

99. G. Attardi, H. Parnas, M. L. H. Hwang, and B. Attardi, Giant-size rapidly labeled nuclear ribonucleic acid and cytoplasmic messenger ribonucleic acid in immature duck erythrocytes, *J. Mol. Biol.* **20**:145–182 (1966).

100. J. F. Houssais and G. Attardi, High molecular weight non ribosomal type nuclear RNA and cytoplasmic messenger RNA in HeLa cells, *Proc. Natl. Acad. Sci.* **56**:616–623 (1966).

101. M. Jacob, J. Stevenin, R. Jund, C. Judes, and P. Mandel, Rapidly-labeled ribonucleic acids in brain, *J. Neurochem.* **13**:619–628 (1966).

102. J. Kempf and P. Mandel, RNA de haut poids moléculaire, rapidement marqué des sarcomes plasmocytaires de la souris, *Bull. Soc. Chim. Biol.* **48**:211–224 (1966).

103. K. L. Scherrer, L. Marcaud, F. Zajdela, I. M. London, and F. Gros, Patterns of RNA metabolism in a differentiated cell; a rapidly labeled, unstable 60 S RNA with messenger properties in duck erythroblasts, *Proc. Natl. Acad. Sci.* **56**:1571–1578 (1966).

104. K. Scherrer, L. Marcaud, F. Zajdela, B. Breckenridge, and F. Gros, Etude des RNA nucléaires et cytoplasmiques à marquage rapide dans les cellules érythropoiétiques aviaires différenciées, *Bull. Soc. Chim. Biol.* **48**:1037–1075 (1966).

105. R. Soeiro, H. C. Birnboim, and J. E. Darnell, Rapidly labeled HeLa cell nuclear RNA II. Base composition and cellular localization of a heterogeneous RNA fraction, *J. Mol. Biol.* **19**:363–372 (1966).

106. J. Warner, R. Soeiro, H. C. Birnboim, M. Girard, and J. E. Darnell, Rapidly labeled HeLa cell nuclear RNA. I. Identification by zone sedimentation of a heterogeneous fraction separate from ribosomal precursor RNA, *J. Mol. Biol.* **19**:349–361 (1966).

107. M. Yoshikawa-Fukada, The nature of the two rapidly labeled ribonucleic acid components of animal cells in culture, *Biochim. Biophys. Acta* **123**:91–101 (1966).

108. T. Balazs and W. A. Cocks, RNA metabolism in subcellular fractions of brain tissue, *J. Neurochem.* **14**:1035–1055 (1967).

109. J. B. Lingrel, Studies on the rapidly labeled RNA of rabbit bone marrow cells, *Biochim. Biophys. Acta* **142**:75–88 (1967).

110. E. Egyhazi and H. Hydén, Biosynthesis of rapidly labeled RNA in brain cells, *Life Sci.* **5**:1215–1223 (1966).

111. S. C. Bondy and S. Roberts, Messenger ribonucleic acid of cerebral nuclei, *Biochem. J.* **105**:1111–1118 (1967).

112. C. Vesco and A. Giuditta, Pattern of RNA synthesis in rabbit brain, *Biochim. Biophys. Acta* **142**:385–402 (1967).

113. L. Casola and B. W. Agranoff, Studies on RNA in goldfish brain. I. Isolation and *in vivo* labeling, *Brain Res.* **10**:227–238 (1968).

114. Z. S. Tencheva and A. A. Hadjiolov, Characterization of rat brain ribonucleic acids by agar gel electrophoresis, *J. Neurochem.* **16**:769–776 (1969).

115. M. Girard, H. Latham, S. Penman, and J. E. Darnell, Entrance of newly formed messenger RNA and ribosomes into HeLa cell cytoplasm, *J. Mol. Biol.* **11**:187–201 (1965).

116. M. Jacob, J. Samec, J. Stevenin, J. P. Garel, and P. Mandel, Polysomes and polysomal RNA of rat brain, *J. Neurochem.* **14**:169–178 (1967).

117. M. Jacob, J. Stevenin, and P. Mandel, Signification et destin des RNA géants du cerveau de rat, *C. R. Ac. Sc.*, *Série D*, **266**:1675–1676 (1968).

118. J. Stevenin, P. Mandel, and M. Jacob, Relationship between nuclear giant-size *d*RNA and microsomal *d*RNA of rat brain, *Proc. Natl. Acad. Sci.* **62**:490–497 (1969).

119. R. W. Shearer and B. J. McCarthy, Evidence for ribonucleic acid molecules restricted to the cell nucleus, *Biochemistry* **6**:283–289 (1967).

120. S. H. Appel, Turnover of brain messenger RNA, *Nature* **213**:1253–1254 (1967).

121. U. N. Singh, Nature of newly synthesized RNA in developing rat brain, *Canad. J. Biochem. Physiol.* **43**:1083–1090 (1965).

122. O. W. Jones, M. Dieckmann, and P. Berg, Ribosome induced dissociation of RNA from a RNA polymerase DNA-RNA complex, *J. Mol. Biol.* **31**:177–189 (1968).

123. M. Revel, M. Herzberg, A. Becarevic, and F. Gros, Role of a protein factor in the functional binding of ribosomes to natural messenger RNA, *J. Mol. Biol.* **33**:231–249 (1968).

124. T. Tamaoki and G. C. Mueller, A rapidly labeled ribonucleoprotein in HeLa cell nuclei, *Biochim. Biophys. Acta* **108**:81–94 (1965).

125. T. Tamaoki, The particulate fraction containing 45 S RNA in L cell nuclei, *J. Mol. Biol.* **15**:624–639 (1966).

126. J. R. Warner and R. Soeiro, Nascent ribosomes from HeLa cells, *Proc. Natl. Acad. Sci.* **58**:1984–1990 (1967).

127. M. Yoshikawa-Fukada, The intermediate state of ribosome formation in animal cells in culture, *Biochim. Biophys. Acta* **145**:651–663 (1967).

128. M. C. Liau and R. P. Perry, Ribosome precursor particles in nucleoli, *J. Cell Biol.* **42**:272–283 (1969).

129. E. C. Henshaw, M. Revel, and H. H. Hiatt, A cytoplasmic particle bearing messenger ribonucleic acid in rat liver, *J. Mol. Biol.* **14**:241–256 (1965).

130. E. H. McConkey and J. W. Hopkins, Subribosomal particles and the transport of messenger RNA in HeLa cells, *J. Mol. Biol.* **14**:257–270 (1965).

131. L. M. Hogan and A. Korner, Ribosomal subunits of Landschütz ascites tumor cells, *Biochem. J.* **100**:74P (1966).

132. R. P. Perry and D. E. Kelley, Buoyant densities of cytoplasmic ribonucleoprotein particles of mammalian cells: distinctive character of ribosome subunits and the rapidly labeled components, *J. Mol. Biol.* **16**:255–268 (1966).

133. H. Ristow and K. Köhler, Polysomal precursors in KB cells, *Biochim. Biophys. Acta* **123**:265–273 (1966).

134. P. Mandel and J. Kempf, Stimulation of protein synthesis by isolated RNA-protein particles, *Life Sci.* **8**:165–172 (1969).

135. C. Quirin-Stricker and P. Mandel, Particules ribonucléoprotéiques légères dans le cytoplasme de foie de rat, *FEBS Letters* **2**:230–232 (1969).

136. J. Samec, M. Jacob, and P. Mandel, Occurrence of light particles carrying DNA-like RNA in the microsomal fraction of adult rat brain, *Biochim. Biophys. Acta* **161**:377–385 (1968).

137. N. V. Belitsina, M. A. Aitkhozhin, L. P. Gavrilova, and A. S. Spirin, Messenger RNA of differentiating animal cells, *Biochemistry* (Engl. Transl.) **29**:315–324 (1964).

138. A. S. Spirin and M. Nemer, Messenger RNA in early sea-urchin embryos: cytoplasmic particles, *Science* **150**:214–217 (1965).

139. F. C. Kafatos, Cytoplasmic particles carrying rapidly labeled RNA in developing insect epidermis, *Proc. Natl. Acad. Sci.* **59**:1251–1258 (1968).

140. O. P. Samarina, E. M. Lukanidin, J. Molnar, and G. P. Georgiev, Structural organization of nuclear complexes containing DNA-like RNA, *J. Mol. Biol.* **33**:251–263 (1968).

141. O. P. Samarina, E. M. Lukanidin, and G. P. Georgiev, On the structural organization of the nuclear complexes containing messenger RNA, *Biochim. Biophys. Acta* **142**:561–563 (1967).

142. J. Stevenin, P. Mandel, and M. Jacob, Forme particulaire du *d*RNA géant dans les noyaux de cerveau de rat, *Bull. Soc. Chim. Biol.* **52**:703–720 (1970).

143. J. Samec, Etude des particules ribonucléoprotéiques cytoplasmiques du cerveau de rat. Caractérisation de particules contenant du RNA messager, Thèse d'Ingénieur-Docteur, pp. 77, Strasbourg (1969).

144. R. Ekholm and H. Hydén, Polysomes from microdissected fresh neurons, *J. Ultrastruct. Res.* **13**:269–280 (1965).

145. M. K. Campbell, H. R. Mahler, W. J. Moore, and S. Tewari, Protein synthesis systems from rat brain, *Biochemistry* **5**:1174–1184 (1966).

146. Y. Takahashi, K. Mase, and H. Sugano, Preparation of polysomes from rat brain tissue, *Biochim. Biophys. Acta* **119**:627–629 (1966).

147. C. E. Zomzely, S. Roberts, D. M. Brown, and C. Provost, Cerebral protein synthesis. I. Physical properties of cerebral ribosomes and polyribosomes, *J. Mol. Biol.* **20**:455–468 (1966).

148. E. C. Henshaw, Messenger RNA in rat liver polyribosomes: evidence that it exists as ribonucleoprotein particles, *J. Mol. Biol.* **36**:401–411 (1968).

149. R. P. Perry and D. E. Kelley, Messenger RNA-protein complexes and newly synthesized ribosomal subunits: analysis of free particles and components of polyribosomes, *J. Mol. Biol.* **35**:37–59 (1968).

150. F. M. Ritossa, K. C. Atwood, D. L. Lindsley, and S. Spiegelman, On the chromosomal distribution of DNA complementary to ribosomal and soluble RNA, National Cancer Institute Monograph n° 23, pp. 449–472, presented symposium (1965).

151. R. H. Burdon, B. T. Martin, and B. M. Lal, Synthesis of low molecular weight ribonucleic acid in tumor cells, *J. Mol. Biol.* **28**:357–372 (1967).

152. R. H. Burdon, Molecular configuration of cytoplasmic transfer RNA precursors, *J. Mol. Biol.* **30**:571–573 (1967).

153. J. E. Darnell Jr., Ribonucleic acids from animal cells, *Bacteriol. Rev.* **32**:262–290 (1968).

154. L. N. Simon, A. J. Glasky, and T. H. Rejal, Enzymes in the central nervous system. I. RNA methylase, *Biochim. Biophys. Acta* **142**:99–104 (1967).

155. T. C. Johnson, Aminoacyl-RNA synthetase and transfer RNA binding activity during early mammalian brain development, *J. Neurochem.* **16**:1125–1131 (1969).

156. For references, see D. B. Roodyn and D. Wilkie, in *The Biogenesis of Mitochondria* (K. Mellanby, ed.) p. 123, Methuen's Monographs on biological subjects, London (1968).

157. P. Mandel, S. Harth, and T. Borkowski, in *Regional Neurochemistry* (S. S. Kety and J. Elkes, eds.) pp. 160–174, Pergamon Press, Oxford (1960).

158. G. E. Vladimirov, M. N. Baranov, L. Z. Pevzner, and W. Tsyn-Yan, in *Regional Neurochemistry* (S. S. Kety and J. Elkes, eds.) pp. 126–134, Pergamon Press, Oxford (1960).

159. P. Mandel, T. Borkowski, S. Harth, and R. Mardell, Incorporation of ^{32}P in ribonucleic acid of subcellular fractions of various regions of the rat central nervous system, *J. Neurochem.* **8**:126–138 (1961).

160. R. Mardell, S. Harth, T. Borkowski, and P. Mandel, Constitution du noyau cellulaire des diverses zones du système nerveux central chez le rat, *C. R. Soc. Biol.* **155**:1096–1099 (1961).

161. M. N. Baranov and L. Z. Pevzner, Microchemical and microspectrophotometric studies on the intralaminar distribution of nucleic acids in the brain cortex under various experimental conditions, *J. Neurochem.* **10**:279–283 (1963).

162. P. Mandel, H. Rein, S. Harth-Edel and R. Mardell, in *Comparative Neurochemistry* (D. Richter, ed.) pp. 149–163, Pergamon Press, Oxford (1964).

163. L. Z. Pevzner, Topochemical aspects of nucleic acid and protein metabolism within the neuron-neuroglia unit of the superior cervical ganglion, *J. Neurochem.* **12**:993–1002 (1965).

164. R. Landolt, H. H. Hess, and C. Thalheimer, Regional distribution of some chemical structural components of the human nervous system. I. DNA, RNA and ganglioside sialic acid, *J. Neurochem.* **13**:1441–1452 (1966).

165. C. Hoch-Ligeti, E. Stutzman, and C. Gruenwald Jr., Differences in the nuclear ribonucleic acid content in different areas of the human adult and foetal brain, *J. Neurochem.* **15**:417–425 (1968).

166. Y. Valotaire, J. Duval, and P. Jouan, Influence de la castration et de la testostérone sur la biosynthèse *in vitro* des acides ribonucléiques dans le cortex cérébral, l'hypothalamus et l'hypophyse du rat, *Bull. Soc. Chim. Biol.* **51**:211–224 (1969).

167. E. Egyhazi, Microchemical fractionation of neuronal and glial RNA, *Biochim. Biophys. Acta* **114**:516–526 (1966).

168. E. Egyhazi and H. Hydén, RNA with high specific activity in neurons and glia, *Brain Res.* **2**:197–200 (1966).

169. B. Daneholt and S.-O. Brattgård, A comparison between RNA metabolism of nerve cells and glia in the hypoglossal nucleus of the rabbit, *J. Neurochem.* **13**:913–921 (1966).

170. P. Volpe and A. Giuditta, Kinetics of RNA labeling in fractions enriched with neuroglia and neurons, *Nature* **216**:154 (1967).

171. P. Volpe and A. Giuditta, Biosynthesis of RNA in neuron- and glia-enriched fractions, *Brain Res.* **6**:228–240 (1967).

172. D. Munoz and P. Mandel, Unpublished data.

173. T. Utakoji and T. C. Hsu, Nucleic acids and protein synthesis of isolated cells from chick embryonic spinal ganglia in culture, *J. Exptl. Zool.* **158**:181–202 (1965).

174. M. Sensenbrenner, Z. Lodin, J. Treska, M. Jacob, and P. Mandel, Différenciation morphologique et biosynthèse des RNA de neurones isolés en culture, *C. R. Ac. Sci. Série D*, **267**:1660–1662 (1968).

175. D. Bodian, A suggestive relationship of nerve cell RNA with specific synaptic sites, *Proc. Natl. Acad. Sci.* **53**:418–425 (1965).

176. E. Koenig, Synthetic mechanisms in the axon. II. RNA in myelin-free axons of the cat, *J. Neurochem.* **12**:357–361 (1965).

177. J. E. Edström, D. Eichner, and A. Edström, The ribonucleic acid of axons and myelin sheaths from Mauthner neurons, *Biochim. Biophys. Acta* **61**:178–184 (1962).

178. N. Miani, A. Di Girolamo, and M. Di Girolamo, Sedimentation characteristics of axonal RNA in rabbit, *J. Neurochem.* **13**:755–759 (1966).

179. H. Rahmann, Zum Vorkommen von RNS im Nervenfaserbereich, *Experientia* **22**:762–766 (1966).

180. J. J. Bray and L. Austin, Flow of protein and ribonucleic acid in peripheral nerve, *J. Neurochem.* **15**:731–740 (1968).

181. A. Edström, J. E. Edström, and T. Hökfelt, Sedimentation analysis of ribonucleic acid extracted from isolated Mauthner nerve fiber components, *J. Neurochem.* **16**:53–66 (1969).

182. E. Koenig, Synthetic mechanisms in the axon. IV. *In vitro* incorporation of (^{3}H) precursors into axonal protein and RNA, *J. Neurochem.* **14**:437–446 (1967).

183. L. Austin, J. J. Bray, and R. J. Young, Transport of proteins and ribonucleic acid along nerve axons, *J. Neurochem.* **13**:1267–1269 (1966).

184. J. A. Peterson, J. J. Bray, and L. Austin, An autoradiographic study of the flow of protein and RNA along peripheral nerve, *J. Neurochem.* **15**:741–745 (1968).

185. L. Casola, G. A. Davis, and R. E. Davis, Evidence for RNA transport in rat optic nerve, *J. Neurochem.* **16**:1037–1041 (1969).

186. A. Haddad, S. Iucif, and A. R. Cruz, Synthesis of RNA in neurons of the hypoglossal nerve nucleus after section of the axon, in mice, *J. Neurochem.* **16**:865–868 (1969).

187. B. Droz, Protein metabolism in nerve cells, *Intern. Rev. Cytol.* **25**:363–390 (1969).

188. E. Koenig, Synthetic mechanisms in the axon. I. Local axonal synthesis of acetylcholinesterase, *J. Neurochem.* **12**:343–355 (1965).

189. E. Koenig, Synthetic mechanisms in the axon. III. Stimulation of acetylcholinesterase synthesis by actinomycin D in the hypoglossal nerve, *J. Neurochem.* **14**:429–435 (1967).

190. A. Edström, Inhibition of protein synthesis in Mauthner nerve fiber components by actinomycin D, *J. Neurochem.* **14**:239–243 (1967).

191. B. Jakoubek and J. E. Edström, RNA changes in the Mauthner axon and myelin sheath after increased functional activity, *J. Neurochem.* **12**:845–849 (1965).

192. H. A. Hartmann, J. Lin and M. C. Shively, RNA of nerve cell bodies and axons after β,β-iminodipropionitrile, *Acta Neuropathol.* **11**:275–281 (1968).

193. W. Sandritter, V. Novakova, J. Pilny, and G. Kieffer, Cytophotometrische Messungen des Nukleinsaüre- und Proteingehaltes von Ganglienzellen der Ratte während der postnatalen Entwicklung und im Alter, *Z. Zellforsch.* **80**:145–152 (1967).

194. Z. Lodin, J. Faltin, and V. Mares, Distribution of nucleic acids in the course of ontogenetic development of the nerve cell, *Folia Morphologica* **16**:171–183 (1968).

195. J. Eschner and P. Glees, Free and membrane-bound ribosomes in maturing neurons of the chick and their possible functional significance, *Experientia* **19**:301–308 (1963).

196. M. Sensenbrenner and P. Mandel, RNA biosynthesis during differentiation of various cell types of chicken embryo in cerebral hemispheres. Histoautoradiographic study, *Z. Zellforsch.* **82**:65–81 (1967).

197. D. H. Adams, The relationship between cellular nucleic acids in the developing rat cerebral cortex, *Biochem. J.* **98**:636–640 (1966).

198. U. N. Singh, Synthesis of ribonucleic acid in developing rat brain *in vitro*, *Nature* **206**:1115–1117 (1965).

199. S. Yamagami, R. R. Fritz, and D. A. Rappoport, Biochemistry of the developing rat brain. VII. Changes in the ribosomal system and nuclear RNA's, *Biochim. Biophys. Acta* **129**:532–547 (1966).

200. M. R. V. Murthy, Protein synthesis in growing rat tissues. II. Polyribosome concentration of brain and liver as a function of age, *Biochim. Biophys. Acta* **119**:599–613 (1966).

201. H. Dellweg, R. Gerner, and A. Wacker, Quantitative and qualitative changes in ribonucleic acids of rat brain dependent on age and training experiments, *J. Neurochem.* **15**:1109–1119 (1968).

202. R. S. Piha, Changes in the activity of amino acid: *t*RNA-ligases in the mouse brain during postnatal development. I. Intern. Meeting Neurochem. Strasbourg, Abstr. p. 167 (1967).

203. J. A. Burdman, Early effects of a nerve growth factor on the RNA content and base ratios of isolated chick embryo sensory ganglia neuroblasts in tissue culture, *J. Neurochem.* **14**:367–371 (1967).

204. E. Howard, Effects of corticosterone and food restriction on growth and on DNA, RNA and cholesterol contents of the brain and liver in infant mice, *J. Neurochem.* **12**:181–191 (1965).

205. R. Balazs, S. Kovacs, P. Teichgräber, W. A. Cocks, and J. T. Eayrs, Biochemical effects of thyroid deficiency on the developing brain, *J. Neurochem.* **15**:1335–1349 (1968).

206. L. Cook, A. B. Davidson, D. J. Davis, H. Green, and E. J. Fellows, Ribonucleic acid: effect on conditioned behavior in rats, *Science* **140**:268–269 (1963).

207. E. L. Bennett and M. Calvin, Failure to train planarians reliably, *Neuroscience Res. Progr. Bull.* **2**(n°4):2–24 (1964).

208. F. R. Babich, A. L. Jacobson, and S. Bubash, Cross-species transfer of learning: effect of ribonucleic acid from Hamsters on rat behavior, *Proc.Natl. Acad. Sci.* **54**:1299–1302 (1965).

209. A. L. Jacobson, F. R. Babich, S. Bubash, and A. Jacobson, Differential-approach tendencies produced by injections of RNA from trained rats, *Science* **150**:636–637 (1965).

210. F. R. Babich, A. L. Jacobson, S. Bubash, and A. Jacobson, Transfer of a response to naïve rats by injection of ribonucleic acid extracted from trained rats, *Science* **149**:656–657 (1965).

211. G. Ungar and C. Oceguera-Navarro, Transfer of habituation by material extracted from brain, *Nature* **207**:301–302 (1965).

212. W. L. Byrne, D. Samuel, E. L. Bennett, M. R. Rosenzweig, E. Wasserman, A. R. Wagner, F. Gardner, R. Galambos, B. D. Berger, D. L. Margules, R. L. Fenichel, L. Stein, J. A. Corson, C. E. Holt, P. H. Schiller, L. Chiappetta, M. E. Jarvik, R. C. Leaf, J. D. Dutcher, Z. P. Horovitz, and P. L. Carlson, Memory transfer, *Science* **153**:658–659 (1966).

213. M. W. Gordon, G. G. Deanin, H. L. Leonhardt, and R. H. Gwynn, RNA and memory: a negative experiment, *Am. J. Psychiat.* **122**:1174–1178 (1966).

214. A. L. Jacobson, F. R. Babich, S. Bubash, and C. Goren, Maze preference in naïve rats produced by injection of ribonucleic acid from trained rats, *Psychon. Sci.* **4**:3–4 (1966).

215. A. L. Jacobson, C. Fried, and S. D. Horowitz, Planarians and memory, *Nature* **209**:599–601 (1966).

216. M. Luttges, T. Johnson, C. Buck, J. Holland, and J. McCaugh, An examination of "transfer of learning" by nucleic acid, *Science* **151**:834–837 (1966).

217. F. Rosenblatt, J. T. Farrow, and W. F. Herblin, Transfer of conditioned responses from trained rats to untrained rats by means of a brain extract, *Nature* **209**:46–48 (1966).

218. F. Rosenblatt, J. T. Farrow, and S. Rhine, The transfer of learned behavior from trained to untrained rats by means of brain extracts. I., *Proc. Natl. Acad. Sci.* **55**:548–555 (1966).

219. F. Rosenblatt, J. T. Farrow, and S. Rhine, The transfer of learned behavior from trained to untrained rats by means of brain extracts. II., *Proc. Natl. Acad. Sci.* **55**:787–792 (1966).

220. G. Ungar, Transfer of learned behavior by brain extracts, *J. Biol. Pyschol.* **9**:12–27 (1967).

221. J. H. Nodine, M. W. Shulkin, J. W. Slap, M. Levine, and K. Freiberg, A double-blind study of the effect of ribonucleic acid in senile brain disease, *Am. J. Psychiat.* **123**:1257–1259 (1967).

222. H. H. Røigaard-Petersen, Th. Nissen, and E. J. Fjerdingstad, Effect of ribonucleic acid (RNA) extracted from the brain of trained animals on learning in rats. III. Results obtained with an improved procedure, *Scand. J. Psychol.* **9**:1–16 (1968).

223. J. Jarlstedt, Functional localization in the cerebellar cortex studied by quantitative determinations of Purkinje cell RNA, *Acta Physiol. Scand.* **67**:243–252 (1966).

224. C. Blomstrand, O. Hallen, A. Hamberger, and J. Jarlstedt, Effect of cerebellectomy upon the cytochemistry of neurons in the lateral vestibular nucleus. II. Effects on the RNA content and succinoxidase activity in Deiters' neurons after warm and cold water calorization, *Brain Res.* **11**:648–654 (1968).

225. G. Attardi, Quantitative behavior of cytoplasmic ribonucleic acid in rat Purkinje cells following prolonged physiological stimulation, *Exptl. Cell Res. Suppl.* **4**:25–53 (1957).

226. R. G. Grenell, H. Hazama, M. Nakazawa, and E. Einberg, Effects of gravitational changes on RNA of cerebral neurons and glia. I. RNA changes of Deiters' cells and glia, *Brain Res.* **9**:115–125 (1968).
227. D. A. Rappoport and H. F. Daginawala, Olfaction and changes in brain nuclear RNA in a teleost, I. Intern. Meeting Neurochem. Strasbourg, Abstr. p. 175 (1967).
228. D. A. Rappoport and H. F. Daginawala, Changes in nuclear RNA of brain induced by olfaction in catfish, *J. Neurochem.* **15**:991–1006 (1968).
229. S. Goswamy and P. Mandel, Role of light in the biosynthesis of ribonucleic acids during ontogenesis of visual pathways, I. Intern. Meeting Neurochem. Strasbourg, Abstr. p. 88 (1967).
230. S. Goswamy, P. Mandel, and P. Karli, in *Biochemistry of the Eye* (M. U. Dardenne and J. Nordmann, eds.) pp. 514–520, Karger, Basel (1968).
231. G. R. Winsberg, RNA synthesis in the retina of xenopus during early dark adaptation, *Exptl. Cell Res.* **52**:555–564 (1968).
232. I. Nir, N. Hirschmann, J. Mishkinsky, and F. G. Sulman, The effect of light and darkness on nucleic acids and protein metabolism of the pineal gland, *Life Sci.* **8**:279–287 (1969).
233. C. Kupper and J. L. DeC. Downer, Ribonucleic acid content and metabolic activity of lateral geniculate nucleus in monkey following afferent denervation, *J. Neurochem.* **14**:257–263 (1967).
234. S. H. Appel, W. Davis, and S. Scott, Brain polysomes: response to environmental stimulation, *Science* **157**:836–838 (1967).
235. G. P. Talwar, B. K. Goel, S. P. Chopra, and B. D'Monte, in *Macromolecules and Behavior* (J. Gaito, ed.) pp. 71–88, Appleton-Century-Crofts, New York (1966).
236. H. Hydén, in *The Cell* (J. Brachet and A. E. Mirsky, eds.) pp. 215–323, Vol. IV, Academic Press, New York (1960).
237. C. Vesco and A. Giuditta, Disaggregation of brain polysomes induced by electroconvulsive treatment, *J. Neurochem.* **15**:81–85 (1968).
238. F. Orrego, Synthesis of RNA in normal and electrically stimulated brain cortex slices *in vitro*, *J. Neurochem.* **14**:851–858 (1967).
239. W. Grampp and J. E. Edström, The effect of nervous activity on ribonucleic acid of the crustacean receptor neuron, *J. Neurochem.* **10**:725–731 (1963).
240. J. E. Edström and D. Eichner, Relation between nucleolar volume and cell body content of ribonucleic acid in supraoptic neurons, *Nature* **181**:619 (1958).
241. C. Prives and J. H. Quastel, Effect of cerebral stimulation on biosynthesis of nucleotides and RNA in brain slices *in vitro*, *Biochim. Biophys. Acta* **182**:285–294 (1969).
242. J. Krivanek, Concerning the mechanism of the inhibitory effect of spreading cortical depression on the cerebral cortex proteosynthesis, Intern. Sympos. "Metabolism of Nucleic acids and proteins and the function of the neuron", Castle Liblice (Czechoslovakia), Abstr. p. 71 (1967).
243. M. Ruscak and D. Ruscakova, in *Macromolecules and the Function of the Neuron* (Z. Lodin and S. P. R. Rose, eds.) pp. 212–220, Excerpta Medica Foundation, Amsterdam (1968).
244. G. P. Talwar, B. Sadasivudu, and V. S. Chitre, Changes in the pentose-nucleic acid content of subcellular fractions of the brain of the rat during "metrazol" convulsions, *Nature* **191**:1007–1008 (1961).
245. V. S. Chitre, S. P. Chopra, and G. P. Talwar, Changes in the ribonucleic acid content of the brain during experimentally induced convulsions, *J. Neurochem.* **11**:439–448 (1964).
246. I. W. Maggiolo Barbato and L. Barbato, The effect of hypoglycaemia and insulin coma on the total RNA content of isolated Purkinje cells, *J. Neurochem.* **12**:60–61 (1965).
247. R. Di Carlo, H. Randrianarisoa, A. Lehmann, and P. Mandel, Variations du métabolisme des RNA du cerveau au cours des crises audiogènes, *J. Physiol.* (*Paris*), **62**:271 (1970).

248. R. Di Carlo, S. Edel, H. Randrianarisoa, and P. Mandel, Amphetamine and cerebral RNA metabolism, Abstr. Congr. Collegium Intern. Neuropsychopharmacologicum, Prague (1970).

249. R. N. Bryan and E. L. Bliss, Incorporation of (^{3}H) uridine into mouse brain RNA during stress, I. Intern. Meeting Neurochem. Strasbourg, Abstr. p. 33 (1967).

250. H. Hydén and E. Egyhazi, Glial RNA changes during a learning experiment in rats, *Proc. Natl. Acad. Sci.* **49**:618–624 (1963).

251. H. Hydén and E. Egyhazi, Changes in RNA content and base composition in cortical neurons of rats in a learning experiment involving transfer of handedness, *Proc. Natl. Acad. Sci.* **52**:1030–1035 (1964).

252. H. Hydén and P. W. Lange, A differentiation in RNA response in neurons early and late during learning, *Proc. Natl. Acad. Sci.* **53**:946–952 (1965).

253. H. Hydén and P. W. Lange, A genetic stimulation with production of adenine-uracil rich RNA in neurons and glia in learning. The question of transfer of RNA from glia to neurons, *Naturwissensch.* **53**:64–70 (1966).

254. J. W. Zemp, J. E. Wilson, K. Schlesinger, W. O. Boggan, and E. Glassman, Brain function and macromolecules. I. Incorporation of uridine into RNA of mouse brain during short-term training experience, *Proc. Natl. Acad. Sci.* **55**:1423–1431 (1966).

255. J. W. Zemp, J. E. Wilson, and E. Glassman, Brain function and macromolecules. II. Site of increased labeling of RNA in brains of mice during a short term training experience, *Proc. Natl. Acad. Sci.* **58**:1120–1125 (1967).

256. L. B. Adair, J. E. Wilson, J. W. Zemp, and E. Glassman, Brain function and macromolecules. III. Uridine incorporation into polysomes of mouse brain during short-term avoidance conditioning, *Proc. Natl. Acad. Sci.* **61**:606–613 (1968).

257. B. Machlus and J. Gaito, Successive competition hybridization to detect RNA species in a shock avoidance task, *Nature* **222**:573–574 (1969).

258. A. J. Glasky and L. N. Simon, Magnesium pemoline: enhancement of brain RNA polymerases, *Science* **151**:702–703 (1966).

259. L. N. Simon and A. J. Glasky, Magnesium pemoline: enhancement of brain RNA synthesis *in vivo*, *Life Sci.* **7**:197–202 (1968).

260. N. Plotnikoff, Magnesium pemoline: enhancement of learning and memory of a conditioned avoidance response, *Science* **151**:703–704 (1966).

261. P. W. Frey and V. J. Polidora, Magnesium pemoline: effect on avoidance conditioning in rats, *Science* **155**:1281–1282 (1967).

262. D. F. Cain, Lack of effect of pemoline on the incorporation of (^{3}H) uridine into RNA in brain slices, *Life Sci.* **6**:1919–1926 (1967).

263. A. Yuwiller, W. Greenouch, and E. Geller, Biochemical and behavioral effects of magnesium pemoline, *Psychopharmacologia* **13**:174–180 (1968).

264. S. H. Barondes and M. E. Jarvik, The influence of actinomycin D on brain RNA synthesis and on memory, *J. Neurochem.* **11**:187–195 (1964).

265. H. D. Cohen and S. H. Barondes, Further studies of learning and memory after intracerebral actinomycin D, *J. Neurochem.* **13**:207–211 (1966).

266. B. W. Agranoff, R. E. Davis, L. Casola, and R. Lim, Actinomycin D blocks formation of memory of shock-avoidance in goldfish, *Science* **158**:1600–1601 (1967).

267. S. H. Appel, Effect of inhibition of RNA synthesis on neural information storage, *Nature* **207**:1163–1166 (1965).

268. T. J. Chamberlain, G. H. Rothschild, and R. W. Gerard, Drugs affecting RNA and learning, *Proc. Natl. Acad. Sci.* **49**:918–924 (1963).

269. R. E. Jewett, J. H. Pirch, and S. Norton, Effect of 8-azaguanine on learning of a fixed-interval schedule, *Nature* **207**:277–278 (1965).

270. O. A. Krylov, R. A. Danylova, and V. S. Tongur, On the participation of RNA in reflectoral activity of white mice, *Life Sci.* **4**:1313–1317 (1965).

Chapter 5

THE TURNOVER OF LIPIDS*†

Robert Main Burton

Department of Pharmacology and the
Beaumont-May Institute of Neurology
Washington University School of Medicine
St. Louis, Missouri

I. INTRODUCTION

Lipids, as a broad category of cellular constituents such as carbohydrates, proteins, vitamins, hormones, and minerals, play varied and assorted roles in living organisms. Some fatty acids are used as energy sources and energy reserves—the triglycerides in adipose tissue represent the energy reserves storing fatty acids as esters and the brown adipose tissue functions to maintain body temperature. Other fatty acids are combined in covalent bonds in complex compounds which form much of the lipid portions of cellular and organelle membranes. Finally, the third major function of lipids is to provide structural elements of which myelin of nerve tissue is an excellent example. In addition, certain lipids may be viewed as vitamins (vitamins A, D, E, and essential polyunsaturated fatty acids) and as hormones (the prostaglandins, steroids, etc.).

While carbohydrates and proteins are largely dependent on covalent bonds to form structural units, the lipids often build structural elements in which covalent bonds play a somewhat minor role. Because of their hydrophobic nature, the lipids are repulsed by water and tend to form a separate phase in which the hydrophobic residues dissolve in each other. This hydrophobic bonding (water lattice repulsion) aided by van der Waal forces gives rise to a very strong association between the lipid molecules. Examples which illustrate the strength of these cooperative forces are soap bubbles and lipid bilayer membranes.[1]

Intuitive reasoning would indicate that the utilization and replacement of fuel lipids would occur at relatively high rates; the specific rates would be

* This work was supported in part by the U.S. Public Health Service Grant NB-01575-12.

† This article is not intended to be a complete review of the literature. Rather, it is an overview of the complex problem of lipid anabolism and catabolism, precursor pools and product pools, the blood–brain barrier, and their relationship to lipid turnover. Lipid turnover is reported as half-life time $(t_{\frac{1}{2}})$. The theoretical points are illustrated by selected, but not comprehensive, data from the literature.

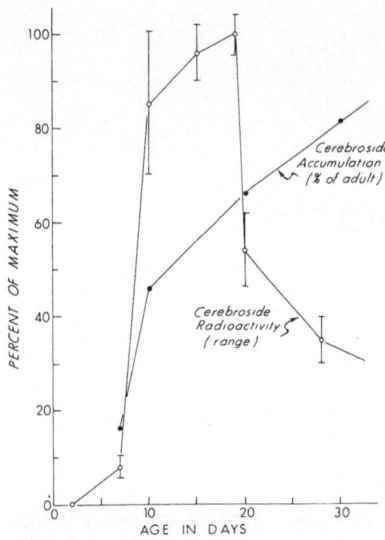

Fig. 1. Rat brain cerebrosides and their rate of biosynthesis (i.e. incorporation of radioactivity) as a function of age. A replot of the data of Burton *et al.*[3]

dependent upon the current nutritional state of the organism. As a corollary, intuition suggests that the turnover of structural lipids would be relatively slow. Whereas the fuel lipid turnover relates to the nutritional state and activities of the organism, the structural lipids will be formed most rapidly during the appropriate phase of fetal and neonatal development. As the organism reaches adult status (for each specific cellular lipid), the turnover of these lipids will be markedly slowed. Figure 1 presents data which illustrate the accumulation in brain of the structural lipid cerebroside with increasing age until adult levels are obtained. The rate of synthesis of cerebrosides is also a function of age and is highest at or near the inflection point of the cerebroside accumulation curve.[2,3]

II. FUNCTIONAL CLASSIFICATION OF LIPIDS

This intuitive principle leads to a classification of lipids based on their metabolic activities and functions.[2] Such a classification is shown in Table I. Three classes of lipids are indicated:

Class I lipids are those with half-lives ($t_{\frac{1}{2}}$) of 1 week or less. These lipids are primarily those used for energy production or storage and as intermediates in synthetic pathways. A typical example is the triglyceride of adipose tissue, whose turnover is rapid and is under hormonal control, and palmitic acid, which serves as an intermediate in the synthesis of longer-chain fatty acids and sphingosine.

Class II comprises those lipids with intermediate $t_{\frac{1}{2}}$, being lipids which form cellular plasma membranes and the membranes of organelles probably including the endoplasmic reticulum.

Class III lipids are components of structural units and have $t_{\frac{1}{2}}$'s of 4 weeks or longer. Myelin is certainly one of the best examples. Thus, if the

TABLE I
Functional Classes of Lipids

Class	Class half-time days	Examples			Observed half-time days
		Lipid	Source	Moiety labeled	
I	7 or less (1 week)	Palmitic acid	Whole brain	Fatty acid	3
		Triglycerides	Adipose tissue	Fatty acid	fast[a]
		Phosphatidyl inositols	Sympathetic neurons	^{32}P	fast[b]
		Phosphatidyl inositols	Whole brain	Fatty acid	$\frac{1}{2}$–1
		Phosphatidyl choline	Mitochondria	Fatty acids	11–23
II	8 to 30 (intermediate)	Sphingomyelin	Mitochondria	Fatty acids	30
		Gangliosides	Whole brain	Glucose galactose	<10
		Gangliosides	Whole brain	N-Acetylgalactosamine	24
				N-Acetyl neuraminic acid	24
III	31 or more (4 weeks or more)	Lignoceric acid	Whole brain	Fatty acid	96
		Phosphatidyl choline	Myelin	Fatty acid	>60
		Sphingomyelin	Myelin	Fatty acid	300
		Cerebrosides	Myelin	Galactose	31
		Cerebrosides	Myelin	Fatty acid	>360
		Cholesterol	Myelin	Cholesterol	60 to 1000

[a] Fast, but under hormonal control.
[b] Fast, but under hormonal and neuronal control.

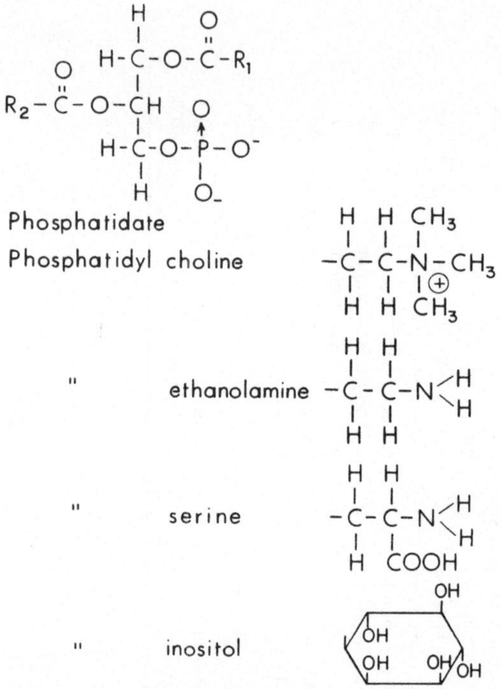

Fig. 2. Formulas of phosphatidyl lipids.

$t_{\frac{1}{2}}$ of a lipid is known, then an educated guess can be made as to the type of its function. Or conversely, if the function of a lipid is known, then it is possible to set approximate lower limits on its half-time.

III. LIPID STRUCTURE-ACTIVITY RELATIONSHIPS

The metabolically active lipids, the triglycerides, are the simple esters of fatty acids with glycerol. The fatty acid profile of the triglycerides centers around palmitic and stearic acids. The triglycerides are the most highly concentrated of the energy stores. Carbon atom for carbon atom, the triglycerides contain about twice the number of calories contained in glucose and the polysaccharide energy reserves. The covalent bonds employed are simple ester bonds which are readily hydrolyzed; the fatty acids are then oxidized to yield the needed energy and the glycerol moiety can be used to form the glucose necessary for the resynthesis of triglycerides.

Phosphatidic acid is the structural foundation for the whole series of phosphoglycerol lipids, i.e., the phosphatidyl lipids. Phosphatidic acid is the phosphoryl ester of a diglyceride. These phospholipids are the primary lipid components of plasma membranes, organelle membranes, and are important in myelin membranes as well.

The triglycerides are insoluble in water and form globules, unless other compounds are present to act as detergents to form micelles containing the hydrophobic triglycerides. However, the replacement of one fatty acid of a triglyceride by phosphoric acid results in an amphipathic lipid capable of being oriented at a water–lipid interface. The specific extent of the polar nature of the phosphatidic acid lipids is the result of forming a second ester of the phosphoric acid moiety. Phosphatidyl choline (lecithin) has the hydrophobic diglyceride on one end of the phosphate, choline on the other (see Fig. 2). Phosphatidyl ethanolamine has a primary amine on the polar end capable of interacting with other ions to form a polar quarternary amine or of not reacting and remaining a neutral hydrophobic moiety. Phosphatidylserine has an anion on its polar end and the phosphatidylinositols form a family of phosphatidic acids containing one, two, or perhaps more, phosphoric acid residues.

The transition from the phosphatidic acids to the lipids required for more structural stability occurs in four stages. The first is the change from a monoglyceride *ester* to an isosteric structure containing a carbon-to-carbon covalent bond replacing the ester bond. The second is the replacement of the hydroxyl group at carbon 2 of the monoglyceride by a primary amine. With the formation of an amide (replacing the *ester* bond), the compound ceramide (N-acyl sphingosine) is formed, which is isosteric with the diglyceride (see Figs. 3 and 4).

The last two steps to increase the structural stability of these lipids are changes in the fatty acid moieties. In line with the Radin principle,[4] the turnover time of fatty acids (at least those longer than palmitic acid) is inversely related to their hydrocarbon chain length. Thus, triglycerides have fatty acids centering around 16:0 and 18:0, while the structural lipids contain fatty acids centering around 24:0. This is illustrated by the data shown in Table II. Compare the fatty acid profiles of sphingomyelin isolated from nerve terminals and synaptic vesicles (89–93 % as 18:0) with the sphingomyelin from the stable structure myelin in which only 31 % is present as stearic acid and 50 % is present as 24:0 and 24:1.[4–6] Another example is the ganglioside present in brain tissue, where it is most likely a membrane lipid (91 % containing 18:0),[5,7] and the red blood cell structural glycolipid, which contains only 10 % as stearic acid and 64 % as 24:0.[6]

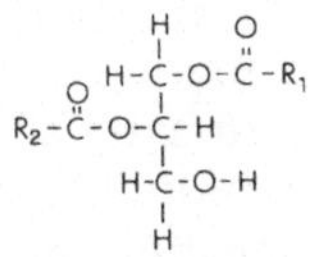

$$
\begin{array}{c}
\text{H} \quad \text{O} \\
\text{H-C-O-C-R}_1 \\
\text{O} \quad \text{|} \\
\text{R}_2\text{-C-O-C-H} \\
\text{H-C-O-H} \\
\text{H}
\end{array}
$$

D-1,2- Diglyceride

$$
\begin{array}{c}
\text{H} \quad \text{H} \\
\text{HO-C-C=C-(CH}_2)_{12}\text{-CH}_3 \\
\text{O} \quad \text{H} \\
\text{R}_2\text{-C-N-C-H} \\
\text{H} \\
\text{H-C-O-H} \\
\text{H}
\end{array}
$$

Fig. 3. Comparison of the formulas of a diglyceride (1,2-diacylglycerol) and ceramide (N-acyl sphingosine). Compare these formulas with the space-filling molecular models in Fig. 4.

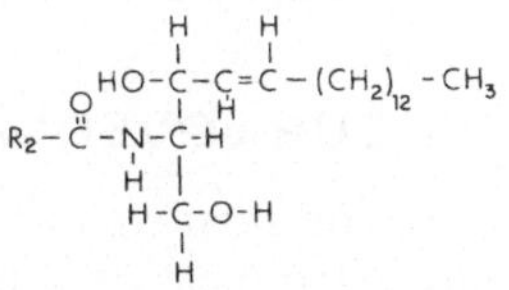

N-acyl trans-D-erythro-1,3-dihydroxy-2-amino -4-octadecene

(N-acyl sphingosine)

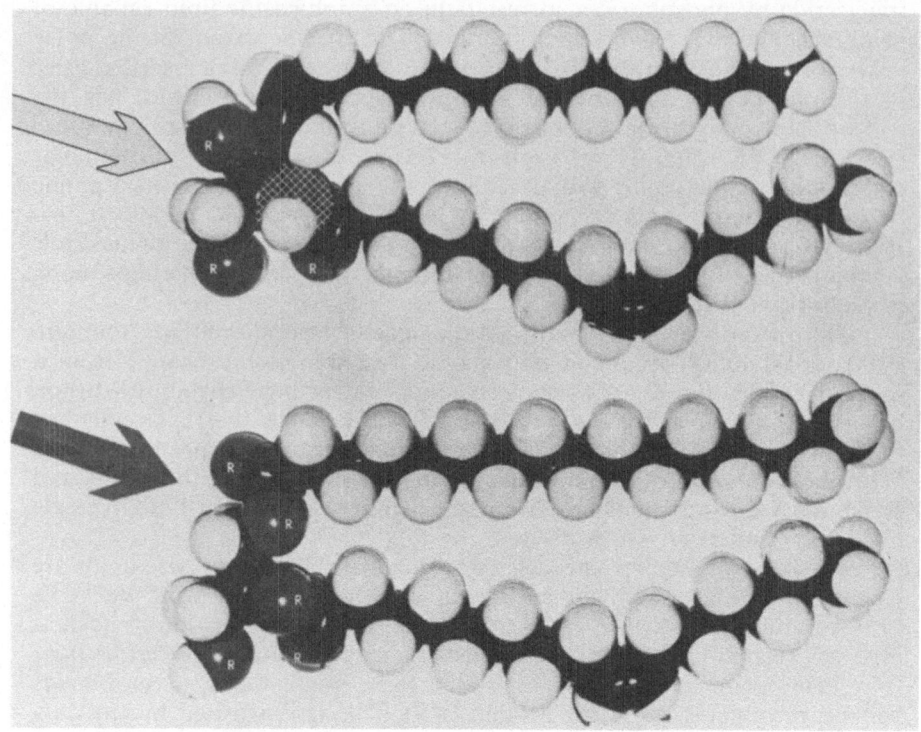

Fig. 4. Space-filling molecular models of a diglyceride and a ceramide (see Fig. 3). Diglyceride, lower model, and ceramide, upper model. The arrows point to the region of difference. Carbon, black atoms; hydrogen, white; oxygen, gray with an R; and nitrogen, cross-hatched. These models have been oriented to facilitate their comparison and to illustrate the change in structure introduced by a *cis*-double bond.

Fourth, even though the Radin principle indicates that the longer chain fatty acids are more stable biologically, additional stability is introduced by the formation of α-hydroxy fatty acids. Unlike the β-hydroxy fatty acids, which is an early step in the metabolism of fatty acid, the α-hydroxy fatty acids are more difficult to metabolize. Note in Table II that cerebrosides, which constitute a major portion of myelin lipids, contain 69% of their fatty acids as α-hydroxy fatty acids. In addition, 53% of the total fatty acids present in cerebrosides are the 24:0 and 24h:0.

IV. TURNOVER OF LIPIDS

The turnover of lipids is measured by the rate of incorporation of a labeled precursor into the lipid under study or by the rate of loss of the label from the lipid. While simple in concept, the procedure itself must be carefully controlled. There are two major areas of concern in experiments of

TABLE II
Fatty Acid Profiles[a]

	Myelin (%)	Nerve terminal (%)	Synaptic vesicles (%)	%
Phosphatidylcholine				
16:0	25	57		
18:0	25	14		
18:1	40	24		
20:4	5	4		
Gangliosides				RBC Glycolipids
18:0		91	92	10
24:0		6	1	64
Sphinomyelin				
16:0	2	4	2	
18:0	31	93	89	
24:0	10	1	4	
24:1	40	0	1	
Cerebrosides				Whole Brain α-OH
18:0				18 0
20:0				9 0
22:0				17 27
24:0				47 59
26:0				4 2
				(31%) (69%)

[a] Data from Radin et al.,[5] Hajra and Radin,[4] and Handa and Yamakawa.[6]

this type. (1) The size of the precursor pools may directly influence the *apparent* rate of turnover of the lipid. (2) The specific part of the lipid labeled must be known. For example, the possible portions of cerebrosides which may be labeled are (a) the galactose, (b) the fatty acid, and (c) the sphingosine moieties. Reference to Table I shows that cerebrosides labeled in the fatty acid moiety have a $t_{\frac{1}{2}}$ for loss of radioactivity of > 360 days.[8] A label in the sphingosine indicates a lower turnover rate.[9] However, when the cerebrosides are labeled in the galactose, the $t_{\frac{1}{2}}$ is about 31 days.[10] The apparent discrepancy in these three $t_{\frac{1}{2}}$ for cerebrosides can be explained in two possible ways. One, the ceramide portion of the cerebroside is relatively fixed in its place in the myelin structure and the galactose moiety is exposed and may be removed and replaced without disturbing the ceramide. Two, and most likely, the entire molecule undergoes total degradation into sphingosine, fatty acid, and galactose (see Fig. 5). The sphingosine and fatty acid enter small pools from which new cerebroside is made. The small pools

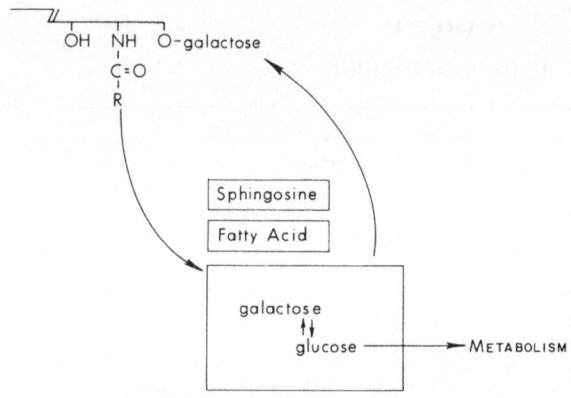

Fig. 5. Effect of precursor pool size on the apparent rate of cerebroside turnover as measured by labeled components.

mean that the radioactivity in the sphingosine and fatty acid is conserved and reincorporated into the cerebroside. The apparent rate of loss of radioactivity does not reflect the true turnover time of cerebrosides. On the other hand, the galactose hydrolyzed from the cerebroside enters a relatively large galactose pool (which is most likely in equilibrium with the glucose pool). Consequently, the radioactive labeled galactose is greatly diluted by the nonradioactive galactose (and glucose) pool. Little radioactivity is reincorporated into the cerebrosides and the apparent rate of loss of the radioactive galactose gives an estimate of the $t_{\frac{1}{2}}$ for cerebroside that is closest to the real turnover time. In experiments of this type, the smallest value obtained for the $t_{\frac{1}{2}}$ is the value most likely to be correct. At the least, it places an upper limit on the turnover time.

A. Fatty Acids

A very important experiment was conducted by N. Radin[4] who administered acetate-1-^{14}C to rats and isolated the fatty acids from their brains. The results are presented in Fig. 6. At 4 hr after the injection, the palmitic acid specific radioactivity was over two times that of the other long-chain fatty acids, e.g., 18:0 to 24:0. The palmitic acid specific radioactivity decreased continuously with increasing time after the injection. All of the longer chain fatty acids increased in specific radioactivity until their specific radioactivities were larger than that of the palmitic acid. After 4 to 14 days, the specific radioactivity of all the fatty acids decreased. A re-plot of Radin's data (Fig. 7) in which the specific radioactivity is presented as a function of the carbon number of the fatty acid, emphasizes the fact that the product appears to be of greater specific radioactivity than the precursors. The fatty acids up to palmitic acid are formed from acetate via malonyl coenzyme A and fatty acid synthetase. The fatty acids of carbon number greater than 16:0 are formed by a chain elongation mechanism.[4,11] Since the total amounts of each of the fatty acids remain the same or increase with time,[4] the decrease in specific radioactivity must reflect a "pulse" label with movement of the isotope from the initial pool—palmitic acid—to the

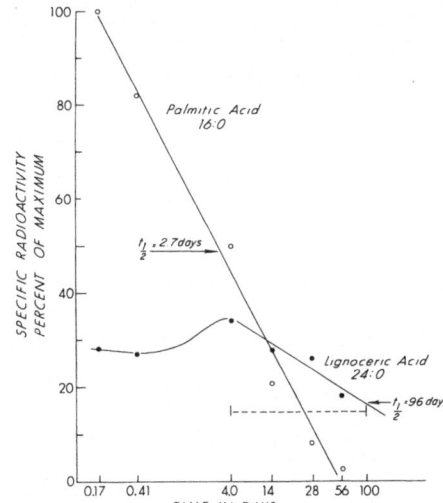

Fig. 6. Fatty acid turnover in rat brain lipids.
Data from Hajra and Radin.[4]

higher fatty acids. The quantity of palmitic acid loss must be replaced by
nonradioactive palmitic acid. That such a flow of fatty acids must occur
was documented by Radin[4] in experiments with tritium-labeled acetate.
In these studies, the radioactivity of the long-chain fatty acids increased
with time. Degradation of the tritium-containing fatty acids results in the
loss of tritium as water (whereas carbon-14 could be recycled). Thus the
radioactive higher fatty acids must be formed from the labeled shorter chain

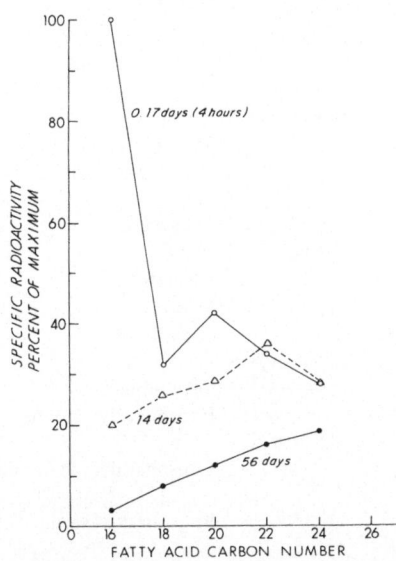

Fig. 7. Specific radioactivity of rat brain fatty
acids as a function of carbon number and time.
A replot of data from Hajra and Radin.[4]

fatty acids without degradation. These observations are schematically shown in Fig. 8.

Figure 8 shows that the initial pulse primarily labels the 16:0. The carbon-14 incorporated into the higher fatty acids are as "C_2" units; therefore the radioactivity per carbon or "C_2" unit is greatest in palmitic acid. With time, and the replacement of acetate-^{14}C by nonradioactive acetate, the highly radioactive 16:0 serves as a precursor "pool" for the longer chain fatty acids. Thus, the ^{14}C-pulse progresses up the fatty acid sequence (16:0 → 24:0) and appears to result in a more radioactive product from a less labeled precursor. This abnormal product-precursor relationship reflects not a departure from traditional kinetics but rather the effect of the choice of time for each fatty acid specific radioactivity measurement.

The kinetics are complex for the incorporation of radioactivity into or out of fatty acids and into the lipids, which contain these fatty acids. The recycling of the specific fatty acids themselves and the recycling of "C_2" units are impediments to the correct estimation of the turnover of the fatty acids and of the lipids containing these fatty acids. As summarized in the last sentence of the preceding section, turnover times for these compounds may provide an upper limit, but may not yield the true $t_{\frac{1}{2}}$ or a lower limit. With this reservation, the half-life of several brain lipids computed from acetate-1-^{14}C incorporation are tabulated in Table III.

TABLE III

Turnover Rates of Rat Brain Lipids[a]

Brain or subcellular fraction	Lipid	Half-life $t_{\frac{1}{2}}$ days	Reference	Functional class
Brain	Fatty acids:			
	Palmitic acid	3	[b]	I
	Lignoceric acid	96	[b]	III
	Neutral:			
	cholesterol	190	[c]	III
Nuclei	Neutral:			
	cholesterol	100	[c]	III
Myelin	Neutral:			
	cholesterol	136	[d]	III
		210–240	[e]	
	Phospholipids:			
	Phosphatidyl choline	60	[e]	III
		NS	[d]	
	Phosphatidyl ethanolamine	210	[e]	III
		NS	[d]	
	Phosphatidyl serine	62	[d]	III
		120	[e]	

TABLE III (continued)

Brain or subcellular fraction	Lipid	Half-life $t_{\frac{1}{2}}$ days	Reference	Functional class
	Phosphatidyl inositol	35	*e*	III
		NS	*d*	
	Sphingomyelin	300	*e*	III
		NS	*d*	
	Glycolipids:			
	Cerebrosides	360	*e*	III
		NS	*d*	
	Cerebroside-3'-sulfate	360	*e*	III
Mitochondria	Neutral:			
	cholesterol	60	*e*	III
		800–1000	*c*	
		NS	*d*	
	Phospholipids:			
	Phosphatidyl choline	11–23	*d*	II
		14	*e*	
	Phosphatidyl ethanolamine	22–35	*d*	II
		30	*e*	
	Phosphatidyl serine	21	*e*	II
		27	*d*	
	Phosphatidyl inositol	2	*e*	I
		29–34	*d*	II
	Sphingomyelin	30	*e*	II
		NS	*d*	
Microsomes	Neutral:			
	cholesterol	60	*c*	III

a Estimated from the loss of radioactivity from the lipid after labeling with acetate-1-^{14}C *in vivo*. The cholesterol was labeled by glucose-U-^{14}C. NS, no significant turnover.
b Hajra and Radin.[4]
c Khan and Folch[31]; data for young rats. Older rats (6 months) showed little turnover in 190 days.
d Cuzner, Davison, and Gregson.[13]
e Smith and Eng.[12]

B. Sphingosine

In addition to serving as the primary unit for the synthesis of the longer chain fatty acids, palmitic acid is a primary unit for the formation of sphingosine. The pathway elucidated by Brady and his co-workers involves the CoA-dependent reduction of palmitic acid to palmitaldehyde and the condensation of the palmitaldehyde and serine with the evolution of carbon dioxide to form dihydrosphingosine.[14] The labeling of sphingosine with serine-3-^{14}C and the subsequent loss of radioactivity indicated a half-life

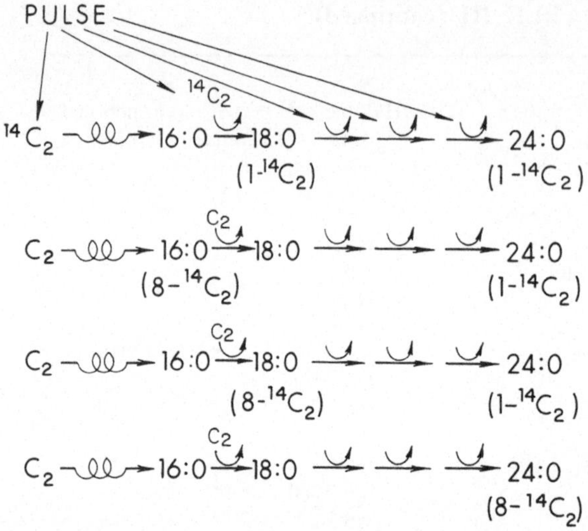

Fig. 8. Schematic diagram illustrating the "pulse" labeling of long chain fatty acids.

of over 50 days in experiments in which radioactivity from acetate had a half-life of 30 days.[10,15,16] Other experiments by Bartsch[9] showed that acetate-^{14}C which would label the fatty acids of sphingolipids failed to incorporate comparable radioactivity into the sphingosine moiety.

C. Carbohydrates

There are two major classes of glycolipids in brain tissue, the cerebrosides and the gangliosides. The cerebrosides constitute about 5% of the wet weight of brain and have relatively simple structures—the primary variant being the fatty acid profile. The turnover of cerebrosides has been considered in Section IV.

Gangliosides have a similarity to the cerebrosides in that both have a sphingosine base and a fatty acid amide (85–95% as stearic acid in the gangliosides); however, the gangliosides have an oligosaccharide residue on carbon 1. These carbohydrate moieties are composed of glucose, galactose, N-acetyl galactosamine, and N-acetyl neuraminic acid. As discussed in detail for cerebrosides, the turnover rate of the ganglioside family depends upon the label used, i.e., glucose-1-^{14}C, galactose-1-^{14}C, or glucosamine-1-[14]C.

As indicated in Table I, the half-life of the total ganglioside function in rat brain has been estimated at about 8–10 days for a glucose or galactose label,[10,16,17] and about 24 days for the amino sugars.[16] Data from the human suggests that certain of the gangliosides are in metabolic equilibrium[18] and that the cerebellum has a faster ganglioside turnover than the cerebrum.[18] While Suzuki[19] reported a half-life for total gangliosides labeled with glucosamine-1-^{14}C in rat brain consistent with Burton et al.,[16] his value for glucose-U-^{14}C labeled gangliosides is about 20 days—twice as long as indicated by Radin[17] and Burton.[16] The temporal pattern of the labeled gangliosides of rat brain appears not to be in equilibrium.[19] The discrepancy in these turnover experiments illustrates again the difficult nature of conducting and interpreting studies on complex molecules with a number of precursor pools.

D. Sulfate and Phosphate

Sulfate incorporated into brain lipids resides exclusively as cerebroside-3'-sulfate.[20,21] Experiments with sulfate-^{35}S suggest that cerebroside sulfates turn over at a very low rate. While extrapolation of the data of Green and Robinson[22] suggests a half-life of approximately 100 days, the value obtained by Davison and Gregson[13,23,24] was greater than 334 days,* and the data of Radin et al.,[17] indicate no turnover of the cerebroside sulfate. However, Radin did note that sulfate-^{35}S still can be incorporated into cerebroside sulfate of old rats.

Experiments performed by Davison and Dobbing[25] in which phosphate-^{32}P was injected intraperitoneally into rats results in the formation of radioactive phospholipids in both central and peripheral nerve tissue. In both tissues (whole brain and sciatic nerve), maximum incorporation was achieved in about 12 to 20 days. An extrapolation of their data indicates a half-life for loss of radioactive phosphate from sciatic phospholipids of approximately 38 days, and similar calculations for the removal of ^{32}P from the phospholipids of whole brain suggest a half-life of about 175 days. Later studies by Davison and his co-workers[8] showed that the specific radioactivities of the various phospholipids in rat brain were similar at each time period studied, i.e., 4, 10, 18, and 27 days after the administration of phosphate-^{32}P. This was true for the myelin phospholipids from both newborn and adult rats. Data of this type suggest that myelin and its component lipids turn over as units rather than as individual phospholipids.

It would be useful to know the pool size for each anion and the ease with which they can enter and leave neuronal tissue. As discussed in detail earlier, the recycling of lipid components through small pools can give rise to turnover numbers of excessive size. The next section discusses hormonal and neuronal control of phosphatidyl inositol.

The in vivo incorporation of phospholipid precursors has been investigated by Ansell and Spanner[26] and Ansell and Chojnacki.[27] Phosphorylcholine-^{32}P, phosphorylethanolamine-^{32}P, CMP-(^{32}P)phosphorylcholine, and CMP-(^{32}P)phosphorylethanolamine were injected into the subarachnoid space of 20-day-old rats. A rapid hydrolysis of these compounds was noted— the phosphorylcholine-^{32}P and phosphorylethanolamine-^{32}P had $t_{\frac{1}{2}}$ values of 15 or 16 min. The nucleotides were rapidly hydrolyzed by pyrophosphatase and the ^{32}P-products hydrolyzed as indicated. Even with the rapid loss of the ^{32}P-precursors, the phospholipid fractions were labeled by these four compounds. The nucleotides injected showed a greater extent of labeling

---

* The $t_{\frac{1}{2}}$ for cerebroside sulfate in rat brain mitochondria is estimated to be 39 days.[13]

of the phospholipids than did the phosphoryl bases. These studies, too, emphasize our lack of information concerning the size of the precursor pools, what pools actually exist *in vivo*, the rate of removal of precursors from these pools by hydrolysis and other metabolic pathways, and the ability of precursors to pass neuronal or glial membranes.

E. Phosphatidyl Inositols

The inositol-containing lipids are distinguished by their different biosynthetic pathway, i.e.,

$$CMP\text{-diglyceride} + \text{inositol} \rightarrow \text{phosphatidyl inositol} + CMP$$

from that for the other phospholipids, e.g.,

$$CDP\text{-choline} + \text{phosphatidic acid} \rightarrow$$

$$\text{phosphatidylcholine} + CMP + \text{pyrophosphate}$$

This difference in biosynthesis could be the result of a biological need to keep these phospholipids metabolically separated because they serve totally different functions. While the phosphatidyl derivations already discussed appear to be concerned with membrane structures, the phosphatidyl inositols are implicated in cellular or synaptic functions. The Hokins have demonstrated increased turnover of phosphatidyl inositols when acetylcholine or norepinephrine was added to media bathing brain slices or sympathetic ganglia.[28,29] Recently, Larrabee[30] has shown that increased turnover of phosphatidyl inositols occurs with transynaptic stimulation *in situ*. Under the conditions of these experiments, no increase in turnover was observed for any of the other phospholipids. While the *in vivo* data reported by Larrabee do not permit the computation of a $t_{\frac{1}{2}}$ for the phosphatidyl inositols, they do provide comparative values. Thus, the increase in turnover of the phosphatidyl-inositols is about 20 % greater than phosphatidylcholine of the same ganglia.

F. Cholesterol

Cholesterol is another major lipid in brain tissue constituting 5 % of the wet weight. In contrast to the other brain lipids which are a simple assembly of common metabolites, e.g., cerebrosides are composed of galactose, a fatty acid, and sphingosine, cholesterol synthesis involves a complex series of reactions to yield the stable compound composed of carbon–carbon bonds. Studies on the turnover of brain cholesterol indicated a long half-life.[13] Recent studies by Khan and Folch-Pi[31] have shown different rates of cholesterol turnover in the different brain subcellular particles (Table III).

V. REFERENCES

1. R. M. Burton (ed.), *Lipid Monolayer and Bilayer Models and Cellular Membranes*, The American Oil Chemists' Society, Chicago, Illinois (1967).

2. R. M. Burton, *in Drugs and Poisons in Relation to the Developing Nervous System* (G. M. McKhann and S. J. Yaffe, eds.) pp. 60–75, Public Health Service Publication 1791, Bethesda, Md. (1967).

3. R. M. Burton, M. A. Sodd, and R. O. Brady, The incorporation of galactose into galactolipids, *J. Biol. Chem.* **233**: 1053–1060 (1958).

4. A. K. Hajra and N. S. Radin, Isotopic studies of the biosynthesis of cerebroside fatty acids in rats, *J. Lipid Res.* **4**: 270–278 (1963).

5. Y. Kishmoto, B. W. Agranoff, M. S. Radin, and R. M. Burton, Comparison of the lipids and fatty acids of subcellular brain fractions, *J. Neurochem.* (in press).

6. S. Handa and T. Yamakawa, Chemistry of lipids of posthemolytic residue or stroma of erythrocytes. XII. Chemical structure and chromatographic behavior of hematosides obtained from equine and dog erythrocytes, *Jap. J. Exptl. Med.*, 5, Vol. 34, 293–304 (1964).

7. E. Klenk, *in Biochemistry of the Developing Nervous System* (H. Waelsch, ed.) Academic Press, pp. 397–410, New York (1955).

8. L. A. Cuzner, A. N. Davison, and N. A. Gregson, Chemical and metabolic studies of rat myelin of the central nervous system in research in demyelinating diseases, *Ann. N.Y. Acad. Sci.* **122**: 86–92 (1965).

9. G. Bartsch, Incorporation of acetate-1-C^{14} into the sphingolipids of the rat brain, VI Int. Congress of Biochem., (1964), p. 564.

10. R. M. Burton and J. M. Gibbons, Ganglioside and cerebroside turnover rates in rat brain, *Fed. Proc.* **23**: 230 (1964).

11. S. J. Wakil, *in Ann. Biochem.* (J. M. Luck, ed.) pp. 369–406, Annual Reviews Inc., Palo Alto (1962).

12. M. E. Smith and L. F. Eng, The turnover of the lipid components of myelin, *J. Am. Oil Chem. Soc.* **42**: 1013–1018 (1965).

13. M. L. Cuzner, A. N. Davison, and N. A. Gregson, Turnover of brain mitochondrial membrane lipids, *Biochem. J.* **101**: 618–626 (1966).

14. R. O. Brady and G. J. Koval, The enzymatic synthesis of sphingosine, *J. Biol. Chem.* **233**: 26–31 (1958).

15. A. N. Davison and M. Wajda, Metabolism of myelin lipids: estimation and separation of brain lipids in the developing rabbit, *J. Neurochem.* **4**: 353–359 (1959).

16. R. M. Burton, *in Lipids and Lipidoses* (G. Schettler, ed.) pp. 122–167, Springer-Verlag, New York (1967).

17. N. S. Radin, F. B. Martin, and J. R. Brown, Galactolipide Metabolism, *J. Biol. Chem.* **224**: 499–507 (1957).

18. R. M. Burton, S. Handa, R. E. Howard, and T. Vietti, Incorporation of Selected Isotopes into Lipids of Humans with Cerebral Lipidoses: Studies on D-glucosamine-l-^{14}C, *Path. Europ.* **3**: 424–430 (1968).

19. K. Suzuki, Formation and turnover of the major brain gangliosides during development, *J. Neurochem.* **14**: 917–925 (1967).

20. T. Taketomi and T. Yamakawa, Further confirmation on the structure of brain cerebroside sulfuric ester, *J. Biochem.* (*Tokyo*) **55**: 87–89 (1964).

21. P. Stoffyn and A. Stoffyn, Structure of sulfatides, *Biochim. Biophys. Acta* **70**: 218–220 (1963).

22. J. P. Green and J. D. Robinson, Jr., Cerebroside sulfate (sulfatide A) in some organs of the rat and in the mast cell tumor, *J. Biol. Chem.* **235**: 1621–1623 (1960).

23. A. N. Davison and N. A. Gregson, The physiological role of cerebron sulphuric acid (sulphatide in the brain), *Biochem. J.* **85**: 558–568 (1962).

24. A. N. Davison and N. A. Gregson, Metabolism of cellular membrane sulpholipids in rat brain, *Biochem. J.* **98**: 915–922 (1966).

25. A. N. Davison and J. Dobbing, Phospholipid metabolism in nervous tissue. I. A reconsideration of brain and peripheral-nerve phospholipid metabolism *in vivo, Biochem. J.* **73**:701–706 (1959).
26. G. B. Ansell and S. Spanner, Incorporation of orthophosphate labeled with phosphorus-32 into cerebral phospholipids *in vivo, Nature* **185**:826–828 (1960).
27. G. B. Ansell and T. Chojnacki, The fate of radioactive phospholipid precursors injected into the subarachnoid space of the rat, *Biochim. Biophys. Acta Library* **1**:425–432 (1963).
28. M. R. Hokin, L. E. Hokin, and W. D. Shelp, The effects of acetylcholine on the turnover of phosphatic acid and phosphoinositide in sympathetic ganglia, and in various parts of the central nervous system *in vitro, J. Gen. Physiol.* **44**:217–226 (1960).
29. M. R. Hokin, Effect of norepinephrine on ^{32}P incorporation into individual phosphatides in slices from different areas of the guinea pig brain, *J. Neurochem.* **16**:127–134 (1969).
30. M. G. Larrabee, Transynaptic stimulation of phosphatidylinoside metabolism in sympathetic neurons *in situ, J. Neurochem.* **15**:803–808 (1968).
31. A. A. Khan and J. Folch-Pi, Cholesterol turnover in brain subcellular particles, *J. Neurochem.* **14**:1099–1105 (1967).

Chapter 6

AMMONIA METABOLISM

Yasuzo Tsukada

Department of Physiology
School of Medicine
Keio University
Tokyo, Japan

I. PHYSIOLOGICAL SIGNIFICANCE OF AMMONIA IN THE NERVOUS SYSTEM

A. Ammonia Content and Nervous Activity

An intimate relationship between nervous activity and ammonia has already been pointed out by many workers.[1-5] It is well known that the liberation of ammonia from excised peripheral nerve is increased by electrical stimulation and is depressed by anesthesia or nerve section.[6,7] When living brain was fixed immediately by freezing in liquid air or liquid nitrogen, it was shown that ammonia was formed rapidly in the brain postmortem.[2,8] Ammonia content in living brain is also elevated significantly before and at the onset of convulsive seizures produced by the injection of several drugs, such as picrotoxin,[2] fluoroacetate,[3] pentamethylene tetrazole,[9] or Telodrin,[57] by electrical stimulation of the brain, by prolonged auditory stimulus,[10] or by anoxia[2,61] (Table I). The rate of increase in the content of ammonia in the brain which is formed by the stimuli is quite high (over 30%). The initial rate of formation of cerebral ammonia following application of convulsive drugs is high, corresponding to 450 μmoles/g/hr. Thus ammonia is regarded as a labile chemical constituent in living brain.

During convulsive seizures, the level of ATP in the brain is generally observed to be decreased.[11] On the other hand, it has been reported that no alteration from normal content of cerebral ATP is found in rats with ammonia acetate-induced precomatose and comatose states. But this finding cannot exclude ATP deficiency in some circumscribed cerebral site.[12]

In the absence of convulsion, the content of ammonia in the brain is also raised by applying electric shock through the paws of animals, by conditioned reflex[13,14] or by injection of ammonium salts[58] or methamphetamine.[15]

On the other hand, a decrease of ammonia in the brain is observed in anesthesia,[2] sleep,[13] and hibernative state.[16]

215

TABLE I

Ammonia and Glutamine Contents in Young Rat Brain[a]

Conditions	Ammonia (μmoles/g)	Glutamine (μmoles/g)	Remarks
Resting state	0.20	5.56	
Picrotoxin injection	0.34	5.49	Convulsion
Electric shock	0.35	5.40	Preconvulsion
Anoxia	0.46	—	Strong convulsion
Anesthesia	0.04	5.65	Nembutal narcosis
Immature	0.47	—	At birth
Immature	0.35	—	14 days old
Immature	0.20	—	28 days old

[a] From Richter and Dawson[2] and Oja, von Bonsdorff, and Lindroos.[27]
[b] Rats of 20–40 g are used and killed by immersion in liquid air.

From these observations, it has been deduced that ammonia level in the brain is closely related to the state of nervous activity; that is, that ammonia level in the brain is increased in the excitatory state of neuronal cells, and decreased in the inhibitory state.

The level of ammonia in the brain *in vivo* can be accurately determined by rapid freezing technique, and the values of 0.20 μmoles and 0.36 μmoles per gram of fresh brain are found in mouse[2] and rat[14] respectively. Postmortem changes dramatically affect the content of ammonia in the brain, and a value of 2 μmoles per gram has been found at 3 min after death.[25,26] This shows that ammonia formation after death is almost explosive in brain tissue.

The concentration of ammonia in an immature rat brain is twice that in a mature one. The ammonia level reaches that of adult brain at the age of about 3 weeks.[27] Electrical activity in the brain of the newborn rats can be recorded at 3–5 days old, and the EEG shows the adult pattern at 2 weeks.[28] In this situation, an increased ammonia level would not correspond to brain activity, but might reflect some metabolic differences in immature brain.

In the case of El mouse, which is a special strain and causes convulsive seizures genetically by the stimulation of postural change "seesawing," ammonia and glutamine content in the brain in the resting state are significantly lower than that of control mice (dd and gpc strains). Nevertheless, ammonia content in the brain is able to increase remarkably during the seizures[29] (Table II).

Ammonia liberation has been also tested *in vitro*. Ammonia is not liberated to any extra degree from the slices[17] or the excised cervical sympathetic ganglion[18] by the application of electric shock, although transmitted action potential can be recorded even from those preparations which are being incubated in the medium. However, greater ammonia

TABLE II
Ammonia and Glutamine Contents in Mouse Brain[a]

Strains	Conditions	Ammonia (μmoles/g)	Glutamine (μmoles/g)
El	Resting	0.13	2.50
	Preconvulsion[b]	0.28	2.46
	Convulsion[b]	0.52	2.53
gpc	Resting	0.21	3.26
	Metrazol-convulsion	0.49	3.42
	Seesawing	0.18	3.39
dd	Resting	0.20	—

[a] From Naruse et al.[29]
[b] Seesawing stimulation was applied.

liberation can be obtained when the brain slices or excised cervical sympathetic ganglia are incubated in the medium without substrate.

From the clinical aspects, interest in the concentration of ammonia in the blood and its relation to neuropsychiatric disturbances has largely centered on liver disease.[21,23] Hyperammonemia frequently demonstrated certain specific defects in motor performances and recognition, and sometimes abnormalities of EEG have been observed to occur in cirrhotic patients demonstrating clinical manifestations of hepatic encephalopathy.[20,22,24] A highly suggestive correlation between blood ammonia levels and alterations of mental status and characteristic involuntary movements in the patients with hyperammonemia has been reported.[19] When the blood ammonia level is uniformly elevated in hepatic disease, the uptake of ammonia by brain is roughly proportional to arterial concentrations. The use of substances (α-ketoglutarate, glutamate) in therapy of hepatic coma has been designed to reduce free ammonia in the blood, but clinical experience has not been uniformly favorable.[25]

As mentioned above, though the important role of ammonium ions upon nervous activities has been realized for many years, the origin of ammonia formed in the brain in vivo has not yet been clarified. It is thus considered that ammonia is being metabolized actively in the nervous tissue and that it contributes a specific physiological function in the brain.

B. Effect of Ammonium Ions on Cerebral Metabolism

The oxygen consumption and aerobic glycolysis of brain slices, more than those of any other tissues, are sensitive to the change in the ionic composition of the incubation medium and particularly to an increase in the concentration of potassium or ammonium ions in the medium.[30] Ammonium ions have the same effect as potassium ions on the enhancement of the respiration and aerobic lactate formation in brain tissue. In contrast, ammonium ions inhibit anaerobic glycolysis. Moreover, ammonium ions are much more active at lower concentration than potassium ions; that is, 40–100 mM potassium ions is required to cause a noticeable metabolic effect[31] whereas only 0.3 mM ammonium is sufficient for that effect.

Recently, the effect of ammonium ions on slice oxygen consumption has been found to be variable; that is, in lower concentrations of ammonium ions, oxygen uptake is stimulated but at higher concentrations it shows no change or depression.[64,65] On mitochondrial preparation from brain cortex, it is reported that oxygen consumption is reduced significantly by the addition of 15 mM NH_4Cl when pyruvate or α-ketoglutarate are used as substrate, but no change occurs with glutamate, GABA, or succinate as substrates.[65] It is suggested that the toxic effect of ammonium ions on the oxidative metabolism of mitochondria is related to the pyruvate and α-keto-glutarate oxidase stages. When sodium fluoroacetate (1 mM) is added together with NH_4 ions, oxygen consumption of rat brain cortex slices accelerated by NH_4 ions is suppressed. However, the free ammonia in brain cortex slices after incubation in the presence of glucose can be increased by the addition of fluoroacetate (1 mM), while the formation of ^{14}C-labeled glutamate from glucose-^{14}C by the brain cortex slices is suppressed.[63]

In patients with hepatic coma and elevated blood ammonia levels, the clinical study also demonstrated marked decreases of cerebral oxygen consumption without comparable changes in cerebral blood flow.[66]

The amount of ammonia accumulated in the brain slices during incubation has been decreased by the addition of increasing potassium concentration in the medium. This means that potassium ions and ammonium ions are complementary to each other for ion transport across the brain cell membranes.[32,33] The active transport of L-histidine into the cells of rat brain cortex slices is inhibited clearly by the addition of NH_4Cl in the medium at the concentration of 27–50 mM.[67] On the other hand, ammonium ions have an effect on acetylcholine metabolism in minced or sliced brain, i.e., they increase the amount of free acetylcholine at the expense of its bound form, just like the potassium ions, whereas ammonium ions inhibit the synthesis of acetylcholine in brain slices.[34,35]

Some of the drugs which have certain effects on the central nervous system usually contain a group of quarternary ammonium salts in their molecules. An ammonium ion itself shows also strong pharmacological effects on the CNS as a powerful irritant.[59] When ammonium salts are injected into the rat in a large dose and their concentration in the brain reaches 9 mg%, convulsive seizures occur.[2] Ammonium ions administered to the animal have a strychninelike action on the medulla and spinal cord, indicating the changes of EEG pattern.[36,37]

II. AMMONIA FORMATION IN BRAIN SLICES

It has been observed that ammonia is actively formed when excised brain tissue, as either homogenate or slice, is incubated in the absence of substrate. In brain homogenates, the initial ammonia formation is quite active compared with that of brain slices.[1,20,38,39] On the other hand, ammonia is formed continuously at a steady rate for several hours in respiring brain slices in a glucose-free medium. These results might mean that cellular integrities, intact mitochondria, and metabolic cofactors are necessary for the continuous production of ammonia in excised brain tissue.

Weil-Malherbe[38,40] has shown that the brain slices respiring in a glucose-free medium produce ammonia at a steady rate. This finding has been also confirmed by other investigators.[32,41] Guinea pig brain cortex slices incubating in a substrate-free medium produce ammonia at a rate of 7.60 μmoles/g/hr. This rate is surprisingly high, and the value of ammonia formed during incubation is almost the same as the total amino acid nitrogen in the tissue. In the presence of glucose or glutamic acid, the ammonia formation is depressed almost completely. When brain slices are incubated in the presence of glutamic acid in the medium, glutamine is formed at a rate corresponding to the decrease of ammonia formation. However, the inhibition of ammonia formation by the addition of glucose is not accompanied by a significant increase of amide synthesis. So the inhibition induced by the addition of glucose is not explained simply by glutamine synthesis. The further addition of iodoacetic acid at a fairly low concentration in the incubation medium again increases the formation of ammonia, even in the presence of glucose. It was thus considered that glucose metabolism itself is necessary to prevent ammonia formation (Table III). Also, it might be partly due to a lower degree of proteolysis and the operation of an ammonia-binding mechanism.

TABLE III

Ammonia Formation in Guinea Pig Brain Cortex Slices[a]

Conditions	O_2 uptake (μmole/g/hr)	NH_3[b] (μmole/g)	Glutamine[b] (μmole/g)
Glucose-free	41.2	7.60 ± 0.92	0.55 ± 0.76
Glucose (11 mM)	58.2	1.51	2.03
Glucose + IAA (5×10^{-5}M)	50.9	3.20	2.80
Glutamate (10^{-2} M)	67.1	1.22	7.74
Glutamate + 2,4 DNP (5×10^{-5}M)	62.5	7.70	2.25
Lactate (10^{-2}M)	—	1.70	1.41
Succinate (10^{-2}M)	—	4.69	1.79

[a] From Takagaki and Tsukada.[41]
[b] Values indicate the contents of ammonia and glutamine after 1 hr incubation.

The endogenous respiration in brain slices of guinea pig or rat amount to 41.2 μmoles/g/hr and it accounts for over 70% of the value for the oxygen consumption in the presence of 11 mM glucose. In this case, the respiratory quotient is almost unity and is not different from the value for brain cortex slices respiring in glucose-containing medium.[41] Along with the endogenous respiration of the brain slices, ammonia formation is increased markedly and the content of glutamic acid in the slices is decreased simultaneously at a rate of 4.59 μmoles/g/hr.

The rate of ammonia formation is strongly and progressively inhibited under anaerobic conditions, as well as in aerobic conditions, in the presence of certain metabolic inhibitors, e.g., 2,4-dinitrophenol (5×10^{-5} M), cyanide (5×10^{-4} M), and arsenite (10^{-3} M). These inhibitors also suppress the endogenous oxygen consumption at the same degree (Table IV). On the other hand, malonate, azide, and iodoacetate do not significantly inhibit the ammonia formation; those agents do not inhibit the respiration either.[41]

TABLE IV

Effect of Metabolic Inhibitors on the Ammonia Formation of Guinea Pig Brain Slices[a]

Inhibitors	Ammonia formed (μmole/g/hr)	Change (%)	O_2 uptake (μmole/g/hr)	Change (%)
Anaerobic	3.90	−48		
Cyanide (5×10^{-4} M)	4.22	−44	11.8	−71
2,4 DNP (5×10^{-5} M)	5.36	−29	22.9	−44
IAA (5×10^{-5} M)	6.60	−13	37.8	−8
NaF (10^{-2} M)	7.68	0	22.7	−45
Malonate (10^{-2} M)	7.75	+2	40.8	−1
Azide (10^{-3} M)	8.24	+8	34.9	−15

[a] From Takagaki and Tsukada.[41]

It might therefore be concluded that ammonia formation and endogenous oxygen consumption have the same origin. These observations also suggested that the origin of endogenous oxygen uptake and ammonia formation is not single, but composite.

The mechanism of ammonia formation in brain slices may connect in some way with that of ammonia liberation resulting from brain activity *in vivo*. It is conceivable that the results from *in vitro* experiments will be useful in finding the physiological significance of ammonia in living brain.

III. THE SOURCE OF AMMONIA IN NEURAL TISSUE

The nature of the origin of ammonia in the nervous tissue is of great interest, but this problem is still unsolved. The possible implication in the ammonia formation of brain slices in a substrate-free medium will be considered here. In this preparation, the tissue can produce ammonia at the rate of 30 μmoles/g for 4 hr, and this value corresponds to about the total amino acid nitrogen of the tissue. It has been shown that brain tissue contains various deaminating enzyme systems such as glutaminase,[47] glutamic dehydrogenase,[40] amine oxidase,[44] adenylic deaminase,[62] adenosine deaminase,[51,62] NAD deaminase,[45] glucosamine-6-phosphate deaminase,[42] guanine and guanosine deaminase,[43] and neutral proteinase.[54] The cyclic deamination reaction through NAD and aspartic acid will be discussed in Chapter 9.

During aerobic as well as anaerobic incubation, the glutamine content in brain slices remains fairly constant. And the ammonia formation of brain slices in a substrate-free medium is unaffected by the addition of D-glutamic acid, which strongly inhibits the deamination of added glutamine by glutaminase.[38] Thus it seems that glutamine can be ruled out as a source of ammonia in excised brain tissue.

It is well known that glutamic acid is contained in brain tissue in high concentration as a free form. When glutamic acid is added to the respiring brain slices, it is oxidized, and it synthesizes glutamine without any liberation of free ammonia. The reaction equilibrium of glutamic dehydrogenase is in favor of the reductive amination of ketoglutarate.[48] Thus it is now generally believed that glutamic acid could not become a candidate for precursor of ammonia. However, it is possible that metabolism of intrinsic glutamate in brain tissue is different from that of added glutamate and part of it can participate in ammonia formation.[41]

Muntz[49] has reported that brain tissue contains an unusual adenylic deaminase which is activated by the addition of ATP. Parnas and Bendall suggested that the ammonia formation of working muscle is attributable to adenylic deaminase.[50] In brain tissue, a similar mechanism for the formation of ammonia could be considered. Accordingly, it has been speculated by Weil-Malherbe[38,51] that ammonia formation of brain slices involves a deamination-reamination cycle of adenosine nucleotides. However, the concentration of preformed adenosine nucleotides in brain is far too low to account for the ammonia formation in brain slices. It has been shown by Tsukada et al.[26] that adenylic compounds in brain slices are not deaminated during an incubation of 60 min and no amination reaction is observed when inosine triphosphate is added to the brain slices as a substrate[28] (Fig. 1).

Since a significant increase of NPN has been demonstrated in incubating slices of brain, an unstable neutral proteinase that was reported by Ansell and Richter[52] might have some role in part upon the formation of ammonia. Vrba[32] suggested that a quarter of the ammonia formed in brain slices incubated without substrate may be accounted for by the splitting of protein-bound amide bonds.

In in vivo experiments, Geiger et al.[53] reported that NPN is also increased in cat brain following the stimulation of the brain. Also Vrba[54] described the decrease of amido nitrogen in the protein fraction of the brain during physical exercise. This amido nitrogen seemed to originate from

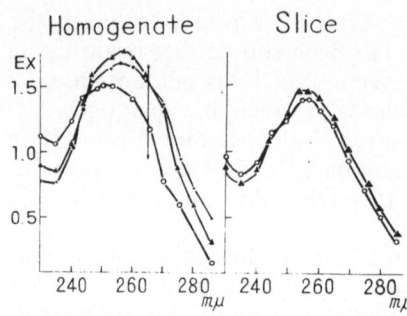

Fig. 1. Absorption spectra of homogenate and slice of guinea pig brain. ▲ = 0, ● = 10 min, ○ = 50 min incubation at 37°C. (From Tsukada and Takagaki.[26]).

bound glutamine, and he assumed that this would be one of the sources of ammonia in brain tissue.

Other deaminating enzymes mentioned before may have important roles in the brain. But their contribution to ammonia formation is presumably small, owing to low concentration of their substrates in the brain. There may be more than one source of ammonia in brain tissue (Fig. 2).

IV. THE FATE OF AMMONIA IN BRAIN TISSUE

The content of ammonia in living brain is kept at a relatively low concentration. It is strongly suggested that a detoxication mechanism for ammonia will operate efficiently in brain, where ammonium ions show a toxic action.

When ammonium salt is injected into the rat, the ammonia level in the brain is elevated up to a value of 0.50 μmoles/g while the rat is alive.[5,14] On the other hand, the glutamine content in the brain increases according to the amount of ammonium salt injected (Table V). A maximum dose of ammonium salt could be administered up to 80 mg N/kg body wt. when ammonium glucuronate is injected intraperitoneally into the adult rat. This

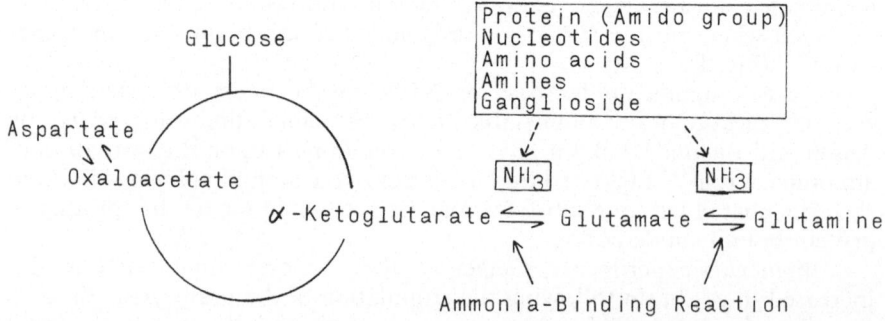

Fig. 2. Map of ammonia metabolism in the brain.

TABLE V

Effect of Ammonium Salt Administration on Ammonia and Glutamine Contents of Rat Brain[a]

Dose	Ammonia (μmoles/g)[b]	Glutamine (μmoles/g)[b]
NH$_4$Cl 200 mg/kg	0.44 (9)	5.57
NH$_4$Cl 50 mg/kg	0.47 (9)	4.09
NH$_4$Cl 20 mg/kg	0.35 (6)	3.34
None	0.38 (11)	3.38

[a] From Tsukada et al.[14]
[b] Ammonia and glutamine contents are measured 1 hr after injection of NH$_4$Cl. Number of experiments indicated in parentheses.

means that the glutamine-forming system is operating quite actively in the brain as a detoxication mechanism of ammonia at the expense of ATP. And it is postulated that the synthesis of glutamine has priority over many other energy-consuming processes for removal of free ammonia in the brain (Fig. 3).

It has been known from the data using ^{14}C-labeled glucose as a tracer that amino acids such as glutamic acid, aspartic acid, γ-aminobutyric acid, and glutamine are synthesized in brain tissue from glucose. And the molecule of ammonia combines with α-ketoglutarate and with glutamate to form glutamate and glutamine respectively.

The mechanism of transformation of glucose into amino acids has been studied using isotopes of glucose-^{14}C and ammonia-^{15}N. The fate of ammonia-^{15}N infused through the carotid artery of the cat has been reported by Berl et al.[55] The incorporation of ^{15}N into several amino acids such as glutamic acid, glutamine, aspartic acid, glutathione, γ-amonobutyric acid,

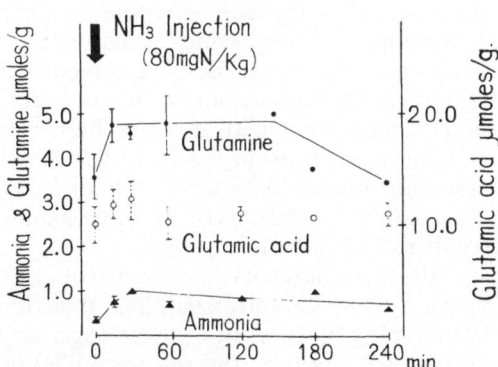

Fig. 3. Changes of glutamine and glutamic acid in rat brain after intraperitoneal injection of ammonium glucuronate. (From Tsukada.[5])

TABLE VI

Labeling of Amino Acids During Ammonia Infusion[a]

	Brain		Liver	
	μmole/g	^{15}N atom % excess	μmole/g	^{15}N atom % excess
Ammonia	4.2	63	18.6	77
Glutamine	7.7		2.2	
Amido-N		34.5·		38.4
α-Amino-N		9.3		9.4
Glutamic acid	8.0	0.69	2.7	27.6
Urea	4.7	0.40	5.3	11.8

[a] 1.0 mmole of ammonium-^{15}N acetate/min (99.6 atom % excess) is infused for 15 min through carotid artery of adult cat. From Berl et al.[55]

and urea in cerebral cortex, liver, and blood was measured. In the case of brain, the highest ^{15}N-concentration was found in the amide group of glutamine followed by its α-amino group (Table VI). Glutamine was the only cerebral amino acid showing a considerable increase of ^{15}N-labeling as a result of the ammonia infusion, and it occurred without a corresponding decrease in the concentration of cerebral glutamic acid. This indicates that glutamic acid is newly formed from glucose during synthesis of glutamine. The α-amino group of cerebral glutamine contained a significantly higher concentration of ^{15}N than that of the cerebral glutamic acid. This result demonstrates that the newly formed glutamine is derived from a small and metabolically very active compartment of glutamic acid, and this compartment does not rapidly reach equilibrium with the total tissue glutamic acid, which is assumed to be slowly metabolized. These findings were obtained specifically in brain, but not in liver tissue. Aspartic acid and γ-aminobutyric acid are also formed from a metabolically active glutamic acid pool. It is suggested that the concept "metabolic compartmentation" provides the operational terms by which structure and metabolism of an organ may be integrated. This result is in agreement with that of the study which shows that intracisternal administration of glutamic acid-^{14}C results in the formation of glutamine with 5 times higher specific activity than in the total glutamic acid in brain tissue. In this experimental situation, the amount of ammonia infused (0.27–0.64 mmole/min/kg body wt. for 25 min) is unusually high, and it is conceivable that this would be an unphysiological condition compared with the normal one.

In our laboratory,[5] ^{15}N-distribution among amino acids was determined by a single intraperitoneal injection of ammonium glucuronate-^{15}N (80 mg N/kg body wt.) to the rat together with glucose-^{14}C. At 60 min after injection of isotopes, the rat was killed in liquid nitrogen, and the specific activities of ^{14}C and ^{15}N in the amino acids in brain and liver tissues were determined. The highest percentage excess of ^{15}N atom is demonstrated in amido nitrogen of glutamine, and the labeling of ^{15}N in α-amido nitrogen of glutamine was much higher than that of glutamic acid. However, ^{14}C-labeling into glutamine from glucose-^{14}C was almost the same degree as that into glutamic acid. In the liver tissue, the highest incorporation of ^{15}N was found in urea instead of glutamine, and ^{15}N in other amino acids was much less than that of the brain.

From these results, the glutamine synthesizing system in the brain could be considered as contributing to the ammonia binding mechanism. Evidently ammonia and glutamine metabolism in the brain must be highly specific in comparison with that of the other tissues.

Chain et al.[56] have investigated the participation of the ammonium ions in the transformation of glucose into amino acids in brain slices incubated with ammonium nitrate-^{15}N and glucose-^{14}C. ^{14}C from glucose was incorporated into amino acids such as glutamic acid, aspartic acid, glutamine, glycine, alanine, and GABA which are produced in brain slices during incubation. At the same time, ^{15}N-isotope from ammonium nitrate was readily incorporated into glutamine, glutamic acid, and aspartic acid, but only to a small extent into GABA, glycine, and alanine. ^{15}N in α-amino nitrogen of glutamine was not higher than that of glutamic acid. It might be assumed that the active metabolic compartment disappears in the case of excised brain slices.

Tower et al.[65] have reported that the exogenous addition of ammonium chloride to the incubation medium in a concentration of 10 mM caused an elevation of free glutamine in cat brain slices and at the same time increased protein-bound glutamine in respiring brain cortex slices at the expense of protein-bound glutamic acid. However, these changes could not be observed in the slices from subcortical white matter. It seems that amidation reaction in brain tissue may be compartmentalized in neurons.

V. BRAIN AMMONIA AND CEREBRAL FUNCTIONS

Now it is believed that ammonium ions play an important role as a cerebral irritant, and glutamine participates in a detoxication process of ammonia in the brain. The ammonia content of the brain in rats or mice can now be estimated accurately using the rapid freezing technique in liquid nitrogen during various states of cerebral activity; the ammonia level in the brain is regarded as a good biochemical indicator, giving an index of the state of physiological activity of the brain.

It has been shown that liberation of cerebral free ammonia occurs in epileptic convulsion induced by electric shock or by administration of pentamethylene tetrazole, picrotoxin, fluoroacetate, or Telodrin. A rise of ammonia in the brain has been observed before the onset of convulsion and throughout its course. In contrast, ammonia level in the brain decreases during anesthetic state or sleep.

In the case of chemically induced convulsion, ammonia metabolism begins before the onset of the seizures. When fluoroacetate is administered to the dog, ammonia content in the brain is already elevated in the pre-convulsive period at 30 min after administration.[3] As another example, acute Telodrin intoxication causes an increase in the content of glutamine and a decrease in glutamic acid at 10 min after administration of the drug. In the early states of Telodrin intoxication, the content of free ammonia in the brain remains unchanged, but an increase is observed later in the seizure pattern at 40 min[57] (Fig. 4). It is assumed that ammonia is liberated in the brain by the administration of Telodrin and then ammonia produced is bound with available glutamate to form glutamine. It is of interest that some animals previously treated with glutamine are resistant to Telodrin convulsions. A close association between convulsion and the metabolism of ammonia and glutamate in brain is strongly suggested. Although ammonia content in the rat brain is increased by electroshock, ammonia levels do not correlate with brain excitability as measured by electroshock seizure threshold or maximal seizure duration. In contrast, acetylcholine content in the brain seems to be influenced by intensity of seizure.[60]

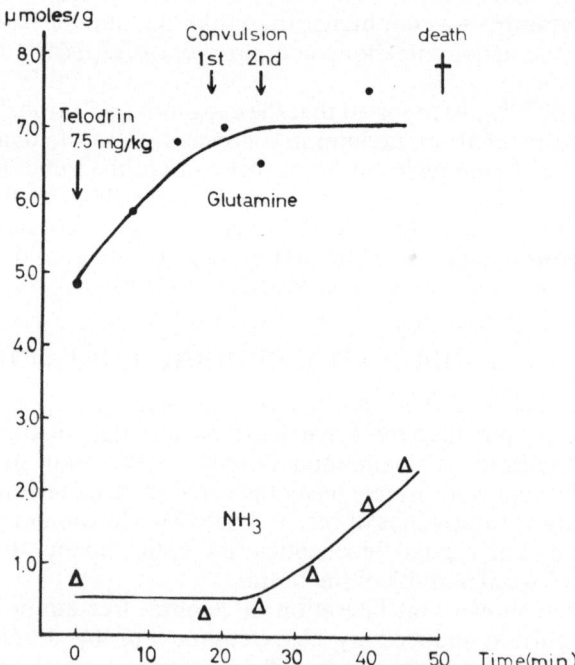

Fig. 4. Changes of ammonia and glutamine content of rat brain after administration of Telodrin. (From Hathway and Mallinson.[57])

The physiological changes of ammonia content of brain have been examined *in vivo* by Vladimirova[13,47] and Tsukada[4,5] under the conditions of active defense reflex and avoidance reflex without convulsive seizures.

In the experimental setup used in our laboratory, a rat is put into the experimental chamber, and an electric shock of 30–50 V ac is applied to the paws through the grid as an unconditioned stimulus, and the light from a 60 W electric lamp is used as a conditioned stimulus. Under these conditions, behavioral changes of the rat have been observed without evoking seizures. To fix the rat brain rapidly, an electric magnet is switched off; the animal then drops into liquid nitrogen and is frozen instantly. The frozen brain is chiseled out, ground to powder and deproteinized in 10% trichloroacetic acid. The mean value of 0.36 μmoles/g is obtained for the ammonia content of rat brain in the resting state. Electrical stimulation for 5 sec, which causes an active defense reflex, results in a significant increase in the concentration of ammonia in the brain, and a value of 0.45 μmoles/g is found. However, with continuous stimulation for longer than 5 sec, i.e., 30, 60, 120 sec, or of much longer duration, ammonia content in the brain shows just the resting level. In these experiments, glutamine content did not change significantly. When the stimulation is applied to the rat for just 1 sec, the ammonia level in the brain is not elevated, and 4 sec later, after 1 sec electric stimulation, the ammonia content is increased to the same extent as after 5 sec stimulation. It is suggested that even 1 sec of stimulation is enough to initiate an increase of the ammonia content in living brain, but it takes a further 4 sec to accumulate the ammonia at a measurable amount.

On the other hand, when the rat is stimulated continuously for 30–60 min with electric current, the glutamine content increases significantly, though the ammonia level stays constant at the resting level. As to the behavior of rat stimulated continuously, the rat assumes a peculiar posture, a fixed rigid posture, to minimize the effect of electric shock. Such posture produced by continuous electric stimulation might correspond to a strong defensive inhibition as a term of conditioned reflex. After prolonged electric stimulation, both the content of glutamine and the specific activity of labeled glutamine in the brain which comes from glucose-^{14}C are increased. It is then explained that the glutamine-synthesizing system is also augmented, accompanied by the continuous liberation of ammonia in living brain during the stimulation. A schematic representation for ammonia metabolism in functioning brain is illustrated in Fig. 5.

In the next step, the physiological effects that are induced by 5-sec electrical stimulation are tested. When the second 5-sec stimulation is applied at any moment within 120 min after the first 5-sec stimulation, ammonia content in the brain does not increase by the application of a second 5-sec stimulation and it tends to lower the level of ammonia. At 150 min after the first stimulation, an increase in ammonia induced by the second stimulation is observed just as in the case of naïve animals. It may be concluded that the processes concerned with ammonia metabolism in the rat brain are sparked to operation by a transient increase of ammonia

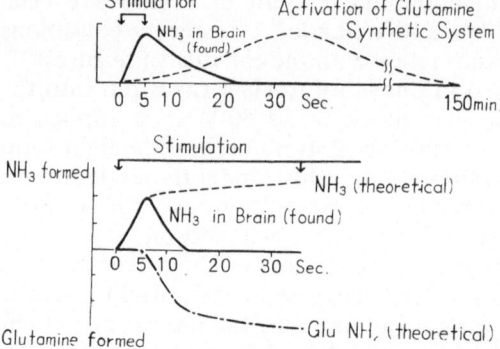

Fig. 5. Schematic representation for ammonia metabolism in brain produced by stimulation. (From Tsukada.[5])

caused by electric shock, and that the glutamine-synthesizing system, which is thought to be a main route of detoxication of ammonia, is activated for 120 min following the first stimulation. Eventually the stimulation produces an after-effect on the chemical process in the brain, which lasts a long time. Experiments are carried out using conditioned rats, enforcing by the lighting of a lamp as a conditioned stimulus instead of electric current. The results obtained on light-conditioned animals are quite similar to those with electric shock (Table VII).

TABLE VII

Effect of Conditioned Stimulus on Ammonia and Glutamine Contents in Rat Brain[a]

Conditions	Ammonia (μmoles/g)[b]	Glutamine (μmoles/g)
Resting state	0.35 (8)	3.41
5-sec Conditioned stimulus	0.42 (8)[c]	3.19
60-sec Conditioned stimulus	0.35 (4)	3.25
Elect. stim. 60 min after cond. stim.	0.41 (5)[c]	3.51
Twice cond. stim. with 60-min interval	0.39 (7)[d]	3.13
Three times cond. stim. with 60-min interval	0.35 (6)	3.18
Elect. stim. 10 min after lighting[e]	0.43 (9)[c]	3.40

[a] From Tsukada.[5]
[b] Numbers in parentheses indicate number of experiments.
[c] $P < 0.01$.
[d] $P < 0.05$.
[e] Unconditioned naïve rats are used.

Furthermore, to ensure that the effect of the electrical stimulation on the changes of ammonia metabolism of *in vivo* brain could be related to a transient increase of ammonia in the brain, ammonium salts are injected into the rate experimentally instead of the application of electrical stimulation. At 60 min after administration of ammonium chloride to the rat, electrical stimulation for 5 sec does not increase the ammonia level in the brain further, although there are found to be no significant changes in the brain ammonia and glutamine levels by the dose of 20 mg NH_4Cl/kg body weight (Table VIII). This shows that ammonia injection does alter the ammonia metabolism in the brain as well as the electrical stimulation does.

TABLE VIII

Effect of Ammonium Chloride Injection on Ammonia and Glutamine Contents in Rat Brain[a]

Injection	Electric shock	Ammonia (μmoles/g)	Glutamine (μmoles/g)
NH_4Cl 20 mg/kg	−	0.35 (6)	3.34
NH_4Cl 20 mg/kg	+	0.36 (6)	3.31
H_2O 0.2 ml	+	0.47 (11)[b]	3.83

[a] From Tsukada et al.[14]
[b] $P < 0.01$.

Utena has reported[15] that ammonia metabolism in the brain changes very gradually in a plateau-shaped course, accompanied by a long-lasting behavioral aberration in the mouse. The ammonia content in the brain of El mouse and chronic methamphetamine-intoxicated mouse in the resting state shows a much lower value than the control mouse (Table IX). However, when they are fed in a revolving wheel as a stimulative living condition for several days, the ammonia levels in the brain increase up to the control level. Also, El mice which have been treated in a revolving wheel never get convulsive seizures from seesawing or tossing stimulation. The threshold of

TABLE IX

Ammonia and Glutamine Contents in the Brain of Methamphetamine-Intoxicated Mice[a]

Substances	Control mouse	Methamphetamine-intoxicated mouse
Ammonia (μmole/g)	0.25	0.15
Glutamine (μmole/g)	5.07	5.54
Ach (mμmole/g)	9.2	11.0
ATP (μmole/g)	2.83	2.47

[a] From Utena.[15]

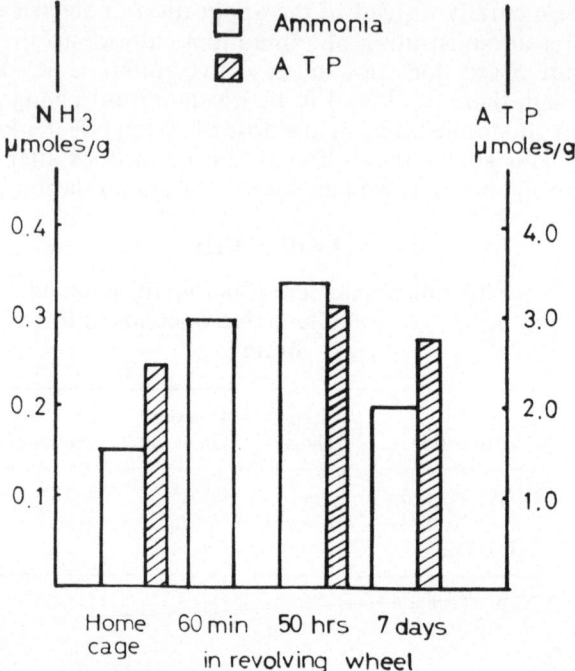

Fig. 6. Changes of ammonia and ATP in the brain of metam-
phetamine-intoxicated mice by transfer from home cage to
revolving wheel. (From Utena.[15])

convulsion seems to rise by a subliminal stimulation caused by running in a
revolving wheel (Fig. 6). These experiments demonstrate that ammonia
metabolism in the brain responds quite rapidly to varying physiological
status of the brain; it can also be changed very slowly by regulating the level
of brain activity.

As for the physiological activity of the brain, the ammonia forming and
binding systems (catabolic and anabolic processes) may be involved. The
source of ammonia liberated in the brain *in vivo* has not yet been fully
clarified.

VI. REFERENCES

1. H. Weil-Malherbe, Significance of glutamic acid for the metabolism of nervous tissue,
   Physiol. Rev. **30**:549–568 (1950).
2. D. Richter and R. M. C. Dawson, The ammonia and glutamine content of the brain, *J.
   Biol. Chem.* **176**:1199–1210 (1948).
3. D. Benitez, G. R. Pscheidt, and W. E. Stone, Effects of fluoroacetate poisoning on citrate,
   lactate and energy rich phosphates in the cerebrum, *Amer. J. Physiol.* **176**:483–487 (1954).
4. H. Weil-Malherbe, Ammonia metabolism in the brain, in *Neurochemistry*, 2nd ed., pp.
   321–330 (K. A. C. Elliott, I. H. Page, J. H. Quastel, ed.) (1962).

5. Y. Tsukada, Amino acid metabolism and its relation to brain functions, *Prog. Brain Res.* **21A**:268–291 (1966).
6. S. Tashiro, Studies on alkaligenesis in tissues. 1. Ammonia production in the nerve fiber during excitation, *Am. J. Physiol.* **60**:519–543 (1922).
7. H. Winterstein, The chemical foundation of nerve excitation, *Naturwissenschaften* **21**:21–23 (1933).
8. M. Bülow and E. G. Hohmes, Die Sauerstoffaufnahme und Ammoniakbildung von Gehirn bei Gegenwart narkotisch wirkender Stoffe, *Biochem. Z.* **245**:459–465 (1932).
9. C. Torda, Ammonium ion content and electrical activity of the brain during the preconvulsive and convulsive phases induced by various convulsants, *J. Pharmacol.* **107**:197–203 (1953).
10. B. E. Leonard, The effect of audiogenic seizures on labile nitrogen and phosphorus containing compounds in the rat brain, *Biochem. Pharmacol.* **14**:1293–1298 (1965).
11. F. N. Minard and R. V. Davis, The effects of electroshock on the acid soluble phosphates of rat brain, *J. Biol. Chem.* **237**:1283–1289 (1962).
12. S. Schenker and J. H. Mendelson, Cerebral adenosine triphosphate in rats with ammonia induced coma, *Am. J. Physiol.* **206**:1173–1176 (1964).
13. E. A. Vladimirova, Ammonia formation in rat cerebral hemispheres induced by conditioned stimuli, *Daklady Akad. Nauk.* (USSR) **XCV**:905–908 (1954).
14. Y. Tsukada, G. Takagaki, S. Sugimoto, and S. Hirano, Changes in the ammonia and glutamine content of the rat brain induced by electric shock, *J. Neurochem.* **2**:295–303 (1958).
15. H. Utena, Behavioral aberrations in methamphetamine-intoxicated animals and chemical correlates in the brain, *Prog. Brain Res.* **21B**:192–207 (1966).
16. Y. Godin, J. Mark, CH. Kayser, and P. Mandel, Ammonia and urea in the brain of garden dormice during hibernation, *J. Neurochem.* **14**:142–144 (1967).
17. E. V. Rowsell, Applied electrical pulses and the ammonia and acetylcholine of isolated cerebral cortex slices, *Biochem. J.* **57**:666–673 (1954).
18. M. G. Larrabee, P. Harowicz, W. Stekiel, and M. Dolivo, Metabolism in relation to function in mammalian sympathetic ganglia, in *Metabolism of the Nervous System*, pp. 208–220 (D. Richter, ed.), Pergamon Press (1957).
19. G. M. Clark and B. Eiseman, Studies on ammonia metabolism. IV. Biochemical changes in brain tissue of dogs during ammonia-induced coma, *New England J. Med.* **259**:178–180 (1958).
20. R. Cohn and D. C. Castell, The effect of acute hyperammonaemia on the electroencephalogram, *J. Lab. Clin. Med.* **68**:195–205 (1966).
21. E. Baraona, A. Salinas, E. Navia, and H. Orrego, Alternations of ammonia metabolism in the cerebral cortex of rats with hepatic damage induced by carbon tetrachloride, *Clin. Sci.* **28**:201–208 (1965).
22. C. B. Booth, J. G. Swadey, B. Klein, and J. C. Hall, Neurologic status of patients with liver disease, *Arch. Neurol.* **8**:257–263 (1963).
23. J. F. Sullivan, H. Linder, P. Holdener, and L. Ortmeyer, Blood ammonia in cerebral dysfunction, *Amer. J. Med.* **30**:893–898 (1961).
24. M. Eichler, Psychological changes associated with induced hyperammonaemia, *Science* **144**:886–888 (1964).
25. S. P. Bessman and A. N. Bessman, The cerebral and peripheral uptake of ammonia in liver disease with an hypothesis for the mechanism of hepatic coma, *J. Clin. Invest.* **34**:622–628 (1958).
26. Y. Tsukada and G. Takagaki, Ammonia formation systems in brain tissue, *Nature* **173**:1138 (1958).
27. S. S. Oja, H. A. R. von Bonsdorff, and O. F. C. Lindroos, Ammonia content of the developing rat brain, *Nature* **212**:937–938 (1966).

28. S. M. Crain, Development of electrical activity in the cerebral cortex of the albino rat, *Proc. Soc. Exp. Biol. Med.* **81**:49–51 (1952).

29. H. Naruse, M. Kato, M. Kurokawa, R. Hara, and T. Yabe, Metabolic defects in a convulsive strain of mouse, *J. Neurochem.* **5**:359–369 (1960).

30. F. Dickens and G. D. Greville, The metabolism of normal and tumor tissue. XIII Neutralsalt effects, *Biochem. J.* **29**:1468–1483 (1935).

31. H. Weil-Malherbe, Observations on tissue glycolysis, *Biochem. J.* **32**:2257–2275 (1938).

32. R. Vrba, J. Folberger, and V. Kanturek, On the mechanism of ammonia formation in guinea pig brain slices, *J. Neurochem.* **2**:187–196 (1958).

33. R. Rybová, The effect of cations on ammonia accumulation in brain cortex slices, *J. Neurochem.* **4**:304–310 (1959).

34. P. J. C. Mann, M. Tennenbaum, and J. H. Quastel, Acetylcholine metabolism in the central nervous system. The effects of potassium and other cations on acetylcholine liberation, *Biochem. J.* **33**:822–835 (1939).

35. C. Torda and H. C. Wolff, Effect of amino acids on acetyl-choline synthesis, *Proc. Soc. Exp. Biol Med.* **59**:181–183 (1945).

36. T. Sollmann, *Manual of Pharmacology*, Saunders, Philadelphia and London (1937).

37. C. Ajimone-Marsan, M. G. F. Fuortes, and F. Marossero, Influence of ammonium chloride on the electrical activity of the brain and spinal cord, *EEG. Clin. Neurophysiol.* **1**:291–298 (1949).

38. H. Weil-Malherbe and R. H. Green, Ammonia formation in brain. I. Studies on slices and suspensions, *Biochem. J.* **61**:210–218 (1955).

39. G. Takagaki, Ammonia formation and glutamic acid in brain tissue, *J. Biochem.* (Tokyo) **42**:131–139 (1955).

40. H. Weil-Malherbe, XCV. Studies on brain metabolism. I. The metabolism of glutamic acid in brain, *Biochem. J.* **30**:665–676 (1936).

41. G. Takagaki and Y. Tsukada, Endogenous respiration and ammonia formation in brain slices, *Arch. Biochem. Biophys.* **68**:196–205 (1957).

42. P. Faulkner and J. H. Quastel, Anaerobic deamination of D-glucosamine by bacterial and brain extracts, *Nature* **177**:1216–1218 (1956).

43. W. K. Jordan, R. March, O. B. Houchin, and E. Popp, Intracellular partition of purine deaminases in rodent brain, *J. Neurochem.* **4**:170–174 (1959).

44. C. E. M. Pugh and J. H. Quastel, Oxidation of aliphatic amines by brain and other tissues, *Biochem. J.* **31**:2306–2321 (1937).

45. G. Kh. Buniatian and S. G. Movsesian, Deamination and reamination of nicotinamide adenine dinucleotides in brain, *Vop. Biokhim. Mozga. Akad. Nauk. Arm. SSR* **2**:5–22 (1966).

46. E. A. Vladimirova, On the initial phase of excitation in absolute differentiation and the content of free ammonia in the greater cerebral hemispheres of rats, *Biull. Eksp. Biol. Med.* **26**:42–45 (1961).

47. H. A. Krebs, The synthesis of glutamine from glutamic acid and ammonia and the enzymic hydrolysis of glutamine in animal tissues, *Biochem. J.* **29**:1951–1969 (1935).

48. H. J. Strecker, Glutamic dehydrogenase, *Arch. Biochem. Biophys.* **46**:128–140 (1953).

49. J. A. Muntz, The formation of ammonia in brain extracts, *J. Biol. Chem.* **201**:221–233 (1953).

50. J. R. Bendall and C. I. Davey, Ammonia liberation during rigor mortis and its relation to changes in the adenine and inosine nucleotides of rabbit muscle, *Biochim. Biophys. Acta* **26**:93–103 (1957).

51. H. Weil-Malherbe and R. H. Green, Ammonia formation in brain. 2. Brain adenylic deaminase, *Biochem. J.* **61**:218–224 (1955).

52. G. B. Ansell and D. Richter, Evidence for a neutral proteinase in brain tissue, *Biochim. Biophys. Acta* **13**:92–97 (1954).

53. A. Geiger, J. Dobkin, and J. Magnes, Accumulation of acid-soluble nitrogen in the brain cotex of cats during stimulation, *Science* **118**:655–657 (1953).

54. R. Vrba, A source of ammonia and changes of protein structure in the rat brain during physical exertion, *Nature* **176**:117–118 (1955).

55. S. Berl, G. Takagaki, D. D. Clarke, and H. Waelsch, Metabolic compartments *in vivo*, *J. Biol. Chem.* **237**:2562–2569 (1962).

56. E. B. Chain, M. Chiozzotto, F. Pocchiari, C. Rossi, and R. Sandman, Participation of the ammonium ion in the transformation of glucose to amino acids in brain tissue, *Proc. Roy. Soc. (B)* **152**:290–297 (1960).

57. D. E. Hathway and A. Mallinson, Chemical studies in relation to convulsive conditions, *Biochem. J.* **90**:51–60 (1964).

58. F. Salvatore, V. Bocchini, and F. Cimino, Ammonia intoxication and its effects on brain and blood ammonia levels, *Biochem. Pharmacol.* **12**:1–6 (1963).

59. F. Rosenthal, P. S. Timiras, and K. E. Roberts, A biometric study of the effects of ammonium acetate on electroshock convulsions, *J. Pharmacol. Exp. Therap.* **133**:364–370 (1961).

60. R. Takahashi, T. Nasu, T. Tamura, and T. Kariya, Relationships of ammonia and acetylcholine levels to brain excitability, *J. Neurochem.* **7**:103–112 (1961).

61. W. Thorn and J. Heimann, Beeinflussung der Ammoniak-Konzentration in Gehirn, Herz, Leber, Niere und Muskulatur durch Ischämie, Anoxie, Asphyxie und Hypothermie, *J. Neurochem.* **2**:166–177 (1958).

62. E. J. Conway and R. Cooke, The deaminases of adenosine and adenylic acid in blood and tissues, *Biochem. J.* **33**:479:492 (1939).

63. S. Lahiri and J. H. Quastel, Fluoroacetate and metabolism of ammonia in brain, *Biochem. J.* **89**:157–163 (1963).

64. M. B. R. Gore and H. McIlwain, Effects of some inorganic salts on the metabolic response of sections of mammalian cerebral cortex to electrical stimulation, *J. Physiol.* **117**:471–480 (1952).

65. D. B. Tower, J. R. Wherrett, and G. M. Mckhann, Functional implications of metabolic compartmentation in the central nervous system, in *Regional Neurochemistry*, pp. 65–88 (S. S. Kety and J. Elkes, eds.) Pergamon Press (1961).

66. R. L. Wechsler, W. Crum, and J. L. A. Roth, The blood flow and O_2 consumption of the human brain in hepatic coma, *Clin. Res. Proc.* **2**:74–75 (1954).

67. D. F. De Almeida, E. B. Chain, and F. Pocchiori, Effect of ammonium and other ions on the glucose dependent active transport of L-histidine in slices of rat brain cortex, *Biochem. J.* **95**:793–796 (1965).

Chapter 7

THE UREA CYCLE

Hrachia Chachatur Buniatian

Institute of Biochemistry
Academy of Sciences of Armenian S.S.R.
Yerevan, Armenia S.S.R.

I. INTRODUCTION

The presence of 22–30 mg% (3.66–5.0 μmoles/g) of urea in brain was reported by Marshall and Davis as early as 1914.[1] Similar amounts were found in cat brain by Tallan *et al.*[2] The values given for rat brain by Sporn *et al.*[3] (4.9 μmoles/g) are in good agreement with the amount found by us (4.5 μmoles/g)[4,5] and by Gershenovitch *et al.* (3.8–4.92 μmoles/g).[6] Roberts and Morelos found the level of urea in cerebral cortex to be equal to 5.49 μmoles/g.[7] According to Shaw and Heine,[8] the content of urea in rat brain cerebral hemispheres was 6.56, in midbrain 5.78, in cerebellum 7.68, and in pons medulla 4.76 μmoles/g. The figures presented show that the brain contains a considerable amount of urea, which is almost equal to that found in liver—5.0–6.0 μmoles/g.

The study of the urea cycle in brain was promoted by the discovery by Allan *et al.*[9] in the urine and CSF of two mentally retarded sibs of large amounts of a ninhydrin-positive substance which was identified by Westall[10,11] as argininosuccinic acid (ASA), an intermediate in urea formation in liver.[12] Later on Cusworth and Dent confirmed that the urine of one of the affected children contained large amounts of ASA.[13]

II. ENZYMES PARTICIPATING IN UREA FORMATION IN BRAIN

Of the enzymes of the urea cycle Walker[14] found argininosuccinase (ASAase) activity in the brain as well as in other organs of the dog. Sporn *et al.*[3,15] and Davies *et al.*[16] administered intracisternally uniformly labeled and guanidino-^{14}C labeled L-Arg and showed that the specific activity of brain urea was at least thirty times as much as the blood urea-specific activity. These results indicated that the brain possessed arginase activity. Tomlinson and Westall[17] demonstrated ASAase activity in homogenates of rat brain, spleen, and blood cells. ASA was formed when homogenates of these tissues were incubated with Arg and fumaric acid. On incubating homogenates with ASA in the presence of added arginase, Orn was formed. The cerebellum and brain stem displayed the highest activity of ASAase, cerebral gray matter having greater activity than white. Arginase activity of these brain regions did not differ significantly in the presence of added Mn^{2+}.[17]

The activity of the enzymes participating in the conversion of citrulline to Arg and that of the latter to urea have been studied by Ratner *et al.*[18] in brain of various mammalian species. In their experiments ASA was formed in brain tissue following the addition of Asp, ATP, and citrulline by a mechanism similar to that observed in liver. An ASA synthetase activity of the same order was demonstrated in the same ammonia sulfate fractions of rat, ox, monkey, and human brain. The ASAase was localized in the supernatant, its acitivity being higher than that of ASA synthetase. Arginase activity was demonstrated in homogenates and its various ammonium sulfate fractions.[18] Its activity was increased considerably following the addition of Mn^{2+}. Highest activity was found for rat brain, lowest for steer.

A block of ASAase activity in brain has been also considered as a possible cause of ASA aciduria by Ratner *et al.*[18] The aforementioned investigations, especially those of Ratner *et al.*,[18] are strong indications

TABLE I

Citrulline Synthesis (μmoles/flask) in Rat Liver and Brain[a]

Number of experiments	Liver, washed residue	Brain	
		Washed residue	Whole homogenate
1	5.44	0.00	0.00
2	5.60	0.00	0.00
3	4.98	0.00	0.00
4	5.84	0.00	0.02
5	6.60	0.02	0.00
6	5.26	0.00	0.01
7	5.78	0.02	0.00
8	6.32	0.00	0.00
9	6.00	0.00	0.00
Mean	5.73	0.01	0.01

[a] Incubation mixture (3.5 ml) was composed of 0.7 ml of washed residue [prepared according to Braunstein *et al.*[21]]; 30.0 μmoles of $NaHCO_3$; 35.0 μmoles of $MgSO_4$; 25.0 μmoles of NH_4Cl; 40.0 μmoles of DL-Orn; 120 μmoles of L-Glu; 7.0–10 μmoles of ATP and 0.017 M potassium phosphate buffer, pH 7.2 (final concentration). Incubation at 37.5°C in an atmosphere of O_2 for 1 hr, reaction stopped by 5 ml of 10 g % $HClO_4$, citrulline determined in the supernatant by the method of Archibald.[22]

pointing to the presence of the three enzymes of the urea cycle—ASA synthetase, ASAase, and arginase in brain. However, the presence of carbamylphosphate synthetase (carb-P synthetase) and that of Orn transcarbamylase in brain remained to be shown.

The investigations of our laboratory (Table I) showed that in contrast to liver, no citrulline was synthesized in brain tissue.[5,19,20]

As seen from the data presented in Table I, an intensive synthesis of citrulline took place in washed residues of liver tissue but not in washed residues or homogenates of brain. Further studies showed that citrulline was not formed in washed residues and homogenates of brain, even following the addition of Orn transcarbamylase (Table II), which indicated the absence of carb-P synthetase in brain tissue.

TABLE II

Carbamylphosphate Synthetase Activity of Rat Brain (Citrulline, μmoles/flask)[a]

Number of experiments	Washed residue	Whole homogenate
1	0.11	0.06
2	0.09	0.04
3	0.08	0.05
4	0.14	0.07
5	0.07	0.03
6	0.08	0.04
7	0.10	0.06
Mean	0.09	0.05

[a] Conditions were the same as in Table I. 25–30 units of Orn transcarbamylase, obtained according to Caravaco and Grisolia[23] were added.[5,19]

No citrulline was formed in brain tissue even after the addition of carb-P and Orn to the above-mentioned brain preparations (Table III). A considerable amount of citrulline was detected on the joint addition of carb-P, Orn, and Orn transcarbamylase. The latter was determined in homogenates of liver and brain also through the arsenolysis of citrulline by the method of Krebs et al.[25] It is known that the arsenolysis of citrulline was catalyzed by Orn transcarbamylase.[25,26] We have shown that while this process did not proceed in brain, it was quite active in liver tissue.[5,19,20]

The results obtained indicated that brain tissue possessed no carb-P synthetase and Orn transcarbamylase activities and confirmed the presence of ASA synthetase, ASAase, and arginase, in this tissue. Kemp and Woodbury[27] studying the intermediates of the urea cycle in rat cerebral cortex

TABLE III

Ornithine Transcarbamylase Activity of Rat Brain Homogenates (Citrulline, μmoles/flask)^a

Number of experiments	Without Orn transcarbamylase	With 30 units Orn transcarbamylase
1	0.02	8.64
2	0.00	5.82
3	0.01	7.50
4	0.02	8.96
5	0.01	8.09
6	0.00	7.98
7	0.00	7.85
Mean	0.01	7.83

^{a} Incubation mixture was composed of 1 ml homogenate; 40 μmoles DL-Orn; 50 μmoles glycylglycine buffer (pH 8.0) and carb-P (obtained biosynthetically by the method of Lowenstein and Cohen.[24] Incubation at 37.5°C, in an atmosphere O_2, for 40 min.

following the intracisternal administration of ureido-^{14}C citrulline had shown that it was easily converted to ASA, Arg, and urea. In preliminary studies with 2-^{14}C Orn they identified succinic acid, Asp, Glu, Gln, and GABA, but no labeled citrulline or other intermediates of the urea cycle. In experiments carried out by us on various brain preparations, Orn was shown to be in-effective in the synthesis of citrulline except following the addition of carb-P and Orn transcarbamylase. The absence of carb-P synthetase in brain tissue was confirmed by Appel and Silberberg.[29] Following the intracarotid administration of ureido-^{14}C citrulline to rabbits, we detected labeled citrulline, Arg, and urea in brain. These results favor the proposition that the blood provides the brain with citrulline. Whether citrulline can be formed in brain via other pathways remains to be shown.

It must be noted that the enzymes catalyzing one or more of the steps of the formation of Arg from citrulline are very widespread in various organisms.[28] Kidneys, for instance, are similar to brain in possessing the enzymes catalyzing the conversion of citrulline to Arg and that of the latter to urea.[28] Moreover, citrulline is converted to Arg in chicken, especially in the kidneys, while the enzymes responsible for the synthesis of citrulline from Orn are absent.[30] In chicken Arg is an essential amino acid and can be replaced by citrulline but not by Orn. Tamir and Ratner,[31] who have found no carb-P synthetase activity in any of the tissues of birds, have established the presence of ASA synthetase and ASAase in kidneys, but not in liver of chicken. Arginase activity was high in kidneys and low in liver.[31] The liver of ureotelic animals has been shown to be the only organ possessing all the enzymes of the urea cycle.[32,33] The above-mentioned facts and the data available led to the proposal that the main difference between ureotelism and uricotelism is the inability of uricotelic animals to synthesize citrulline in liver thus having a greater requirement for Arg. The synthesis of Arg from citrulline in mammalian brain and kidneys is similar to the formation of Arg from citrulline in kidneys of chicken.

In this respect the presence of enzymes of the urea cycle in the brain of chicken, frogs (*Rana asculenta*), rat [5,19] and turtle is quite interesting. As is shown in Table IV, the content of urea in brain of rats and frogs equals 4.36 and 4.52 μmoles/g respectively; in aquatic turtle only traces were found while the brain of terrestrial turtle and chicken contain none. No carp-P synthetase and Orn transcarbamylase have been found in the brains of all the species studied. The enzymes converting citrulline to Arg were very active in rat brain and exhibited low activity in frogs and chicken. The results presented indicate that the brain of birds also contained enzymes, although of low activity, which catalyze the conversion of citrulline to Arg, which is essential in protein synthesis. It is interesting that arginase activity is higher in the frog and especially in the chicken but is low in rats and turtle. A higher arginase activity together with an insignificant content of urea (0.4–0.5 μmoles/g) has been found by us in brain of chick embryos, which decreases gradually in the process of development, remaining at high levels in 1–3-day-old chicken. The high arginase activity of chicken brain and the absence of urea there remained enigmatic.

It is interesting to note that according to Mora et al.[33] K_m for the arginase of liver of ureotelic animals in respect to Arg is considerably higher than the K_m for the arginase of liver of uricotelic animals. Besides, it was found that the arginase from ureotelic animals was inhibited by excess of substrate whereas such was not the case with the arginase from uricotelic animals. They have shown[34] that the arginase present in the liver of ureotelic animals differs from that of uricotelic animals not only in respect to its K_m, inhibition by an excess of substrate, but its antigenicity and molecular weight. However, both enzymes are highly specific for L-Arg, their activity is enhanced by Mn^{2+} and the optimum pH for their action is about the same. The results obtained make the authors conclude that the arginase of uricotelic animals is functionally different from the arginase of liver of ureotelic animals.

The investigations of Davtian[35] from our laboratory have shown that arginase of chicken brain is also activated by Mn^{2+}, its optimum pH being about 9.5. The brain of rat embryo was found by him to contain 4.16 μmoles/g of urea, that of 1–3-day-old rats 8.63 μmoles/g while in adult rats its concentration decreased to 4.43 μmoles/g. The arginase activity of brain is high in the embryo and low in the newborn and adult—i.e., there is no parallelism between arginase activity and urea content during development.[35] The data obtained by Davtian concerning the arginase activity of chicken, rat, aquatic, and terrestrial turtle are of interest in that they all have a higher K_m than that of rat liver arginase and their activity is not inhibited by excess of substrate, i.e., the arginase activity of brain, independent of the type of

TABLE IV

Urea Content and Activity of Enzymes of the Urea Cycle in Various Vertebrate Brains

Animal	Urea content (prior to incubation)	Carb-P synthetase. Citrulline increase[a]	Orn transcarbamylase. Citrulline increase[b]	ASA synthetase and ASAase. Urea synthesis from citrulline and Asp in presence of excess arginase. Increase of urea[c]	Arginase. Increase of urea[d]
Rat	4.36 ± 0.11[e]	0	0	3.2 ± 0.12	42.3 ± 4.0
Frog	4.52 ± 0.61	0	0	0.69 ± 0.046	173 ± 8.2
Chicken	0	0	0	0.56 ± 0.044	185 ± 14
Aquatic turtle	traces	—	—	—	67 ± 6.1
Terrestrial turtle	0	0	—	—	64 ± 4.6

[a] The conditions were the same as in Table II.

[b] The conditions were the same as in Table III.

[c] 2 ml brain homogenate (200 mg tissue) prepared in 0.05 M potassium phosphate buffer (pH 7.4) were incubated with: L-Asp, 6.7 mmoles; L-Cit, 6.7 mmoles; ATP, 3.4 mmoles; $MgSO_4$, 1.7 mmoles; succinic acid, 6.7 mmoles and 20.0 units arginase (Sigma). Incubation at 37.5°C, in an atmosphere of oxygen, for 1 hr.

[d] Arginase activity was determined by incubating brain homogenates at 37.5°C for 1 hr, at pH 9.0–9.1 in 0.08 M tris or 0.4 M glycine buffer in the presence of L-ArgHCl—(50 μmoles and $MnCl_2 \cdot H_2O$ — 5.0 μmoles with the subsequent estimation of urea by the urease method.

[e] The results are given in μmoles/g fresh tissue and are means ± S.E.M. of six to eight experiments.

nitrogen excretion, belongs to the type of arginases characteristic of the liver of uricotelic animals.

Davtian has shown that the arginase of brain is mainly localized in the nuclear fraction. He succeeded in purifying this enzyme from rat brain by 1470-fold and demonstrated that the purified arginase was homogeneous (by electrophoresis and ultracentrifugation) and in its molecular weight (62,000), amino acid composition, and other properties differed markedly from liver arginase.[35a]

The results of numerous investigations indicating the universal distribution of arginase in vertebrate tissues[36] and the data presented make us conclude that the arginase of nonhepatic tissue, including the brain, differs from hepatic arginase of ureotelic animals and, apparently, has its specific, as-yet-unestablished role in metabolic processes.

III. PARTICIPATION OF AMINO ACIDS IN THE SYNTHESIS OF UREA IN BRAIN

The investigations conducted by us[4,19] indicated (Table V) that in brain tissue slices Orn was similar to other amino acids in having no noticeable effect on the synthesis of urea, while in liver slices it increased its

TABLE V

Amino Acids Participating in Urea Synthesis in Rat Brain and Liver Slices[a]

	Brain	Liver
Incubation mixture (prior to incubation)	4.46 ± 0.29^b	5.48 ± 0.31
Incubation mixture (after incubation)	6.82 ± 0.19	7.64 ± 0.26
Incubation mixture + Orn	7.56 ± 0.23	16.81 ± 0.59
Incubation mixture + Cit	9.83 ± 0.14	18.36 ± 0.98
Incubation mixture + Glu	7.18 ± 0.22	8.29 ± 0.19
Incubation mixture + Asp	7.74 ± 0.26	8.94 ± 0.24
Incubation mixture + L-Ala	7.12 ± 0.28	8.64 ± 0.18
Incubation mixture + GABA	7.82 ± 0.22	8.86 ± 0.22
Incubation mixture + β-Ala	7.28 ± 0.24	8.58 ± 0.26
Incubation mixture + Lys	7.35 ± 0.16	8.50 ± 0.16
Incubation mixture + Gly	7.21 ± 0.18	8.32 ± 0.12

[a] Brain cortex slices (200 mg) were incubated in 3 ml of 0.05 M potassium phosphate buffer (pH 7.4) containing: ATP, 10 μmoles; $MgSO_4$, 10 μmoles; glucose, 14 μmoles; amino acids, 20 μmoles and 1 mg (20 units) arginase (Sigma). Incubation was carried out at 37.5°C, in an atmosphere of O_2, for 1 hr. The urea formed was determined by the urease method.

[b] The results are given in μmoles/g fresh tissue and are means $\pm$ S.E.M. of nine to ten experiments.

synthesis considerably. Citrulline enhanced urea formation in brain tissue slices, which was more pronounced in liver slices. These data point to the absence of enzymes affecting the conversion of Orn to citrulline in brain and to the high activity of those participating in the synthesis of Arg and in its cleavage to urea and Orn.

Other amino acids: Glu, Asp, Ala, GABA, β-Ala, Lys, and Gly exhibited nonsignificant activity in urea production, their effect being of the same magnitude in brain and liver slices. Investigations carried out on brain homogenates showed that all the amino acids mentioned, except Asp, participated in the synthesis of urea through transamination. They all induced an increase of urea on their simultaneous addition with citrulline. Hydroxylamine abolished completely their effect and only in experiments with added citrulline and Asp was there a noticeable increase of urea both in the presence and absence of hydroxylamine.[4,19]

The data presented show that GABA and other γ-amino acids did not differ in their effect on urea formation from other amino acids. In this respect the results obtained by us do not agree with those of Gerschenovitch et al.,[6] according to whom GABA increases the synthesis of urea in rat brain slices. They proposed that GABA through transamidination with Arg can be converted to γ-guanidinobutyric acid which according to them, increases the urea content of brain slices.[37] Pisano et al.[38] reported that for the synthesis of γ-guanidinobutyric acid from canavanine and GABA it was necessary to prepare brain extracts low in Gly, inasmuch as the latter is ten times more active than GABA as an acceptor of amidine groups. On incubation of homogenates of rat, rabbit, mouse, guinea pig, and dog brain with canavanine, detectable amounts of guanidinoacetic acid were formed in the absence of added Gly, no additional increase of guanidinoacetic acid or formation of γ-guanidinobutyric acid occurring upon the addition of Gly or GABA, indicating that the enzyme was already saturated with Gly. They found γ-guanidinobutyric acid not to be effectively utilized by brain homogenates.[38]

In our experiments no increase of urea was noticed following the addition of γ-guanidinobutyric acid to rat, rabbit, and chicken brain slices and homogenates.[5,20] According to Gerschenovitch et al.,[37] the joint addition of GABA and carb-P enhance urea formation in rat brain slices. We failed to observe the formation of any ureido compound from Gly, β-Ala, Lys, and GABA in rat brain tissue.[20]

The liver, kidney, and intestinal mucous membrane of rabbits, and liver and kidney of guinea pigs[39,38] have been reported to contain an enzyme splitting guanidinobutyric acid to urea. Rat tissues do not possess such activity.[39,38] Mora et al.[34] have demonstrated the presence of an enzyme hydrolyzing γ-guanidinobutyric acid and a number of guanidino compounds in frog, turtle, lizard, and chicken liver. Such enzymatic activity has not been demonstrated for brain tissue. Some other enzymes present in tissues of certain birds and fish catalyzing the oxidative deamination and subsequent decarboxylation of Arg to γ-guanidinobutyric acid are absent from brain.

IV. UREA LEVEL OF BRAIN IN VARIOUS FUNCTIONAL STATES

Agrawal *et al.*[39] have demonstrated that the level of urea is high in the brains of newborn and 1-day-old rats (11.04–15.233 μmoles/g wet wt.) and that it decreases with age, reaching a level of 4.090 μmoles/g in the adult. These results are in keeping with the results of Davtian.[35]

A number of studies indicate that the content and formation of urea in rat liver increase following a protein-rich diet, fasting, administration of corticosteroids,[40] and in alloxan diabetes.[41] It was interesting to study the concentration of urea and the activity of enzymes affecting its synthesis in brain. The studies of Yeghian and Hagopian[42] in our laboratory (Table VI) have shown that during alloxan diabetes and fasting the concentration of urea increased considerably both in brain, liver, and blood. One to 2 hr following the injection of insulin, the urea content of the brain was not affected while that of liver and blood was considerably increased (Table VI). Show and Heine[43] found that the injection of zinc-insulin induced a significant decrease of urea in rat cerebral hemispheres and some decrease in the cerebellum and other parts of the brain, the decrease in pons medulla being nonsignificant.

TABLE VI

Urea Level of Rat Brain, Liver and Blood in Alloxan Diabetes, Starvation, and Insulin Hypoglycemia

	Brain	Liver	Blood
Control	3.28 ± 1.04^a	3.25 ± 0.09	3.62 ± 0.21
Alloxan diabetes	9.39 ± 1.34	14.44 ± 2.64	12.38 ± 1.25
Starvation	5.97 ± 0.22	7.81 ± 0.47	7.85 ± 0.59
Insulin hypoglycemia	3.78 ± 0.30	5.18 ± 0.39	4.93 ± 0.44

^{a} The results are given in μmoles/g fresh tissue, μmoles/ml blood and are means $\pm$ S.E.M. of II to 38 experiments.

Having in mind that the increased synthesis of urea following a protein-rich diet,[40] fasting,[40,44] and alloxan diabetes[41] is accompanied by a considerable activation of all of the enzymes of the urea cycle in liver, it was interesting to study their activity in rat brain in alloxan diabetes and starvation during which the urea content of the brain is increased. The investigations of Yeghian in our laboratory confirmed once more the absence of carb-P synthetase and Orn transcarbamylase in brain both in normal and alloxan diabetic rats. ASA synthetase and ASAase activity increased about eight times in brain and twice in liver, the arginase activity of both organs increasing to a smaller extent. The results obtained indicate that the enzymes of the urea cycle present in brain are also adaptive. As is shown in Table VI, in alloxan diabetes the content of urea in brain increased about three times. The sources of citrulline in brain for such an increase of urea formation remains to be shown. In alloxan diabetes and fasting the NH_3 and especially the urea content of the blood was increased. In brain no citrulline is formed from NH_3. Following $^{15}NH_3$ administration ^{15}N was recovered in brain mainly in Gln and only negligibly in urea.[45] In the experiments of Flock et al.,[46] $2\frac{1}{2}$, 6, and 24 hr after hepatectomy the NH_3 and especially Gln content of brain was markedly increased (two to five times), while the content of urea was not changed considerably. The possibility cannot be excluded that the increase of brain urea observed in alloxan diabetes is due to a certain extent to the uptake by brain of urea, which is known to pass the blood–brain barrier freely. The preliminary experiments of Yeghian in our laboratory on dogs have shown that in alloxan diabetes the content of urea is higher in arterial blood than in the venous blood coming out of the brain.

The experiments carried out in our laboratory by B. G. Dallakian have shown a marked increase of urea in brain, liver, and blood of rats with brain edema induced by trauma.

What is the role of urea formation in brain? There is no reason to consider it a specific compound for the brain inasmuch as it is absent in the brain of chicken and certain uricotelic animals. Its synthesis in brain is not connected with NH_3 neutralization, since the brain has no carb-P synthetase and Orn transcarbamylase activities. Possibly the synthesis of urea in brain has a role in the neutralization of toxic ASA, the accumulation of which in brain and CSF induces serious mental disturbances.[9,11] In this respect the cyclic forms of ASA,[11,28] which can have more toxic effects, may be of importance.

It is worth mentioning that in the urine and in body fluids of a mentally retarded child, citrulline was found in large amounts while the level of urea in blood and urine of this patient was normal.[47] Kemp and Woodbury[27] found certain unknown compounds following the intracisternal injection of citrulline which are worth studying. A number of authors are of the opinion that the synthesis of Arg from citrulline in nonhepatic tissue is important for the replenishment of the Arg deficit, necessary for protein synthesis.

The denaturation of proteins by urea is well known. The possibility that the small amounts of urea found in brain have an effect on the structure and properties of proteins remains to be shown.

There are indications that even low concentrations of urea (0.148– 0.8 M) induce considerable changes in the activity of various enzymes.[48–50] According to Martinson and Tjakhepield,[51] subcutaneous administrations of urea in 2 g/kg amounts induced a change of brain proteins characteristic of reversible denaturation. Stevenson et al.[52] subsequent to administration of 20% urea via the femoral vein, observed the appearance of spindling and seizure discharges in the electrocorticogram. It was demonstrated that urea administered intravenously in amounts producing blood NPN values commensurate with those observed in patients with uremia, alter cortical activity and excitability. Administrations of hypertonic solutions of urea during edema of brain induced a considerable decrease of intracranial pressure. This beneficial effect of urea cannot be solely ascribed to changes in osmotic pressure inasmuch as hypertonic solutions of glucose or sucrose were less effective.[53]

According to Rose and Dekker,[54] urea-^{15}N is utilized for the synthesis of amino acids in rats maintained on diets containing a mixture of the essential amino acids at their minimal levels. Cys, Gly, and Asp isolated from the animals contained the isotope in high concentrations. There are indications that the rate of urea[55] excretion is much lower than that of its production, which indicates the existence of considerable urea catabolism.

The data presented are reasons for thinking that urea may be viewed as a metabolically and physiologically active substance. Clarification of the biological role of urea and of its formation in brain must await further studies.

V. REFERENCES

1. E. K. Marshall and D. M. Davis, Urea: its distribution in and elimination from the body, *J. Biol. Chem.* **18**:53–80 (1914).
2. H. H. Tallan, S. Moore, and W. H. Stein, Studies on the free amino acids and related compounds in the tissues of the cat, *J. Biol. Chem.* **211**:927–939 (1954).
3. M. B. Sporn, N. Dingman, A. Defalco, and R. K. Davies, The synthesis of urea in the living rat brain, *J. Neurochem.* **5**:62–67 (1959).
4. H. Ch. Buniatian and M. A. Davtian, Urea synthesis in brain slices and homogenates, *in* *Problems of Brain Biochemistry* Vol. 1, pp. 97–104, Akademiia Nauk Armianskoi, S.S.R., Yerevan (1964).
5. H. Ch. Buniatian and M. A. Davtian, Urea synthesis in brain, *J. Neurochem.* **13**:743-753 (1966).
6. Z. S. Gershenovitch, A. A. Krichevskaya, and W. S. Shugalei, Synthesis of urea by brain slices, *Dokl. Akad. Nauk SSSR*, **157**:464–468 (1964).
7. S. Roberts and B. S. Morelos, Regulation of cerebral metabolism of amino acids, *J. Neurochem.* **12**:373–387 (1965).
8. R. K. Shaw and J. D. Heine, Ninhidrin positive substances present in different areas of normal rat brain, *J. Neurochem.* **12**:151–155 (1965).
9. J. D. Allan, D. C. Cusworth, C. E. Dent, and V. K. Wilson, A disease, probably hereditary characterised by severe mental deficiency and constant gross abnormality of amino acid metabolism, *Lancet* **1**:182–187 (1958).
10. R. G. Westall, Abstr. Comm. 4-th Int. Congr. Biochem. N13-14, p. 168, Vienna (1958).
11. R. G. Westall, Argininosuccinic acid-uria: Identification and reactions of the abnormal metabolite in a newly described form of mental disease, with some preliminary metabolic studies, *Biochem. J.* **77**:135–144 (1960).
12. S. Ratner, B. Petrack, and O. Rochovansky, Biosynthesis of urea. V. Isolation and properties of argininosuccinic acid, *J. Biol. Chem.* **204**:95–113 (1953).
13. D. C. Cusworth and C. E. Dent, Renal clearance of amino acids in normal adults and in patients with aminoaciduria, *Biochem. J.* **74**:550–561 (1960).
14. J. B. Walker, Role of pancreas in biosynthesis of creatine, *Proc. Soc. Exp. Biol. Med.* **98**:7–9 (1958).
15. M. B. Sporn, W. Dingman, A. J. Defalco, and R. K. Davies, Formation of urea from arginine in the brain of the living rat, *Nature* **183**:1520–1521 (1959).

16. R. K. Davies, A. J. Defalco, D. Shander, A. Kopelman, and J. Kiyasu, Urea synthesis in the living rat brain, *Nature* **191**:288 (1961).
17. S. Tomlinson and R. G. Westall, Argininosuccinase activity in brain tissue, *Nature* **188**:235–236 (1960).
18. S. Ratner, H. Morell, and E. Carvalho, Enzymes of arginine metabolism in brain, *Arch. Biochem. Biophys.* **91**:280–289 (1960).
19. H. Ch. Buniatian and M. A. Davtian, The synthesis of citrulline in brain, *in Problems of Brain Biochemistry* Vol. 1, pp. 105–111, Akademiia Nauk Armianskoi, S.S.R., Yerevan (1964).
20. H. Ch. Buniatian and M. A. Davtian, Urea synthesis in brain, *in Chemistry and Function of the Nervous System*, Proc. of the International Symposium, September, 1965, pp. 78–88, Leningrad (1967).
21. A. E. Braunstein, I. S. Severina, and H. E. Babskaya, The inhibition of ornithine cycle of urea formation by α-methyl-DL-aspartic acid, *Biokhimiya* **21**:738–745 (1956).
22. R. M. Archibald, Determination of citrulline and allantoin and demonstration of citrulline in blood plasma, *J. Biol. Chem.* **156**:121–142 (1944).
23. J. Caravaco and S. Grisolia, Synthesis of citrulline with animal and bacterial enzymes, *J. Biol. Chem.* **235**:684–693 (1960).
24. J. M. Lowenstein and P. P. Cohen, Studies on the biosynthesis of carbamylaspartic acid, *J. Biol. Chem.* **220**:57–70 (1956).
25. H. A. Krebs, L. V. Eggleston, V. A. Knivett, Arsenolysis and phosphorolysis of citrulline in mammalian liver, *Biochem. J.* **59**:185–193 (1955).
26. P. Reichard, Ornithine carbamyl transferase from rat liver, *Acta Chem. Scand.* **11**:523–536 (1957).
27. J. W. Kemp and D. M. Woodbury, Synthesis of urea cycle intermediates from citrulline in brain, *Biochim. Biophys. Acta* **111**:22–31 (1965).
28. S. Ratner, Urea synthesis and metabolism of arginine and citrulline, *Advanc. Enzymol.* **15**:319–387 (1954).
29. S. H. Appel and D. H. Silberberg, Pyrimidine synthesis in tissue culture, *J. Neurochem.* **15**:1437–1443 (1968).
30. H. Tamir and S. Ratner, A study of ornithine, citrulline and arginine synthesis in growing chicks, *Arch. Biochem. Biophys.* **102**:259–269 (1963).
31. H. Tamir and S. Ratner, Enzymes of arginine metabolism in chicks, *Arch. Biochem. Biophys.* **102**:249–258 (1963).
32. M. E. Jones, A. D. Anderson, A. Anderson, and S. Hodes, Citrulline synthesis in rat tissues, *Arch. Biochem. Biophys.* **95**:499–507 (1961).
33. J. Mora, J. Martuscelli, J. Ortiz-Pined, and G. Soberon, The regulation of urea-biosynthesis enzymes in vertebrates, *Biochem. J.* **96**:28–35 (1965).
34. J. Mora, R. Tarrab, J. Martuscelli, and G. Soberon, Characteristics of arginase from ureotelic and non-ureotelic animals, *Biochem. J.* **96**:588–594 (1965).
35. M. A. Davtian, The arginase activity of brain tissue, *in Problems of Brain Biochemistry* Vol. 3, pp. 273–278, Akademiia Nauk Armianskoi, S.S.R., Yerevan (1967).
35a. M. A. Davtian and H. Ch. Buniatian, Isolation and properties of arginase of rat brain, *Biokhimyia* **35**:412–418 (1970).
36. G. W. Brown, Jr. and P. P. Cohen, Comparative biochemistry of urea synthesis. 3. Activities of urea-cycle enzymes in various higher and lower vertebrates, *Biochem. J.* **75**:82–91 (1960).
37. Z. S. Gershenovitch, A. A. Krichevskaya, J. I. Weksler, I. M. Agafonova, N. G. Atabekova, I. V. Gotlober, V. A. Kheruvimova, W. S. Shungalai, and L. A. Shtherbina, Urea and amides in brain metabolism in normal and external conditions, *in Chemistry and Function of Nervous System*, Proc. of the Int. Symposium, September 1965, pp. 90–96, Leningrad (1967).

38. J. J. Pisano, D. Abraham, and S. Udenfriend, Biosynthesis and disposition of y-guanidino-butyric acid in mammalian tissues, *Arch. Biochem. Biophys.* **100**:323–329 (1963).
39. H. C. Agrawal, J. M. Davis, and W. A. Himwich, Postnatal changes in free amino acids pool of rat brain, *J. Neurochem.* **13**:607–615 (1966).
40. R. T. Schimke, Studies on factors affecting the levels of urea cycle enzymes in rat liver, *J. Biol. Chem.* **238**:1012–1018 (1963).
41. P. McLean and F. Novello, Influence of pancreatic hormones on enzymes concerned with urea synthesis in rat liver, *Biochem. J.* **94**:410–422 (1965).
42. V. B. Yeghian and G. E. Hagopian, Changes in nitrogen metabolism in rat brain in diabetes, starvation and insulin hypoglycemia, in *Problems of Brain Biochemistry*, Vol. 3, pp. 329–339, Akademiia Nauk Armianskoi, S.S.R., Yerevan (1967).
43. R. K. Shaw and J. D. Heine, Effect of insulin on nitrogenous constituents of rat brain, *J. Neurochem.* **12**:527–532 (1965).
44. R. T. Schimke, Differential effects of fasting and protein-free diets on levels of urea cycle enzymes in rat liver, *J. Biol. Chem.* **237**:1921–1928 (1962).
45. D. Duda and P. Handler, Kinetics of ammonia metabolism *in vivo*, *J. Biol. Chem.* **232**:303–314 (1958).
46. E. V. Flock, G. M. Tyce, and Ch. A. Owen, Jr., Utilization of [U-^{14}C] glucose in brain after total hepatectomy in the rat, *J. Neurochem.* **13**:1389–1406 (1966).
47. N. C. McMurray, F. Mohyddin, R. J. Rossiter, J. C. Rathbun, and D. E. Zarfas, An abnormality of citrulline metabolism in a mentally retarded child, *Biochem. J.* **84**:106P (1962).
48. B. N. Manukhin, A study of the mechanism of the inhibitory action of urea upon thiol enzymes as exemplified by monoamino oxidase, *Biokhimyia* **22**:225–229 (1958).
49. T. Tursky, The effect of ammonia ion and urea on glutamate decarboxylase of brain, *Biology (Czéchóslóvákia)* **16**:831–837 (1961).
50. R. Jaenicke and G. Pfleiderer, Über den Einfluss auf die molekulare Strucktur von Lactat-Dehydrogenase aus dem Sweineherz, *Biochim. Biophys. Acta* **60**:615–629 (1962).
51. E. E. Martinson and L. J. Tjakhepild, The effect of ammonia, glutamic acid and urea upon intravitam changes in the macrostructure of cerebral proteins with reference to functional state, *Biokhimiya* **26**:984–992 (1961).
52. G. C. Stevenson, J. R. Campbell, W. Ross, W. F. Collins, and C. T. Randt, Effect of urea on central nervous activity in the cat, *Amer. J. Physiol.* **197**:141–144 (1959).
53. A. E. Richardson, Clinical (pharmacotherapeutical) aspects of cerebral oedema, *Proc. Roy. Soc. Med.* **58**:604 (1965).
54. W. C. Rose and E. E. Dekker, Urea as a source of nitrogen for the biosynthesis of amino acids, *J. Biol. Chem.* **223**:107–121 (1956).
55. E. Regolczi, L. Irons, A. Koj, and A. S. McFarlane, Isotopic Studies of urea metabolism in rabbits, *Biochem. J.* **95**:521–532 (1965).

Chapter 8

FREE NUCLEOTIDES

Paul Mandel

Centre de Neurochimie du Centre National de la Recherche Scientifique
Strasbourg, France

I. INTRODUCTION

The various types of free nucleotides and their synthesis and utilization have been the subject of many papers and several reviews.[1-8] In this article, we will discuss the main problems concerning free nucleotides of brain, and the importance of such information in studies of the regulation of metabolic processes and behavior. To avoid repetition, we have neglected several aspects, namely, pathways related to oxidative phosphorylations, RNA, polysaccharide and phospholipid biosynthesis, which are treated elsewhere in this Handbook.

The introduction of ion-exchange chromatography of nucleotides by Cohn[9] adapted by Hurlbert, Schmitz, and Potter,[10-13] and others[14] as well as paper chromatography and electrophoresis techniques contributed to a tremendous development of our knowledge on free nucleotides. The ability to detect these nucleotides by their UV absorption[15-17] greatly facilitated this progress.

II. CHEMISTRY

Nucleotides are sugar-*o*-phosphate esters of nucleoside compounds in which a purine or pyrimidine base is linked glycosidically to a carbohydrate, mainly ribose or deoxyribose.

The purine and pyrimidine bases are adenine, guanine, uracil, cytosine, and thymidine. Substituted bases exist in nucleic acids, but their free nucleotide precursors have not been found in the free nucleotide pool.

Among the free nucleotides, nucleoside triphosphates are precursors of natural macromolecules, RNA and DNA, and of homopolyribo- and deoxyribonucleotides like polyA, polyC, polyADPR, polyAT.

Some nucleotides, such as NAD, NADP, FAD, and coenzyme A, also contain a vitamin molecule. Some others may contain in addition to a nucleoside diphosphate a sugar, ethanolamine, choline, and so on. Only nucleotides which exist in nervous systems will be discussed here.

III. METHODS OF SEPARATION AND IDENTIFICATION OF FREE NUCLEOTIDES

A valid assay of the amount of free nucleotides depends on the method of preparation of the tissue.

A. Preparation of the Tissue Extract

The technique must minimize the degradation of the tissue-free nucleotides between the time when the animal is killed until degrading enzymes are inactivated. Even with a short delay, ATP is degraded very rapidly to ADP and AMP. The procedure of removing the brain and transferring it immediately to liquid air or cold trichloracetic acid is much too slow. In the case of rat brain, a dissection lasting only 20 sec is sufficient to produce a decrease in ATP content of 70%, even if the brain on removal is immediately immersed in liquid air and then acid-treated at low temperature.[18,19] If the animal is as large as an adult rat, even when it is plunged into liquid air, the delay before the brain becomes frozen allows a significant decrease of the various nucleoside triphosphates to occur. Several other organs[18,20,21] need similar caution.[8]

Thus, to avoid any degradation or any secondary changes in nucleoside triphosphate distribution, appropriate means of killing and of extraction are necessary. Decapitation of the animal by a guillotine after conditioning or a slight and short chloral anesthesia, may prevent stress and a rapid breakdown of high energy phosphates caused by a generalized state of cerebral excitation, without effect on the free nucleotide pool.[22] The whole head has to be frozen, then split and lyophilized. Impregnation of the brain *in situ* by anhydrous methanol–acetone (4:1) followed by a slight drying may permit keeping intact the overall anatomical structure of the brain and isolation of different brain regions.[19] Similar results are obtained with chloroform.[23] The latter procedure had the advantage of allowing subsequent embedment for isolation of single nerve cells and studies of their free nucleotide pool. When the head is of great volume, it is useful to remove the hair in a preliminary step and to make some holes in the skull.

B. Chemical Techniques

Free nucleotides are separated from high polymers either by 10% trichloracetic acid, by 0.6 N perchloric acid, by acidic acetone at −78°C, or by 50% alcohol followed by dilute acids. The last two procedures seem to provide the best results.[19,24]

Several methods are available for the purification, identification, and assay of free nucleotides. The classical procedure, which consisted in the precipitation of the nucleotides as salts of heavy metals such as silver, zinc, barium, and lead,[25-27] may be useful for isolation of mononucleotide, nucleoside diphosphate, or nucleoside triphosphate fractions as well as for elimination of sugar esters.

Ion-exchange chromatography offers the most accurate method for isolation and determination of many tissue nucleotides. Introduced by Cohn et al.,[9,28-31] the method was elaborated by Hurlbert, Schmitz et al.[10-13] for different tissues. Many variants with different ion-exchange adsorbants or elutions have been devised.[30-43] The nucleotide elution is followed by UV absorption. Column rechromatography, paper chromatography,[38,44-60] iontophoresis,[49,61-72] thin-layer chromatography,[58,73-91] or thin-layer electrophoresis[92,93] are necessary for isolation of single nucleotides. The last four methods are very useful when small quantities of nucleotides are available, but these methods applied directly to an acid-soluble tissue extract do not afford a satisfactory separation of the 22 free nucleotides which exist in every tissue. It is better to begin with either column chromatography or salt precipitation.

The pure nucleotides can be assayed by UV absorption at 2600 or 2800 Å, a highly sensitive method. However, measurements of phosphorus, ribose, or deoxyribose, and of associated residues (sugars, amines, etc.) are necessary for precise identification and quantitative evaluation.

Certain nucleotides can be assayed enzymatically[94-99]: NAD and NADP,[100] 5-deoxycytidylic acid.[101] The use of the cycling procedure developed by Lowry et al. through reactions in which NAD and NADP are involved for determination in a whole homogenate of microquantities at a cellular level is a very useful technique.[102,103] Coupling in succession several enzyme systems, ADP can also be determined.[104] However, the enzymatic procedures, which are very sensitive and can be used without preliminary separations of the nucleotides, are not always strongly specific and are sensitive to inhibitors that may exist in the medium. Fluorometric assays of pyridine nucleotides enable determinations in very small amounts.[105-110] A microfluorometric technique to measure the intracellular oxidation-reduction state of pyrimidine nucleotides was applied to the isolated neuron preparations; the method allows the simultaneous recording of impulse activity. The behavior of the fluorescence with continuous UV exposure can be studied, to investigate cell metabolism.[111] Small amounts of ATP can be determined using the light produced by firefly luciferase.[112] Recently, specific and sensitive procedures for measuring UDP-glucuronic acid and UDP-glucose have been applied to the brain of several species.[113] The separation of all free nucleotides by the microelectrophoretic method[114] seems unlikely. Only the four RNA nucleotides and a small number of free nucleotides can be isolated and measured by this method.[115]

In studies on RNA biosynthesis, it is convenient to know the specific activity of the precursors either of α-phosphate of free nucleotides or the specific activity of the nucleotides which correspond to the bases or nucleosides used as radioactive tracers. In this respect, it is necessary to proceed to a hydrolysis of nucleotide di- and triphosphates. After barium hydrolysis[116] or pyrophosphatase treatment, separation of mononucleotides can be performed by ion-exchange chromatography, paper or thin-layer chromatography, or microelectrophoresis, and for nucleosides after dephosphorylation by gas chromatography[117,118] or column chromatography on polyacrylamide gel.[119]

IV. BIOSYNTHESIS OF FREE NUCLEOTIDES

A. Pyrimidine Nucleotides

All studies suggest that the pathways of pyrimidine or purine nucleotide biosynthesis are the same in the nervous system as in most other tissues.[120–124]

Soluble enzymes from cat spinal cord or rat brain convert orotic acid into orotidine-5'-phosphate.[125,126] Decarboxylation of orotidine-5'-monophosphate gives uridine-5'-monophosphate. Injection of 5-azauridine into the fourth ventrical of cat inhibits by 75–89% incorporation of orotic acid into acid-soluble pyrimidines and produces an accumulation of orotidine-5'-phosphate.[125] In total, five reaction steps and two ATP molecules are required for the synthesis of UMP, the parent pyrimidine nucleotide. The latter converted to UTP is aminated to give cytidine-5'-triphosphate.

B. Purine Mononucleotides

The sources of the atoms in the purine ring, first established by Buchanan and Hartman[127] by administrating suitable labeled precursors,

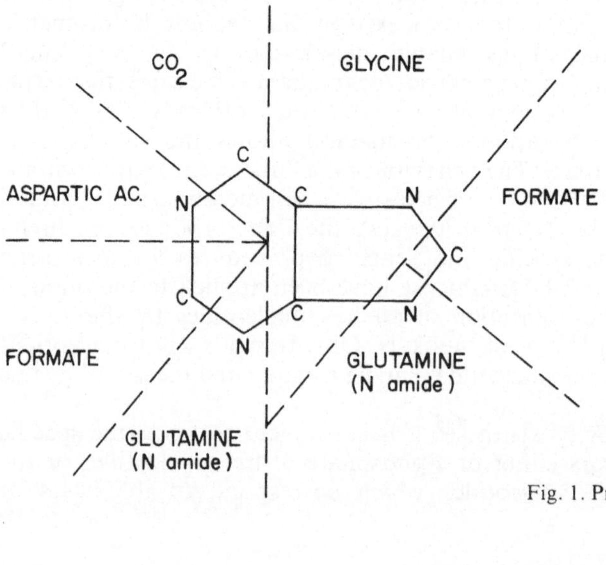

Fig. 1. Precursors of the purine cycle.

are shown in Fig. 1. It was also demonstrated by incorporation experiments with intracerebrally injected labeled precursors such as formate-^{14}C,[128–130] glycine,[129,131] inosine-^{3}H.[132]

The formation of GMP from IMP is also a two-stage reaction in which xanthosine-5′-monophosphate is initially formed and then aminated to give GMP. IMP dehydrogenase is present in very low amounts in the soluble fraction of rat cortex.[131]

Among various tissues studied, brain cortex exhibited the highest adenyl cyclase and cyclic 3′,5′-AMP phosphodiesterase activities.[133] The latter, which controls the level of 3′,5′-AMP, seems to exist *in vivo* in a greatly inhibited state; both ATP and PP are effective inhibitors.[134] Most of the adenyl cyclase activity was found in the mitochrondrial fraction, with a slightly smaller amount present in the microsomal fraction. Cyclic phosphodiesterase is primarily present in fractions containing nerve endings, although the highest specific activity was found in the soluble fraction.[135]

C. Phosphorylation of Mononucleotides

The mononucleotides are phosphorylated by kinases through the diphosphate stage to give nucleoside triphosphates.[131,136–141]

Phosphotransferases, from ATP to AMP, GMP, CMP, and UMP, were shown in brain. In the hog brain, there are different enzymes for GMP and AMP phosphorylation.[131,141] In hog brain and in rat brain,[131] ATP:CMP and ATP:UMP phosphotransferase activities may be due to a single enzyme.

D. The Control of Nucleotide Biosynthesis

There are many data suggesting that the regulation of nucleotide biosynthesis occurs by feedback control or repression. It is also suggested that activation and inhibition are caused in many cases by allosteric modifers, that is, by substances of low molecular weight that modify the catalytic activity of the enzyme by combining with it at a site other than the catalytic site.[142–144]

Carbamyl phosphate synthesis by carbamyl phosphate synthetase, which represents the first step of pyridine nucleotide biosynthesis,[145,146] is inhibited by UTP.[146]

Aspartate transcarbamylase, the second enzyme of the pyridine nucleotide synthesis pathway, seems rather insensitive to feedback inhibition in animal cells.[146,147] In contrast, feedback inhibition of orotidylate decarboxylase by pyrimidine nucleotides[147,148] appears to provide a possible alternative means of regulating *de novo* pyrimidine synthesis.

The first enzyme of *de novo* purine biosynthesis, phosphoribosylpyrophosphate amidotransferase, is inhibited by AMP, IMP, and their analogs.[149] It was also reported that IMP rather than adenine or guanine nucleotides may be responsible for the control of the rate of purine biosynthesis.

The other enzymes of the purine pathway are apparently not controlled by feedback inhibition. For IMP transformylase, which catalyzes the last step in IMP synthesis, a product repression was demonstrated.[149a,149b] It may be noted that while purine nucleotides participate in regulation of the biosynthesis of pyrimidine nucleotides, the reverse has not been observed.

Adenylate deaminase catalyzes the first step in the conversion of AMP to GMP via IMP and XMP. This enzyme, found in brain,[150] is activated by ATP[151-153] and strongly inhibited by GTP.[153,154] The enzyme purified from calf brain is stimulated by alkali metal cations and, in particular, by lithium, both in the presence and absence of ATP.[155] ATP and, to a smaller extent, tripolyphosphate and AMP protect the enzyme against thermal inactivation.[155]

A model in which the enzyme exists in equilibrium between an active form in which the catalytic site is freely accessible to AMP, and an inactive form in which the catalytic site is inaccessible, was suggested.[155]

Ribonucleoside or ribonucleotide reduction seems the main biosynthetic pathway to deoxynucleotides.[156,157] Little information is available on regulation of ribonucleotide reductase synthesis in animal cells, although it is known to occur.[158]

E. The Salvage Pathway

The maintenance of purine and pyrimidine ribonucleotide pools in the brain is essential to many metabolic processes, including RNA synthesis. Whether the rate of *de novo* synthesis of purine and pyrimidine nucleotides in the central nervous system *in vivo* is sufficient for the need of RNA synthesis remains to be determined.

TABLE I

Uptake of Purine and Pyrimidine Bases, Nucleosides and Nucleotides in Brain after Intravenous Injection of Labeled Compounds[a]
(Percentage uptake per gram of brain)

	2 min	15 min	30 min
Adenine-^{14}C	0.028	0.032	0.034
Adenosine-^{3}H	0.18	0.20	0.21
Adenosine-5′-phosphate adenine-^{3}H	0.17	0.19	0.25
Adenosine-5′-phosphate-^{32}P	0.014	0.016	0.022
Uracil-^{14}C	0.032	0.042	0.048
Uridine-^{3}H	0.074	0.130	0.146
UMP uracil-^{3}H	0.048	0.102	0.126
UMP-^{32}P	0.010	0.016	0.018

[a] From Mandel and Mandel.[164]

Despite their capacity for *de novo* nucleotide synthesis, cells of most organisms also synthesize nucleotides from purine, pyrimidine bases, and nucleosides. The enzymes associated with this pathway, including mono-nucleotide pyrophosphorylases, nucleoside phosphorylases and nucleo-side kinases,[140,159] may play a vital role in maintaining neuronal function. Their presence or absence may determine the sensitivity or resistance to growth inhibition by various purine and pyrimidine analogs.[160,161]

Neuronal tissue contains relatively large amounts of the enzymatic activities associated with the salvage pathways. The contribution of these pathways to RNA synthesis in cortical slices seems of the same order of magnitude as the contribution by *de novo* synthesis.[131] Moreover, adenine, uracil, and their nucleosides and nucleotides injected intravenously penetrate the free nucleotide pool and the RNA of brain tissue. The permeability of the blood–brain barrier is the highest for nucleosides or nucleoside moieties of mononucleotides and the lowest for nucleoside monophosphates (Table I).

1. *Nucleotide Biosynthesis Involving Purine Bases*

Formation of purine mononucleotides from the appropriate purine and phosphoribosyl pyrophosphate appears to be a function of at least two enzymes which are only specific for adenine and for guanine and hypoxan-thine.[160] Purine phosphoribosyl transferases are inhibited by purine nucleotides. The immediate products of the enzymes are the most potent inhibitors.[162,163] The role of adenine phosphoribosyl transferase in main-taining the purine nucleotide pool in brain remains obscure. If adenine is carried to the brain by the blood stream, and there is good evidence for such a possibility,[164] the role of this enzyme is obvious.

The importance of hypoxanthine-guanine phosphoribosyl transferase in nucleotide metabolism is demonstrated by the observation that a genetic deficiency of this enzyme is associated with a neurological disorder.[165] Adenine phosphoribosyl transferase may function as a means of intracellular translocation or intercellular transfer of adenine nucleotides. Since nucleo-tides are not known to cross cell membranes intact, and since the reaction catalyzed by adenine phosphoribosyl transferase is readily reversible,[166] this enzyme can carry out the net transfer of AMP from one cell compartment to another or from one cell to another with a minimum expenditure of energy. Uptake of preformed adenine-^{14}C and guanine-^{14}C into brain RNA after intrathecal injections was observed.[132,167–169] There is, however, a greater uptake of glycine-^{14}C than adenine.[132] Azaserine reduces markedly the uptake of glycine-^{14}C into AMP. In contrast, the utilization of adenine-^{3}H via the salvage pathway was not affected.[170] *De novo* pyrimidine precursors are preferentially utilized over preformed bases.[132] In contrast, the orotic acid pathway played a more significant role in the maintenance of pyrimidine nucleotides in the feline neuraxis than the utilization of preformed pyrimi-dines.[125] The incorporation of orotic acid may occur five times faster via the *de novo* than by the preformed base route.[169]

Guanine is rapidly deaminated to xanthine; deamination of adenine and xanthine oxidase activity[131] could not be demonstrated.

2. Nucleotide Biosynthesis Involving Nucleosides

ATP nucleoside phosphotransferase activity was demonstrated in brain for cytidine.[131] The product of enzymatic hydrolysis of brain RNA, the nucleoside-3' or 5'-monophosphates may be converted to nucleosides by nucleotidases[171,171b] and phosphatases, and then to nucleoside-5'-phosphates by phosphoribosyl transferases.[131,171a] It may be a key function in the recycling of purine and pyrimidine derivatives from the breakdown products of RNA. However, the occurrence of an adenosine phosphorylase or hydrolase in brain is not yet established.

Thymidine kinase,[140] which exists in brain, is inhibited by dTTP, the end product of the pathway.[172] It has been suggested that thymidine kinase is important in the control of DNA synthesis in cells.[173,174]

CTP and UTP inhibit both uridine and cytidine phosphorylation.[175,176] This inhibition is partially reversed by high concentrations of either uridine or ATP,[175] or by GTP and deoxyGTP(dGTP).[177]

3. Uptake of Nucleotide Precursors in Brain Slices

By comparing the incorporation of aspartate, glycine, adenine, and adenosine in brain slices, it was demonstrated that these compounds or their metabolites are concentrated above the medium concentration. Hypoxanthine, inosine, guanine, guanosine, orotate, uridine, cytidine, uracil, and cytosine diffuse into the slices, but they or their metabolites are not concentrated by the slices.[131]

Adenine, hypoxanthine, and adenosine as well as inosine, orotate, uridine, or cytidine, are actively converted to nucleotides.[125,131] As well as azauracil-6-^{14}C, azauridine-^{14}C underwent conversion into acid-soluble nucleotides and was incorporated in feline hind-brain RNA.[125] Similarly, in cortical slices incubated with glycine-^{14}C, 1–2% of the radioactivity is in the form of TCA-soluble nucleotides. No conversion of aspartate to pyrimidine nucleotides could be demonstrated, but this may be due to the low rate of the conversion.[131]

In rat brain cortical slices, glycine and adenine are incorporated faster into RNA than hypoxanthine, adenosine, and inosine. Orotic acid and cytidine are incorporated to approximately the same extent.[131]

V. THE METABOLICAL ROLE OF VARIOUS NUCLEOTIDES

Many important biological compounds are nucleotides. They can be classified according to the nucleoside base which they contain.

A. Adenine Nucleotides

Lipmann[178] recognized and clarified the fundamental role of adenosine triphosphate as the most important energy-rich phosphate compound which drives biochemical processes. This free nucleotide is the energy source for such diverse physiological phenomena as muscular contraction, bioelectrical processes, and bioluminescence. *In vivo* and *in vitro* studies have shown that the biosynthesis of proteins, nucleic acids, carbohydrates, and lipids is ATP dependent. Many enzymes require ATP directly as energy source or substrate. These reactions may be placed into two categories, the first involving the cleavage of ATP to provide the driving force, the second the transfer of some portion of ATP to a suitable acceptor molecule.

In transfer reactions the terminal phosphate or pyrophosphate may be cleaved. γ-Phosphate transfer occurs in the phosphorylation of sugars which then enter metabolic cycles, of nucleosides which yield the corresponding monophosphates, and during the biosynthesis of riboflavine-5'-phosphate, pyridoxal phosphate, nicotinamide coenzymes, and coenzyme A. γ-Phosphorylations are also important in the metabolism of glycerol, choline, and ethanolamine involved in lipid biosynthesis.

Thus the first reaction of glucose metabolism is the phosphorylation by hexokinase or glucokinase of glucose. Glucose-6-phosphate is the starting point for anaerobic aerobic glycolysis, for the pentose cycle, and glycogenesis. A high hexokinase activity has been found in brain.[136,179–180] A part of the brain hexokinase is bound to particles.[181,182] The particulate and soluble forms of hexokinase differ in their kinetic and electrophoretic properties.[183] Several phosphotransferases producing metabolites involved in different metabolic pathways were found in nervous tissue (see the first volume of this series, pp. 325–468).

A very active creatine kinase was described in brain by Nachmansohn *et al.*[184,185]; a dual subcellular localization in soluble and particulate fractions[186–189] and also the existence of isozymic forms were demonstrated.[190] Transphosphorylations have been demonstrated in brain.[137] A similar series of reactions was established for the deoxyribose analogues of the nucleotides.[191,192] ATP regeneration through oxidative phosphorylation, very active in brain mitochondria,[193] is discussed elsewhere in this Handbook.

The most important compounds formed by pyrophosphate transfer through pyrophosphokinase are the synthesis of phosphothiamine and phosphoribosyl pyrophosphate. The enzyme which hydrolyzes thiamine pyrophosphate is active in brain.[194,195]

Phosphoribosyl pyrophosphate also produced by a pyrophosphate transfer is involved in the first step of purine biosynthesis and salvage reactions for nucleotide synthesis.[127,131]

The transfer of a nucleoside monophosphate group (i.e., adenylation) to substrates such as nicotinamide monophosphate, flavin mononucleotide, and phosphopantotheine yields the corresponding coenzymes. This type of reaction was discovered by Kornberg in his studies of NAD synthesis.[196] It involves the formation of a covalent bond between two substrate molecules coupled with the cleavage of the pyrophosphate linkage of ATP; the pyrophosphate group is then mineralized.

$$\text{NMN} + \text{ATP} \rightarrow \text{nicotinamide-ribose-P} \overset{O}{\underset{O^-}{\overset{\diagdown}{\underset{|}{P}}}} \overset{O\text{-Adenine}}{\overset{\diagup}{\cdots}} \text{OPO}_3 \cdot \text{PO}_3 \rightleftharpoons$$

$$\text{NAD} + \text{P}_2\text{O}_7$$

Reactions in which the attacking group is a carboxyl group of an amino acid produce amino acyl adenylates as the first step of protein biosynthesis.[191] These ATP-dependent amino acid activations are catalyzed by specific enzymes for every amino acid.[197-199] Since the brain has a rather high protein turnover,[200,201] the amino acid adenylate production is of great importance. By an analogous process adenyl acetate is produced in a reaction coupled with CoA, finally producing acetyl-CoA.[203]

$$\text{ATP} + \text{acetate} \rightleftharpoons \text{adenylacetate} + \text{PP}$$

$$\text{adenylacetate} + \text{CoA} \rightleftharpoons \text{acetyl-CoA} + \text{AMP}$$

An enzymatic reaction discovered by Bachhawat and Coon[203] produces adenylic acid-CO_2. This enzyme appears particularly active in brain tissue[204] and is of importance in the biosynthesis of malonyl CoA.[205] The brain apparently is the second most active source of the CO_2-activating enzyme.[204]

Sulfate may be activated through the transfer of an AMP molecule producing phosphoadenosine phosphosulfate. AMP can also act as a donor of an $=N-C$ linkage, which is utilized in the synthesis of imidazole phosphate, a precursor of histidine and therefore of histamine.[206]

ATP reacts with methionine[207] as a donor of the adenosine moiety, producing S-adenosylmethionine, a methyl donor in the biosynthesis of creatine,[208] adrenaline,[209] phosphatidyl choline from phosphatidyl ethanolamine, and many other compounds or metabolites.

ATP is also a substrate for 3′,5′-cyclic AMP by adenyl cyclase.[210-213] This enzyme has the highest activity in brain. Cyclic 3′,5′-AMP-involved glycogenolysis[214] plays probably an important role in several metabolic pathways, although the mechanisms are not yet clear.

The role and metabolism of certain adenine nucleotides discovered in different tissues but not yet in brain are still unsolved. This is the case for adenosine tetraphosphate,[215] diadenosine pentaphosphate,[216] adenosine pentaphosphate,[217,218] adenosine 5′-pyrophosphate-2-phosphoryl-3-glyceric acid,[219,220] and histamine adenine dinucleotide.[221-224]

B. Uracil Nucleotides

The recognition of the role of uridine nucleotide coenzymes is one of the most important recent advances in the understanding of carbohydrate metabolism. Since the initial studies by Leloir,[225] numerous uridine nucleotides have been isolated from a variety of natural sources,[12,226–228] and much has been learned of the role of these nucleotides in the metabolism of polysaccharides, mucopolysaccharides, and nucleic acids.

Besides a possible participation in transphosphorylation reactions,[229] UTP intervenes in the synthesis of acid anhydrides of uridine-5'-phosphate and various sugar phosphates, coenzymes in carbohydrate metabolism.[225, 226,230,231] UDP sugars are produced in enzymatic reactions between UTP and the sugar-1-phosphate.[227] Enzymes catalyzing such reactions are called UDP sugar pyrophosphorylases or uridyltransferases. A variety of such enzymes catalyze the synthesis of several UDP sugars.

The presence in rat, rabbit, and human brain of UTP α-D-glucose-1-phosphate uridyltransferase has been shown[227,232–240]:

$$\text{UTP} + \text{glucose-1-phosphate} \rightleftharpoons \text{UDP-glucose} + \text{pyrophosphate}$$

Adenosine and guanosine triphosphates are not active in this reaction.[241] Galactose-1-phosphate cannot replace glucose-1-phosphate. However the UDP-glucose produced may act with α-D-galactose-1-phosphate, yielding UDP-galactose as illustrated by the following reaction[234,235,240]:

$$\text{UDP-glucose} + \text{galactose-1-phosphate} \rightleftharpoons \text{glucose-1-phosphate}$$
$$+ \text{UDP-galactose}$$

UTP-acetylglucosamine uridyltransferase was found in brain.[242] This enzyme synthesizes UDP-N-acetylglucosamine (UDPAG),[13,243] which exists in nervous tissue, as follows:

$$\text{UTP} + \alpha\text{-D-}N\text{-acetylglucosamine-1-phosphate} \rightleftharpoons \text{UDPAG}$$
$$+ \text{pyrophosphate}$$

UDP-N-acetylglucosamine was identified in several tissues, including brain.[10,228,243–246] This compound and also UDP-galactose are involved in the biosynthesis of glycoproteins and glycolipids. UDP-acetylglucosamine, through UDP-acetylglucosamine pyrophosphatase, may give acetyl-glucosamine-1-phosphate + UMP.

The role of UDP sugar coenzymes in enzymatic catalysis is of two types. In one type of reaction, they are involved in glycosyl transfer, for example, lengthening of the glycogen chain by addition of a terminal glucose, synthesis of complex polysaccharides like hyaluronic acids from UDP-acetylglucosamine and UDP-glucuronic acids.

UDP-Galactose is involved in the synthesis of psychosine[247] and in the incorporation of galactose into glycolipids.[248,249]

The other type of UDP sugar reaction is characterized by chemical transformation of the sugar. The epimerization of glucose to galactose occurs when the sugar is present in glycosyl linkage with the uridine nucleotide. The enzyme catalyzing this reaction is named UDP-galactose-4-epimerase. The enzyme requires NAD; NADP is inhibitory.[232,250-252]

C. Cytosine Nucleotides

Cytosine nucleotides are involved in nucleic acid and polyC synthesis[253-255] and in group transfer reactions at the alcohol level of oxidation. CDP-choline,[256,257] CDP-ethanolamine,[257] and CDP-glycerol[258,259] were discovered as essential intermediates in the biosynthesis of phospholipids.[256,257,260,261] These nucleotides are formed biosynthetically from CTP and alcohol-1-phosphate by a transfer from phosphoalcohol to CTP catalyzed by specific transferase[259,262]:

CTP + choline phosphate = pyrophosphate + CDP choline

CDP-alcohols transfer an alkyl phosphate group to a variety of acceptors to yield lecithin,[257,262-267] phosphatidylethanolamine,[257,262,267] sphingomyelin,[268] or phosphatidyl inositol.[269-271]

CTP-N-acetylneuraminic acid transferase produces CMP-acetylneuraminic acid[272,273] which through transferase reaction is involved in ganglioside biosynthesis.[249,274] (See Chapters 13 and 14 of Volume 3 in this series.)

D. Guanine Nucleotides

Cabib and Leloir[275] isolated a guanine nucleotide: guanosine diphosphomannose, similar to uridine diphosphoglucose.[225] The role of guanosine nucleotides as coenzymes was discovered by Sanadi et al.[276] and Kaufman and Alivisatos.[277] GDP-mannose has since been identified in several types of cells[245,278-282] and in brain.[283] GTP reacts with mannose-1-phosphate through GTP-mannose-1-phosphate GDP transferase in a manner similar to the reaction between UTP and glucose-1-phosphate, to yield GDP-mannose as an intermediate for transfer of mannosyl groups.[280,284]

In brain[285] as well as in other tissues, GTP and ITP are formed from corresponding nucleoside diphosphates through an oxidation reaction of α-oxylglutarate.

GTP acts also as an energy donor in the formation of adenylosuccinate from IMP and α-aspartate, a key step in the biosynthesis of AMP. Oxalacetate decarboxylase requires ITP or ATP.[286,287]

The role of GTP in the biosynthesis of proteins has been extensively investigated.[288-291] It seems probable that the energy required for moving the mRNA relative to the ribosomes and positioning the amino acyl-tRNA is provided by the hydrolysis of GTP. In the binding of tRNA to a specific codon, according to Schweet et al.[292-294] a transfer factor and GTP are involved. It seems that the first step of the binding is nonenzymatic and requires a relatively high magnesium concentration. In the second step, there occurs a conformational alteration in the ribosome at the expense of GTP and requiring the transfer factor. Hence the amino-acyl-*t*RNA moves to a newly exposed second site, freeing the first one for the nonenzymatic binding of a second molecule of amino-acyl-*t*RNA.[292–294]

VI. DISTRIBUTION OF FREE NUCLEOTIDES

A. In Whole Brain

Since the early work on the distribution of the free nucleotides in brain was essentially concerned only with adenine nucleotides, such studies have been extended considerably.[8,18,19,50,283,295–304] Mono-, di-, and triphosphates of adenosine, guanine, uridine, and cytidine are present in the central nervous system and the different regions, as are the nucleotide coenzymes, NAD, NADP, uridine diphosphate sugars, cytidine diphosphocholine, and ethanolamine. The distribution of free nucleotides has a common feature in the central nervous system in various mammals[18,299,303] (Table II).

TABLE II

Free Nucleotides in the Cerebral Hemispheres and Cerebellum
in Different Mammal Species
(μmoles/100 g wet weight)

Nucleotides	Mouse hemispheres[a]	Rat[b]		Guinea pig[c]		Rabbit[c]	
		Hemi-spheres	Cere-bellum	Hemi-spheres	Cere-bellum	Hemi-spheres	Cere-bellum
CDP-choline	13.3	13.7	12.7	8.8	8.3	9.4	9.4
CMP	Traces	—	—	—	—	—	—
NAD	15.1	12.0	15.1	2.3	3.0	5.4	5.7
AMP	—	2.1	—	5.0	3.4	5.4	4.6
ADP	10.9	12.2	15.5	20.0	18.7	18.4	23.3
ATP	194.0	208.2	260.0	201.4	216.66	213.0	210.1
GMP	—	—	1.8	Traces	Traces	Traces	Traces
GDP	2.2	4.4	7.1	5.5	15.7	9.8	12.1
GTP	21.8	29.9	38.6	32.2	34.2	35.7	30.7
UMP	1.2	1.7	1.5	Traces	Traces	2.3	4.0
UDP	1.7	2.2	6.2	3.1	4.9	0.8	Traces
UDP-Co	14.9	11.4	18.5	20.7	22.6	22.3	29.8
UTP	17.4	22.0	19.0	26.0	27.1	25.6	29.7
IMP	4.8	5.1	4.7	2.6	5.5	4.7	9.3

[a] Mandel and Harth.[18]
[b] Mandel, Harth and Rebel.[299]
[c] Mandel, Edel and Poirel.[303]

The adenine nucleotides predominate and account for 60–65% of the total nucleotides. The ATP/ADP ratio, which is an index of the state of oxidative phosphorylation, has a value up to 22 in mouse brain, 10 to 12 in rat, guinea pig, and rabbit brain. The amount of guanine nucleotides, about 12%, is quite high[18,303,305,306] in contrast to a very low amount of this type of nucleotide in skeletal muscle.[51,307] GTP accounts for 91% of the guanine nucleotides. The very high concentration of guanine nucleotides in brain relative to the other nucleotides may perhaps be a reflection of the active protein synthesis in nervous tissue. The amount of uracil nucleotides is close to that of guanine nucleotides. UTP and UDP coenzymes predominate among the uridine nucleotides. The cytosine nucleotides are very scarce, accounting for about 4% of the total nucleotides. There are very small amounts of mononucleotides.

The amount of nucleotides expressed as a ratio to the diploid amount of DNA, in order to have an indication of content per cell, is very low in fish and much higher in chicken and mammalian brain.

The nucleoside triphosphates, which amount to 80% of total free nucleotides of mammalian brain, are lower in inferior vertebrates (Table III). The

TABLE III

Distribution of Nucleoside Triphosphates in the Brain of Different Species of Vertebrates
(mμmoles/100 mg wet weight)

	Trout[a]	Carp[a]	Turtle[a]	Grass snake[b]	Lizard[b]	Chicken[b]	Rat[b]
ATP	88.5	108.0	114.0	159.0	189.0	160.5	189.8
GTP	14.4	16.2	16.4	18.8	25.1	21.0	23.8
UTP	9.3	10.4	10.5	12.3	16.1	13.4	18.8

[a] Rein, Harth-Edel, and Mandel.[302]
[b] Edel and Rein-Lovtrup.[283]

amount of ATP in the total brain nucleotides seems lower in inferior vertebrates. ATP ranges from 85 to 95% of adenylic nucleotides, according to the species examined.[302,303] Similarly, the ATP/ADP and GTP/GDP ratios are lower in inferior vertebrates than in mammals. The amount of CTP found in all adult animals is extremely low, suggesting that it may be a limiting metabolite.[308] The reciprocal relations between various nucleotides indicate an energy pattern similar to that of muscle fibers, which is very different from that of liver or kidney. The high levels of ATP are without doubt related to the functional activity of the nervous system and explain its extreme sensitivity to hypoxia.

B. Regional Distribution of Free Nucleotides

1. *Brain*

Per unit of fresh weight, the highest amounts of free nucleotides were found in the olfactory lobe, different areas of cerebral hemispheres (corpus striatum, hippocampus, hypothalamus), and cerebellar gray matter; the lowest, as could be expected, in the white matter. When the whole free nucleotides are expressed as its ratio to the diploid amount of DNA in order to have an indication of content per cell, the highest values were still found in the cortical gray matter, the corpus striatum, the hippocampus, and the thalamus, and the lowest in the olfactory lobe, the cerebellum gray matter, and the white matter (Table IV).

TABLE IV

Regional Distribution of Free Nucleotides in Rat Central Nervous System[a,c]

	100 g fresh weight[b]	diploid nucleus DNA $\cdot 10^{12}$
Olfactory lobe	424 ± 50	10.8
Gray matter	414 ± 55	28.4
White matter	244 ± 18	14.4
Corpus striatum	361 ± 42	24.5
Hippocampus	354 ± 27	23.8
Thalamus	296 ± 28	23.1
Hypothalamus	333 ± 38	18.5
Mesencephalon	253 ± 14	19.2
Cerebellum white matter	249 ± 27	14.5
Cerebellum gray matter	341 ± 44	4.0
Medulla oblongata	212 ± 30	17.8

[a] From P. Mandel and S. Edel, unpublished data.
[b] Average values of 6 experiments each of a pool of 8 brains.
[c] Values are in μmoles.

A special technique of dissection of different areas of the central nervous system has to be used in order to avoid the degradation of nucleoside triphosphates.[19] Thus, it appears that the amount of nucleoside polyphosphates is particularly high in the cortex as compared to cerebellum, mesencephalon, and medulla oblongata. However, the distribution of bases within the nucleotides is the same for the different areas: adenine nucleotides account for 70%; the amount of guanine nucleotides is slightly higher in cortex. Uridine nucleotides are slightly higher in mesencephalon and medulla oblongata.

2. Retina

The amount of free nucleotides in terms of wet weight is lower than that found in brain. The principal characteristic is the much higher amount of NAD, guanine nucleotides, and UDP sugars and a lower amount of adenine nucleotides (48 % of the total).[309]

Profiles of nicotinamide adenine nucleotide levels in different layers of retina from rabbit and monkey were comparable. Both total NADP and NADPH were the lowest in the outer segments of the retina and the highest in the inner layers. The ratio of total NADP/NADPH in the mitochondrial layer was much higher than expected from studies with isolated mitochondria, and the amount of total NADP was surprisingly high in nonmitochondrial layers.[304]

3. Brain Slices

Rat brain slices prepared according to McIlwain[310] and incubated for 10 min in a phosphate–glucose buffer at 37°C, contain more nucleoside mono- and diphosphates and much smaller amounts of nucleoside triphosphates than normal brain. Since these energy-rich compounds are required for the active transport and intermediate metabolism, changes in the biochemical functioning of the nerve cells in brain slices might be expected.[311]

4. Subcellular Distribution of Free Nucleotides

The content of ATP which remains relatively stable after rapid initial postmortem fall has been determined by the firefly method in some subcellular fractions.[312] In guinea pig brain, as well as in the gray matter of dog cerebral cortex, the highest amount of stable ATP was found in the nerve ending fraction. Reserpine failed to deplete the brain of stable ATP in doses that almost completely depleted it of hydroxytryptamine. This is in a marked contrast to the effect of reserpine on the adrenal medulla[313] and duodenal mucosa.[314] The amount of adenosine triphosphate in the nerve ending fraction was found to be greatly in excess of the 1 to 4 ratio of ATP to pressor amines expected, if the function of ATP there was to bind amines, as it is believed to be in other storage particles.

In order to determine the free nucleotide pool size in mammalian cell nuclei, it is necessary to use a nonaqueous medium for isolation of these organelles. A simple technique derived from that of Behrens[315] and Siebert[316] was described by Mandel and Mandel.[317] Since the nucleotide amount was very small and during the isolation of the nuclei hydrolysis of nucleoside triphosphates may occur, all the nucleotides were transformed in mononucleotides by barium hydroxide hydrolysis. It appears that the quantity of free nucleotides present in the nuclei is extremely low compared to the velocity of RNA biosynthesis. Only 10 % of total cellular nucleotides are present in the brain nuclei. AMP counts for 70 % of the whole nucleotide pool. The lowest values are obtained for CMP (Table V).

TABLE V

Distribution of Free Nucleotides in
Rat Brain Nuclei and Cytoplasm[a]

Nucleotide	Total (μmoles/mg DNA)		Percent of total	
	Nuclei	Cytoplasm	Nuclei	Cytoplasm
—	140.4	1415		
AMP			69.2	70.0
UMP			11.5	11.2
GMP			12.5	13.2
CMP			6.8	5.6

[a] From Mandel and Mandel.[317]

VII. CHANGES IN FREE NUCLEOTIDE POOLS IN VARIOUS PHYSIOLOGICAL AND EXPERIMENTAL CONDITIONS

A. In Development

The free nucleotide changes in the central nervous system during post-natal development were investigated[301,318] by the classical technique of LePage,[319] by column chromatography,[308,320] and by enzymatic determination.[321] Since the brain weight and its development depend upon the number of animals per litter, only rats born in a litter of less than five have been taken into account.[308,321a] (Table VI). Owing to this selection, the brain weight for the same age is almost identical and the curve of changes during development is found to be regular. In rat, the values of the total nucleotides expressed in terms of wet weight indicate a reduction on the second day as compared to the first day after birth. Following this, there is a progressive increase up to 14 to 20 days of age; the values become almost constant up to 35 days of age. However, these changes represent a ratio of varying amounts of free nucleotides to an increasing mass of brain constituents. Therefore it is wise to relate the results to the quantity of DNA which expresses the amount of free nucleotides in terms of cell content. Hence, when the values are related to the amount of DNA, the fall of total nucleotides, as related to wet weight observed between the first and second day after birth, is not seen. There is a regular and progressive increase of total nucleotides

TABLE VI

Nucleoside Triphosphates in the Rat Brain
During Postnatal Growth
($m\mu$moles/100 mg wet weight)

Day after birth	1[a]	2[a]	8[a]	15[a]	21[a]	30[a]	35[a]	Adults[a,b]
ATP	189.7	146.2	165.1	183.6	210.0	216.1	223.0	189.8
GTP	34.8	23.8	30.3	32.3	36.3	34.4	33.7	23.8
UTP	49.5	33.3	30.4	33.8	28.8	29.1	26.7	18.6
CTP	12.8	7.0	5.5	5.6	2.4	1.3	3.1	Traces

[a] Mandel and Edel-Harth.[308]
[b] Jacob and Mandel.[321a]

from birth to the adult age. At each stage of development, there is a very small quantity of nucleoside monophosphates. The nucleoside triphosphates constitute 70–80 % of the total nucleotides. The ratio of ATP to DNA increases significantly up to 14 days of age, followed by a constant but slower increase. Thus, there are two phases in the increase of the ATP. One phase is a rapid increase which is observed in the first 14 days and the other phase is a slow increase between 14 days and adult age. The absolute quantity of GTP, UTP, and UDPG increases until 14 or 21 days and then a slight decrease appears. The decrease of UTP and CTP is in all probability due to the arrest of growth of the cytoplasmic mass and the myelination.

The relatively high values of GTP maintained until adult age may be explained by its active role in the biosynthesis of proteins, which becomes intense in the adult. Thus, the postnatal evolution of the free nucleotides in rat brain is characterized by a progressive increase of adenylic nucleotides, which is in good agreement with the type of energetic distribution of the nucleotides in brain. The rise of ATP corresponds to the energy requirements, while the evolution of GTP, UTP, CTP, and the nucleoside diphosphate complexes parallels a greatly increasing cytoplasmic mass in the initial postnatal development. The low quantity of CTP in the adult suggests that it is a limiting factor in the biosynthesis of ribonucleic acid and eventually phosphatides.[308]

During development of the chick brain[322,323] the total amount of nucleotides increases regularly from the sixth to the eighteenth day and then remains constant until hatching. In contrast, the values for total nucleotides in terms of dry weight regularly diminishes during the whole development period and also during the postnatal period up to 2 years, which indicates that the deposition of other solid material occurs to a greater extent than deposition of nucleotides. CTP is present only in small traces at the hatching moment. The relative increase of adenylic nucleotides in comparison with other nucleotides can be related here again to the increased utilization of this type of nucleotides to produce energy by an ionic transport and by mobility.[324] The ATP/ADP ratio decreases during embryonic development and remains constant after hatching up to adult age.[325] A decrease of UDPG with increase in age was also observed in chicken, and in mice and dogs.[113] A decrease in UDPAG also observed can be related to an increase in hyaluronic acid synthesis.[323]

From 2 to 10 days following birth the pyridine nucleotide content of rat brain is relatively constant on a wet weight basis and increases about 50% until 21 days and then remains constant.[326] Measurements of oxidized and reduced nicotinamide adenine dinucleotides have shown that the amount of the oxidized forms per wet weight increased while the level of the reduced coenzymes underwent a progressive decline.[327] In normal developing retina of the rat, the NAD/NADP ratio decreases and that of NADP/NADPH increases.[327]

The NAD nucleotidase (NADase) activity of rat brain tissue increases from about 2 units in the 2-day-old animals to almost 28 units per gram of wet weight tissue in the 20-day-old rats. Thereafter the enzyme level is relatively constant.

B. Effects of Ischemia

Many authors have observed a depletion of ATP by anoxia or ischemia. However, phosphocreatine decreases more rapidly than ATP.[328–333] The well-known loss of spontaneous electrical activity of the brain during early stages of anoxia or ischemia may be correlated to the splitting of phosphocreatine rather than to a breakdown of nucleoside triphosphates, which occurs more slowly. According to Gerlach et al.,[334] 30 sec after interruption of the blood flow by simple compression of the neck blood vessels, more than 40% of the creatine phosphate disappeared without a decrease in the level of ATP, while the amount of the nucleotide was reduced to about 50% after 2 min.[334] In the brain of rats maintained under hypoxia for 10 min, the content of ATP diminished over 80%, while the amount of guanosine triphosphate decreased only to one-quarter. The labeling of the β-phosphate of ATP relative to that of the γ-phosphorus is also reduced under these conditions.[335] When the blood supply is cut off and the brain is forced to use its endogenous sources of energy, the depletion of ATP in 10-day-old animals is less marked than in adults. In anesthetized mice the decrease is even less.[331] The changes are smaller in the Ammon's horn than in other areas (parietal cortex, cerebellar cortex, or medulla).[336] Slower disappearance of rat brain ATP was noted during a second exposure to anoxia.[337]

Hypercapnea produced in rats exposed continuously to gas mixtures containing 6–8% CO_2 causes an exponential loss of ATP up to the twelfth day.[338]

In rats suffering from chronic anemia a marked decrease of ATP in the cerebral and cerebellar hemispheres was noted.[299]

C. Effects of Anesthetics, Tranquilizers, and Sleep

Anesthetics produce changes in brain metabolism which are reflected in nucleotide metabolism, but there are some contradictory results in the literature. The available data[333,334,339–347] are mainly confined to adenine nucleotides and have been systematically reviewed.[297] Rats anesthesized with either ether or pentobarbitone for a short time have the same brain levels of AMP, ADP, and ATP as those of control animals.[22,50,348] A wide variety of anesthetics administered for a period of 10 min preceding the decapitation, always produced a "stabilization" of ATP and creatine phosphate levels.[334] Under very deep phenobarbitone anesthesia, ATP was unchanged in mouse brain, although this drug caused a progressive depression in metabolic rate (apparent even with lighter dosages) in the whole brain or in separate regions, such as the parietal cortex, Ammon's horn, cerebellar cortex, and medulla. There is, however, accumulation of phosphocreatine.[336,339–346,349]

During habituation to barbital and also for a short period after withdrawal, a fall in ATP and creatine phosphate was found in the brain of two substrains of albino rats.[353]

The primary pharmacological action of ether and pentobarbitone seems to be neither a depression of oxidative metabolism or ATP synthesis,[350] nor an uncoupling of oxidative phosphorylation.[351,352] Conceivably, however, an inhibition of ATP synthesis might occur in certain areas of the brain, resulting in a depression of brain function locally and secondary depression of overall brain function with accumulation of labile phosphate compounds elsewhere from lack of utilization. This would be reflected in elevation of the concentrations of labile phosphate compounds if the whole brain was analyzed and could be one explanation for the results obtained in various studies.[340,348,354] Turnover studies on phosphorus metabolism either *in vitro* or *in vivo* do not give a satisfactory picture of the actual sequence of events occurring in drug-treated animals.[352]

Effects of tranquilizers have been extensively studied with contradictory results.[24,355–358] Depressants of cerebral activity have been reported either to increase or to decrease the level of ATP and phosphocreatine in the brain of treated animals. This effect has been observed, for example, with chlorpromazine[355,357] in rats and with reserpine in mice or rats.[356,359] Changes in ATP content varying with time in the hypothalamus as well as in whole rat brain were reported.[358] However, with an improved extraction for nucleotides, Minard and Davis[24] were unable to confirm these observations. An explanation for this discrepancy might be that depressant drugs inhibit the very rapid degradation of ATP and phosphocreatine by brain tissue which occurs immediately following decapitation.

It has been found that the increase in NAD after intrathecal injection of nicotinamide is greatly delayed when the animals are pretreated with reserpine.[360] This drug seems to interfere with NAD turnover in brain.[361] Nicotinamide with reserpine prolongs pentobarbitone-induced anesthesia in mice longer than reserpine does alone, whereas nicotinamide itself is without effect.[326] There are, however, no clear-cut changes in the nucleotide levels during sleep or sleep deprivation.[362]

D. Effects of Stimulations

1. Electrical Stimulations and Seizures

A lowering of brain adenosine triphosphate and phosphocreatine levels as an effect of electrical stimulation both *in vivo*[127,363–370] and *in vitro*[366] was reported.

Adenine nucleotides showed rapid changes followed by a return to the control values. GTP exhibited minor changes and a relatively slow return to its starting concentration. Using tritiated uridine, dramatic changes in pyrimidine nucleotide metabolism could be observed.[370]

When tonic extensor seizures were induced in rats, ATP and phosphocreatine reached minimal levels in 10 sec, the ATP decrease being about 50% and that of phosphocreatine about 70%. ATP was again at its normal level 75 sec after the electroshock; phosphocreatine was slower in reaching its normal level.[127] Anoxia is a concomitant of electroshock seizures, but after 10 sec in a chamber of 95% N_2 and 5% CO_2,[371] the decrease of ATP values was only about 10% and that of phosphocreatine about 50%.[127]

In experiments performed by Wallgren,[368] levels of ATP, ADP, AMP, and creatine phosphate in slices of rat cerebral cortex did not change after electrical stimulation. However, using ^{32}P, a decrease in the specific activity of the γ-ATP-phosphate caused by electrical stimulation and blocked by ethanol was observed.[368] The application of electrical pulses to slices of guinea pig cerebral cortex leads to an increase in the levels of adenosine-3′,5′-phosphate of more than elevenfold which returns to normal within 10 min.[372] This phenomenon cannot be ascribed to the release and action of norepinephrine and histamine,[372] but may be caused either by alteration of the ionic environment of adenyl cyclase or by a release of other active endogenous substances.

In audiogenic seizures in mice, there were no changes of ATP after the clonic phase,[373] although a great decrease of phosphocreatine was noticed. Only a slight decrease of ATP (15%) was observed after the tonic phase (2 sec) of the seizure. There are no significant changes of other nucleotides.

When adenine nucleotides and creatine phosphate were measured from 15 to 180 sec after exposure of mice to a convulsant, Indoklon [bis(2,2,2-trifluoroethyl)ether], creatine phosphate fell to half its original value after the initiation of convulsion.[374] In contrast, in the epileptogenic focus induced by topical application of mescaline to cat brain cortex, the level of ATP and the incorporation of ^{32}P into ATP is markedly higher in the focal area compared to the contralateral one.[375,376]

2. Stimulation by Drugs

There is a correlation between brain hyperactivity and ATP content. Thus Palladin[346] and Heim *et al.*[377] showed an increase of ATP in the central nervous system after injection of small doses of pervitin, which produce a psychomotor excitation without convulsions. Similarly, after administration of $\beta\beta'$-iminodipropionitrile (IDPN) which causes a psychomotor agitation in the mouse, the ATP increases progressively in parallel with the development of the psychomotor activity. The total activity of radioactive P incorporated into ATP compared to the specific activity of PO_4 appears higher in mice having had IDPN injections.[299] A significant increase in the retinal ATP content was also obtained after IDPN injection.[377a]

There is only a slight alteration of ATP content, if any, in mouse brain under amphetamine treatment.[378–380] However, a high increase of ^{32}P in α, β, and γ-phosphate of ATP was noted.[380] It should be taken into account when RNA turnover is studied. Some antidepressive drugs seem to increase the ATP level in rat brain.[381]

E. Effects of Nucleotides and Their Analogues

Geiger and Yamasaki found that by adding small amounts of cytidine and uridine to a perfusion fluid of cat brain, the electrocorticogram, which was practically extinguished after 1 hr perfusion could be restored.[382] Uridylic acid, cytidylic acid, guanylic acid, uridine, guanosine, uracil, and guanine, injected intraperitoneally, caused definite electroencephalographic changes both in the voltage and the wave frequency of the rat EEG. An apparent graduation of the effects was observed when injecting uracil, uridine, and uridylic acid.[383] It is not clear whether the electrographic differences are related to a differential permeability of the blood–brain barrier or to a differential reactivity of the nervous system to a group of compounds not discriminated by the blood–brain barrier.

Flavine adenine dinucleotide elicited an increase in voltage and the appearance of 10–15 cps high voltage activity in spindle form in the ECG of guinea pigs.[384] Flavine adenine dinucleotide partially reversed the effect of chlorpromazine on the EEG.[385]

Nicotinamide, which elevates tissue NAD levels, appears to exert an influence on the central nervous system, interacting with drugs such as reserpine and chlorpromazine[386] and blocking audiogenic seizure in mice.[387] Humans suffering from severe nicotinic acid deficiency exhibit neurasthenia syndromes, organic psychosis often manifested as memory impairment and disorientation, and encephalopathic syndromes.[386]

Intraventricular injection of high doses of ATP and methacholine into rats elicited clonic-tonic convulsions, or an akinetic state, or both. Cholinergiclike properties of ATP and adenosine were partially antagonized by concomitant perfusion with methylene blue.[388] ATP and other nucleoside phosphates prevent or abolish some neutral effects of calcium deficiency in concentrations which do not affect normal nerve function.[389] This stabilizing effect may be associated with replacement of nucleotide lost from the excitable membrane into the Ca^{2+}-deficient medium.[390]

Fluorinated pyrimidines administrated intrathecally produce structural, physiological, and biochemical alterations in brain and spinal cord.[125,391]

Lethargy appears several weeks after azauridine injections followed by hyperaflexia, mental obtundation, hallucination, semicoma, and electro-encephalographic abnormality. Conversion of a tracer dose of orotic acid to acid-soluble nucleotides is inhibited by azauridine, whereas conversion of uridine is unaffected and orotidine-3'-phosphate accumulates.

F. Miscellaneous

The formation of cyclic 3',5'-AMP by broken-cell preparations from central nervous system was observed to be increased by catecholamines.[213] It was also found that catecholamines as well as histamine induced the accumulation of cyclic nucleotides in slices of rabbit cerebellum and cerebral cortex.[392,393] Changes in cyclic 3',5'-AMP levels of brain slices induced by neurohormones were not accompanied by an increase of phosphorylase levels.[393]

Thyroxine and hydrocortisone have no effect on brain adenine nucleotide levels nor is there a change after thyroidectomy.[394] However, a reduced ^{32}P incorporation into brain ATP in several species was observed after thyroxine administration.[395] Adrenalectomy produces in rat brain a significant decrease of ATP and ATP/ADP ratio, suggesting a decrease of oxidative phosphorylation rate.[394]

The hydrolysis of CDP choline, CDP ethanolamine, CDP and UDP glucose, but not of UDP, is significantly increased in homogenates of rat sciatic nerve after transection.[396]

Intracisternal injection of serotonin to rabbits increased the level of free nucleotides in the cerebral hemispheres and in the cerebellum.[397] No alteration of ATP level was obtained in rats in coma induced by ammonium acetate.[398,399] In rats cooled to 15°, brain ATP levels were normal.[400] A slight increase occurs in deeper hypothermia (personal data). Apparently hypothermia and especially conditioned hypothermia prevent the breakdown of high energy phosphates in brain.[400]

Intravenous injection of methanol in white mice produced a fall in brain ATP content, a result of impaired biosynthesis of this nucleotide.[401] ATP levels were lowered in brain slices by ouabain, glutamate, or omission of Na from the medium.[402]

VIII. CONCLUSIONS

Acid-soluble nucleotides play a role in the metabolism of small molecules as well as in the biosynthesis of large polymers. They constitute a group of active compounds that are encountered in all metabolic cycles. We do not know yet in what measure their chemical structure is related to the functions they undertake.

Although the rationale of the specific intervention of nucleotides is little understood, it is possible to propound a theory on the significance of their action. Thus, a biosynthetic pathway benefits from proceeding through a nucleotide derivative because of the reaction cycle necessary for the reconstitution of the nucleotide derivative, which prevents the reversion as in simple esterifications.

The adaptation of the nucleotide pool to biochemical function, as in the energetic or metabolic distribution, or the adjustment of the free nucleotide pool must lead us to seek in the distribution and the changes of free nucleotides an expression of regulation that implicates high polymers and certain biological aspects of cell normal or pathological life. This shows the importance of exploring free nucleotide pools in tissues and the manner of their regulation.

IX. REFERENCES

1. J. Gregoire, *Bull. Soc. Chim. Biol.* **40**:1245–1275 (1958).
2. J. Henderson and G. A. LePage, *Chem. Revs.* **58**:645–688 (1958).
3. V. R. Potter, *Fed. Proc.* **17**:691–697 (1958).
4. J. L. Strominger, *Physiol. Rev.* **40**:55–111 (1960).
5. Symposium of the section on Neurochemistry, American Academy of Neurology, *The Neurochemistry of Nucleotides and Amino Acids* (R. O. Brady and D. B. Tower, eds.), 292 pp., John Wiley & Sons, New York (1960).
6. Colloquium der Gesellschaft für Physiologische Chemie, *Zur Bedeutung der freien Nucleotide* (Mosbach) Vol. 11, 176 pp., Springer Verlag, Berlin (1961).
7. A. M. Michelson, *Ann. Rev. Biochem.* **30**:133–164 (1961).
8. P. Mandel, in *Progress in Nucleic Acid and Molecular Biology* (W. E. Cohn and J. N. Davidson, eds.) **3**, pp. 299–334, Academic Press, New York (1964).
9. W. E. Cohn, *J. Am. Chem. Soc.* **72**:1471–1478 and 2811–2812 (1950).
10. H. Schmitz, V. R. Potter, and R. B. Hurlbert, *Z. Krebsforsch.* **60**:419–432 (1955).
11. R. B. Hurlbert, H. Schmitz, A. F. Brumm, and V. R. Potter, *J. Biol. Chem.* **209**:23–39 (1954).
12. H. Schmitz, R. B. Hurlbert, and V. R. Potter, *J. Biol. Chem.* **209**:41–54 (1954).
13. H. Schmitz, V. R. Potter, R. B. Hurlbert, and D. M. White, *Cancer Res.* **14**:66–74 (1954).
14. J. Klethi and P. Mandel, *Biochim. Biophys. Acta* **24**:642–643 (1957).
15. E. R. Holiday and E. A. Johnson, *Nature* **163**:216–217 (1949).
16. R. Markham and J. D. Smith, *Nature* **163**:250–251 (1949).
17. C. E. Carter, *J. Am. Chem. Soc.* **72**:1466–1471 (1950).
18. P. Mandel and S. Harth, *J. Neurochem.* **8**:116–125 (1961).
19. S. Edel, C. Judes, M. Ramuz, and P. Mandel, *C.R. Acad. Sci.* (*Paris*) **261**:3695–3697, Groupe 13 (1965).
20. C. Bishop, D. M. Rankine, M. B. Klein, and J. H. Talbott, *Proc. Soc. Exptl. Biol. Med.* **104**:754–756 (1960).
21. K. Trautner and F. Bramstedt, *Arch. Fischereiwiss.* **13**:130–138 (1962).
22. S. Harth and P. Mandel, *Bull. Soc. Chim. Biol.* **43**:969–980 (1961).
23. S.-O. Brattgård, H. Løvtrup-Rein, and M. L. Moss, *J. Neurochem.* **13**:1257–1260 (1966).
24. F. N. Minard and R. V. Davis, *Nature* **193**:277–278 (1962).
25. P. A. Levene and L. W. Bass, *Nucleic Acids*, American Chem. Soc. Monogr. New York, Ser. 56, 337 pp. (1931).
26. G. A. LePage, *Am. J. Physiol.* **146**:267–281 (1946).
27. W. W. Umbreit, R. H. Burris, and J. F. Stauffer, *Manometric Methods*, 3rd ed., 338 pp., Burgess Publishing, Minneapolis, Minn. (1957).

28. E. R. Tomkins, J. X. Khym, and W. E. Cohn, *J. Am. Chem. Soc.* **69**:2769–2777 (1947).
29. W. E. Cohn and C. E. Carter, *J. Am. Chem. Soc.* **72**:4273–4275 (1950).
30. W. E. Cohn, *Ann. N.Y. Acad. Sci.* **57**:204–224 (1953).
31. W. E. Cohn, in *Chromatography* (E. Heftmann, ed.) pp. 554–583, Reinhold, New York (1961).
32. G. Schoen, *Acta Vitaminol.* **7**:151–167 (1953).
33. R. Bergkvist and A. Deutsch, *Acta Chem. Scand.* **8**:1877–1879 (1954).
34. H. Schmitz, *Biochim. Biophys. Acta* **14**:160–161 (1954).
35. R. K. Crane, *Science* **127**:285–286 (1958).
36. H. G. Pontis and N. L. Blumson, *Biochim. Biophys. Acta* **27**:618–624 (1958).
37. E. D. Korn, *Biochim. Biophys. Acta* **32**:554–555 (1959).
38. E. Lederer, *Chromatographie*, Collection Monographies de Chimie organique, Masson, Paris, Vol. I (1959), Vol. II (1960).
39. R. L. Stambaugh and D. W. Wilson, *J. Chromatog.* **3**:221–224 (1960).
40. A. Ishikawa, *Folia Psychiat. Neurol. Jap.* **15**:120–132 (1961).
41. J. Ingle, *Biochim. Biophys. Acta* **61**:147–149 (1962).
42. S. A. Morell, V. E. Ayers, and T. J. Greenwalt, *Anal. Biochem.* **3**:285–297 (1962).
43. N. G. Anderson, J. G. Green, M. L. Barber, and SR. F. C. Ladd, *Anal. Biochem.* **6**:153–169 (1963).
44. R. Markham and J. D. Smith, *Biochem. J.* **52**:552–557 (1952).
45. A. Bourdet and P. Mandel, *C.R. Acad. Sci.* (*Paris*) **237**:530–532 (1953).
46. H. Boser, *Z. Physiol. Chem.* **298**:145–150 (1954).
47. K. Matsuzaki, *Vitamins* (*Japan*) **7**:26–29 (1954).
48. E. Gerlach, E. Weber, and H. J. Döring, *Arch. Exptl. Pathol. Pharmakol.* **226**:9–17 (1955).
49. R. Bergkvist, *Acta Chem. Scand.* **11**:1465–1472 (1957).
50. H. J. Döring and E. Gerlach, *Arch. Exptl. Pathol. Pharmakol.* **232**:271–273 (1957).
51. C. J. Threlfall, *Biochem. J.* **65**:694–699 (1957).
52. J. Stockx, *Naturwiss.* **45**:570 (1958).
53. G. Hem, *Arch. Biochem. Biophys.* **82**:485–487 (1959).
54. K. K. Tsuboï and T. D. Price, *Arch. Biochem. Biophys.* **81**:223–237 (1959).
55. M. A. Moscarello, B. G. Laine, and C. S. Hanes, *Canad. J. Biochem.* **39**:1755–1764 (1961).
56. J. Stockx and R. Van Parijs, *Arch. Int. Physiol. Biochim.* **69**:263–264 (1961).
57. K. Randerath, *J. Chromatog.* **10**:235–236 (1963).
58. F. Pavelka, *Mikrochimica Acta*, N° 5–6:1059–1064 (1965).
59. H. Verachtert, S. T. Bass, J. Wilder, and R. G. Hansen, *Anal. Biochem.* **11**:497–509 (1965).
60. J. F. Morrison, *Anal. Biochem.* **24**:106–111 (1968).
61. H. E. Wade and D. M. Morgan, *Biochem. J.* **56**:41–43 (1954).
62. A. Wollenberger, *Nature* **173**:205–207 (1954).
63. R. Bergkvist, *Acta Chem. Scand.* **12**:555–560 (1958).
64. R. Bergkvist, *Acta Chem. Scand.* **12**:572–575 (1958).
65. A. B. Banerjee and N. C. Ganguli, *J. Electroanal. Chem.* **2**:501–505 (1961).
66. E. W. Busch, H. Scheitza, and W. Thorn, *Biochem. Z.* **335**:62–68 (1961/62).
67. H. M. Klouwen, *J. Chromatog.* **7**:216–222 (1962).
68. T. R. Sato, J. F. Thomson, and W. F. Danforth, *Anal. Biochem.* **5**:542–547 (1963).
69. Z. Stransky, *J. Chromatog.* **10**:456–462 (1963).
70. C. Cocito and P. Laduron, *Anal. Biochem.* **7**:429–438 (1964).
71. E. Sulkowski and M. Laskowski, *Anal. Biochem.* **20**:94–101 (1967).
72. B. Zak, B. A. Welsh, and L. M. Weiner, *J. Chromatog.* **34**:275–277 (1968).
73. K. Randerath, *Angew. Chem.* **74**:484–488 (1962). In English **1**:435–439 (1962).
74. G. Weinmann and K. Randerath, *Experientia* **19**:49–50 (1963).
75. K. Randerath and E. Randerath, *J. Chromatog.* **16**:111–125 (1964).

76. E. Randerath and K. Randerath, *J. Chromatog.* **16**:126–129 (1964).
77. S. Fahn, R. W. Albers, and G. J. Koval, *Anal. Biochem.* **10**:468–471 (1965).
78. J. Neuhard, E. Randerath, and K. Randerath, *Anal. Biochem.* **13**:211–222 (1965).
79. K. Randerath, *Nature* **205**:908 (1965).
80. E. Randerath and K. Randerath, *Anal. Biochem.* **12**:83–93 (1965).
81. I. W. F. Davidson and W. G. Drew, *J. Chromatog.* **21**:319–323 (1966).
82. R. J. Morin, *Clin. Chim. Acta* **13**:395–397 (1966).
83. G. Pataki and A. Kunz, *J. Chromatog.* **23**:465–467 (1966).
84. E. Bancher and J. Washüttl, *Mikrochimica Acta* N° 2:223–227 (1967).
85. A. P. Remenchik and J. Bernsohn, *Anal. Biochem.* **18**:1–9 (1967).
86. J. Washüttl and E. Bancher, *Mikrochimica Acta* N° 2:395–398 (1967).
87. J. E. Ciardi and E. P. Anderson, *Anal. Biochem.* **22**:398–408 (1968).
88. P. Hastings and J. T. Wong, *Anal. Biochem.* **22**:169–171 (1968).
89. E. Königk, *J. Chromatog.* **37**:128–131 (1968).
90. R. Bondivenne, J. Simond, and N. Busch, *J. Chromatog.* **41**:205–219 (1969).
91. M. Cashel, R. A. Lazzarini, and B. Kalbacher, *J. Chromatog.* **40**:103–109 (1969).
92. H. A. Ramsey, *Anal. Biochem.* **5**:83–91 (1963).
93. F. M. De Filippes, *Science* **144**:1350–1351 (1964).
94. O. Warburg, W. Christian, and A. Griese, *Biochem. Z.* **282**:157–205 (1935).
95. H. J. Hohorst, F. H. Kreutz, and T. Bücher, *Biochem. Z.* **332**:18–46 (1959/60).
96. R. Scholz, H. Schmitz, T. Bücher, and J. O. Lampen, *Biochem. Z.* **331**:71–86 (1959).
97. G. Siebert and R. Beyer, *Biochem. Z.* **335**:14–24 (1961/62).
98. H. U. Bergmeyer, in *Methoden der enzymatischen Analysen* (H. U. Bergmeyer, ed.) pp. 1–42, Verlag Chemie, Weinheim, Germany (1962).
99. H. Holzer, St. Goldschmidt, W. Lamprecht, and E. Helmreich, *Z. Physiol. Chem.* **297**:1–18 (1954).
100. J. A. Bassham, L. M. Birt, R. Hems, and U. E. Loenig, *Biochem. J.* **73**:491–499 (1959).
101. E. Scarano, M. Talarico, L. Bonaduce, and B. De Petrocellis, *Nature* **186**:237–238 (1960).
102. O. H. Lowry, J. V. Passonneau, D. W. Schulz, and M. K. Rock, *J. Biol. Chem.* **236**:2746–2755 (1961).
103. O. H. Lowry, J. V. Passonneau, and M. K. Rock, *J. Biol. Chem.* **236**:2756–2759 (1961).
104. S.-C. Cheng, *J. Neurochem.* **7**:271–277 (1961).
105. N. O. Kaplan, S. P. Colowick, and C. C. Barnes, *J. Biol. Chem.* **191**:461–472 (1951).
106. B. Chance, *Nature* **169**:215–221 (1952).
107. O. H. Lowry, N. R. Roberts, and J. J. Kapphahn, *J. Biol. Chem.* **224**:1047–1064 (1957).
108. T. Bücher and M. Klingenberg, *Angew. Chem.* **70**:552–570 (1958).
109. R. W. Estabrook, *Anal. Biochem.* **4**:231–245 (1962).
110. O. H. Lowry, Harvey Lectures, Ser. 58:1 (1962/63).
111. W. Sickel, *Science* **148**:648–651 (1965).
112. B. L. Strehler and J. R. Totter, in *Methods of Biochemical Analysis* (D. Glick, ed.) Vol. 1, pp. 341–356, Interscience, New York (1954).
113. K. P. Wong and T. L. Sourkes, *J. Neurochem.* **15**:201–203 (1968).
114. J. E. Edström, in *Methods in Cell Physiology* (D. M. Prescott, ed.) Vol. 1, pp. 417–447, Academic Press, New York (1964).
115. S. Edel and P. Mandel, unpublished data.
116. L. Berger, in *Methods in Enzymology* (S. P. Colowick and N. O. Kaplan, eds.) Vol. III, pp. 866–867, Academic Press, New York (1957).
117. R. L. Hancock and D. L. Coleman, *Anal. Biochem.* **10**:365–367 (1965).
118. L. P. Turner, *Anal. Biochem.* **28**:288–294 (1969).
119. M. Carrara and G. Bernardi, *Biochim. Biophys. Acta* **155**:1–7 (1968).
120. H. Koenig, *J. Biophys. Biochem. Cytol.* **4**:664–666 (1958).

121. H. Koenig, *J. Biophys. Biochem. Cytol.* **4**:785–792 (1958).
122. H. Koenig, in *Progress in Neurobiology* (S. Korey, ed.) Vol. 4, 241, Hoeber, New York (1959).
123. P. Reichard, *Adv. Enzymol.* **21**:263–294 (1959).
124. G. Schmidt, *Ann. Rev. Biochem.* **33**:667–728 (1964).
125. W. Wells, D. Gaines, and H. Koenig, *J. Neurochem.* **10**:709–723 (1963).
126. Wm. T. Murakami, D. W. Visser, and H. E. Pearson, *Proc. Soc. Exptl. Biol. N.Y.* **100**:463–467 (1959).
127. J. M. Buchanan and S. C. Hartman, *Adv. Enzymol.* **21**:199–261 (1959).
128. W. A. Manell and R. J. Rossiter, *Biochem. J.* **61**:418–424 (1955).
129. J. F. Henderson and G. A. LePage, *J. Biol. Chem.* **234**:2364–2368 (1959).
130. G. Ricceri and A. M. Guiffrida, *Boll. Soc. Ital. Biol. Sper.* **36**:1881–1884 (1960).
131. J. N. Santos, K. W. Hempstead, L. E. Kopp, and R. P. Miech, *J. Neurochem.* **15**:367–376 (1968).
132. I. Held and W. Wells, *J. Neurochem.* **16**:529–536 (1969).
133. R. W. Butcher and E. W. Sutherland, *J. Biol. Chem.* **237**:1244–1250 (1962).
134. W. Y. Cheung, *Biochem. Biophys. Res. Commun.* **23**:214–219 (1966).
135. E. De Robertis, G. Rodriguez De Lores Arnaiz, and M. Alberici, *J. Biol. Chem.* **242**:3487–3493 (1967).
136. H. Weil-Malherbe and A. D. Bone, *Biochem. J.* **49**:339–347 (1951).
137. H. A. Krebs and R. Hems, *Biochem. J.* **61**:435–441 (1955).
138. I. T. Oliver, *Biochem. J.* **61**:116–122 (1955).
139. R. M. Smillie, *Arch. Biochem. Biophys.* **67**:213–224 (1957).
140. S. M. Weissman, R. M. S. Smellie, and J. Paul, *Biochim. Biophys. Acta* **45**:101–110 (1960).
141. R. P. Miech and R. E. Parks, Jr., *J. Biol. Chem.* **240**:351–357 (1965).
142. D. E. Atkinson, *Ann. Rev. Biochem.* **35**, Part I, 85–124 (1966).
143. E. R. Stadtman, *Adv. Enzymol.* **28**:41–154 (1966).
144. R. L. Blakley and E. Vitols, *Ann. Rev. Biochem.* **37**:201–224 (1968).
145. M. E. Jones and S. E. Hager, *Science* **154**:422 (1966).
146. M. Tatibana and K. Ito, *Biochem. Biophys. Res. Commun.* **26**:221–227 (1967).
147. S. H. Appel and P. Pettis, *Fed. Proc.* **26**:292, N° 204 (1967).
148. D. G. R. Blair and V. R. Potter, *J. Biol. Chem.* **236**:2503–2506 (1961).
149. R. J. McCollister, W. R. Gilbert, Jr., D. M. Ashton, and J. B. Wyngaarden, *J. Biol. Chem.* **239**:1560–1563 (1964).
149a. A. P. Levin and B. Magasanik, *J. Biol. Chem.* **236**:184–188 (1961).
149b. H. Momose, H. Nishikawa, and I. Shiio, *J. Biochem.* **59**:325–331 (1966).
150. J. A. Muntz, *J. Biol. Chem.* **201**:221–233 (1953).
151. H. Weil-Malherbe and R. H. Green, *Biochem. J.* **61**:218–224 (1955).
152. J. Mendicino and J. A. Muntz, *J. Biol. Chem.* **233**:178–183 (1958).
153. B. Cunningham and J. M. Lowenstein, *Biochim. Biophys. Acta* **96**:535–537 (1965).
154. B. Setlow, R. Burger, and J. M. Lowenstein, *J. Biol. Chem.* **241**:1244–1245 (1966).
155. B. Setlow and J. M. Lowenstein, *J. Biol. Chem.* **242**:607–615 (1967).
156. I. A. Rose and B. S. Schweigert, *J. Biol. Chem.* **202**:635–645 (1953).
157. A. Larsson and J. B. Neilands, *Biochem. Biophys. Res. Commun.* **25**:222–226 (1966).
158. R. L. P. Adams, R. Abrams, and I. Lieberman, *J. Biol. Chem.* **241**:903–905 (1966).
159. H. L. A. Tarr and J. Roy, *Canad. J. Biochem. Physiol.* **44**:1435–1446 (1966).
160. R. W. Brockman, C. S. Debavadi, P. Stutts, and D. J. Hutchinson, *J. Biol. Chem.* **236**:1471–1479 (1961).
161. L. L. Bennett, Jr., H. P. Schneibli, M. H. Vail, P. W. Allan, and J. A. Montgomery, *Mol. Pharmacol.* **2**:432–443 (1966).

162. A. W. Murray, *Biochem. J.* **100**:671–674 (1966).
163. A. W. Murray, *Biochem. J.* **103**:271–279 (1967).
164. P. Mandel and L. Mandel, unpublished data.
165. J. E. Seegmiller, F. M. Rosenbloom, and W. N. Kelley, *Science* **155**:1682–1683 (1967).
166. J. G. Flaks, in *Methods in Enzymology* (S. P. Colowick and N. O. Kaplan, eds.) Vol. VI, pp. 136–162, Academic Press, New York (1963).
167. H. Koenig, *Proc. Soc. Exptl. Biol. N.Y.* **97**:255–260 (1958).
168. L. L. Bennett, Jr., L. Simpson, and H. E. Skipper, *Biochim. Biophys. Acta* **42**:237–243 (1960).
169. D. H. Adams, *J. Neurochem.* **12**:783–790 (1965).
170. I. Held, W. Wells, and H. Koenig, *J. Neurochem.* **16**:537–542 (1969).
171. N. Robinson and B. M. Phillips, *Clin. Chim. Acta* **10**:414–419 (1964).
171a. P. Mandel and L. Mandel, unpublished data.
171b. P. L. Ipata, *Biochem. Biophys. Res. Commun.* **27**:337–343 (1967).
172. E. Bresnick, U. B. Thompson, and K. Lyman, *Arch. Biochem. Biophys.* **114**:352–359 (1966).
173. G. F. Maley, M. G. Lorenson, and F. Maley, *Biochem. Biophys. Res. Commun.* **18**:364–370 (1965).
174. P. Eker, *J. Biol. Chem.* **241**:659–662 (1966).
175. E. P. Anderson and R. W. Brockman, *Biochim. Biophys. Acta* **91**:380–386 (1964).
176. Z. J. Lucas, *Science* **156**:1237–1240 (1967).
177. A. Orengo, *Exptl. Cell Res.* **41**:338–349 (1966).
178. F. Lipmann, *Adv. Enzymol.* **1**:99–162 (1941).
179. S. Ochoa, *J. Biol. Chem.* **141**:245–251 (1941).
180. C. Long, *Biochem. J.* **50**:407–415 (1952).
181. R. K. Crane and A. Sols, *J. Biol. Chem.* **203**:273–292 (1953).
182. M. K. Johnson, *Biochem. J.* **77**:610–618 (1960).
183. H. S. Bachelard, *Biochem. J.* **104**:286–292 (1967).
184. D. Nachmansohn, R. T. Cox, C. W. Coates, and A. L. Machado, *J. Neurophysiol.* **6**:383–396 (1943).
185. D. Nachmansohn, C. W. Coates, M. A. Rothenberg, and M. W. Brown, *J. Biol. Chem.* **165**: 223–231 (1946).
186. T. Wood, *Biochem. J.* **89**:210–220 (1963).
187. H. Jacobs, H. W. Heldt, and M. Klingenberg, *Biochem. Biophys. Res. Commun.* **16**:516–521 (1964).
188. T. Wood and P. S. Swanson, *J. Neurochem.* **11**:301–307 (1964).
189. R. J. Sullivan, O. Neal Miller, and O. Z. Sellinger, *J. Neurochem.* **15**:115–121 (1968).
190. A. Burger, R. Richterich, and H. Aebi, *Biochem. Z.* **339**:305–314 (1963/64).
191. H. Klenow and E. Lichtler, *Biochim. Biophys. Acta* **23**:6–12 (1957).
192. H. Klenow and B. Andersen, *Biochim. Biophys. Acta* **23**:92–97 (1957).
193. T. M. Brody and J. A. Bain, *J. Biol. Chem.* **195**: 685–696 (1952).
194. H. G. K. Westenbrink, E. P. Steyn-Parve, and J. Goudsmit, *Enzymologia* **11**:26–45 (1943).
195. O. E. Pratt, *Biochem. J.* **55**:140–145 (1953).
196. A. Kornberg and W. E. Pricer, Jr., *J. Biol. Chem.* **182**:763–778 (1950).
197. M. B. Hoagland, E. B. Keller, and P. Zamecnik, *J. Biol. Chem.* **218**:345–358 (1956).
198. F. Lipmann, *Fed. Proc.* **8**:597–602 (1949).
199. J. A. DeMoss and G. D. Novelli, *Biochim. Biophys. Acta* **18**:592–593 (1955).
200. M. K. Gaitonde and D. Richter, *Proc. Roy. Soc.* (*B*) **145B**:83–99 (1956).
201. D. H. Clouet and D. Richter, *J. Neurochem.* **3**:219–229 (1959).
202. P. Berg, *J. Am. Chem. Soc.* **77**:3163–3164 (1955).
203. B. K. Bachhawat and M. J. Coon, *J. Am. Chem. Soc.* **79**:1505–1506 (1957).

204. B. K. Bachhawat and M. J. Coon, *J. Biol. Chem.* **231**:625–635 (1958).
205. J. V. Formica and R. O. Brady, *J. Am. Chem. Soc.* **81**:752 (1959).
206. H. S. Moyed and B. Magasanik, *J. Am. Chem. Soc.* **79**:4812–4813 (1957).
207. G. L. Cantoni, *J. Biol. Chem.* **204**:403–416 (1953).
208. G. L. Cantoni and P. J. Vignos, Jr., *J. Biol. Chem.* **209**:647–659 (1954).
209. J. Axelrod, *Science* **126**:400–401 (1957).
210. E. W. Sutherland, T. W. Rall, and T. Menon, *J. Biol. Chem.* **237**:1220–1227 (1962).
211. T. W. Rall and E. W. Sutherland, *J. Biol. Chem.* **237**:1228–1232 (1962).
212. F. Murad, Y. M. Chi, T. W. Rall, and E. W. Sutherland, *J. Biol. Chem.* **237**:1233–1238 (1962).
213. L. M. Klainer, Y. M. Chi, S. L. Freidberg, T. W. Rall, and E. W. Sutherland, *J. Biol. Chem.* **237**:1239–1243 (1962).
214. E. W. Sutherland, Harvey Lectures, Ser. 57, 17–33 (1961/62).
215. I. Lieberman, *J. Am. Chem. Soc.* **77**:3373–3375 (1955).
216. G. Embden, *Arch. Exptl. Pathol. Pharmakol.* **167**:50–52 (1932).
217. J. Sacks, L. Lutwak, and P. D. Hurley, *J. Am. Chem. Soc.* **76**:424–426 (1954).
218. J. Sacks, *Biochim. Biophys. Acta* **16**:436 (1955).
219. T. Hashimoto and H. Yoshikawa, *Biochem. Biophys. Res. Commun.* **5**:71–74 (1961).
220. T. Hashimoto and H. Yoshikawa, *J. Biochem.* **53**:219–226 (1963).
221. S. G. A. Alivisatos, F. Ungar, L. Lukacs, and L. LaMantia, *J. Biol. Chem.* **235**:1742–1750 (1960).
222. S. Muraoka, M. Inoue, and H. Yamasaki, *Nature* **190**:532–533 (1961).
223. S. Muraoka, M. Sugiyama, and H. Yamasaki, *Nature* **196**:441–443 (1962).
224. S. G. A. Alivisatos, A. A. Abdel-Latif, F. Ungar, and G. A. Mourkides, *Nature* **199**:907–908 (1963).
225. R. Caputto, L. F. Leloir, C. E. Cardini, and A. C. Paladini, *J. Biol. Chem.* **184**:333–350 (1950).
226. A. C. Paladini and L. F. Leloir, *Biochem. J.* **51**:426–432 (1952).
227. A. Munch-Petersen, H. M. Kalckar, E. Cutolo, and E. E. B. Smith, *Nature* **172**:1036–1037 (1953).
228. E. E. B. Smith and G. T. Mills, *Biochim. Biophys. Acta* **13**:386–400 (1954).
229. L. A. Heppel, J. L. Strominger, and E. S. Maxwell, *Biochim. Biophys. Acta* **32**:422-430 (1959).
230. C. E. Cardini, A. C. Paladini, R. Caputto, and L. F. Leloir, *Nature* **165**:191–192 (1950).
231. L. F. Leloir, *Proc. 3rd Intern. Congr. Biochem. Brussels*, pp. 154–162 (1955) (Publ. 1956).
232. E. S. Maxwell, H. M. Kalckar, and R. M. Burton, *Biochim. Biophys. Acta* **18**:444–445 (1955).
233. H. M. Kalckar, E. P. Anderson, and K. J. Isselbacher, *Biochim. Biophys. Acta* **20**:262–268 (1956).
234. K. J. Isselbacher, *J. Biol. Chem.* **232**:429–444 (1958).
235. K. Kurahashi and E. P. Anderson, *Biochim. Biophys. Acta* **29**:498–501 (1958).
236. C. Villar-Palasi and J. Larner, *Arch. Biochem. Biophys.* **86**:270–273 (1960).
237. D. K. Basu and B. K. Bachhawat, *J. Neurochem.* **7**:174–179 (1961).
238. B. M. Breckenridge, S. Scott, J. L. Strominger, and E. J. Crawford, *J. Neurochem.* **7**:228–233 (1961).
239. B. M. Breckenridge and E. J. Crawford, *J. Neurochem.* **7**:234–240 (1961).
240. D. Bertoli and S. Segal, *J. Biol. Chem.* **241**:4023–4029 (1966).
241. H. M. Kalckar and E. S. Maxwell, *Physiol. Rev.* **38**:77–90 (1958).
242. T. N. Pattabiraman and B. K. Bachhawat, *J. Neurochem.* **11**:55–60 (1964).
243. R. B. Hurlbert and V. R. Potter, *J. Biol. Chem.* **209**:1–21 (1954).
244. E. Cabib, L. F. Leloir, and C. E. Cardini, *J. Biol. Chem.* **203**:1055–1070 (1953).

245. J. L. Strominger, *Biochim. Biophys. Acta* **17**:283–285 (1955).
246. J. L. Strominger, *J. Biol. Chem.* **237**:1388–1392 (1962).
247. W. W. Cleland and E. P. Kennedy, *J. Biol. Chem.* **235**:45–51 (1960).
248. R. M. Burton, M. A. Sodd, and R. O. Brady, *J. Biol. Chem.* **233**:1053–1060 (1958).
249. B. Kaufman, S. Basu, and S. Roseman, in *Inborn Disorders of Sphingolipid Metabolism* (S. M. Aronson and B. W. Volk, eds.) pp. 193–213, Pergamon Press, Oxford (1967).
250. R. Caputto, L. F. Leloir, R. E. Trucco, C. E. Cardini, and A. C. Paladini, *J. Biol. Chem.* **179**:497–498 (1949).
251. K. J. Isselbacher, E. P. Anderson, K. Kurahashi, and H. M. Kalckar, *Science* **123**:635–636 (1956).
252. E. S. Maxwell, *J. Am. Chem. Soc.* **78**:1074 (1956).
253. S. B. Weiss and L. Gladstone, *J. Am. Chem. Soc.* **81**:4118–4119 (1959).
254. J. Hurwitz, A. Bresler, and R. Diringer, *Biochem. Biophys. Res. Commun.* **3**:15–19 (1960).
255. S. Ochoa, D. P. Burma, H. Kröger, and J. D. Weill, *Proc. Natl. Acad. Sci. U.S.* **47**:670–679 (1961).
256. E. P. Kennedy and S. B. Weiss, *J. Am. Chem. Soc.* **77**:250–251 (1955).
257. E. P. Kennedy and S. B. Weiss, *J. Biol. Chem.* **222**:193–214 (1956).
258. J. Baddiley, J. G. Buchanan, and G. R. Greenberg, *Biochem. J.* **66**:51P–52P (1957).
259. D. R. D. Shaw, *Biochem. J.* **66**:56P (1957).
260. M. Sribney and E. P. Kennedy, *J. Am. Chem. Soc.* **79**:5325–5326 (1957).
261. L. F. Borkenhagen and E. P. Kennedy, *J. Biol. Chem.* **227**:951–962 (1957).
262. E. P. Kennedy, *Canad. J. Biochem. Physiol.* **34**:334–347 (1956).
263. S. B. Weiss, S. W. Smith, and E. P. Kennedy, *Nature* **178**:594–595 (1956).
264. H. G. Williams-Ashman and J. Banks, *J. Biol. Chem.* **223**:509–521 (1956).
265. R. J. Rossiter, W. C. McMurray, and K. P. Strickland, *Fed. Proc.* **16**:853–855 (1957).
266. S. B. Weiss, S. W. Smith, and E. P. Kennedy, *J. Biol. Chem.* **231**:53–64 (1958).
267. E. P. Kennedy, *Am. J. Clin. Nutr.* **6**:216–219 (1958).
268. M. Sribney and E. P. Kennedy, *J. Biol. Chem.* **233**:1315–1322 (1958).
269. B. W. Agranoff, R. M. Bradley, and R. O. Brady, *J. Biol. Chem.* **233**:1077–1083 (1958).
270. W. Thompson, K. P. Strickland, and R. J. Rossiter, *Biochem. J.* **87**:136–142 (1963).
271. R. J. Rossiter and F. B. Palmer, in *Lipoide, 16.Coll.Ges. für Physiol. Chem. in Mosbach* (E. Schütte, ed.), pp. 40–48, Springer Verlag, Berlin (1966).
272. M. Shoyab, T. N. Pattabiraman, and B. K. Bachhawat, *J. Neurochem.* **11**:639–646 (1964).
273. M. Shoyab and B. K. Bachhawat, *Indian J. Biochem.* **2**:6–9 (1965).
274. A. Arce, H. F. Maccioni, and R. Caputto, *Arch. Biochem. Biophys.* **116**:52–58 (1966).
275. E. Cabib and L. F. Leloir, *J. Biol. Chem.* **206**:779–790 (1954).
276. H. Hift, L. Ouellet, J. W. Littlefield, and D. R. Sanadi, *J. Biol. Chem.* **204**:565–575 (1953).
277. S. Kaufman and S. G. A. Alivisatos, *J. Biol. Chem.* **216**:141–152 (1955).
278. P. Mandel and J. Klethi, unpublished data.
279. J. L. Strominger, *Fed. Proc.* **13**:307 (1954).
280. A. Munch-Petersen, *Arch. Biochem. Biophys.* **55**:592–593 (1955).
281. P. Mandel and E. Kempf, *Biochim. Biophys. Acta* **51**:184–186 (1961).
282. G. T. Mills and E. E. B. Smith, in *Zur Bedeutung der freien Nucleotide, Colloq. Ges. Physiol. Chem. in Mosbach*, Vol. 11, pp. 44–54, Springer Verlag, Berlin (1961).
283. S. Edel and H. Rein-Løvtrup, *Bull. Soc. Chim. Biol.* **48**:915–933 (1966).
284. M. F. Utter, in *The Enzymes* (P. D. Boyer, H. Lardy, and K. Myrbäck, eds.), Vol. 2, 2nd ed., pp. 75–88, Academic Press, New York (1960).
285. D. R. Sanadi, in *The Neurochemistry of Nucleotides and Amino Acids* (R. O. Brady and D. B. Tower, eds.), pp. 13–27, John Wiley and Sons, New York (1960).
286. R. Abrams and M. Bentley, *J. Am. Chem. Soc.* **77**:4179–4180 (1955).

287. I. Lieberman, *J. Biol. Chem.* **223**:327–339 (1956).

288. E. B. Keller and P. C. Zamecnik, *J. Biol. Chem.* **221**:45–59 (1956).

289. S. Slapikoff, J. M. Fessenden, and K. Moldave, *J. Biol. Chem.* **238**:3670–3676 (1963).

290. K. Moldave, Adv. Chem. Ser. **44**:41–53 (1964).

291. Y. Nishizuka and F. Lipmann, *Proc. Natl. Acad. Sci. U.S.* **55**:212–219 (1966).

292. B. Hardesty, R. Miller, and R. Schweet, *Proc. Natl. Acad. Sci. U.S.* **50**:924–931 (1963).

293. B. Hardesty, J. J. Hutton, R. Arlinghaus, and R. Schweet, *Proc. Natl. Acad. Sci. U.S.* **50**:1078–1085 (1963).

294. A. R. Williamson and R. Schweet, *J. Mol. Biol.* **11**:358–372 (1965).

295. P. Mandel and S. Harth, *C.R. Acad. Sci. (Paris)* **245**:1843–1845 (1957).

296. W. Koransky, *Arch. Exptl. Pathol. Pharmacol.* **234**:46–65 (1958).

297. P. J. Heald, *Phosphorus Metabolism of Brain*, Pergamon Press, Oxford (1960).

298. P. Mandel and S. Harth, *J. Physiol. (Paris)* **52**:166 (1960).

299. P. Mandel, S. Harth, and G. Rebel, in *Chemical Pathology of the Nervous System*, Proc. 3rd Intern. Neurochem. Symp. Strasbourg (J. Folch-Pi, ed.), pp. 551–561, Pergamon Press, Oxford (1961).

300. G. B. Ansell and B. J. Bayliss, *Biochem. J.* **78**:209–213 (1961).

301. P. Mandel and J. D. Weill, *J. Physiol. (Paris)* **54**:199–268 (1962).

302. H. Rein, S. Harth-Edel, and P. Mandel, *Bull. Soc. Chim. Biol.* **46**:206 (1964).

303. P. Mandel, S. Edel, and G. Poirel, *J. Neurochem.* **13**:885–886 (1966).

304. F. M. Matschinsky, *J. Neurochem.* **15**:643–657 (1968).

305. S. E. Kerr, *J. Biol. Chem.* **145**:647–656 (1942).

306. P. J. Heald, *Biochem. J.* **67**:529–536 (1957).

307. R. Bergkvist and A. Deutsch, *Acta Chem. Scand.* **7**:1307–1308 (1953).

308. P. Mandel and S. Edel-Harth, *J. Neurochem.* **13**:591–595 (1966).

309. P. F. Urban, J. Klethi, and P. Mandel, unpublished data.

310. H. McIlwain, *Biochem. J.* **78**:24–32 (1961).

311. M. Ramuz, C. Judes, and P. Mandel, *J. Neurochem.* **11**:826–827 (1964).

312. M. Nyman and V. P. Whittaker, *Biochem. J.* **87**:248–255 (1963).

313. H. J. Schümann, *Arch. Exptl. Pathol. Pharmakol.* **233**:237–249 (1958).

314. W. H. Prusoff, *Brit. J. Pharmacol.* **17**:87–91 (1961).

315. M. Behrens, in *Handbuch der Biologischen Arbeitsmethod* (E. Abderhalden, ed.) Vol. 10, pp. 1363, Urban and Schwarzenberg, Berlin and Vienna (1938).

316. G. Siebert, in *Biochemisches Taschenbuch* (H. M. Rauen, ed.) Vol. II, 2nd ed., pp. 541–546, Springer Verlag, Berlin (1964).

317. L. Mandel and P. Mandel, *J. Physiol. (Paris)* **62**:184 (1970).

318. P. Mandel, R. Bieth, and J. D. Weill, in *Metabolism of the Nervous System*, Proc. 2nd Intern. Symp. Neurochem. Aahrus (D. Richter, ed.), pp. 291–296, Pergamon Press, Oxford (1957).

319. G. A. LePage, in *Manometric Techniques* (W. W. Umbreit, R. H. Burris, and J. F. Stauffer, eds.), 3rd ed., pp. 268–281, Burgess Publishing, Minneapolis (1957).

320. M. M. Cohen and S. Lin, *J. Neurochem.* **9**:345–352 (1962).

321. A. Devi, M. A. Mukundan, U. Srivastava, and N. K. Sarkar, *Exptl. Cell Res.* **32**: 242–250 (1963).

321a. M. Jacob and P. Mandel, in *Protides of the Biological Fluids* (H. Peeters, ed.), pp. 63–80, Elsevier Publishing, Amsterdam (1965).

322. I. Wegelin and F. A. Manzoli, *J. Neurochem.* **14**:1161–1165 (1967).

323. F. A. Manzoli and I. Wegelin, *J. Neurochem.* **16**:829–831 (1969).

324. V. Hamburger and R. Oppenheim, *J. Exptl. Zool.* **166**:171–204 (1967).

325. J. M. Milstein, J. G. White, and K. F. Swaiman, *J. Neurochem.* **15**:411–415 (1968).

326. R. M. Burton, *J. Neurochem.* **2**:15–20 (1957).

327. R. Guarneri and V. Bonavita, *Brain Res.* **2**:145–150 (1966).
328. W. Thorn, G. Pfleiderer, R. A. Frowein, and I. Ross, *Pflüger's Arch. Ges. Physiol.* **261**:334–360 (1955).
329. H. J. Döring, A. Knoff, and Th. Martin, *Pflüger's Arch. Ges. Physiol.* **269**:375–391 (1959).
330. H. McIlwain, *Biochemistry and the Central Nervous System*, 2nd ed., 288pp., Little, Brown and Co., Boston (1959).
331. O. H. Lowry, J. V. Passonneau, F. X. Hasselberger, and D. W. Schulz, *J. Biol. Chem.* **239**:18–30 (1964).
332. F. W. Schmahl, *Acta Neurol. Scand.* **41**, Suppl 14:156–159 (1965).
333. F. W. Schmahl, E. Betz, H. Talke, and H. J. Hohorst, *Biochem. Z.* **342**:518–531 (1965).
334. E. Gerlach, H. J. Döring, A. Fleckenstein, H. Fikentscher, and G. Wirth, *Pflüger's Arch. Ges. Physiol.* **266**:266–291 (1958).
335. T. N. Ivanova, N. I. Pravdina, and L. N. Rubel, *Biokhimiya* **27**:293–305 (1962) [English Translation, *Biochemistry* **27**:250–260 (1962)].
336. P. D. Gatfield, O. H. Lowry, D. W. Schulz, and J. V. Passonneau, *J. Neurochem.* **13**:185–195 (1966).
337. N. A. Dahl and W. M. Balfour, *Am. J. Phys.* **207**:452–456 (1964).
338. S. Navon and A. Agrest, *Am. J. Phys.* **205**:957–958 (1963).
339. W. E. Stone, *Biochem. J.* **32**:1908–1918 (1938).
340. W. E. Stone, *J. Biol. Chem.* **135**:43–50 (1940).
341. D. Richter and R. M. C. Dawson, *Am. J. Phys.* **154**:73–79 (1948).
342. H. McIlwain, *Brit. Med. Bull.* **6**:301 (1950).
343. H. McIlwain, *Biochem. Soc. Sympos.* **8**:27–43 (1952).
344. D. Richter, *Biochem. Soc. Sympos.* **8**:62–76 (1952).
345. H. G. Albaum, W. K. Noell, and H. Chinn, *Am. J. Phys.* **174**:408–412 (1953).
346. A. V. Palladin, *Ukrain. Biochem. J.* **26**:112–127 (1954).
347. I. Pechan and P. Marko, *Physiol. Bohemoslovenica* **12**:458–462 (1963).
348. N. Weiner, *J. Neurochem.* **7**:241–250 (1961).
349. H. McIlwain, L. Buchel, and I. D. Cheschire, *Biochem. J.* **48**:12–20 (1951).
350. W. J. Johnson and J. H. Quastel, *Nature* **171**:602–605 (1953).
351. J. A. Bain, *Fed. Proc.* **11**:653–658 (1952).
352. J. A. Bain, in *Progress in Neurobiology* (S. R. Korey and J. I. Nurnberger, eds.), Vol. 2, pp. 139–157, P. B. Hoeber, New York (1957).
353. B. E. Leonard, *Biochem. Pharmacol.* **15**:255–262 (1966).
354. H. McIlwain, *Biochem. J.* **53**:403–412 (1953).
355. R. G. Grenell, *Ann. N.Y. Acad. Sci.* **66**:826–835 (1957).
356. K. Schwemmle and F. Heim, *Med. Experimentalis* **1**:351–356 (1959).
357. C. L. Kaul and J. J. Lewis, *Biochem. Pharmacol.* **12**:1279–1282 (1963).
358. C. L. Kaul, J. J. Lewis, and S. D. Livingstone, *Biochem. Pharmacol.* **14**:165–175 (1965).
359. C. L. Kaul and J. J. Lewis, *J. Pharmacol. Exptl. Ther.* **140**:111–116 (1963).
360. N. Bonasera, G. Mangione, and V. Bonavita, *Biochem. Pharmacol.* **12**:633–636 (1963).
361. N. Bonasera, G. Mangione and V. Bonavita, *Int. J. Neuropharmacol.* **4**:71–77 (1965).
362. P. Mandel and Y. Godin, in *Aspects anatomo-fonctionnels de la physiologie du sommeil*, N° 127, pp. 1–36, Editions du Centre National de la Recherche Scientifique, Paris (1964).
363. J. R. Klein and N. S. Olsen, *J. Biol. Chem.* **167**:747–756 (1947).
364. P. F. Minaev and T. P. Kurokhtina, *Ukrain. Biokhim. Zhur.* **21**:359 (1949).
365. R. M. C. Dawson and D. Richter, *Am. J. Phys.* **160**:203–211 (1950).
366. H. McIlwain, *Biochem. J.* **50**:132–140 (1952).
367. D. B. Tower, *Neurochemistry of Epilepsy*, C. C. Thomas, Springfield, Ill. (1960), 335 pp.
368. H. Wallgren, *J. Neurochem.* **10**:349–362 (1963).

369. L. J. King, O. H. Lowry, J. V. Passonneau, and V. Venson, *J. Neurochem.* **14**:599–611 (1967).
370. F. Piccoli, R. Camarda, and V. Bonavita, *J. Neurochem.* **16**:159–169 (1969).
371. E. W. Davis, W. S. McCulloch, and E. Roseman, *Am. J. Psychiat.* **100**:825 (1944).
372. S. Kakiuchi, T. W. Rall, and H. McIlwain, *J. Neurochem.* **16**:485–491 (1969).
373. P. Mandel, S. Simler, and H. Randrianarisoa, unpublished data.
374. B. Sacktor, J. E. Wilson, and C. G. Tiekert, *J. Biol. Chem.* **241**:5071–5075 (1966).
375. I. Dosseva, Z. Tencheva, and S. Dimov, *Compt. Rend. Acad. Bulgare Sci.* **18**:283–286 (1965).
376. I. Dosseva, Z. Tencheva, and S. Dimov, *Compt. Rend. Acad. Bulgare Sci.* **19**:313–316 (1966).
377. F. Heim, F. Leuschner, and C. J. Estler, *Experientia* **13**:462–471 (1957).
377a. S. Williams, R. A. Paterson, and H. Heath, *J. Neurochem.* **15**:227–233 (1968).
378. T. Osuka and T. Shimizu, *Seishin Shinkeigaku Zasshi* **68**:501–511 (1966).
379. C. J. Estler and H. P. T. Ammon, *J. Neurochem.* **14**:799–805 (1967).
380. S. Edel, *J. Physiol. (Paris)* **59**:234–235 (1967).
381. J. J. Lewis and G. R. Van Petten, *Brit. J. Pharmacol. Chemother.* **20**:462–470 (1963).
382. A. Geiger and S. Yamasaki, *J. Neurochem.* **1**:93–100 (1956/57).
383. V. Bonavita, N. Bonasera, M. Zito, and E. Scarano, *J. Neurochem.* **10**:155–164 (1963).
384. T. Muramatsu, M. Ando, T. Nagatsu, and K. Yagi, *J. Neurochem.* **4**:229–233 (1959).
385. K. Yagi, T. Ozawa, M. Ando, and T. Nagatsu, *J. Neurochem.* **5**:304–306 (1960).
386. R. M. Burton, in *Inhibition of the Nervous System and γ-Aminobutyric Acid*, pp. 249–259, Pergamon Press, Oxford (1960).
387. A. Lehmann, S. Simler, and P. Mandel, *J. Physiol. (Paris)* **59**:446 (1967).
388. F. N. Minard and R. V. Davis, *J. Biol. Chem.* **237**:1283–1289 (1962).
389. M. Okamoto, A. Askari, and A. S. Kuperman, *J. Pharmacol. Exptl. Ther.* **144**:229–235 (1964).
390. A. S. Kuperman, M. Okamoto, and E. Gallin, *J. Cell Physiol.* **70**:257–264 (1967).
391. H. Koenig, in *Intern. Rev. Neurobiology* (C. C. Pfeiffer and J. R. Smythies, eds.), Vol. 10, pp. 199–230, Academic Press, New York (1967).
392. S. Kakiuchi and T. W. Rall, *Mol. Pharmacol.* **4**:367–378 (1968).
393. S. Kakiuchi and T. W. Rall, *Mol. Pharmacol.* **4**:379–388 (1968).
394. M. Ramuz and P. Mandel, *C.R. Acad. Sci. (Paris)* **256**, Groupe 13:4523–4524 (1963).
395. I. Potop, G. Mreana, and C. Neacsu, *Rev. Sci. Med. Acad. Rep. Populaire Roumaine* **8**:3–4 (1963).
396. J. F. Berry and J. D. Coonrad, *J. Neurochem.* **14**:245–255 (1967).
397. M. D. Kurs'kii and O. M. Zryakov, *Ukrain. Biokhim. Zh.* **36**:679–684 (1964).
398. S. Schenker, J. H. Mendelson, and C. Corbett, *Am. J. Phys.* **206**:1173–1176 (1964).
399. S. Schenker, D. W. McCandless, E. Brophy, and M. S. Lewis, *J. Clin. Invest.* **46**:838–848 (1967).
400. I. R. Petrov and N. V. Korostovtseva, *Fiziol. Zh. SSSR* **49**:1155–1162 (1963).
401. F. Heim, C. J. Estler, H. P. T. Ammon, and U. D. Wiedemann, *J. Neurochem.* **13**:1495–1500 (1966).
402. S. P. R. Rose, *Biochem. Pharmacol.* **14**:589–601 (1965).

Chapter 9

THE TRICARBOXYLIC ACID CYCLE*

Sze-Chuh Cheng

New York State Research Institute for
Neurochemistry and Drug Addiction
Ward's Island, New York

I. INTRODUCTION

This chapter serves as a tiepoint to intermediate metabolism in the nervous tissue. Pyruvate is chosen as the starting point for this discussion in order to cover the interconversion of three-carbon and four-carbon acids. Consequently, the acetyl-coenzyme A (acetyl-CoA, AcCoA, $CoAS \sim Ac$) pathway which contributes a two-carbon element to citrate is also included here. A discussion of individual enzymatic reactions will be the principal scheme of this chapter. Since our knowledge of these individual reactions in the nervous tissue is extremely scanty,[1,2] much of the discussion has to rely on knowledge gained in other tissues, particularly liver, heart, and microorganisms.[3-6] There is no reason to suspect that these enzymatic mechanisms could not apply to the enzymes of nervous tissue. The material is organized in such a way that it could be used as a starting point for future work in this area. The known concentrations of intermediates and activities of enzymes are given under each enzymatic reaction in Section II as well as in Tables I and II. To avoid duplication, many areas covered in other parts of this treatise are deleted. Thus, the isolation and characterization of the mitochondrial element together with its properties are already covered in another chapter[7] and will not be included here. Several other closely related subjects are also covered elsewhere in this treatise, such as the glutamate-γ-aminobutyrate shunt,[8] the oxidative phosphorylation of substrates,[9] the compartmentalization of α-ketoglutarate and glutamate[10] and the metabolism of glutamate, aspartate, and acetyl-aspartate.[11]

The tricarboxylic acid cycle (TCAC), variously known as the Krebs cycle or the citric acid cycle, occupies a central position in metabolic activities. It is the main site to produce metabolically useful energy, primarily in the form of reduced diphosphopyridine nucleotide (NADH) or "reducing

* This work was supported in part by a grant from the U.S.P.H.S. NB 5869-02.

TABLE I

The Activity of Individual Enzymes of the TCAC in Brain

Reaction No.	Name	Rate (μmoles/ g/hr)	Animal	References
1	Pyruvate dehydrogenase complex	30–50	Sheep, rabbit, rat	2, 20
2	Condensing enzyme	0.76	Sheep	52
		820	Rat	51
	Citrate cleavage enzyme	1.8	Sheep	52
		3.9	Pig	53
		8–11	Rat, rabbit, beef	53
3	Aconitase	0.60	Rabbit	70
		90	Rat	2
4	Isocitric dehydrogenase	0.5–56	Rabbit	70, 85
		400	Rat	2
5	α-Ketoglutarate dehydro-genase complex	—		
6	Succinate thiokinase	—		
7	Succinic dehydrogenase	200	Rat	2
8	Fumarase	710	Sheep	53, 105
		1400–2800	Rabbit	70, 103, 104
		2000	Rat	2
9	Malate dehydrogenase	30–40	Rabbit or monkey retina	113
		5300	Rat	2
		6000–88,000	Rabbit	85, 111, 112

power" which upon oxidative phosphorylation gives rise to adenosine triphosphate (ATP).[12] Its intermediates also provide "carbon skeletons" for the syntheses of various biological compounds. The scope of this chapter starts from pyruvate which is produced from sugar breakdown.[13] In the simplified scheme (Fig. 1), pyruvate is converted to citrate by dismutation. One pyruvate molecule is decarboxylated to AcCoA, and another pyruvate molecule is carboxylated to oxaloacetate. Then the two products condense together into citrate. Oxaloacetate could be, and is, regenerated from the cyclic operation of the TCAC. If so, a net decarboxylation of pyruvate coupled to a reduction of diphosphopyridine nucleotide (NAD) to NADH takes place.

Citrate is isomerized via cis-aconitate to isocitrate which is then decarboxylated to α-ketoglutarate with the gain of another NADH. α-Ketoglutarate serves as an important branching point for the glutamate-γ-amino-butyrate shunt which rejoins the TCAC in the form of succinate. This shunt is covered in another chapter.[8] Here, α-ketoglutarate is decarboxylated very much like pyruvate to succinyl-coenzyme A (succinyl-CoA) with the gain of

TABLE II

The Concentration of the Intermediates in the TCAC

Compound	μmoles/g	Source	References
Pyruvate	0.004–0.06	Cat brain	41
	0.06	Rat brain	39
	0.08	Mouse brain	23, 42
	2	Rat brain	40
Acetyl-CoA	0.005	Rat brain	61
Oxaloacetate	0.0015	Rabbit sciatic nerve	63
	0.004	Rat and mouse brain	23
	0.08	Lobster nerve	62
	0.64	Rat brain	40
Citrate	0.08	Rat sciatic nerve	60
	0.25–0.4	Rat brain	40, 53, 60, 73, 74
	0.1–0.3	Mouse brain	23, 75, 96
	0.3	Rabbit sciatic nerve	60, 62
	0.46	Lobster nerve	60, 63
cis-Aconitate	<0.01	Rat brain	40
Isocitrate	0.02	Mouse brain	23
	0.03	Rat brain	40
α-Ketoglutarate	0.001–0.016	Cat brain	41
	0.012	Rat brain	39
	0.13	Mouse brain	23
	1.3	Rat brain	40
Succinyl-CoA	—		
Succinate	0.34	Rat brain	40
	0.69	Mouse brain	23
Fumarate	0.07	Mouse brain	23
	1.2	Rat brain	40
Malate	0.24	Rat brain	40
	0.44	Mouse brain	23
	4.3	Rabbit sciatic nerve	63
	6	Lobster nerve	62, 116, 117

another NADH. Succinate is formed from succinyl-CoA, gaining a "high energy phosphate bond" ($\sim$P) as a guanosine triphosphate (GTP), and is oxidized to fumarate with a concurrent reduction of a flavin-adenine-dinucleotide (FAD). Fumarate is hydrated to malate which is then oxidized to oxaloacetate with a fourth NAD reduced to NADH and the cycle is completed with a new oxaloacetate molecule to replace the one used up in the previous cycle.

In all, from each pyruvate, four NAD are reduced to four NADH, one FAD to $FADH_2$ and one GTP formed from GDP (guanosine diphosphate). If we calculate, based on the commonly accepted $\sim$P yield of 2 and 3 respectively for $FADH_2$ and NADH, a maximum of 15 $\sim$P could be formed

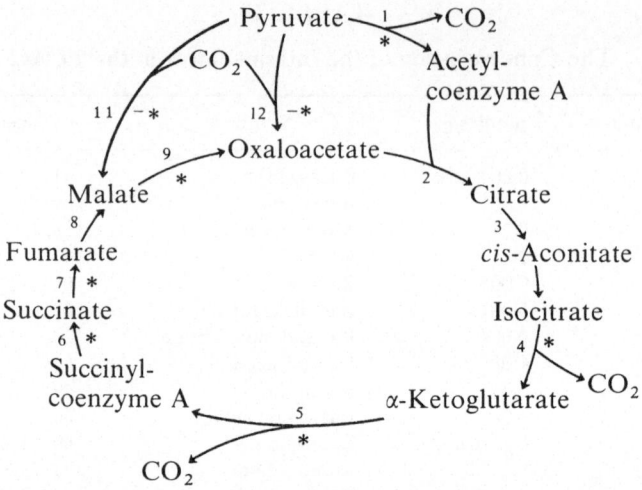

Fig. 1. Simplified diagram of the tricarboxylic acid cycle in its forward direction, showing the reactions involving CO_2 as well as those which produce and consume metabolic energy (* and −*). Individual reactions are numbered according to those in Section II. They are: 1, pyruvate dehydrogenase system; 2, condensing enzyme and citrate cleavage enzyme; 3, aconitase; 4, isocitric dehydrogenase; 5, α-ketoglutarate dehydrogenase system; 6, succinate thiokinase; 7, succinic dehydrogenase; 8, fumarase; 9, malate dehydrogenase; 11, malic enzyme; and 12, pyruvate carboxylase.

from the oxidation of each pyruvate to three carbon dioxide (see Section II, 10).

Normally, and particularly during intense anabolism such as regeneration, some of the intermediates of the TCAC are tapped off. The regeneration of oxaloacetate is then reduced. In order to keep the TCAC in operation, the carboxylation of a three-carbon compound (pyruvate or phosphoenolpyruvate), i.e., the carbon dioxide fixation pathway, will commence to a significant degree. In brain, carbon dioxide fixation accounted for about 10% of the pyruvate utilized as compared to the decarboxylation of pyruvate,[14,15] while in the isolated lobster nerve, the two processes were nearly equal in magnitude.[15]

II. INDIVIDUAL REACTIONS AND ENZYMES

1. The Pyruvate Dehydrogenase System (Pyruvate: Lipoate Oxidoreductose (Acceptor-acetylating), E.C. 1.2.4.1)

Overall reaction:

$$\text{pyruvate} + \text{CoASH} + \text{NAD} \rightarrow \text{CoAS} \sim \text{Ac} + \text{NADH} + CO_2$$

Mechanism:

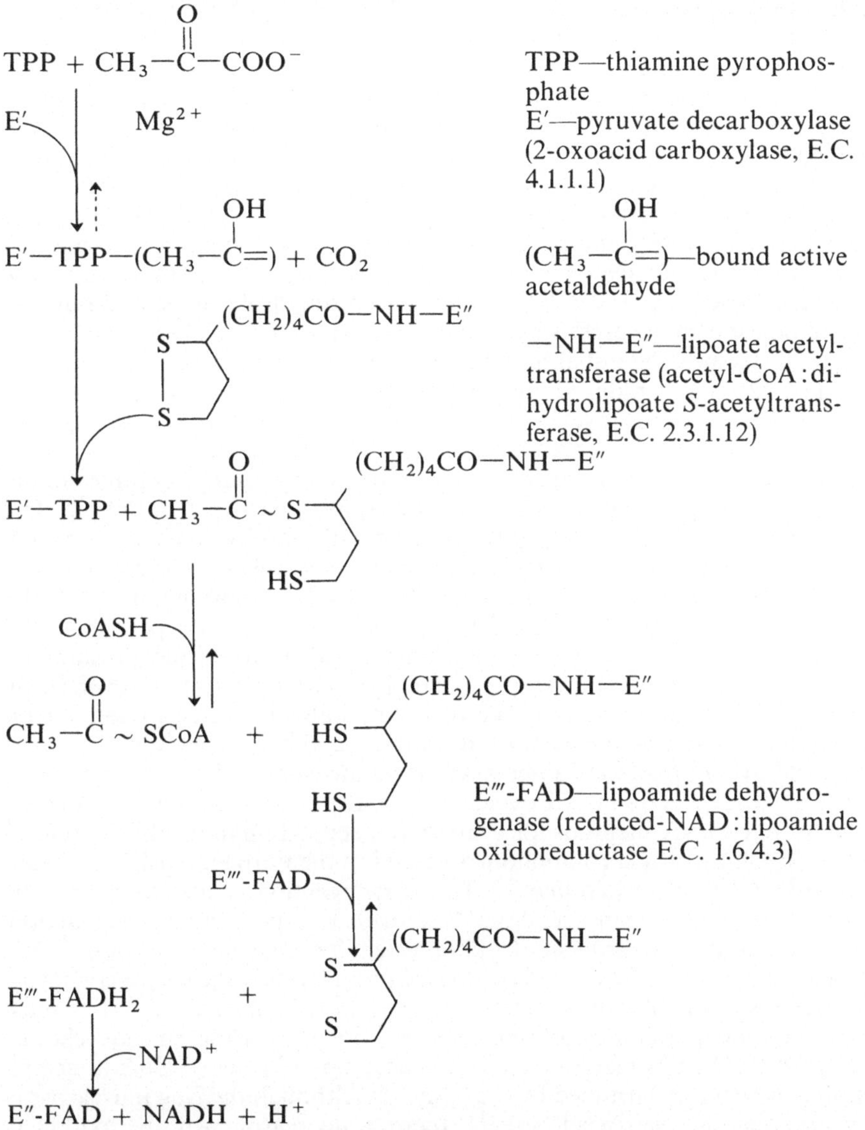

$$TPP + CH_3-\overset{\overset{\displaystyle O}{\|}}{C}-COO^-$$

TPP—thiamine pyrophosphate

E'—pyruvate decarboxylase (2-oxoacid carboxylase, E.C. 4.1.1.1)

E'⟍ | Mg²⁺

$$E'-TPP-(CH_3-\overset{\overset{\displaystyle OH}{|}}{C}=) + CO_2$$

$$(CH_3-\overset{\overset{\displaystyle OH}{|}}{C}=)\text{—bound active acetaldehyde}$$

$(CH_2)_4CO-NH-E''$

—NH—E''—lipoate acetyltransferase (acetyl-CoA:dihydrolipoate S-acetyltransferase, E.C. 2.3.1.12)

$(CH_2)_4CO-NH-E''$

$$E'-TPP + CH_3-\overset{\overset{\displaystyle O}{\|}}{C} \sim S$$

HS

CoASH⟍

$$CH_3-\overset{\overset{\displaystyle O}{\|}}{C} \sim SCoA \quad + \quad HS$$

$(CH_2)_4CO-NH-E''$

HS

E'''-FAD—lipoamide dehydrogenase (reduced-NAD:lipoamide oxidoreductase E.C. 1.6.4.3)

E'''-FAD⟍

$(CH_2)_4CO-NH-E''$

S

$E'''-FADH_2 \quad + $

S

⟍NAD⁺

$E''-FAD + NADH + H^+$

The keto acid oxidation mechanism[16,17] is rather complicated, requiring three sets of enzyme-bound reactions leading ultimately to CoAS ∼ Ac and NADH. The enzymes are tightly bound together and are hard to separate. The isolated enzymes do tend to reassociate with the transacetylase as the central piece.[18,19] Mg^{2+}, TPP, ferredoxin, lipoate, and FAD are all involved.

The reaction starts with pyruvate decarboxylase, which is reversibly linked to the quarternary nitrogen of the thiazole ring of thiamine pyrophosphate. Carbon number 2 of pyruvate, in the presence of Mg^{2+}, reacts with the carbon between the nitrogen and the sulfur atoms of the thiazole, forming a lactyl intermediate. This intermediate loses a carbon dioxide and leaves behind, still bound, an active acetaldehyde. Somewhere in this reaction sequence, ferredoxin is involved. The lipoate, bound to the enzyme lipoate acetyltransferase, functions as an acceptor for the active aldehyde group with the opening of the —S—S— bond to form a —S—Ac linkage. The decarboxylase–TPP complex is regenerated. The acetyl-lipoyl enzyme will now donate the active acetaldehyde to a reduced coenzyme A and the resultant CoAS~Ac is cast free. The reduced dihydrolipoyl enzyme undergoes oxidation, in the presence of the third enzyme with FAD as its prosthetic group, to recover the initial lipoate acetyltransferase complex. The enzyme-bound FAD is now reduced but is recovered through the reduction of NAD to NADH. Thus the overall reaction produces CoAS~Ac and NADH. This mechanism has not been studied in the nervous tissue.

The overall reaction rate of the pyruvate dehydrogenase complex in the nervous tissue is about 30–50 μmoles per gram (fresh) per hour.[2,20] The Michaelis constant, K_M, for pyruvate varies with the electron acceptor used. With indophenol, it is 2×10^{-5} M for the pigeon breast muscle enzyme. The reaction at neutral pH is far in favor of CoAS~Ac formation and $\Delta G^{0'}$, the change of Gibbs free energy at 25°C and pH 7, is -9.38 kcal per mole. The reaction is inhibited by arsenite and other trivalent arsenicals, presumably acting on the enzyme-bound lipoic acid. The inhibition can be reversed by BAL (2,3-dimercaptoethanol). The reaction is also inhibited by long-chain fatty acids,[21] and by the reaction product, acetyl-coenzyme A.[22]

The overall reaction can be assayed by measuring either the remaining pyruvate with lactate dehydrogenase[23] or the acetyl-coenzyme A formed. The primary assay methods for acetyl-CoA depend upon its ability to form a C—N bond with a chromophore attached to the nitrogen, such as sulfanilamide[24–26] or p-nitroaniline.[27] The latter reaction is enzymatic and the purity of arylamine transacetylase[28] controls the specificity of the reaction. Other methods, primarily spectrophometric after enzymatic coupling,[29,30] depend either on the formation of NADH or on its cleavage to free coenzyme A. Free coenzyme A can be determined by many methods.[31–36] The most sensitive, and possibly most specific, method utilizes the enzyme choline acetylase.[37] In this reaction, choline is acetylated by acetyl-CoA to acetylcholine which is determined by bioassay.[38] Although the frog muscle assay for acetylcholine was used here,[37] there is no reason why the extremely sensitive method of leech muscle or clam heart could not be employed.[38]

The concentration of pyruvate in rat brain has been variously estimated from 0.06[39] to 2 μmoles[40] per gram of fresh tissue. That in cat brain was reported to be 0.004–0.06 μmoles per gram,[41] and in mouse brain 0.08 μmoles per gram.[23,42] Assay of pyruvate is commonly performed by reacting its keto group to 2,4-dinitrophenylhydrazine to form the corresponding hydrazone.[43,44] It can also be decarboxylated easily and the resultant carbon dioxide measured manometrically.[43] Or it can be measured using the lactate dehydrogenase reaction.[23]

2. Condensing Enzyme (Citrate Oxaloacetate-lyase (CoA-Acetylating), E.C. 4.1.3.7)

Overall reaction:

$$CoAS \sim Ac + Oxaloacetate + H_2O \rightarrow Citrate + CoASH + H^+$$

Mechanism:

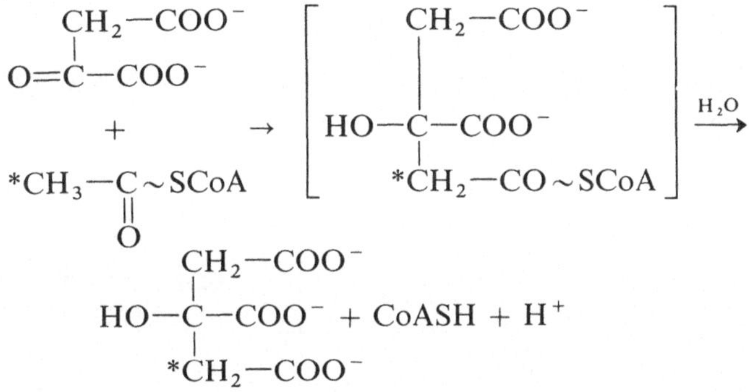

This enzymatic reaction[45,46] condenses carbon 2 (marked with an asterisk and carried through reaction 6) of acetyl-coenzyme A to carbon 2 of oxaloacetate. The proposed intermediate is then hydrolyzed to give citrate and free coenzyme A. The reverse reaction is catalyzed by the citrate cleavage enzyme [ATP:citrate oxaloacetate-lyase (CoA-acetylating and ATP-dephosphorylating), E.C. 4.1.3.8].[47,48] This latter reaction requires an ATP for citrate activation, an ADP (adenosine diphosphate) and an inorganic phosphate (P_i) are formed. Presumably, both reactions involve an enzyme-bound intermediate of citryl-coenzyme A. Recently, citryl-1-phosphate was shown to be a substrate, and presumably an intermediate, in the reverse direction.[49,50] This mechanism has not been shown using nervous tissue.

The overall reaction rates are approximately 820 and 0.76 μmoles per gram per hour in the forward direction for rat brain[51] and sheep brain[52] respectively, and 2–11 μmoles per gram per hour in the reverse direction in brains of various animals.[52,53] The K_M is $(0.5 - 1) \times 10^{-5}$ M for oxaloacetate,[53,54] 5.8×10^{-6} M for CoAS $\sim$ Ac[54] and 5.8×10^{-4} M for citrate.[55] The equilibrium is far to the citrate side for the forward reaction and $\Delta G^{0\prime}$ is -9.08 kcal per mole. This reaction is inhibited by long-chain acyl-CoA compounds,[56] fluoroacetyl-CoA,[57] and adenine nucleotides, particularly ATP.[58]

The reaction could be assayed by measuring: (1) citrate formation with the pentabromide method,[29,59] (2) citrate synthesized by an aconitase-isocitrate dehydrogenase preparation where NADP is reduced to

NADPH,[60] (3) NADH formation after coupling to malate dehydrogenase for the reverse reaction,[29] and (4) CoASH liberation in the reverse reaction with choline acetylase[20] or other methods (see Section II, 1).

The concentration of acetyl-coenzyme A has been determined in the rat brain to be 5 mμmoles per gram.[61] That of oxaloacetate in the rat and mouse brain is only 4 mμmoles per gram[23] and in the rat brain 640 mμmoles per gram.[40] There are 80 and 1.5 mμmoles per gram in lobster and rabbit sciatic nerves respectively.[62,63] In addition to the assay using malate dehydrogenase,[23,29] oxaloacetate could be determined as its 2,4-dinitrophenylhydrazone,[43,44] as CO_2 after decarboxylation,[43,64] or as labeled citrate after enzymatic coupling.[65]

3. Aconitase (Citrate (Isocitrate) Hydro-lyase, E.C. 4.2.1.3)

Overall reaction:

$$\text{citrate} \rightarrow \textit{cis}\text{-aconitate} + H_2O \rightarrow \text{isocitrate}$$

Mechanism:

This enzyme-bound reaction[66,67] is readily reversible, and at equilibrium there are 91% citrate, 3% cis-aconitate and 6% isocitrate.[68] This ratio is affected by magnesium ions.[69] The initial reaction involves the association of citrate to the ferrous ion on the enzyme. The hydroxyl group migrates to the Fe^{2+} ion, thus making it possible to transfer a proton (labeled as H^0 for clarity) from carbon 2 to the apoenzyme. With the formation of the double bond between carbons 2 and 3 as a result of losing a proton, the entire molecule regains electric neutrality by releasing the hydroxyl ion to the medium. The enzyme-bound cis-aconitate can be released as free cis-aconitate, or be reconverted back to citrate or be converted in the forward direction to isocitrate. For the latter, the bound cis-aconitate presumably tumbles over, so that C-1 and C-2 exchange their positions with C-6 and C-3. The positions of these carbons in the diagrammatic scheme should really be viewed upside down to give the correct picture. This is not done so that the relative positions of all the carbons can be shown. In this reaction, the enzyme-bound hydrogen (H^0) is returned to carbon 6 which now occupies the position which was held by carbon 2 and the double bond is broken. An hydroxyl group from the medium becomes reattached to the ferrous ion and migrates back to its original position which is now occupied by carbon 2. The final step involves the liberation of this isocitrate molecule. In the hydration of the double bond, the protein is held by the enzyme but the hydroxyl is exchanged with the medium. The addition of the proton and the hydroxyl is trans with respect to the double bond. Again, no such mechanism has been studied using the nervous tissue.

The reaction is readily reversible with a $\Delta G^{0'}$ of 1.59 kcal per mole and, at equilibrium, citrate accounts for over 90% of the total tricarboxylic acids. The overall reaction rates in the brain of rabbit and rat are 0.6 and 90 μmoles per gram per hour respectively.[2,70] With the pig heart enzyme, the K_M values are 3.6×10^{-3} M, 1.2×10^{-4} M, and 4.8×10^{-4} M for citrate, cis-aconitate, and isocitrate respectively. The reaction can be inhibited by iron chelators (such as α,α'-bipyridyl), fluorocitrate, trans-aconitate (competitively) and sulfhydryl reagents (reversible by BAL or glutathione) excluding iodoacetate and arsenic compounds. The enzyme is inactivated in phosphate or veronal buffers.

The assays for this reaction are based on (1) citrate formation as measured by the pentabromide method,[43,59,71] (2) the increase in absorption at 240 mμ of cis-aconitate,[72] or (3) the synthesis of NADPH after coupling to isocitrate dehydrogenase.[60]

The concentration of citrate in rat brain is between 0.25–0.4 μmoles per gram[40,53,60,73,74]; in mouse it is 0.1–0.3 μmoles per gram.[23,75,76] In nerves, it is 0.08, 0.3, and 0.46 μmoles per gram respectively for rat sciatic, rabbit sciatic, and lobster nerves.[60,62,63] The concentration of cis-aconitate and isocitrate are 0.01 and 0.03 μmoles per gram respectively in the rat brain[40] and 0.016 μmoles isocitrate per gram in mouse brain.[23]

Citrate can be determined by the pentabromide method.[43,59,71,77] It can also be determined by coupling the citrate cleavage enzyme[53] or citrate lyase[78] and the malate dehydrogenase to produce NADH. Isocitrate can be assayed using the isocitrate dehydrogenase[71,79] with the formation of NADPH. The NADH and NADPH formed can be followed spectrophotometrically or fluorometrically. The tricarboxylic acids can be separated chromatographically and then each determined separately using an aconitase–isocitrate dehydrogenase preparation.[23,60]

4. *Isocitric Dehydrogenase* (threo-D$_s$-*Isocitrate:*
   NAD oxidoreductase (decarboxylating), E.C. 1.1.1.41)

Overall reaction:

$$\text{isocitrate} + \text{NAD} \rightarrow \alpha\text{-ketoglutarate} + CO_2 + \text{NADH}$$

Mechanism:

$$\begin{array}{l} \text{HO}-\text{CH}-\text{COO}^- \\ \quad\;\; | \\ \quad\;\; \text{CH}-\text{COO}^- \quad + \text{NAD}^+ \\ \quad\;\; | \\ \quad *\text{CH}_2-\text{COO}^- \end{array}$$

$\parallel$ Mg^{2+} or Mn^{2+}

$$\left[\begin{array}{l} \text{HO}-\text{C}-\text{COO}^- \\ \quad\;\; \| \\ \quad\;\; \text{C}-\text{COO}^- \\ \quad\;\; | \\ \quad *\text{CH}_2-\text{COO}^- \end{array}\right] + \text{NADH} + \text{H}^+$$

$\parallel$ Mg^{2+} or Mn^{2+}

$$\begin{array}{l} \text{O}{=}\text{C}-\text{COO}^- \\ \quad\;\; | \\ \quad\;\; \text{CH}_2 \qquad\qquad\qquad\quad + CO_2 \\ \quad\;\; | \\ \quad *\text{CH}_2-\text{COO}^- \end{array}$$

This reaction[80,81] presumably involves an enzyme-bound intermediate,[82,83] oxalosuccinate. The isocitrate is supposed to be bound to the enzyme through the cofactor (Mg^{2+} or Mn^{2+}) at both the carboxyl and α-hydroxyl groups. The enzyme is supposedly bound to the substrate until the final product is released. Thus, the enzyme-bound isocitrate is oxidized by NAD to the enol form of oxalosuccinate which is decarboxylated to the enol form of α-ketoglutarate (2-oxo-glutarate). Upon release from the enzyme, this enol form of α-ketoglutarate is immediately isomerized to the normal keto form. The requirement for the divalent metal ion is definite for the initial reaction involving NAD, but its participation in the decarboxylation reaction is disputed.[82] The addition of the hydrogen to NAD is on the α-side of the pyridine ring. This enzyme in the nervous tissue has not been isolated and characterized.

There are two kinds of isocitrate dehydrogenase.[81] The NAD-specific one is localized in the mitochondria. Its affinity for isocitrate is much larger than that of the NADP enzyme (threo-D_s-isocitrate: NADP oxidoreductase (decarboxylating), E.C. 1.1.1.42). It is very labile and exhibits certain kinetic peculiarities especially toward ATP and ADP. ADP enhances its stability, probably through an allosteric effect.[84] The NADP-specific enzyme is found in the high-speed supernatant of a tissue homogenate. It is relatively more stable. The evidence for the enzyme-bound intermediate is obtained using this enzyme.

For the NADP enzyme, the overall reaction in brain is about 400 (rat)[2] and 0.5–56 (rabbit)[70,85] μmoles per gram per hour. Its K_M for isocitrate is 2.6×10^{-6} M with the pig heart enzyme.[81] The equilibrium favors decarboxylation and $\Delta G^{0\prime}$ is -1.70 kcal per mole for both NAD and NADP enzymes. It is inhibited by sulfhydryl inhibitors and the inhibition can be partially reversed by BAL.[81,83,86] Its activity can be stimulated by ADP and isocitrate and inhibited by ATP and NADH.

The activity of these enzymes can easily be followed by the reduction of NAD or NADP.[23,80,87,88]

The concentration of isocitrate in the nervous tissue and its assay have been discussed in the previous section.

5. α-Ketoglutarate Dehydrogenase System (2-Oxoglutarate:Lipoate Oxidoreductase (Acceptor-Acylating), E.C. 1.2.4.2)

Overall reaction:

$$\alpha\text{-ketoglutarate} + \text{CoASH} + \text{NAD} \rightarrow \text{succinyl}\sim\text{SCoA} + \text{NADH} + CO_2$$

Mechanism:

$$\begin{array}{c} \text{O=C--COO}^- \\ | \\ \text{CH}_2 \\ | \\ \text{*CH}_2\text{--COO}^- \end{array} + \text{NAD}^+ + \text{CoASH} \xrightarrow[\substack{\text{lipoate} \\ \text{FAD}}]{\substack{\text{TPP} \\ \text{Mg}^{2+}}} \begin{array}{c} \text{O=C}\sim\text{SCoA} + CO_2 \\ | \\ \text{CH}_2 \\ | \\ \text{*CH}_2\text{--COO}^- \end{array} + \text{NADH}$$

The α-ketoglutarate dehydrogenase reaction system[16–18,89] is similar to the pyruvate dehydrogenase discussed earlier. The system also includes three enzymes and the corresponding cofactors and coenzymes. It has not been studied in the nervous tissue.

The equilibrium is far toward succinyl-coenzyme A formation with a $\Delta G^{0\prime}$ of -8.82 kcal per mole. The K_M for α-ketoglutarate is 2×10^{-4} M. As with the pyruvate dehydrogenase system, arsenite and trivalent arsenicals are inhibitors of this reaction. This reaction can be assayed by the formation of NADH like other dehydrogenases.[90]

α-Ketoglutarate can be determined as its 2,4-dinitrophenylhydrazone,[43,44,91] as NAD with glutamate dehydrogenase,[23] or as NADPH through further enzymatic coupling.[92] Its concentration in the brain is approximately 1–16 mμmoles per gram[41] in the cat and 130 mμmoles in the mouse. That in the rat has been variously reported as $0.012^{(39)}$ or $1.3^{(40)}$ μmoles per gram.

6. Succinate Thiokinase (Succinate:CoA Ligase (GTP), E.C. 6.2.1.4)

Overall reaction:

$$\text{succinyl} \sim \text{CoA} + \text{GDP} + \text{Pi} + \text{H}_2\text{O} \rightarrow \text{succinate} + \text{CoASH} + \text{GTP}$$

Mechanism:

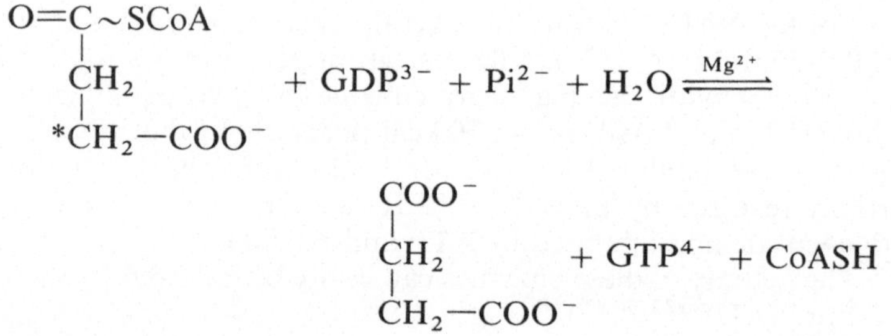

$$
\begin{array}{l}
\text{O=C} \sim \text{SCoA} \\
\quad | \\
\quad \text{CH}_2 \qquad\qquad + \text{GDP}^{3-} + \text{Pi}^{2-} + \text{H}_2\text{O} \xrightarrow{\text{Mg}^{2+}} \\
\quad | \\
\text{*CH}_2 - \text{COO}^-
\end{array}
$$

$$
\begin{array}{l}
\text{COO}^- \\
\quad | \\
\quad \text{CH}_2 \qquad\qquad + \text{GTP}^{4-} + \text{CoASH} \\
\quad | \\
\text{CH}_2 - \text{COO}^-
\end{array}
$$

The energy in the thioester bond is utilized to synthesize a high energy phosphate bond in this reaction.[93] The phosphate acceptor can either be GDP or IDP (inosine diphosphate). The energy in the synthesized GTP or ITP (inosine triphosphate) can easily be transferred to ATP in the presence of nucleotide diphosphate kinase. This phosphorylation associated with the oxidation of α-ketoglutarate is an example of substrate level phosphorylation different from the phosphorylation of NADH oxidation. The significance of this reaction is speculative. Particularly in the brain where glutamate is abundant, one wonders if the glutamate-α-ketoglutarate-succinate pathway could serve as a quick energy source under certain conditions.

There is another important feature in this reaction. Succinate is a symmetrical molecule and the C-4 of α-ketoglutarate (labeled with * throughout the above reactions) becomes randomized between C-2 and C-3 of succinate. The carboxyl carbon, C-5, and carbonyl carbon C-2 of α-ketoglutarate are similarly randomized between C-1 and C-4 of succinate. Thus, all the four-carbon dicarboxylic acids (in reactions to follow) should have the same randomized configuration.

The reaction is in favor of succinate formation with a $\Delta G^{0\prime}$ of -2.12 kcal per mole. It is inhibited by hydroxylamine. Succinyl-CoA can be determined by the hydroxamate reaction[24,26] and can be utilized to follow this reaction. Alternatively, free coenzyme A formation can be followed (see section II, 1). Neither the activity of this enzyme nor the concentration of succinyl-CoA has been determined in the nervous tissue.

7. Succinic Dehydrogenase [Succinate: (Acceptor) Oxidoreductase, E.C. 1.3.99.1]

Overall reaction and mechanism:

$$
\begin{array}{ccc}
\underset{H\ H}{\overset{H\ H\ COO^-}{\diagdown\,\vert\,\diagup}}\!\!\!\!\!\!\!\!\!\!\!\!\!\!
\begin{matrix} C \\ \vert \\ C \end{matrix}
\underset{\diagup\,\vert\,\diagdown}{}
\ H\ H\ COO^-
& +\ E\cdot FAD \rightleftharpoons &
\overset{H\quad COO^-}{\diagdown\ \diagup}
\begin{matrix} C \\ \Vert \\ C \end{matrix}
\underset{^-OOC\quad H}{\diagup\ \diagdown}
\ +\ E\cdot FADH_2
\end{array}
$$

The oxidation of succinate to fumarate is catalyzed by this enzyme[94,95] with bound FAD as a coenzyme. Iron as a function group on the enzyme has been discussed for a long while. Although the EPR signals have been assigned to changes in the iron, it was also thought to originate from flavin or some other part of the complex. It is possible that the iron–flavin–protein functions as a single unit and accounts for the EPR signal of the relatively stable free radical intermediate. A sulfhydryl group was also thought to be involved. The fumarate formed in this reaction is trans in configuration, as shown in the scheme above. The enzyme is tightly bound to the mitochondria and is used as a marker enzyme in subcellular fractionation procedures.

The overall rate of reaction is approximately 200 μmoles per gram per hour for the rat brain.[2] The K_M for succinate of the beef heart enzyme is 5.2×10^{-4} M, and of the yeast enzyme 1×10^{-3} M. The reaction is reversible with $\Delta G^{0\prime}$ approximately zero. An unusual property of this enzyme is its ability to be activated by phosphate through an intramolecular rearrangement. The reaction is inhibited by malonate, oxaloacetate, and certain sulfhydryl inhibitors. Iron chelators do not react with the enzyme-bound iron unless the enzyme is denatured. Also, the brain enzyme is not inhibited by cyanide in contrast to that of the heart muscle.[96]

The reaction can be assayed with an electron acceptor such as certain dyes[94,97,98] or phenazine methosulfate.[99] Solubilized succinic dehydrogenase may behave quite differently from its parent multienzyme complex toward these different electron acceptors, so the actual enzyme concentration in any tissue becomes difficult to evaluate.

Succinate is present in the rat[40] and mouse[23] brain at 0.34 and 0.69 μmoles per gram respectively. It can be determined using the succinic dehydrogenase assays or using the succinoxidase complex.[100] A rather extensive enzymatic method to circumvent experimental difficulties was devised.[23] A microdetermination method appeared recently.[101]

8. Fumarase (L-Malate Hydro-lyase, E.C. 4.2.1.2)

Overall reaction and mechanism:

$$
\overset{H\quad COO^-}{\diagdown\ \diagup}
\begin{matrix} C \\ \Vert \\ C \end{matrix}
\underset{^-OOC\quad H}{\diagup\ \diagdown}
\ +\ H_2O \Longrightarrow
\underset{^-OOC\ \ OH\ H}{\overset{H\ H\ COO^-}{\diagdown\,\vert\,\diagup}}
\begin{matrix} C \\ \vert \\ C \end{matrix}
$$

The seemingly simple reaction from fumarate to malate catalyzed by fumarase[102] was actually rather complicated, with two postulated intermediates and a number of possible related complexes. The mechanism is somewhat analogous to the upper half of the aconitase reaction (section II, 3) where the bottom acetyl group is replaced by hydrogen. The Michaelis constants for fumarate and malate are related through the equilibrium constant, and

$$K_{eq} = \frac{V_F K_m}{V_m K_F}$$

Here K_{eq} is the equilibrium constant, V_F and K_F are maximum initial reaction velocity and Michaelis constant with fumarate as the substrate and V_m and K_m are the corresponding values with malate as the substrate. All these constants are affected by the pH and buffer ions in the medium. The $\Delta G^{0'}$ is -0.88 kcal per mole. It is inhibited by *meso*-tartrate and some other structurally closely related compounds. The addition of water has been shown to be trans in configuration using deuterium water. It should be noted that free fumarate is a symmetrical molecule very much like succinate while malate is not. This enzyme has not been studied in the nervous tissue.

The overall reaction rates in the rabbit, sheep, and rat brain are 1.4–2.8, 0.71 and 2 mmoles per gram per hour respectively.[2,52,70,103–105] The concentration of fumarate in the rat brain has been reported to be 1.2[40] μmoles per gram, and that in the mouse brain 0.073.[23]

Fumarase activity can be measured spectrophotometrically,[106] or by the fluorescence of malate-β-naphthol complex in concentrated sulfuric acid.[103,107] Other methods are also available.[23,108,109]

9. *Malate Dehydrogenase* (L-*Malate:NAD Oxidoreductase, E.C. 1.1.1.37*)

Overall reaction and mechanism:

$$\begin{array}{c} CH_2-COO^- \\ | \\ ^-OOC-CH \\ | \\ OH \end{array} + NAD^+ \rightleftharpoons \begin{array}{c} CH_2-COO^- \\ | \\ C\!-\!\!-\!\!-\!COO^- \\ \| \\ O \end{array} + NADH + H^+$$

The oxidation of L-malate to oxaloacetate catalyzed by malate dehydrogenase[110] causes the loss of the asymmetric carbon in malate with a concurrent addition of hydrogen on the α-side of the pyridine ring in NAD. The formation of the oxaloacetate from this reaction completes the tricarboxylic acid cycle with the net oxidation of pyruvate to three CO_2. The mechanism of this enzyme in the nervous tissue has not been explored.

The overall rate in brain is approximately 5300 and 2500–88,000 μmoles per hour per gram for rat[2] and rabbit[85,111,112] respectively. That in retina is 30–40 μmoles per hour per gram for both rabbit and monkey.[113] The K_M for malate and oxaloacetate are 1–9×10^{-4} and 1.8×10^{-4} M respectively. The equilibrium at neutral pH is far toward malate and the $\Delta G^{0\prime}$ is 6.69 kcal per mole. This equilibrium is pH-dependent, like many other dehydrogenases. It is inhibited by high oxaloacetate concentration, and by fumarate, fluoromalate, β-fluoro-oxaloacetate and sulfhydryl inhibitors.

The reaction is best measured in the reverse direction and the oxidation of NADH can be followed by the decrease in optical density at 340 mμ.[114,115] Since oxaloacetate is unstable and decomposes to pyruvate and CO_2, the method is not specific unless the oxaloacetate used is uncontaminated or freshly generated by another enzymatic reaction such as the glutamate-oxaloacetate transaminase using aspartate and α-ketoglutarate as the substrates. It can also be determined in the forward direction after reacting oxaloacetate with a derivative of hydrazine to form a colored compound.[112]

The concentration of malate is approximately 0.24, 0.44, 4.3, and 6 μmoles per gram wet weight for rat brain[40] and mouse brain,[23] rabbit sciatic nerve,[63] and lobster nerve,[62,116,117] respectively. Malate can be determined by the reduction of NADP using the malic enzyme or by NAD reduction using the malate dehydrogenase.[23,107,112,118] A fluorometric assay is also available, but it is difficult to use.[112] It can be decarboxylated by a bacterial enzyme and the CO_2 measured.[109,119]

10. The Reactions of the Tricarboxylic Acid Cycle and the Replenishment of Oxaloacetate

Reactions 2 to 9 can be summed up as follows:

$$CoAS \sim Ac + 3\,NAD + GDP + Pi + (FAD) + 2\,H_2O$$

$$\rightarrow 2\,CO_2 + 3\,NADH + GTP + (FADH_2) + 2\,H^+$$

Oxaloacetate, which is one of the substrates for the condensing enzyme in reaction 2, is regenerated in reaction 9, so a net metabolism of acetyl-coenzyme A is obtained. If the reaction sequence includes reaction 1 where acetyl-Coenzyme A is produced, the overall reaction becomes

$$pyruvate + 4\,NAD + GDP + Pi + (FAD) + 2\,H_2O$$

$$\rightarrow 3\,CO_2 + 4\,NADH + GTP + (FADH_2) + 2\,H^+$$

Assuming the oxidative phosphorylation of NADH and $(FADH_2)$ yields 3 and $2 \sim P$ respectively, then $CoAS \sim Ac$ and pyruvate will yield a maximum of 12 and $15 \sim P$ respectively. The combustion energy of pyruvate is -273 kcal per mole. If the $\sim P$ bond energy is assumed to be -7.5 kcal per mole,[3] then the biochemical oxidation of pyruvate could yield -112.5 kcal per mole, or a maximum efficiency of 41 %.

However, in the biological system, this is not always achieved. One of the reasons is commonly referred to as the loss of carbon skeletons of the TCAC intermediates. Acetyl and succinyl-coenzyme A, oxaloacetate and α-ketoglutarate are well-known precursors for various metabolites in the body. If any of them were removed from the TCAC, a lower efficiency in terms of energy production would take place. Furthermore, if any of the latter three were removed, oxaloacetate concentration would decrease and eventually could stop the operation of the TCAC. The replenishment of the lost oxaloacetate, in order to ensure the proper operation of the TCAC, depends upon the carboxylation of a three-carbon compound (phospho-enolpyruvate or pyruvate) which is supplied by glycolysis.[13] Four possible reactions could be involved and are discussed in the next four paragraphs.[120-122] None of them has been studied in the nervous tissue.

11. Malic Enzyme (L-Malate: NADP Oxidoreductase (Decarboxylating), E.C. 1.1.1.40)

$$\begin{matrix} CH_3 \\ | \\ O{=}C{\cdots}COO^- \end{matrix} \quad \begin{matrix} +\ CO_2 \\ +\ NADPH \end{matrix} \xrightleftharpoons{Mn^{2+}} \quad \begin{matrix} CH_2{-}COO^- \\ | \\ HO{-}CH{-}COO^- \end{matrix} +\ NADP^+$$

The carboxylation of pyruvate at the expense of a NADPH to form malate[82,123,124] has been thought of as the enzymatic mechanism to replenish the loss of oxaloacetate. However, this is not considered likely anymore. It is now thought that the reverse reaction leading further to the formation of acetyl-coenzyme A is more important. This enzyme is not mitochondrial as all the TCAC enzymes. The reaction mechanism is similar to that of the NADP-linked isocitrate dehydrogenase (reaction 4), and the addition of hydrogen is on the α-side of the pyridine ring. The cofactor Mn^{2+} can be replaced by Co^{2+}. Aspartate increases its activity and this effect is considered to be allosteric.[125]

The overall reaction favors pyruvate formation and is pH dependent. The $\Delta G^{0\prime}$ is 0.36 kcal per mole. The K_M is also pH dependent and at pH 7.5 it is 3.9×10^{-4} M for pyruvate.

The assay of this enzyme utilizes the formation of NADPH which can be followed spectrophotometrically.[124]

12. Pyruvate Carboxylase (Pyruvate: Carbon Dioxide Ligase (ADP), E.C. 6.4.1.1)

$$\begin{matrix} CH_3 \\ | \\ O{=}C{-}COO^- \end{matrix} \quad \begin{matrix} +\ CO_2 \end{matrix} +\ ATP^{4-} +\ H_2O \xrightarrow{Mg^{2+}}$$

$$\begin{matrix} CH_2{-}COO^- \\ | \\ O{=}C{-}COO^- \end{matrix} +\ ADP^{3-} +\ Pi^{2-} -\ 2\ H^+$$

This ATP-dependent carboxylation reaction[126-130] is now thought to be the main contributor for the replenishment of oxaloacetate. An enzyme-bound biotin is supposed to bind a CO_2 before adding it to pyruvate. An excess of ADP inhibits the enzyme.[131] It is also an allosteric enzyme in the sense that it is activated by acetyl-CoA via a lysine residue on the enzyme.[132] This activation can be counteracted by aspartate. When there is an over-abundance of acetyl-CoA, this enzyme is activated and more pyruvate is diverted into oxaloacetate which condenses with the excess acetyl-CoA. The oxidation of the resultant citrate via the TCAC in turn provides the necessary ATP for this carboxylation reaction.

It has been speculated[131] that the synthesis of phosphoenolpyruvate involves this reaction with the subsequent formation of malate and aspartate which, after diffusing out of the mitochondria, give rise to phosphoenolpyruvate via oxaloacetate.

As a biotin enzyme, it is inhibited by avidin. It is also inhibited by sulfhydryl inhibitors. The $\Delta G^{0\prime}$ of this reaction is -0.5 kcal per mole. The reaction can be assayed by coupling to the malate dehydrogenase and following the oxidation of NADH.[115]

13. Phosphoenolpyruvate Carboxykinase (GTP:Oxaloacetate Carboxy-lyase (Transphosphorylating), E.C. 4.1.1.32)

$$
\begin{array}{c}
CH_2 \\
\| \\
C-COO^- \\
| \\
OPO_3^{2-}
\end{array}
+ CO_2
\quad + GDP^{3-} \xrightarrow{Mn^{2+}}
\begin{array}{c}
CH_2-COO^- \\
| \\
C-COO^- \\
\| \\
O
\end{array}
+ GTP^{4-}
$$

This mitochondrial enzyme[120,133,134] utilizes the energy in phosphoenolpyruvate to synthesize a $\sim P$ bond as well as fixing a CO_2. The $\Delta G^{0\prime}$ is 4.0 kcal per mole, i.e., the reverse reaction is favored. The guanosine nucleotides can be substituted by inosine nucleotides. It has been speculated that this reaction could be coupled in reverse in two different manners to yield phosphoenolpyruvate which is important in the synthesis of carbohydrates.

(1) $OAA + GTP \rightarrow PEP + CO_2 + GDP$

$Pyruvate + CO_2 + ATP \rightarrow OAA + Pi + 2\,H^+$

$\underline{ATP + GDP \rightarrow ADP + GTP}$

$Pyruvate + 2\,ATP \rightarrow PEP + 2\,ADP + Pi + 2\,H^+$

(2) $OAA + GTP \rightarrow PEP + CO_2 + GDP$

$\underline{Succinyl \sim SCoA + GDP \rightarrow succinate + GTP + CoASH}$

$OAA + succinyl \sim SCoA \rightarrow PEP + CO_2 + succinate + CoASH$

The first reactions are important in glucogenesis,[135] but the latter reactions are of particular interest since both enzymes are mitochondrial and use the same coenzymes. It is also interesting that aspartate inhibits the forward reaction constituting a feedback mechanism.[129]

The reaction can be assayed by determining the ^{14}C incorporation into oxaloacetate from $^{14}CO_2$[136] or by following NADH disappearance after coupling to malate dehydrogenase. It is inhibited by sulfhydryl inhibitors. Phosphoenolpyruvate can be determined as inorganic phosphate which is easily liberated.[137]

14. *Phosphoenolpyruvate Carboxylase [Orthophosphate:Oxaloacetate Carboxylyase (Phosphorylating), E.C. 4.1.1.31]*

$$
\begin{array}{l}
CH_2 \\
\parallel \\
C-COO^- \\
\mid \\
OPO_3^{2-}
\end{array}
\; + CO_2 \; + H_2O \xrightarrow{Me^{2+}}
\begin{array}{l}
CH_2-COO^- \\
\mid \\
CH-COO^- \\
\mid \\
OH
\end{array}
\; + Pi^{2-} + H^+
$$

A divalent metal cofactor is required for this enzymatic carboxylation of phosphoenolpyruvate to oxaloacetate.[120,138] The reaction is essentially irreversible with a large $\Delta G^{0'}$ of -6.8 kcal per mole. It has not been reported in animals.

III. THE PATH OF CARBONS IN THE TCAC

The greatly simplified diagram (Fig. 2) is drawn to show the number of carbons in each intermediate. Each numeral stands for a carbon atom. The arrow indicates the reactions in the forward direction. Carbon-3 of pyruvate is labeled to give an example of the pathways of a carbon atom, particularly with respect to the carboxylation and decarboxylation reactions. In succinate, the labeling of its carboxyl and methylene carbons is randomized because it is a symmetrical molecule. Such randomization also takes place if malate is dehydrated to fumarate via the reversal of the TCAC. The randomization of the dicarboxylic acids, sometimes referred to as the shunt, can be rather fast, particularly in the brain. The oxaloacetate formed here from CO_2 fixation could be randomized before its condensation with acetyl-coenzyme A to form citrate. This was demonstrated in brain[14,139] and the sciatic nerve of the rabbit.[63]

Citrate appears as a symmetrical molecule in Fig. 2, but it does not behave like one in the TCAC. The origin and fate of the C-1 and C-5 carboxyl group are quite different. In Fig. 2, the fate of each carbon in citrate can be traced. C-1 is decarboxylated. C-5 enters into succinate, gets randomized and decarboxylated in the next cycle. Citrate occupies a central position in the TCAC because it contains all six carbons and represents over 90% of the tricarboxylic acids.[68] It can be seen that when aspartate is compared with citrate, it lacks the carbons of the acetyl-CoA. Since aspartate is normally abundant and easy to handle, its analysis together with citrate could give some kind of estimation of the relative contribution of pyruvate molecules via either the carboxylation or decarboxylation pathways.

In Table III, an assessment of the fate of each carbon of pyruvate is listed. Two sets of numbers are given. They are different in the so-called "dicarboxylic acid shunt and randomization" discussed above. The first set of numbers refers to the condition whereby a four-carbon dicarboxylic acid after its formation from carboxylation is immediately randomized before condensing with acetyl-CoA to form citrate. The second set of numbers refers to carboxylation without randomization. Each cycle starts with

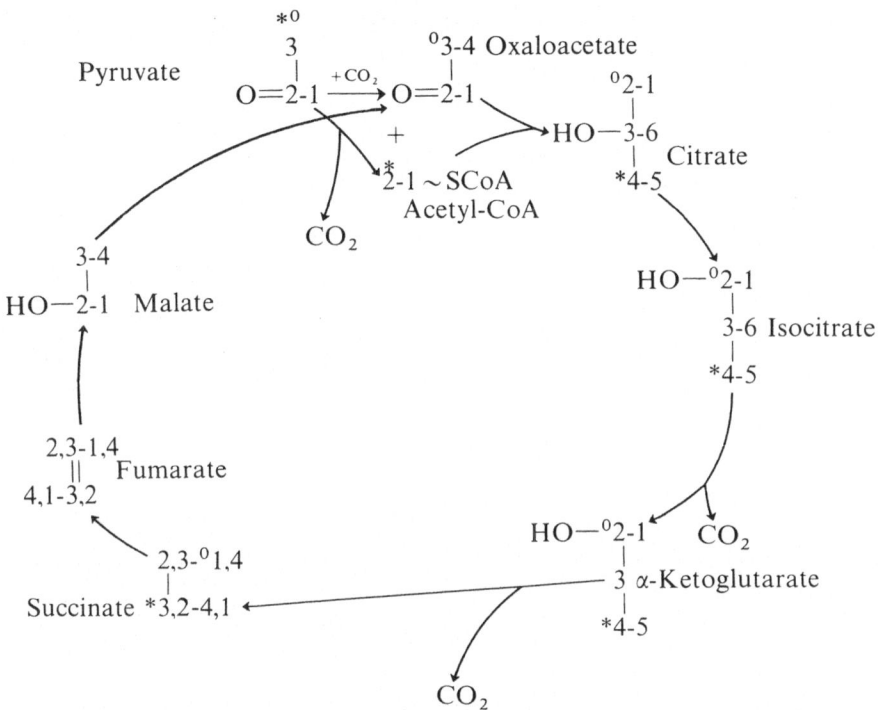

Fig. 2. Schematic of the TCAC showing the numbering of the carbons in each intermediate and the paths of carbons. Each carbon is labeled with a numeral. *—C-3 of pyruvate similarly labeled in Eqs. 2–6 in the text. °—The same carbon going through the carboxylation pathway. Arrows denote the forward reactions of the TCAC.

aspartate with respect to the cycle number. Several different conditions can now be discussed with the help of this table.

Under all conditions using pyruvate-1-^{14}C as a substrate, the total radioactivity in, or the specific activities of, aspartate or citrate is a measure of the carboxylation pathway. Since carbon 1 is decarboxylated, it does not enter into the TCAC via the decarboxylation pathway.

If the contribution of the carboxylation and decarboxylation pathways are about equal and the TCAC is turning over very slowly, then the ratio of the incorporated radioactivity from pyruvate-1-, 2-, or 3-^{14}C should be 1:1:1 for aspartate and 1:2:2 for citrate. This type of result has been obtained in the lobster nerve at 15°C (approximately the ocean temperature) using pyruvate-1- and 2-^{14}C as the substrates.[15] On the other hand, the TCAC can be turning rapidly, such as that in the brain at 37°C. The above ratio of the incorporated radioactivity in the absence of randomization would be 1:5:5 for aspartate and 1:6:6 for citrate respectively, and in the presence of randomization 1:4:6 and 1:5:7 for aspartate and citrate respectively.

TABLE III

The Fate of Each Carbon in Pyruvate as Appearing in Aspartate[a]

Labeled C in pyruvate	±CO$_2$	Cycle number	Labeled C in aspartate				
			A-1	A-2	A-3	A-4	ΣA
P-1	−	1	—	—	—	—	—
	+	1	$\frac{1}{2}$, 0	—	—	$\frac{1}{2}$, 0	1, 1
		2	—	—	—	—	—
		Σ	$\frac{1}{2}$, 0	—	—	$\frac{1}{2}$, 0	$\dfrac{1, 1}{1, 1}$
P-2	−	1	—	—	—	—	—
		2	$\frac{1}{2}$, $\frac{1}{2}$	—	—	$\frac{1}{2}$, $\frac{1}{2}$	1, 1
		3	—	—	—	—	—
		Σ	$\frac{1}{2}$, $\frac{1}{2}$	—	—	$\frac{1}{2}$, $\frac{1}{2}$,	1, 1
	+	1	—	$\frac{1}{2}$, 1	$\frac{1}{2}$, 0	—	1, 1
		2	$\frac{1}{4}$, 0	$\frac{1}{4}$, $\frac{1}{2}$	$\frac{1}{4}$, $\frac{1}{2}$	$\frac{1}{4}$, 0	1, 1
		3	$\frac{1}{8}$, $\frac{1}{4}$	$\frac{1}{8}$, $\frac{1}{4}$	$\frac{1}{8}$, $\frac{1}{4}$	$\frac{1}{8}$, $\frac{1}{4}$	$\frac{1}{2}$, 1
		n	$(\frac{1}{2})^n, (\frac{1}{2})^{n-1}$	$(\frac{1}{2})^n, (\frac{1}{2})^{n-1}$	$(\frac{1}{2})^n, (\frac{1}{2})^{n-1}$	$(\frac{1}{2})^n, (\frac{1}{2})^{n-1}$	$(\frac{1}{2})^{n-2}, (\frac{1}{2})^{n-3}$
		Σ	$\frac{1}{2}$, $\frac{1}{2}$	1, 2	1, 1	$\frac{1}{2}$, $\frac{1}{2}$	$\dfrac{3, 4}{4, 5}$
P3	−	1	—	—	—	—	—
		2	—	$\frac{1}{2}$, $\frac{1}{2}$	$\frac{1}{2}$, $\frac{1}{2}$	—	1, 1
		3	$\frac{1}{4}$, $\frac{1}{4}$	$\frac{1}{4}$, $\frac{1}{4}$	$\frac{1}{4}$, $\frac{1}{4}$	$\frac{1}{4}$, $\frac{1}{4}$	1, 1
		4	$\frac{1}{8}$, $\frac{1}{8}$	$\frac{1}{8}$, $\frac{1}{8}$	$\frac{1}{8}$, $\frac{1}{8}$	$\frac{1}{8}$, $\frac{1}{8}$	$\frac{1}{2}$, $\frac{1}{2}$
		n	$(\frac{1}{2})^{n-1}, (\frac{1}{2})^{n-1}$	$(\frac{1}{2})^{n-1}, (\frac{1}{2})^{n-1}$	$(\frac{1}{2})^{n-1}, (\frac{1}{2})^{n-1}$	$(\frac{1}{2})^{n-1}, (\frac{1}{2})^{n-1}$	$(\frac{1}{2})^{n-3}, (\frac{1}{2})^{n-3}$
		Σ	$\frac{1}{2}$, $\frac{1}{2}$	1, 1	1, 1	$\frac{1}{2}$, $\frac{1}{2}$	3, 3
	+	1	—	$\frac{1}{2}$, 0	$\frac{1}{2}$, 1	—	1, 1
		2	$\frac{1}{4}$, $\frac{1}{2}$	$\frac{1}{4}$, 0	$\frac{1}{4}$, 0	$\frac{1}{4}$, $\frac{1}{2}$	1, 1
		3	$\frac{1}{8}$, 0	$\frac{1}{8}$, 0	$\frac{1}{8}$, 0	$\frac{1}{8}$, 0	$\frac{1}{2}$, 0
		n	$(\frac{1}{2})^n$, 0	$(\frac{1}{2})^n$, 0	$(\frac{1}{2})^n$, 0	$(\frac{1}{2})^n$, 0	$(\frac{1}{2})^{n-2}$, 0
		Σ	$\frac{1}{2}$, $\frac{1}{2}$	1, 0	1, 1	$\frac{1}{2}$, $\frac{1}{2}$	$\dfrac{3, 2}{6, 5}$

[a] P-x, A-x, and C-x denote compounds pyruvate, aspartate, and citrate respectively. When the
1. Of each pair of numbers, the first one considers the case when a complete randomization
to form citrate. The second number assumes no randomization.

TABLE III (continued)

		Labeled C in citrate				
C-1	C-2	C-3	C-4	C-5	C-6	ΣC
—	—	—	—	—	—	—
$\frac{1}{2},1$	—	—	—	—	$\frac{1}{2},0$	$1,1$
—	—	—	—	—	—	—
$\frac{1}{2},1$	—	—	—	—	$\frac{1}{2},0$	$1,1$
						$1,1$
—	—	—	—	$1,1$	—	$1,1$
$\frac{1}{2},\frac{1}{2}$	—	—	—	—	$\frac{1}{2},\frac{1}{2}$	$1,1$
—	—	—	—	—	—	—
$\frac{1}{2},\frac{1}{2}$	—	—	—	$1,1$	$\frac{1}{2},\frac{1}{2}$	$2,2$
—	$\frac{1}{2},0$	$\frac{1}{2},1$	—	—	—	$1,1$
$\frac{1}{4},0$	$\frac{1}{4},\frac{1}{2}$	$\frac{1}{4},\frac{1}{2}$	—	—	$\frac{1}{4},0$	$1,1$
$\frac{1}{8},\frac{1}{4}$	$\frac{1}{8},\frac{1}{4}$	$\frac{1}{8},\frac{1}{4}$	—	—	$\frac{1}{8},\frac{1}{4}$	$\frac{1}{2},1$
$(\frac{1}{2})^n,(\frac{1}{2})^{n-1}$	$(\frac{1}{2})^n,(\frac{1}{2})^{n-1}$	$(\frac{1}{2})^n,(\frac{1}{2})^{n-1}$	—	—	$(\frac{1}{2})^n,(\frac{1}{2})^{n-1}$	$(\frac{1}{2})^{n-2},(\frac{1}{2})^{n-3}$
$\frac{1}{2},\frac{1}{2}$	$1,1$	$1,2$	—	—	$\frac{1}{2},\frac{1}{2}$	$3,4$
						$5,6$
—	—	—	$1,1$	—	—	$1,1$
—	$\frac{1}{2},\frac{1}{2}$	$\frac{1}{2},\frac{1}{2}$	—	—	—	$1,1$
$\frac{1}{4},\frac{1}{4}$	$\frac{1}{4},\frac{1}{4}$	$\frac{1}{4},\frac{1}{4}$	—	—	$\frac{1}{4},\frac{1}{4}$	$1,1$
$\frac{1}{8},\frac{1}{8}$	$\frac{1}{8},\frac{1}{8}$	$\frac{1}{8},\frac{1}{8}$	—	—	$\frac{1}{8},\frac{1}{8}$	$\frac{1}{2},\frac{1}{2}$
$(\frac{1}{2})^{n-1},(\frac{1}{2})^{n-1}$	$(\frac{1}{2})^{n-1},(\frac{1}{2})^{n-1}$	$(\frac{1}{2})^{n-1},(\frac{1}{2})^{n-1}$	—	—	$(\frac{1}{2})^{n-1},(\frac{1}{2})^{n-1}$	$(\frac{1}{2})^{n-3},(\frac{1}{2})^{n-3}$
$\frac{1}{2},\frac{1}{2}$	$1,1$	$1,1$	$1,1$	—	$\frac{1}{2},\frac{1}{2}$	$4,4$
—	$\frac{1}{2},1$	$\frac{1}{2},0$	—	—	—	$1,1$
$\frac{1}{4},\frac{1}{2}$	$\frac{1}{4},0$	$\frac{1}{4},0$	—	—	$\frac{1}{4},\frac{1}{2}$	$1,1$
$\frac{1}{8},0$	$\frac{1}{8},0$	$\frac{1}{8},0$	—	—	$\frac{1}{8},0$	$\frac{1}{2},0$
$(\frac{1}{2})^n,0$	$(\frac{1}{2})^n,0$	$(\frac{1}{2})^n,0$	—	—	$(\frac{1}{2})^n,0$	$(\frac{1}{2})^{n-2},0$
$\frac{1}{2},\frac{1}{2}$	$1,1$	$1,1$	—	—	$\frac{1}{2},\frac{1}{2}$	$3,2$
						$7,6$

carbon number x in the number of cycles n becomes infinite, the sums of series $\Sigma(1/2)^n$ becomes of the four carbon dicarboxylic acids took place before condensing with acetyl coenzymes-A

These ratios are seldom obtained because, in case of rapid cycling of the TCAC, an equal contribution of the carboxylation and decarboxylation pathways seldom occurs. Rather, the carboxylation pathways will contribute much less than the decarboxylation pathways.[14,15]

The joint use of C-1 and C-2 labeled pyruvate can help to elucidate the situation. Since C-1 labeled pyruvate is a measure of carboxylation pathway only, the difference between pyruvate-1- and 2-^{14}C experiments will give a measure of the decarboxylation pathway plus the effect of cycling. The larger the difference, the more contribution there is from the decarboxylation pathway, since the effect of cycling is relatively constant after several cycles (Table III). The radioactivity in aspartate and citrate reinforce each other in the estimation of the relative contribution of the two pathways. Since aspartate and other dicarboxylic acids could be lost from their TCAC pools[140] before reaching citrate, there should be a certain amount of difference in their radioactivity. Thus, in the rat brain slices,[15] a definite drop in radioactivity from aspartate to citrate was found. The percentage of the carboxylation pathway can be estimated to be 7.5 and 17.5%, respectively, based on aspartate or citrate analyses. The formulas used for these approximate estimations are

$$\text{for aspartate} \quad \frac{P_2 - P_1}{P_1} = 1 + \frac{1}{\%}$$

$$\text{for citrate} \quad \frac{P_2 - P_1}{P_1} = \frac{2}{\%}$$

where $\%$ is the percentage of carboxylation reaction, and P_1 and P_2 are the experimental radioactivity in aspartate and citrate respectively using pyruvate-1- and 2-^{14}C as the substrate.

Use of double labeling with ^{14}C at C-1 or C-2 and ^{3}H at C-3 of pyruvate does not help to elucidate the mechanisms of TCAC as a whole. The normal exchange of ^{3}H with water and its unknown degree of exchange with water in the aconitase and fumarase reactions both contribute to the uncertainty in determining exactly the amount of ^{3}H that should be present in any intermediate.

IV. THE TCAC AS A WHOLE

The occurrence of the TCAC has been assumed to be universal in all nervous tissue. In most cases, its presence is implied by measuring only the oxygen consumption using a TCAC intermediate.[2,141–144] In other instances, mitochondria have been isolated as a subcellular fraction and studied.[7,145] The brain as a whole was the subject of intensive study using the subcellular fractionation technique.[146,147] Although the nerve endings were of primary concern, these methods did lead to a better mitochondrial fraction. But extensive purification of enzyme from this fraction has not been done. As a matter of fact, practically no data on the mechanism of action of TCAC enzymes from nervous tissue have been reported.

Metabolically speaking, the most interesting results came from two directions. The micromethods of Lowry[107,148] opened up new avenues of study, and perfusion experiments are developed in addition to the use of slices. The "dicarboxylic acid shunt and randomization" was demonstrated clearly in the perfused cat brain,[41,139,140] in guinea pig brain slices,[149] in rat brain slices,[39,150] and in the rabbit sciatic nerve.[63,151] The reversibility of α-ketoglutarate to oxaloacetate and acetyl-CoA was only recently demonstrated.[62,63,151–153] The reversibility of the last link, from α-keto-glutarate to succinate, has not been demonstrated. Thus, the complete reversibility of the whole TCAC is still in question.

With respect to the rate of operation of the TCAC, it is well known that oxygen consumption increases upon activity. This is true not only for the brain[2,154] but also for the nerve.[155] The precise mechanism is not known but one of the control points seems to lie in the $Pi + ADP \rightleftharpoons ATP$ reaction as well as the change of cation concentrations.[156–158] Other control points were supposedly located at the succinate dehydrogenase and isocitrate dehydrogenase level as well as somewhere between pyruvate and citrate.[23,159] It is known that succinate dehydrogenase is inhibited by oxaloacetate, and isocitrate dehydrogenase is both inhibited by ATP and NADH and stimulated by ADP. An important factor, of course, is the substrate concentrations and the Michaelis constants of various enzymes. For this, one should consult special papers on this subject.[160,161] Other related controls such as that of glutamate dehydrogenase, glutamate-oxaloacetate transaminase, pyruvate kinase, and acetate thiokinase are discussed elsewhere in this treatise and are purposely omitted here.

The control of the replenishment of carbon skeletons, which accounts for 10 % in the brain[14,15] and 50 % in the lobster nerve,[15] involves more than the reactions shown on the top part of Fig. 1. One must also think of the acetyl-CoA carboxylase which synthesizes malonyl-CoA, and the reversal of pyruvate to phosphoenolpyruvate. In some cases, acetate thiokinase also comes into the picture. This part of metabolism represents a crucial division of substrate for energy production, glucogenesis, and lipogenesis (see also Ref. 162). An increase in acetyl-CoA concentration would stimulate the carboxylation of pyruvate, thus favoring citrate synthesis. An increase in citrate concentration would stimulate acetyl-CoA carboxylation, thus draining more acetyl-CoA toward malonyl-CoA for lipid synthesis and reducing the synthesis of citrate. Long-chain acyl-CoA inhibits both malonyl-CoA and citrate synthesis. This pushes acetyl-CoA back toward pyruvate which is converted to phosphoenolpyruvate via oxaloacetate since an increased acetyl-CoA concentration would increase pyruvate carboxylation.[2,3,21,162–165] Many other factors also affect this complicated system, such as nutritional state, which is particularly important for the liver.[163–165]

The compartmentalization in brain of intermediates of the TCAC has been discussed in another chapter of this treatise.[10] In addition to the evidence that glutamate and α-ketoglutarate are compartmentalized,[39,41, 139,149,166,167] the dicarboxylic acids of the TCAC intermediates also seem to be compartmentalized.[149] Unpublished results[151] in this laboratory suggest a compartmentalization of acetyl-CoA and citrate, thus making the compartmentalization of all TCAC intermediates complete. Does this mean that there are at least two kinds of mitochondria? Furthermore, if so, are they located in different cellular (neurons *vs.* glia) or subcellular (cell bodies *vs.* nerve endings) structures? Some evidence is available for two populations of mitochondria[168–170] as well as neuronal and glial differences.[171] Much more work is needed to clarify this problem.

The transfer of ions, particularly of the intermediates of the TCAC, across the mitochondrial membrane is the most intensely studied subject at present in animal tissues. Unfortunately, no studies have been done on the mitochondria from the brain tissue. The subject is outside the scope of this chapter. A few selected references[172–175] are given here as a starting point for further study.

In closing, amazingly little is known of the biochemistry of the TCAC in the nervous tissue. It is reasonably definite that it exists, possibly in more than one form, as suggested by the compartmentalization studies. But there is hardly any study on each individual enzyme or on the transport of any of the intermediates in and out of the mitochondria. Existing studies dealt only with the concentrations of TCAC intermediates and their changes under various conditions.[2,23,41,51,60–63,75,76,116,117,152,176]

V. SEPARATION AND DEGRADATION OF THE TCAC INTERMEDIATES

The methods of separation are based primarily on chromatography. Column methods using an anion-exchange resin were mostly employed,[60, 116,117,176–178] although Celite[179] or silica gel[180] has also been used. Chromatography on paper[178,180,181] or thin layer[41,166,182–189] has also been successful, as is a gas-liquid chromatographic method.[149]

The degradation of the TCAC intermediates, in order to study the radio-activity in each carbon in isotope experiments, has been described in detail elsewhere.[179] Several more recent approaches exist. A number of procedures are available for the breaking down of citrate.[190–192] The carbons of α-ketoglutarate could be determined individually after its conversion to glutamate.[193] Other methods for partial degradation exist using different enzymes for citrate,[60,194] glutamate,[149,195] and aspartate.[149,196,197]

VI. REFERENCES

1. G. D. Greville, Mechanism of carbohydrate metabolism in the brain, *in Neurochemistry* (K. A. C. Elliott, I. H. Page, and J. H. Quastel, eds.), pp. 238–266, C. C. Thomas, Springfield, Ill., 2nd ed. (1962).

2. H. McIlwain, *Biochemistry of the Central Nervous System*, Little, Brown, and Co., Boston, 3rd ed. (1966).

3. H. R. Mahler and E. H. Cordes, *Biological Chemistry*, Harper and Row, New York (1966).

4. A. White, P. Handler, and E. L. Smith, *Principles of Biochemistry*, McGraw-Hill Book Co., New York (1954).

5. P. D. Boyer, H. A. Lardy, and K. Myrbäck, eds., *The Enzymes*, 2nd ed., 8 vols., 1959–1963, Academic Press, New York.

6. S. P. Colowick and N. O. Kaplan, eds., *Methods in Enzymology*, Vols. I–VII, Academic Press, New York (1955–1965).

7. L. G. Abood, Brain mitochondria, *in Handbook of Neurochemistry* (A. Lajtha, ed.), Vol. 2, pp. 303–326, Plenum Press, New York (1969).

8. C. F. Baxter, γ-Aminobutyric acid, *in Handbook of Neurochemistry* (A. Lajtha, ed.), Vol. 3, pp. 289–353, Plenum Press, New York (1969).

9. C. Moore and P. M. Strasberg, Cytochromes, oxidative phosphorylation, *in Handbook of Neurochemistry*, (A. Lajtha, ed.), Vol. 3, pp. 53–56, Plenum Press, New York (1969).

10. S. Berl and D. D. Clarke, Compartmentation of amino acid metabolism, *in Handbook of Neurochemistry* (A. Lajtha, ed.), Vol. 2, pp. 447–472, Plenum Press, New York (1969).

11. C. J. van den Berg, Glutamate and glutamine in nervous tissue, *in Handbook of Neurochemistry* (A. Lajtha, ed.), Vol. 3, pp. 355–379, Plenum Press, New York (1969).

12. H. A. Krebs, Mitochondrial generation of reducing power, *in Biochemistry of Mitochondria* (E. C. Slater, Z. Kaniuga, and L. Wojtczak, eds.), pp. 105–113, Academic Press, New York (1967).

13. R. Balázs, Carbohydrate metabolism, *in Handbook of Neurochemistry* (A. Lajtha, ed.), Vol. 3, pp. 1–35, Plenum Press, New York (1969).

14. S. Otsuki, A. Geiger, and G. Gombos, The metabolic pattern of the brain in brain perfusion experiments *in vivo*. I. The quantitative significance of CO_2 assimilation in the metabolism of the brain, *J. Neurochem.* **10**:397–404 (1963).

15. S.-C. Cheng, R. Nakamura, and H. Waelsch, Relative contribution of carbon dioxide fixation and acetyl-CoA pathways in two nervous tissues, *Nature* **216**:928–929 (1967).

16. D. R. Sanadi, Pyruvate and α-ketoglutarate oxidation enzymes, *in The Enzymes* (P. D. Boyer, H. A. Lardy, and K. Myrbäck, eds.), Vol. 7, pp. 307–344, Academic Press, New York (1963).

17. F. H. Pettit and L. J. Reed, α-Keto acid dehydrogenase complexes VIII. Comparison of dihydrolipoyl dehydrogenases from pyruvate and α-ketoglutarate dehydrogenase complexes of *Escherichia coli*, *Proc. Nat. Acad. Sci.* **58**:1126–1130 (1967).

18. E. R. Schwartz and L. Reed, α-Keto acid dehydrogenase complexes. IX. Effects of iodination of the tyrosyl residues on the properties of the dihydrolipoyl transacetylase of *Escherichia coli*, *J. Biol. Chem.* **243**:639–643 (1968).

19. M. Koike, I. J. Reed, and W. R. Carroll, α-Keto acid dehydrogenase complexes. IV. Resolution and reconstitution of the *E. Coli* pyruvate dehydrogenation complex, *J. Biol. Chem.* **238**:30–39 (1963).

20. S. Tuček, The use of choline acetyltransferase for measuring the synthesis of acetyl-coenzyme A and its release from brain mitochondria, *Biochem. J.* **104**:749–756 (1967).

21. G. V. Jagow, B. Westermann, and O. Wieland, Suppression of pyruvate oxidation in liver mitochondria in the presence of long-chain fatty acid, *Europ. J. Biochem.* **3**:512–518 (1968).

22. P. R. Garland and P. J. Randle, Control of pyruvate dehydrogenase in the perfused rat heart by the intracellular concentration of acetyl-coenzyme A, *Biochem. J.* **91**:6C–7C (1964).

23. N. D. Goldberg, J. V. Passonneau, and O. H. Lowry, Effects of changes in brain metabolism on the levels of citric acid cycle intermediates, *J. Biol. Chem.* **241**:3997–4003 (1966).

24. F. Lipmann and L. C. Tuttle, A specific micromethod for the determination of acyl phosphates, *J. Biol. Chem.* **159**:21–28 (1945).

25. N. O. Kaplan and F. Lipmann, The assay and distribution of coenzyme A, *J. Biol. Chem.* **174**:37–44 (1948).

26. E. R. Stadtman, Preparation and assay of acyl coenzyme A and other thiol esters; use of hydroxylamine, *in Methods of Enzymology* (S. P. Colowick and N. O. Kaplan, eds.), Vol. III, pp. 931–941, Academic Press, New York (1957).

27. K. Decker, Acetyl coenzyme A, *in Methods of Enzymatic Analysis* (H. U. Bergmeyer, ed.), pp. 419–424, Academic Press, New York (1965).

28. H. Tabor, A. H. Mehler, and E. R. Stadtman, The enzymatic acetylation of Amines, *J. Biol. Chem.* **204**:127–138 (1953).

29. S. Ochoa, Crystalline condensing enzyme from pig heart, *in Methods in Enzymology* (S. P. Colowick and N. O. Kaplan, eds.), Vol. I, pp. 685–694, Academic Press, New York (1955).

30. R. W. Von Korff, A. Rapid spectrophotometric assay for coenzyme A, *J. Biol. Chem.* **200**:401–405 (1953).

31. G. D. Novelli, N. O. Kaplan, and F. Lipmann, The liberation of pantothenic acid from coenzyme A, *J. Biol. Chem.* **177**:97–107 (1949).

32. E. R. Stadtman, G. D. Novelli, and F. Lipmann, Coenzyme A function in an acyl transfer by the phosphotransacetylase system, *J. Biol. Chem.* **191**:365–376 (1951).

33. E. R. Stadtman, The enzymatic synthesis of acyl coenzyme A compounds, *J. Cell. Comp. Physiol.* **41S**:89–107 (1953).

34. L. Jaenicke and F. Lynen, Coenzyme A, *in The Enzymes* (P. D. Boyer, H. A. Lardy, and K. Myrbäck, eds.), Vol. 3, pp. 3–103, Academic Press, New York (1960).

35. S. J. Wakil and G. Hübscher, Quantitative determination of coenzyme A by sorbyl coenzyme A formation, *J. Biol. Chem.* **235**:1554–1558 (1960).

36. G. Michal and H. U. Bergmeyer, Coenzyme A, *in Methods of Enzymatic Analysis* (H. U. Bergmeyer, ed.), pp. 512–527, Academic Press, New York (1965).

37. S. Tuček, Observations on the subcellular distribution of choline acetyltransferase in the brain tissue of mammals and comparisons of acetylcholine synthesis from acetate and citrate in homogenates and nerve-ending fractions, *J. Neurochem.* **14**:519–529 (1967).

38. V. P. Whittaker, Identification of acetylcholine and related esters of biological origin, *in Handbook Exp. Pharmakol.* (G. B. Koelle, ed.), Vol. 15, pp. 1–39, (1963).

39. M. K. Gaitonde, Rate of utilization of glucose and "compartmentation" of α-oxoglutarate and glutamate in rat brain, *Biochem. J.* **95**:803–810 (1965).

40. C. E. Frohman, J. M. Orten, and A. H. Smith, Chromatographic determination of the acids of the citric acid cycle in tissue, *J. Biol. Chem.* **193**:277–283 (1951).

41. H. Waelsch, S. Berl, C. A. Rossi, D. D. Clarke, and D. P. Purpura, Quantitative aspects of CO_2 fixation in mammalian brain *in vivo, J. Neurochem.* **11**:717–728 (1964).

42. O. H. Lowry and J. V. Passonneau, The relationships between substrates and enzymes of glycolysis in brain, *J. Biol. Chem.* **239**:31–42 (1964).

43. W. W. Umbreit, R. H. Burris, and J. F. Stauffer, *Manometric Technique and Tissue Metabolism*, Burgess Publ. Co., Minneapolis (1964).

44. T. E. Friedmann, Determination of α-keto acids, *in Methods of Enzymology* (S. P. Colowick and N. O. Kaplan, eds.), Vol. III, pp. 414–418, Academic Press, New York (1957).

45. J. R. Stern, Oxaloacetate transacetylase (condensing enzyme, citrogenase), *in The Enzymes* (P. D. Boyer, H. Lardy, and K. Myrbäck, eds.), Vol. 5, pp. 367–380, Academic Press, New York (1961).

46. P. A. Srere, Citrate-condensing enzyme-oxaloacetate binary complex. Studies on its physical and chemical properties, *J. Biol. Chem.* **241**:2157–2165 (1966).

47. P. A. Srere, Citrate cleavage enzyme, *in Methods in Enzymology* (S. P. Colowick and N. O. Kaplan, eds.), Vol. V, pp. 641–644, Academic Press, New York (1962).

48. M. Wollermann, Die Rolle des Zitrats beider Entstehung des Azetyl-Coenzyma in Tierischen Geweben, *Acta Physiol. (Acad. Sci. Hung.)* **10**:171–183 (1956).

49. P. A. Srere and A. Bhaduri, The citrate cleavage enzyme. III. Citryl coenzyme A as a substrate and the stereospecificity of the enzyme, *J. Biol. Chem.* **239**:714–718 (1964).

50. C. T. Walsh, Jr. and L. B. Spector, Citryl-1-phosphate and the citrate cleavage reaction, *J. Biol. Chem.* **243**:446–448 (1968).

51. R. Balázs, Control of glutamate metabolism. The effect of pyruvate, *J. Neurochem.* **12**:63–76 (1965).

52. S. Tuček, Subcellular distribution of acetyl-CoA synthetase, ATP citrate lyase, citrate synthase, choline acetyltransferase, fumarate hydratase and lactate deyhydrogenase in mammalian brain tissue, *J. Neurochem.* **14**:531–545 (1967).

53. P. A. Srere, The citrate cleavage enzyme. I. Distribution and purification, *J. Biol. Chem.* **234**:2544–2547 (1959).

54. N. O. Jangaard, J. Unkeless, and D. E. Atkinson, The inhibition of citrate synthase by adenosine triphosphate, *Biochim. Biophys. Acta* **151**:225–235 (1968).

55. H. Inoue, F. Suzuki, K. Fukunishi, K. Adachi, and Y. Takeda, Studies on ATP citrate lyase of rat liver. I. Purification and some properties, *J. Biochem.* **60**:543–553 (1966).

56. O. Wieland, L. Weiss, and I. Eger-Neufeldt, Enzymatic regulation of liver acetyl-CoA metabolism in relation to ketosis, *Adv. in Enzyme Regulation* **2**:85–99 (1964).

57. R. O. Brady, Fluoroacetyl coenzyme A, *J. Biol. Chem.* **217**:213–224 (1955).

58. J. A. Hathaway and D. E. Atkinson, Kinetics of regulatory enzymes: Effect of adenosine triphosphate on yeast citrate synthase, *Biochem. Biophys. Res. Comm.* **20**:661–665 (1965).

59. S. Natelson, J. B. Pincus, and J. K. Logovoy, Microestimation of citric acid; A new colorimetric reaction for pentabromoacetone, *J. Biol. Chem.* **175**:745–750 (1948).

60. H. Naruse, S-C. Cheng and H. Waelsch, Microdetermination of citric acid in nervous tissue, *Exp. Brain Res.* **1**:40–47 (1966).

61. J. Schuberth, J. Sollenberg, A. Sundwall, and B. Sörbo, Acetylcoenzyme A in brain. The effect of centrally active drugs, insulin coma and hypoxia, *J. Neurochem.* **13**:819–822 (1966).

62. H. Naruse, S-C. Cheng, and H. Waelsch, CO_2 Fixation in the nervous tissue. IV. CO_2 fixation and citrate metabolism in lobster nerve, *Exp. Brain Res.* **1**:284–290 (1966).

63. H. Naruse, S-C. Cheng, and H. Waelsch, CO_2 Fixation in the nervous system. V. CO_2 fixation and citrate metabolism in rabbit nerve, *Exp. Brain Res.* **1**:291–298 (1966).

64. S. P. Bessman, Preparation and assay of oxaloacetate, *in Methods in Enzymology* (S. P. Colowick and N. O. Kaplan, eds.), Vol. III, pp. 418–421 (1957).

65. W. Prinz, W. Schoner, U. Haag, and W. Seubert, On the mechanism of glucogenesis and its regulation. IV. An isotope assay for the determination of microquantities of acetyl-CoA, *Biochem. Z.* **346**:206–211 (1966).

66. S. R. Dickman, Aconitase, *in The Enzymes* (P. D. Boyer, H. A. Lardy, and K. Myrbäck, eds.), Vol. 5, pp. 495–510, Academic Press, New York (1961).

67. I. A. Rose and E. L. O'Connell, Mechanism of aconitase action. I. The hydrogen transfer reaction, *J. Biol. Chem.* **242**:1870–1879 (1967).

68. H. A. Krebs, The equilibrium constants of the fumarase and aconitase systems, *Biochem. J.* **54**:78–82 (1953).

69. P. J. England, R. M. Denton, and P. J. Randle, The influence of magnesium ions and other bivalent metal ions on the aconitase equilibrium and its bearing on the binding of magnesium ions by citrate in rat heart, *Biochem J.* **105**:32C–33C (1967).

70. J. A. Shepherd and G. Kalnitsky, Intracellular distribution of fumarase, aconitase and isocitric dehydrogenase in rabbit cerebral cortex, *J. Biol. Chem.* **207**:605–611 (1954).

71. J. R. Stern, Assay of tricarboxylic acids, *in Methods in Enzymology* (S. P. Colowick and N. O. Kaplan, eds.), Vol. III, pp. 425–431, Academic Press, New York (1957).

72. C. B. Anfinsen, Aconitase from pig heart muscle, *in Methods in Enzymology* (S. P. Colowick and N. O. Kaplan, eds.), Vol. I, pp. 695–698, Academic Press, New York (1955).

73. P. Buffa and R. A. Peters, The *in vivo* formation of citrate induced by fluoroacetate and its significance, *J. Physiol.* **110**:488–500 (1949).

74. R. M. C. Dawson and R. A. Peters, Observations upon the behavior of some phosphate esters in brain at the start of convulsions induced by fluorocitrate and fluoroacetate, *Biochim. Biophys. Acta* **16**:254–257 (1955).

75. B. Sacktor, J. E. Wilson, and C. G. Tiekert, Regulation of glycolysis in brain, *in situ*, during convulsion, *J. Biol. Chem.* **241**:5071–5075 (1966).

76. B. K. Koe and A. Weissman, Convulsions and elevation of tissue citric acid levels induced by 5-fluorotryptophan, *Biochem. Pharmacol.* **15**:2134–2136 (1966).

77. G. B. Jones, Estimation of microgram quantities of citrate in biological fluids, *Anal. Biochem.* **21**:286–292 (1967).

78. H. Moellering and W. Gruber, Determination of citrate with citrate lyase, *Anal. Biochem.* **17**:369–376 (1966).

79. S. Ochoa, Biosynthesis of tricarboxylic acids by carbon dioxide fixation. III. Enzymatic mechanisms, *J. Biol. Chem.* **174**:133–157 (1948).

80. G. W. E. Plaut and S-C. Sung, Diphosphopyridine nucleotide isocitric dehydrogenase from animal tissues, *in Methods in Enzymology* (S. P. Colowick and N. O. Kaplan, eds.), Vol. I, pp. 710–714 (1955).

81. G. W. E. Plaut, Isocitric dehydrogenases, *in The Enzymes* (P. D. Boyer, H. A. Lardy, and K. Myrbäck, eds.), Vol. 7, pp. 105–126 (1963).

82. J. Moyle, Some properties of purified isocitric enzyme, *Biochem. J.* **63**:552–558 (1958).

83. G. Siebert, M. Carsiotis, and G. W. E. Plaut, The enzymatic properties of isocitric dehydrogenase, *J. Biol. Chem.* **226**:977–991 (1957).

84. A. M. Stein, S. K. Kirkman, and J. H. Stein, Diphosphopyridine nucleotide specific isocitric dehydrogenase of mammalian mitochondria. II. Kinetic properties of the enzyme of the Ehrlich ascites carcinoma, *Biochem.* **6**:3197–3203 (1967).

85. O. H. Lowry, Enzyme concentrations in individual nerve cell bodies, *in Metabolism of the Nervous System* (D. Richter, ed.), p. 323, Pergamon Press, New York (1957).

86. W. D. Lotspeich and R. A. Peters, The action of sulphydryl inhibitors upon isocitric dehydrogenase with especial reference to the behavior of some trivalent arsenicals, *Biochem. J.* **49**:704–709 (1951).

87. S. Ochoa, Isocitric dehydrogenase system (TPN) from pig heart, *in Methods in Enzymology* (S. P. Colowick and N. O. Kaplan, eds.), Vol. I, pp. 699–704 (1955).

88. G. W. E. Plaut, Isocitric dehydrogenase (TPN-linked) for pig heart (revised procedure), *in Methods in Enzymology* (S. P. Colowick and N. O. Kaplan, eds.), Vol. V, pp. 645–651 (1962).

89. M. Hirachima, T. Hayakawa, and M. Koike, Mammalian α-keto acid dehydrogenase complexes. II. An improved procedure for the preparation of 2-oxoglutarate dehydrogenase complex from pig heart muscle, *J. Biol. Chem.* **242**:902–907 (1967).

90. S. Kaufman, α-Ketoglutaric dehydrogenase system and phosphorylating enzyme from heart muscle, *in Methods in Enzymology* (S. P. Colowick and N. O. Kaplan, eds.), Vol. I, pp. 714–722 (1955).

91. H. Weil-Malherbe, Studies on Brain Metabolism. I. The Metabolism of Glutamic Acid in Brain, *Biochem. J.* **30**:665–676 (1936).

92. G. A. Stoner and J. W. Evans, The enzymatic determination of α-ketoglutaric acid, *Anal. Biochem.* **21**:313–316 (1967).

93. L. P. Hager, Succinyl CoA synthetase, in *The Enzymes* (P. D. Boyer, H. A. Lardy, and K. Myrbäck, eds.), Vol. 6, pp. 387–399, Academic Press, New York (1962).

94. W. D. Bonner, Succinic dehydrogenase, in *Methods in Enzymology* (S. P. Colowick and N. O. Kaplan, eds.), Vol. I, pp. 722–729 (1955).

95. T. P. Singer and E. B. Kearney, Succinate dehydrogenase, in *The Enzymes* (P. D. Boyer, H. A. Lardy, and K. Myrbäck, eds.), Vol. 7, pp. 383–445, Academic Press, New York (1963).

96. A. Giuditta and T. P. Singer, Studies on succinate dehydrogenase. XI. Inactivation by cyanide and its mechanism, *J. Biol. Chem.* **234**:666–671 (1958).

97. T. P. Singer and E. B. Kearney, Determination of succinic dehydrogenase activity, in *Methods of Biochemical Analysis* (D. Glick, ed.), Vol. 4, pp. 307–333 (1957).

98. P. Bornath and T. P. Singer, Succinic dehydrogenase, in *Methods in Enzymology* (S. P. Colowick and N. O. Kaplan, eds.), Vol. V, pp. 597–614 (1962).

99. E. B. Kearney and T. P. Singer, Studies on succinate dehydrogenase. I. Preparation and assay of the soluble dehydrogenase, *J. Biol. Chem.* **219**:963–975 (1956).

100. H. A. Krebs, The role of fumarate in the respiration of *Bacterium coli* Commune, *Biochem. J.* **31**:2095–2124 (1937).

101. E. Kmetec, Spectrophotometric method for the enzymic microdetermination of succinic acid, *Anal. Biochem.* **16**:474 (1966).

102. R. A. Alberty, Fumarase, in *The Enzymes* (P. D. Boyer, H. A. Lardy, and K. Myrbäck, eds.), Vol. 5, pp. 531–544 (1961).

103. O. H. Lowry, N. R. Roberts, M-L. Wu, W. S. Hixon, and E. J. Crawford, The quantitative histochemistry of the brain. II. Enzyme measurement, *J. Biol. Chem.* **207**:19–37 (1954).

104. O. H. Lowry, N. R. Roberts, K. Y. Leiner, M-L. Wu, A. L. Farr, and R. W. Albers, The quantitative histochemistry of the brain. III. Ammon's horn, *J. Biol. Chem.* **207**:39–49 (1954).

105. S. Tuček, On subcellular localization and binding of choline acetyltransferase in the cholinergic nerve endings of the brain, *J. Neurochem.* **13**:1317–1327 (1966).

106. V. Massey, Fumarase, in *Methods in Enzymology* (S. P. Colowick and N. O. Kaplan, eds.), Vol. I, pp. 729–785 (1955).

107. O. H. Lowry, Micromethods for the assay of enzymes, in *Methods in Enzymology* (S. P. Colowick and N. O. Kaplan, eds.), Vol. IV, pp. 366–381 (1957).

108. H. A. Krebs, D. H. Smyth, and E. A. Evans, Determination of fumarate and malate in animal tissues, *Biochem. J.* **34**:1041–1045 (1940).

109. P. M. Nossal, Estimation of L-malate and fumarate by malic decarboxylase of *Lactobacillus arabinosus*, *Biochem. J.* **50**:349–355 (1950).

110. E. Kun, Malate dehydrogenase, in *The Enzymes* (P. D. Boyer, H. A. Lardy, and K. Myrbäck, eds.), Vol. 7, pp. 149–160 (1963).

111. O. H. Lowry, N. R. Roberts, and M.-L. W. Chang, The analysis of single cells, *J. Biol. Chem.* **222**:97–107 (1956).

112. J. L. Strominger and O. H. Lowry, The quantitative histochemistry of brain. IV. Lactic, malic and glutamic dehydrogenases, *J. Biol. Chem.* **213**:635–646 (1955).

113. O. H. Lowry, N. R. Roberts, and C. Lewis, The quantitative histochemistry of the retina, *J. Biol. Chem.* **220**: 879–892 (1952).

114. A. M. Mehler, A. Kornberg, S. Grisolia, and O. Ochoa, The enzymatic mechanism of oxidation-reductions between malate or isocitrate and pyruvate, *J. Biol. Chem.* **174**:961–977 (1948).

115. S. Ochoa, Malic dehydrogenase from pig heart, in *Methods in Enzymology* (S. P. Colowick and N. O. Kaplan, eds.), Vol. I, pp. 735–739 (1955).

116. S-C. Cheng and H. Waelsch, Some quantitative aspects of the fixation of carbon dioxide by the lobster nerve, *Biochem. Z.* **338**:643–653 (1963).

117. S-C. Cheng and P. Mela, CO_2 fixation in the nervous system. II. Environmental effects on CO_2 fixation in lobster nerve, and III. Effect of acetylcholine on CO_2 fixation in lobster nerve, *J. Neurochem.* **13**:281–287 and 289–292 (1966).

118. H. J. Hohorst, F. H. Krentz, and Th. Bücher, Uber Metabolitgehalte und Metabolit-Kinzentrationen in der Leber der Ratte, *Biochem. Z.* **332**:18–46 (1959).

119. S. Korkes, Isolation and assay of L-malate, in *Methods in Enzymology* (S. P. Colowick and N. O. Kaplan, eds.), Vol. III, pp. 435–437 (1957).

120. M. F. Utter, Non-oxidative carboxylation and decarboxylation, in *The Enzymes* (P. D. Boyer, H. A. Lardy, and K. Myrbäck, eds.), Vol. 5, pp. 319–340 (1961).

121. D. A. Walker, Pyruvate carboxylation and plant metabolism, *Biol. Rev.* **37**:215–256 (1962).

122. H. G. Wood and M. F. Utter, The role of CO_2 fixation in metabolism, in *Essays in Biochemistry* (P. N. Campbell and G. D. Greville, eds.), Vol. 1, pp. 1–27 (1965).

123. S. Ochoa, Enzymatic mechanism of CO_2 fixation, in *The Enzymes* (J. B. Sumner and K. Myrbäck, eds.), Vol. II, pp. 929–1032 (1952).

124. S. Ochoa, "Malic" enzyme, in *Methods in Enzymology* (S. P. Colowick and N. O. Kaplan, eds.), Vol. I, pp. 739–753 (1955).

125. K. Takeo, T. Murai, J. Nagai, and H. Katsuki, Allosteric activation of DPN-linked malic enzyme from *Escherichia coli* by aspartate, *Biochem. Biophys. Res. Comm.* **29**:717–722 (1967).

126. H. A. Lardy, V. Paetkau, and P. Walter, Paths of carbon in gluconeogenesis and lipogenesis: The role of mitochondria in supplying precursors of phosphoenolpyruvate, *Proc. Nat. Acad. Sci.* **53**:1410–1415 (1965).

127. M. Ruiz-Amil, G. de Torrontegui, E. Palacián, L. Catalina, and M. Losada, Properties and function of yeast pyruvate carboxylase, *J. Biol. Chem.* **240**:3485–3492 (1965).

128. A. S. Mildvan and M. C. Scrutton, Pyruvate carboxylase. X. The demonstration of direct coordination of pyruvate and α-ketobutyrate by the bound manganase and the formation of enzyme-metal-substrate bridge complexes, *Biochem.* **6**:2978–2994 (1967).

129. J. J. Cazzulo and A. O. M. Stoppani, Purification and properties of pyruvate carboxylase from baker's yeast, *Arch. Biochem. Biophys.* **121**:596-608 (1967).

130. M. C. Scrutton and M. F. Utter, Pyruvate carboxylase. IX. Some properties of the activation by certain acyl derivatives of coenzyme A, *J. Biol. Chem.* **242**:1723–1735 (1967).

131. P. Walter, V. Paetkau, and H. A. Lardy, Paths of carbon in gluconeogenesis and lipogenesis. III. The role and regulation of mitochondrial processes involved in supplying precursors of phosphoenolpyruvate, *J. Biol. Chem.* **241**:2523–2532 (1966).

132. D. B. Keech and R. K. Farrant, The reactive lysine residue at the allosteric site of sheep kidney pyruvate carboxylase, *Biochim. Biophys. Acta* **151**:493–503 (1968).

133. H-C. Chang, H. Maruyama, R. S. Miller, and M. D. Lane, The enzymatic carboxylation of phosphoenolpyruvate. III. Investigation of the kinetics and mechanism of the mitochondrial phosphoenolpyruvate carboxykinase-catalyzed reaction, *J. Biol. Chem.* **241**:2421–2430 (1966).

134. H-C. Chang and M. D. Lane, The enzymatic carboxylation of phosphoenolpyruvate. II. Purification and properties of liver mitochondrial phosphoenolpyruvate carboxykinase, *J. Biol. Chem.* **241**:2413–2420 (1966).

135. E. Shrago, H. A. Lardy, R. C. Nordlie, and D. O. Foster, Metabolic and hormonal control of phosphoenolpyruvate carboxykinase and malic enzyme in rat liver, *J. Biol. Chem.* **238**:3188–3192 (1963).

136. M. F. Utter and K. Kurahashi, Oxaloacetate synthesizing enzyme, in *Methods in Enzymology* (S. P. Colowick and N. O. Kaplan, eds.), Vol. I, pp. 758–763 (1955).

137. G. Schmidt, Preparation of phosphoenolpyruvate, in *Methods in Enzymology* (S. P. Colowick and N. O. Kaplan, eds.), Vol. III, pp. 223–228 (1957).

138. H. Maruyama, R. L. Easterday, H-C. Chang, and M. D. Lane, The enzymatic carboxylation of phosphoenolpyruvate. I. Purification and properties of phosphoenolpyruvate carboxykinase, *J. Biol. Chem.* **241**:2405–2412 (1966).

139. S. Berl, G. Takagaki, D. D. Clarke, and H. Waelsch, Carbon dioxide fixation in the brain, *J. Biol. Chem.* **237**:2570–2573 (1962).

140. G. Gombos, A. Geiger, and S. Otsuki, The metabolic pattern of the brain in brain perfusion experiments *in vivo*. II. Pyruvate and lactate formation from ^{14}C-labelled aspartate, *J. Neurochem.* **10**:405–413 (1963).

141. A. Hamberger, Oxidation of tricarboxylic acid cycle intermediates by nerve cell bodies and glial cells, *J. Neurochem.* **8**:31–35 (1961).

142. G. McKhann and D. B. Towers, Ammonia toxicity and cerebral oxidative metabolism, *Amer. J. Physiol.* **200**:420–424 (1961).

143. R. E. Basford, W. L. Stahl, D. S. Beattie, H. R. Sloan, J. C. Smith, and L. M. Napolitano, Enzymatic properties of brain mitochondria, in *Morphological and Biochemical Correlates of Neural Activity* (M. M. Cohen and R. S. Snider, ed.), pp. 192–211, Hoeber Med. Div., Harper and Row, New York (1964).

144. H. P. Cohen, Some factors involved in the use of glucose and hexokinase as a trap for ATP in cerebral mitochondrial studies, in *Morphological and Biochemical Correlates of Neural Activity* (M. M. Cohen and R. S. Snider, eds.), pp. 212–224, Hoeber Med. Div., Harper and Row, New York (1964).

145. T. M. Brody and J. A. Bain, A mitochondrial preparation from mammalian brain, *J. Biol. Chem.* **195**:685–696 (1952).

146. V. P. Whittaker, The application of subcellular fractionation technique to the study of brain function, *Progr. Biophys. Mol. Biol.* **15**:39–91 (1965).

147. E. DeRobertis, G. Rodriguez de Lores Arnaiz, L. Salganicoff, A. Pellegrino de Iraldi, and L. M. Zieher, Isolation of synaptic vesicles and structural organization of the acetylcholine system within brain nerve endings, *J. Neurochem.* **10**:225–235 (1963).

148. E. Robins, N. R. Roberts, K. M. Eydt, O. H. Lowry, and D. E. Smith, Microdetermination of α-keto acids with special reference to malic, lactic and glutamic dehydrogenases in brain, *J. Biol. Chem.* **218**:897–909 (1956).

149. G. Simon, J. B. Drori, and M. M. Cohen, Mechanism of conversion of aspartate into glutamate in cerebral-cortex slices, *Biochem. J.* **102**:153–162 (1967).

150. G. Simon, M. M. Cohen, and J. F. Berry, Conversion of glutamate into aspartate in guinea-pig cerebral-cortex slices, *Biochem. J.* **107**:109–111 (1968).

151. R. Nakamura and S.-C. Cheng, unpublished.

152. H. Waelsch, S.-C. Cheng, L. J. Côté, and H. Naruse, CO_2 fixation in the nervous system, *Proc. Nat. Acad. Sci.* **54**:1249–1253 (1965).

153. A. F. D'Adamo and C. N. Powell, Citrate metabolism in the developing rat brain, *Fed. Proc.* **24**:2365 (1965).

154. H. McIlwain, *Chemical Exploration of the Brain*, Elsevier Publishing Co., New York (1963).

155. F. Brink, Nerve metabolism, in *Metabolism of the Nervous System* (D. Richter, ed.), pp. 187–207, Pergamon Press, New York (1957).

156. B. Chance, Control of energy metabolism in mitochondria, in *Control of Energy Metabolism* (B. Chance, R. W. Estabrook, and J. R. Williamson, eds.), pp. 415–431, Academic Press, New York (1965).

157. M. Klingenberg, Control characteristics of the adenine nucleotide system, in *Control of Energy Metabolism* (B. Chance, R. W. Estabrook, and J. R. Williamson, eds.), pp. 149–155, Academic Press, New York (1965).

158. H. Rasmussen and E. Ogata, Cation flux across the mitochondrial membrane as a possible pacemaker of tissue metabolism, *in Control of Energy Metabolism* (B. Chance, R. W. Estabrook, and J. R. Williamson, eds.), pp. 209–216, Academic Press, New York (1965).

159. N. D. Goldberg, J. V. Passonneau, and O. H. Lowry, The effects of altered brain metabolism on the levels of Krebs cycle intermediates, *in Control of Energy Metabolism* (B. Chance, R. W. Estabrook, and J. R. Williamson, eds.), pp. 321–327, Academic Press, New York (1965).

160. D. Garfinkel, Computer-based analysis of biochemical data, *in Control of Energy Metabolism* (B. Chance, R. W. Estabrook, and J. R. Williamson, eds.), pp. 49–54, Academic Press, New York (1965).

161. D. Garfinkel, A simulation study of the metabolism and compartmentation in brain of glutamate, aspartate, the Krebs cycle, and related metabolites, *J. Biol. Chem.* **241**:3918–3929 (1966).

162. F. J. Ballard, R. W. Hanson, and D. S. Kronfeld, Factors controlling the concentration of mitochondrial oxaloacetate in liver during spontaneous bovine ketosis, *Biochem. Biophys. Res. Comm.* **30**:100–104 (1968).

163. M. S. Kornacker and J. M. Lowenstein, Citrate and the conversion of carbohydrate into fat. Activities of citrate-cleavage enzyme and acetate thiokinase in livers of normal and diabetic rats, *Biochem. J.* **95**:832–837 (1965).

164. A. Spencer, L. Corman, and J. M. Lowenstein, Citrate and conversion of carbohydrate into fat. A comparison of citrate and acetate incorporation into fatty acids, *Biochem. J.* **93**:378–388 (1964).

165. A. F. Spencer and J. M. Lowenstein, Citrate and the conversion of carbohydrate into fat. Citrate cleavage in obesity and lactation, *Biochem. J.* **99**:760–765 (1966).

166. S. Berl, G. Takagaki, D. D. Clarke, and H. Waelsch, Metabolic compartments *in vivo*, *J. Biol. Chem.* **237**:2562–2569 (1962).

167. R. M. O'Neal, R. E. Koeppe, and E. I. Williams, Utilization *in vivo* of glucose and volatile fatty acids by sheep brain for the synthesis of acidic amino acids, *Biochem. J.* **101**:591–597 (1966).

168. J. M. G. Van Kempen, C. J. van den Berg, H. J. Van der Helm, and H. Veldstra, Intracellular localization of glutamate decarboxylase, γ-Aminobutyrate transaminase, and some other enzymes in brain tissue, *J. Neurochem.* **12**:581–588 (1965).

169. C. J. van den Berg, P. Mela, and H. Waelsch, On the contribution of the tricarboxylic acid cycle to the synthesis of glutamate, glutamine and aspartate in brain, *Biochem. Biophys. Res. Comm.* **23**:479–484 (1966).

170. C. J. van den Berg, A. Neidle, and A. Grynbaum, personal communication.

171. S. P. R. Rose, Metabolism of isolated neuronal and glial cells, *in First International Meeting of the International Society for Neurochemistry*, p. 182, Strasbourg (1967).

172. J. B. Chappell and K. N. Haarhoff, The penetration of the mitochondrial membrane by anions and cations, *in Biochemistry of Mitochondria* (E. C. Slater, Z. Kaniuga, and L. Wojtczak, eds.), pp. 75–91, Academic Press, New York (1967).

173. E. J. DeHaan and J. M. Tager, Evidence for a permeability barrier for α-ketoglutarate in rat-liver mitochondria, *Biochim. Biophys. Acta* **153**:98–112 (1968).

174. E. Quagliariello and F. Palmieri, Control of succinate oxidation by succinate-uptake by rat-liver mitochondria, *European J. Biochem.* **4**:20–27 (1968).

175. B. H. Robinson and J. B. Chappell, An apparent energy requirement for oxaloacetate penetration into mitochondria provided by permeant cations, *Biochem. J.* **105**:18P (1967).

176. F. N. Minard and I. K. Mushahwar, The effect of periodic convulsions induced by 1,1-dimethylhydrazine on the synthesis of rat brain metabolites from (2-^{14}C)Glucose, *J. Neurochem.* **13**:1–11 (1966).

177. H. Busch, R. B. Hurlbert, and V. R. Potter, Anion exchange chromatography of acids of the citric acid cycle, *J. Biol. Chem.* **196**:717–727 (1952).
178. D. Wang and D. Mancini, Anion exchange chromatography of organic acids and nucleotides: An improved gradient elution system, *Contr. Boyce Thompson Inst.* **23**:93–100 (1965).
179. H. E. Swim and M. F. Utter, Isotopic experimentation with intermediates of the tricarboxylic acid cycle, *in Methods in Enzymology* (S. P. Colowick and N. O. Kaplan, eds.), Vol. IV, pp. 584–609 (1957).
180. J. E. Varner, Chromatographic analyses of organic acids, *in Methods in Enzymology* (S. P. Colowick and N. O. Kaplan, eds.), Vol. III, pp. 397–404 (1957).
181. A. R. Jones, E. J. Dowling, and W. J. Skraba, Identification of some organic acids by paper chromatography, *Anal. Chem.* **25**:394–396 (1953).
182. R. D. Hartley and G. J. Lawson, R_F values of organic acids in selected solvents, *J. Chromatog.* **7**:69–76 (1962).
183. C. Passera, A. Fedrotti, and G. Ferrari, Thin-layer chromatography of carboxylic acids and ketoacids of biological interest, *J. Chromatog.* **14**:289–291 (1964).
184. I. P. Ting and W. M. Dugger, Jr., Separation and determination of organic acids on silica gel, *Anal. Biochem.* **12**:571–578 (1965).
185. E. Stahl, *Thin-Layer Chromatography*, Academic Press, New York (1965).
186. K. Randerath, *Thin-Layer Chromatography*, pp. 256–258, Academic Press, New York, 2nd ed. (1966).
187. H. Higgins and T. Von Brand, Separation of lactic acid and some Krebs cycle acids by thin-layer chromatography, *Anal. Biochem.* **15**:122–126 (1966).
188. W. F. Myers and K-Y. Huang, Separation of intermediates of the citric acid cycle and related compounds by thin-layer chromatography, *Anal. Biochem.* **17**:210–213 (1966).
189. P. Nygaard, Two-way separation of carboxylic acids by thin-layer electrophoresis and chromatography, *J. Chromatog.* **30**:240–243 (1967).
190. E. H. Mosbach, E. F. Phares, and S. F. Carlson, Degradation of isotopically labeled citric, α-ketoglutaric, and glutamic acids, *Arch. Biochem. Biophys.* **33**:179–185 (1951).
191. K. F. Lewis and S. Weinhouse, Studies on the mechanism of citric acid production in *Aspergillus niger*, *J. Amer. Chem. Soc.* **73**:2500–2503 (1951).
192. S. H. Pomerantz, Citric acid cycle in young rat skin, *J. Biol. Chem.* **236**:2863–2867 (1961).
193. A. J. Finlayson, Degradation of the carbon skeleton of glutamic acid, *Can. J. Biochem.* **44**:397–401 (1966).
194. D. C. Hardwick, The incorporation of carbon dioxide into milk citrate in the isolated perfused goat udder, *Biochem. J.* **95**:233–237 (1965).
195. E. F. Gale, Determination of amino acids by use of bacterial amino acid decarboxylases, *in Methods of Biochemical Analysis* (D. Glick, ed.), Vol. 4, pp. 285–306 (1954).
196. L. V. Crawford, Studies on the aspartic decarboxylase of *Nocardia globerula*, *Biochem. J.* **68**:221–225 (1958).
197. H. A. Krebs and D. Bellamy, The interconversion of glutamate and aspartate in respiring tissues, *Biochem. J.* **75**:523–529 (1960).

Chapter 10

METABOLISM AND FUNCTION IN NERVE FIBERS

P. Greengard and J. M. Ritchie

Department of Pharmacology
Yale University School of Medicine
New Haven, Connecticut

I. INTRODUCTION

The preparations of intact cells that have been used in the past to study metabolism and function in nervous tissue vary in complexity from the human brain *in vivo*[1,2] to isolated single nerve fibers of invertebrates.[3-7] The extent to which results obtained with any one type of preparation are applicable to the others is an open question because there is considerable evidence of differences in the biochemistry of different parts of the nervous system.[8] However, as will be discussed below in comparing myelinated and nonmyelinated fibers, some seemingly qualitative differences in metabolism, such as the enormous difference in sensitivity to glucose deprivation, may be more apparent than real.

II. BRAIN TISSUE AND GANGLION CELLS

A. Human Brain *in vivo*

Many studies have been made on the metabolism of the human brain *in vivo*, particularly by Kety and his colleagues.[8,9] Thus, the use of arterial-venous difference methods combined with measurements of blood flow through the brain has allowed the accurate determination of the quantity of a number of substances consumed or produced by the brain per unit time. These methods have established the very high oxygen consumption of the brain *in vivo*, and have shown that the disappearance of glucose is adequate to account for the oxygen consumed. The average respiratory quotient is close to unity, which is in agreement with the conclusion that glucose is the almost exclusive physiologically available fuel of the brain; it should be noted, however, that under certain experimental conditions other sugars, particularly mannose, can equally well support function (H.

Sloviter, personal communication). Normal variations of functional activity of the brain such as sleep, resting state, or the performance of mental arithmetic are not accompanied by any detectable changes in the cerebral oxygen consumption. Similarly, there is no change in cerebral oxygen consumption in patients sedated by chlorpromazine or even in subjects during hallucinations induced by LSD. In extreme physiological states, however, one does see variations in oxygen consumption; for example, both in coma and in anesthesia, the cerebral oxygen consumption decreases, whereas during convulsions it increases.

B. Animal Brain and Brain Slices

The advantages of studying the human brain in the ways described above are obvious; but the amount of information that can be obtained on intermediary metabolism is necessarily restricted. To overcome this difficulty, a variety of approaches have been used. A common procedure has been to place animals in different states of cerebral activity (anesthesia, convulsions, etc.) and then to quick-freeze and analyze their brains for the resulting chemical changes.[10] Another approach to the study of brain metabolism is that of McIlwain[1,2] who subjected slices of mammalian cerebral cortex to electrical stimulation in the hope of simulating the conditions of continuous electrical activity that exist in the living brain. In this way the respiration of brain slices can be increased nearly to that of the brain *in situ*. This enhanced respiration associated with electrical pulses is more sensitive than the normal respiration to a number of anesthetics, depressants, and anticonvulsants.

C. Ganglion Cells

Another approach to the study of metabolism of nervous tissue is that of Larrabee and his colleagues, who have used the excised superior cervical ganglion.[11,12] This preparation permits the study both of the metabolic events associated with impulse conduction along the preganglionic fibers as well as the corresponding events associated with transmission across the ganglionic synapse. With this preparation it has been shown that some general anesthetics interfere with ganglionic transmission at concentrations as low as those that depress the central nervous system during general anesthesia. It is especially significant therefore that the resting oxygen consumption is not disturbed at concentrations of anesthetic that block ganglionic transmission. These experiments cast serious doubt on the theory that the primary action of general anesthetics is on respiration of nervous tissue. This conclusion is supported by recent work on isolated nonmyelinated nerve fibers, which has shown that local anesthetics have little or no effect on energy metabolism in the concentrations required to produce local anesthesia.[13]

III. NERVE FIBERS

A. General

Perhaps the most important contribution to our understanding of the molecular basis of axonal function has come from the studies of Hodgkin, Huxley, Keynes and their colleagues on the giant axons of invertebrates[14-16] as a result of which the ionic basis of nervous activity has become firmly established. As in most other tissues, nerve cells and axons have a higher concentration of potassium than does the extracellular environment and a lower concentration of sodium ions. During the nerve impulse, sodium ions enter and potassium ions leave the fibers. These movements of ions are downhill in relation to the existing electrochemical gradient so that, thermodynamically, no source of chemical energy is required. However, after the impulse the nerve fibers are left with an excess of sodium and a deficit of potassium. The restoration of the nerve fibers to their initial state, which is the recovery process, involves the coupled movement of sodium ions out of, and potassium ions back into, the axons against electrochemical gradients. This recovery process must use energy and indeed most of the metabolism of peripheral nerve is devoted to the maintenance of the ionic distribution across the nerve membrane. This description of ion movement is of the utmost importance to neurochemists because it provides the basis for a considerably more exact formulation of the problem of the relationship between function and energy requirements of nerve axons than was previously possible. Instead of the vague question: "How is metabolism related to function in nerve axons?" it is now possible to put the more specific question: "What are the energy requirements for moving sodium ions out of, and potassium ions back into, the nerve axons following impulse conduction?"

B. Comparison of Myelinated and Nonmyelinated Nerve Fibers

That nonmyelinated fibers are more suitable than myelinated fibers for studying the relationship between metabolism and function was already suggested by the work of Larrabee and Bronk.[17] They demonstrated that the ability of the nonmyelinated fibers of the excised superior cervical ganglion preparation to conduct impulses is considerably more dependent on an adequate supply of glucose than that of the myelinated fibers in the same preparation. Thus, they demonstrated that the compound action potential of the nonmyelinated fibers at rest in a glucose-free solution falls by 50% in 2 hr, a time that is much shorter than the corresponding time of $2\frac{1}{2}$ days that has been found in numerous laboratories for myelinated fibers. Nonmyelinated fibers thus seem to have considerably higher energy demands than do myelinated fibers; and direct measurements of the oxygen consumption of these nerves shows that this is indeed the case. For example, a nerve trunk composed of nonmyelinated fibers consumes about 100 times as much oxygen per impulse as does a nerve trunk of the same size composed of myelinated fibers.[18] In a theoretical analysis, Rushton[19] has pointed out that from the teleological point of view, nonmyelination of nerve fibers may be desirable because it allows more fibers to run in a given nerve trunk without too great a sacrifice in conduction velocity. This arrangement might well have proven uneconomical because of the much larger amount of nerve membrane participating in conduction in nonmyelinated than in myelinated fibers where ionic interchange is restricted to the nodes of Ranvier; thus in a 2-mm stretch of nerve, containing just one node, a single large myelinated fiber has only between 6 and 60 sq. microns of active membrane[20] whereas a single vagal nonmyelinated fiber of average diameter 0.8μ[21] has $5 \times 10^3 \mu^2$, i.e., it has about 100–1000 times more than the myelinated fiber. Therefore, since about 100 times as many nonmyelinated fibers of diameter 0.8μ can be packed into the space occupied by a single myelinated fiber of diameter 8μ, the amount of active membrane in a nerve trunk composed of non-myelinated fibers would be 10,000–100,000 times the amount in a nerve of the same size composed only of myelinated fibers. The difference in the oxygen consumptions of the two types of fiber, however, is not as great as the above considerations would suggest and there is only a 100-fold difference between them. Indeed, although the oxygen consumption per impulse of non-myelinated nerve is greater than for myelinated nerve, expressed per unit weight of nerve, the cost of transmitting a single impulse in a single fiber over a given distance is about the same in the two types of fiber,[18] which seems to provide an interesting example of economy in nature.

Other studies of oxygen consumption and of the utilization of high-energy phosphates can also be interpreted as indicating that myelinated fibers do not have a very high rate of metabolism. For example, in addition to the low respiration of myelinated fibers, Brink and his colleagues[22,23] have shown that the increase in oxygen consumption associated with electrical stimulation of frog myelinated nerve fibers could be abolished by azide, with little effect on the resting respiration or ionic movements and no detectable effect on the ability to conduct impulses at high frequency for many hours. Furthermore, the papers of Gerard of 30 years ago[24] reveal the enormous difficulty in demonstrating changes in the content of high energy phosphate esters as the result of activity in myelinated nerve fibers. This contrasts with the situation in muscle, where it is relatively easy to demonstrate a depletion of high energy phosphate compounds as a result of activity.

The theoretical reason for the superiority of nonmyelinated nerve fibers in metabolic studies seems clear. For such a study, a preparation is required that shows the closest relation between activity and metabolism, i.e., a preparation in which: (1) altering activity would readily produce altered metabolism; and (2) altering the metabolism would readily result in changes in function. This seems to suggest a preparation in which large changes in ion concentration of tissue occur during activity. This idea rules out myeli-nated fibers because it can be readily calculated that the ionic concentration changes even during intense activity are very small—small enough in fact to be able to explain why myelinated fibers are very much less dependent on metabolism than are other types of vertebrate nervous tissue. This is why, for example, it is necessary to use 2 hr of stimulation at frequencies of 100 per second to find any change in the level of high energy phosphate compounds in myelinated fibers[24]; for except under these severe conditions of stimulation, the metabolism of the nerve fibers is adequate to resynthesize the high energy phosphate required for the restorative processes that follow the exceedingly small ion movements that occur in these myelinated nerve fibers during activity. In a nonmyelinated nerve fiber, on the other hand, even a single impulse produces a significant change in the sodium and potassium concentration. Thus, the average increase per impulse of sodium concentration $\delta[C]$, which equals the average decrease per impulse in potassium concentration, is given by the expression:

$$\delta[C] = \frac{Q \cdot Al}{V} = \frac{Q\pi dl}{0.25\pi d^2 l} = \frac{4Q}{d}$$

where $\delta[C]$ is the concentration change in mole/ml; Q is the influx of sodium or efflux of potassium in mole/impulse $\times$ cm^2 surface area; A is the surface area of the fiber in cm^2; V is the volume of the fiber in cm^3; d is the diameter in cm; and l the length of the fiber in cm. For nonmyelinated fibers at about 20° C, Q equals 1 pmole/cm^2 $\times$ impulse[21]; thus, $\delta[C]$ equals $(4 \times 10^{-12})/d$ mole/ml. By expressing $\delta[C]$ in millimoles per liter and d in microns, we obtain the numerical relationship $\delta[C]$ (mM) $= (0.04)/[d(\mu)]$. Thus, for an average nonmyelinated fiber of 0.8 μ diameter,[21] $\delta[C]$ equals 0.05 mM, i.e., there will be a decrease in potassium concentration and an equal increase in sodium concentration of 0.05 mM for each impulse. For a myelinated fiber, on the other hand, the corresponding changes are much smaller; for not only is d much larger, but the ionic interchange occurs only at that part of the membrane that is at the nodes.

C. Metabolic Aspects

Mammalian nonmyelinated fibers have proven to be eminently suitable for metabolic studies because, as pointed out above, these fibers, with such a large surface to volume ratio, have very much larger ion movements per impulse per unit volume than do the myelinated fibers and a correspondingly larger loss of electrochemical energy in the form of downhill ion movements. Thus, following just a few impulses in nonmyelinated fibers, there are well-defined changes attributable to the recovery process in: the high-energy phosphate compounds required for the recovery process[25-27]; the electrical activity that reflects the operation of the sodium pump[28-32]; their oxygen consumption[18,33]; and heat production.[34,35] These various aspects of the metabolic activity of mammalian C fibers will be discussed separately below.

1. *ATP and Creatine Phosphate*

In spite of their greater energy turnover, mammalian nonmyelinated fibers were little used until recently because of the lack of analytical methods of adequate sensitivity, the amount of tissue available being rather small. But with the advent of highly sensitive and specific fluorimetric techniques for measuring high energy phosphates, methods vastly more sensitive than the old spectrophotometric methods have become available. These methods, based on the difference between the fluorescence properties of the oxidized and reduced forms of the pyridine nucleotides,[36,37,25] have proven more than adequate to demonstrate high energy phosphate changes on stimulation of mammalian nonmyelinated nerve fibers. In such experiments Greengard and Straub[25] mounted vagus nerves of the rabbit on stimulating electrodes with flexible leads, stimulated them for 15 sec, and while still being stimulated, plunged them into a boiling buffer solution. The nerves were then extracted for analysis and examined for eight different phosphate compounds. Four of these, glucose-6-phosphate, 1,3-diphosphoglycerate, 3-phosphoglycerate, and phosphoenolpyruvate could not be detected in the nerve bundles whether extracted at rest or during electrical stimulation; had these substances been present in a concentration higher than 3×10^{-11} mole/mg wet weight they would have been detected. The values obtained for the other four phosphate compounds, creatine phosphate (CrP), adenosine triphosphate (ATP), adenosine diphosphate (ADP), and adenosine monophosphate (AMP) in nerve bundles extracted at rest are shown in Table I. Recently, Chmouliovsky,

TABLE I

Concentrations of CrP, ATP, ADP, and AMP in Resting Cervical Vagus Nerves of Rabbit[a]

CrP	1.96 ± 0.28
ATP	2.11 ± 0.24
ADP	0.28 ± 0.02
AMP	0.08 ± 0.01

[a] Concentrations are expressed as nmoles/mg fresh weight $\pm$ S.E.M. Means of eight, and in the case of CrP of nine experiments. From Greengard and Straub.[25]

Schorderet, and Straub,[27] using completely different analytical methods from those used by Greengard and Straub,[25] found values for the four phosphate compounds in good agreement with those given in Table I. The effect of 15 sec of electrical stimulation on the concentration of these compounds is shown in Tables II and III. When nerves were stimulated either at a frequency of 6 per second or at 15 per second, the sum, CrP + ATP,

TABLE II

Effect of Electrical Stimulation on the Levels of (CrP + ATP) and of (ATP) Alone in Rabbit Cervical Vagus Nerves

Stimulation frequency (per sec)	Number of expts.	Percentage change in total $\sim P^a$ (CrP + ATP) Mean $\pm$ S.E.	Percentage change in ATP^a Mean $\pm$ S.E.
6	3	$+2.0 \pm 1.7$	$+3.3 \pm 1.8$
15	5	$+1.2 \pm 1.2$	$+3.4 \pm 3.3$
50	9	21.2 ± 2.6	-23.1 ± 2.2

^{a} The mean values represent the percent change in concentration in the stimulated nerves with respect to that in the nonstimulated contralateral nerves. The values have been calculated with respect to equal weights of resting and stimulated nerves. Modified from Greengard and Straub.[25]

representing nearly all the known high energy phosphate of the nerve, was virtually the same as that in the nonstimulated nerve (Table II). However, it has been shown by Folkow[38] and by Douglas and Ritchie[39] that the maximum physiological frequency of firing in these autonomic non-myelinated fibers is about 10/sec. Thus, these results are in agreement with the expectation that resynthesis is fully capable of coping with the energetic demands at physiological frequencies of firing. However, when a frequency of stimulation of 50/sec was used, there was a 21 % decrease in the concentration of the high energy phosphate. Assay of the ATP alone gave similar

TABLE III

Recovery of ATP as ADP and AMP in Cervical Vagus Nerve Stimulated at 50 per sec for 15 sec^a

	Rest	Stimulated	Absolute change	Content of stimulated nerve (as percentage of that of resting nerves)
ATP	12.95	10.29	-2.66	79.4
ADP	0.32	2.29	$+1.97$	716.0
AMP	0.13	0.31	$+0.18$	238.0
Total	13.40	12.89	-0.51	96.2

^{a} The resting values were obtained from an equal length of the contralateral nerve. Values expressed as mμmole per nerve. From Greengard and Straub.[25]

results, stimulation at 6 or 15/sec being ineffective, whereas stimulation at 50/sec resulted in a 23% decrease in the ATP concentration.

In some experiments, one of which is shown in Table III, the loss of ATP could largely be accounted for by the formation of ADP and AMP. Thus, the ATP decreased by 2.66 nmole upon stimulation, but the ADP increased by 1.97 and the AMP by 0.18 nmole. The total adenosine phosphate was almost unchanged, agreeing within 4% for the two nerves. In other experiments the ATP did not appear as ADP and AMP, suggesting that the newly formed ADP and AMP had been further metabolized.

Whereas in myelinated nerve fibers 2 hr of activity was necessary to produce a detectable change in the level of high energy phosphate esters, in nonmyelinated nerve fibers a period of 15 sec was adequate.[25] The relative ease with which the high energy phosphate of the nonmyelinated fibers can be depleted appears reasonable from calculation of the ionic movement involved per impulse in those two types of fiber. Thus, it was calculated earlier that in nonmyelinated fibers the average increase in sodium concentration per impulse was 0.05 mM. The experiments described above (Table I) suggest that the total amounts of ATP and of creatine phosphate in these fibers were 2.1 and 2.0 mmoles/kg respectively. Thus, if no resynthesis were to take place, and if one molecule of ATP were utilized for each sodium ion extruded, recovery from about 40 impulses would deplete the entire store of ATP and recovery from an additional 40 impulses would deplete the creatine phosphate reservoir as well. Thus, less than 100 impulses would be adequate, in the absence of resynthesis, to deplete the known high energy phosphate reservoirs of nonmyelinated nerve fibers. On the other hand, in myelinated fibers it can be calculated that because of the smaller surface to volume ratio, as well as because of the presence of the insulating myelin layer, the increase in sodium concentration per impulse is orders of magnitude smaller than that occurring in nonmyelinated fibers of average diameter.[25] It is, therefore, reasonable to expect that the chance of demonstrating an effect of electrical stimulation on the concentrations of phosphate esters is enormously greater in nonmyelinated nerve fibers than in the myelinated fibers previously used for such studies.

It can be calculated that the free energy change resulting from ion movements in mammalian nonmyelinated nerve fibers is large enough to provide a thermodynamic possibility for an extensive depletion of the high-energy phosphates of these nerve fibers. In addition to this thermodynamic requirement, however, there is still a kinetic requirement that has to be fulfilled in order for such a change to occur. In the experiments of Greengard and Straub[25] the time that elapsed between the onset of stimulation and the fixation of the nerve bundles was 15 sec. The observed decrease of ATP and of creatine phosphate in these experiments shows that the utilization of energy associated with the breakdown of these compounds must have occurred within these 15 sec. The total breakdown of ATP and creatine phosphate (after 15 sec stimulation at 50 shocks per second) was found to be 900 pmoles/g impulse (Table I and III of reference 25) which, for a heat of hydrolysis of ATP of 5000 cal/mole,[40] would correspond to about 4.5 μcal/g impulse. These are minimum values, for if any resynthesis occurred during the period of stimulation, the true value for the high-energy phosphate breakdown would have been correspondingly greater than the net value which was measured. Moreover, there would certainly have been a further heat production, from the heat of protonation of buffer by the hydrogen ion liberated during the hydrolysis of ATP. In experiments in which the recovery heat of the same type of nerve fiber was measured, Howarth, Keynes, and Ritchie[35] found that the total amount of heat evolved in the first 15 sec was about 16 μcal/g impulse. Thus, taking into account the difference in the rate of stimulation (50 per sec for the phosphate experiment; 2.5 per sec in the recovery heat experiment), the correspondence between the chemical and thermal data seems satisfactorily close.

Recently, Montant and Chmouliovsky[26] and Chmouliovsky, Schorderet, and Straub,[27] by altering the experimental conditions were able to demonstrate a much more extensive depletion of ATP following stimulation of mammalian nonmyelinated nerve fibers than occurred in the earlier experiments of Greengard and Straub.[25] Thus, by stimulating at 50 shocks per second for two 15-sec periods separated by a 25-sec interval, Montant and Chmouliovsky[26] found that the ATP concentration decreased to 5% of that observed in unstimulated nerves. As pointed out by the authors, the almost complete depletion of ATP suggests that, in addition to the nonmyelinated axons, myelinated fibers as well as Schwann cells may participate in the increased utilization of ATP after electrical activity. These conclusions seem to follow from the fact that only a small fraction (about 1/4) of the rabbit cervical vagus nerve is occupied by nonmyelinated fibers because of dilution by myelinated fibers and by Schwann cells.[21]

2. *Electrical Experiments*

Probably the most convenient way to study the recovery process in nerve is to examine the hyperpolarization that follows a period of stimulation. Such a post-tetanic hyperpolarization occurs in many tissues such as sensory receptors[41] and myelinated nerve fibers,[42-44] but it is particularly marked in nonmyelinated fibers. Thus Ritchie and Straub[28] and Holmes[30] found that this hyperpolarization in mammalian nonmyelinated fibers could be greatly reduced by ouabain, dinitrophenol, cyanide, iodoacetate, and azide. From these and other experiments, rather convincing evidence was accumulated that this post-tetanic hyperpolarization was the reflection of an increased active extrusion of sodium ions that had entered the axons during the period of repetitive stimulation. The value of this post-tetanic hyperpolarization as an index of metabolic activity has been further strengthened by the finding that in a Locke solution in which the chloride has been removed (being replaced by isethionate), a much larger and more reliable response is obtained: this is because the electrogenic component of the sodium pump, which is normally short-circuited by chloride, becomes fully manifest in the absence

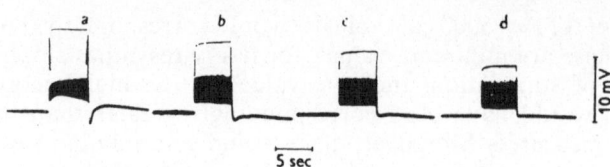

Fig. 1. Effect of antimycin A (1.8×10^{-6} M) on the post-tetanic hyperpolarization recorded from a desheathed bundle of nerve fibers from the rabbit cervical sympathetic trunk. Record (a) obtained before and record (b) obtained 15 min after addition of antimycin A. Three minutes after this the nerve was restored to the control Locke solution and records (c) and (d) were taken 13 and 23 min later. Frequency of stimulation, 15 shocks per sec. Temp. 30°C. Top of spike potential at end of tetanus in record (a) cut off by 6%. (From Greengard and Straub.[31])

of that anion. This post-tetanic response is readily measured and provides a reliable means of mapping metabolic pathways associated with the active extrusion of sodium ions in nerve axons.[31,45,46]

Figure 1 shows records of the post-tetanic hyperpolarization of the non-myelinated fibers of the rabbit's cervical sympathetic trunk. In this experiment (in chloride-Locke solution) the nerve was treated with 1.8 μM antimycin A, a respiratory inhibitor acting on the cytochrome system. It can be seen that the post-tetanic hyperpolarization was abolished rapidly and irreversibly (in agreement with the known irreversible biochemical action of this substance) as the ability to extrude sodium and reabsorb potassium is lost. The action potential was also affected by antimycin A; however, the post-tetanic hyperpolarization was affected much more rapidly than was the height of the action potential itself. Thus, by the time the final records in Fig. 1 were taken (d) the energy production of the nerve fibers had been drastically reduced; but some fibers were still able to maintain the action potential response to repetitive stimulation, presumably by using the energy previously stored in the form of intermediary metabolites and ionic gradients. Even this response eventually failed; and a few minutes after the last record in Fig. 1 was taken, the response to repetitive stimulation could no longer be elicited. Many other metabolic inhibitors with known sites of action (including azide, cyanide, 2,4-dinitrophenol, fluoride, fluoroacetate, hydroxylamine, and salyrganic acid) also cause a selective decrease in the amplitude of the post-tetanic hyperpolarization.[28,30]

Cardiac glycosides also interfere specifically with the production of the post-tetanic hyperpolarization. Thus the large post-tetanic hyperpolarization that follows stimulation of the nonmyelinated fibers of the rabbit cervical vagus in isethionate-Locke solution [Fig. 2(a)] is rapidly abolished when ouabain is added at the peak of the post-tetanic response [Fig. 2(b)], the membrane potential rapidly returning to near its original value; subsequently [Fig. 2(c)] it is impossible to elicit a post-tetanic hyperpolarization.

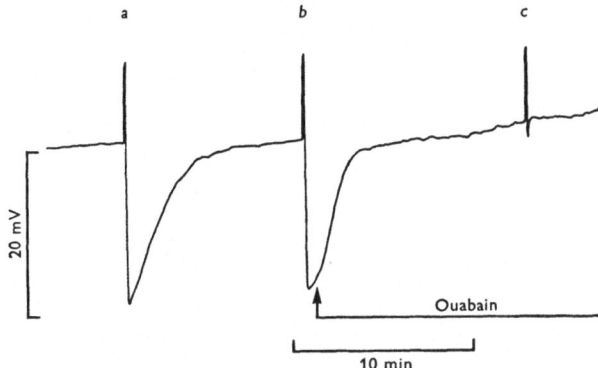

Fig. 2. The effect of ouabain on the post-tetanic hyperpolarization of the nonmyelinated fibers of rabbit desheathed vagus
nerve in isethionate-Locke solution. The records were made on
a potentiometric pen recorder so that the action potentials (the
upward deflexion) were lost; the slower post-tetanic hyperpolarization (downwards deflexion) was, however, faithfully
recorded. Note that the time base is much slower than in Fig. 1.
(From Rang and Ritchie.[32])

As is the case with antimycin A, the action potential is unaffected by this
procedure, provided the fiber is not run down by excessive stimulation.

The results with antimycin A and the other metabolic inhibitors, and
with ouabain, illustrate that the post-tetanic hyperpolarization, reflecting
the rate of active extrusion of sodium ions after activity, is more sensitive
to changes in the metabolic state of the fibers than is the compound action
potential, the latter being merely a reflection of the ionic concentrations
inside and outside the nerve axon that exist at the time of the action potential.
A similar conclusion was reached from a study of the effect on the post-
tetanic hyperpolarization of removing glucose from the solution perfusing
the nerve bundles. Thus, Larrabee and Bronk[17] found that nonmyelinated
fibers are quite sensitive to glucose withdrawal, and that at higher frequencies
of stimulation the action potential falls to 50% of its initial value in about
20 min. Greengard and Straub[31] similarly found that the post-tetanic
hyperpolarization following stimulation of the rabbit cervical sympathetic
nonmyelinated fibers is extremely sensitive to glucose withdrawal. A pharmacological procedure, equivalent to lowering the glucose concentration in the
bathing medium, is the addition of deoxy-D-glucose, which is a competitive
inhibitor of glucose metabolism and which thus markedly reduces the utilization of glucose by the fibers. Recent experiments have shown that this
inhibitor markedly reduces the size of (and slows) the post-tetanic hyperpolarization.[45,46] In keeping with the competitive nature of the inhibition,
less deoxy-D-glucose is required to affect metabolism in glucose-free media

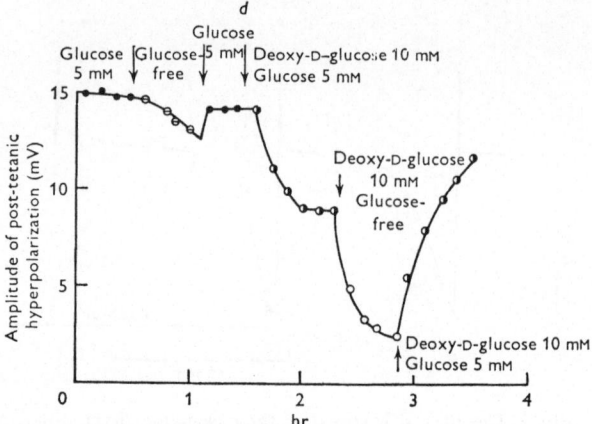

Fig. 3. The effect of deoxy-D-glucose on the amplitude of the post-tetanic hyperpolarization of the nonmyelinated fibers of a rabbit desheathed vagus nerve. The Locke solution contained: ●, glucose 5 mM; ⊖, no glucose; ◐, deoxy-D-glucose 10 mM and glucose 5 mM; ○, deoxy-D-glucose 10 mM in the absence of glucose. *d* marks the beginning of exposure to deoxy-D-glucose. (From den Hertog, Greengard, and Ritchie.[45])

than in Locke solution containing glucose. Also in keeping with the competitive nature, the inhibition can be reversed by glucose itself, as is illustrated in Fig. 3. It can also be reversed by fructose (Fig. 4), which means that these fibers possess fructokinase activity, and by pyruvate and acetate.

These results with inhibitors and with substrates are thus in accordance with what would be expected on the basis of the conventional picture that

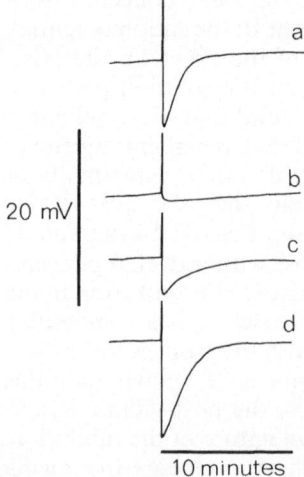

Fig. 4. Records of the post-tetanic hyperpolarization (5 sec, 30 shocks/sec) of the nonmyelinated fibers of a rabbit desheathed cervical vagus nerve in isethionate-Locke solution (recording conditions as in Fig. 2).

Record (a) was obtained from nerves in normal Locke solution. All subsequent records were obtained from nerves that were bathed in Locke solution in which the glucose normally present was removed and replaced by 1 mM deoxy-D-glucose; these deoxy-D-glucose solutions also contained: (b), nothing added; (c) 10 mM fructose; (d) 10 mM glucose.

glucose is metabolized through the glycolytic pathway to pyruvate, which is then oxidized by the tricarboxylic acid cycle. However, in sympathetic non-myelinated fibers, malonate, which is a Krebs cycle inhibitor, rather than decreasing the post-tetanic hyperpolarization actually increases it. Malonate is known to inhibit succinic dehydrogenase and, thus, prevent the formation of oxaloacetate. Greengard and Straub[31] therefore suggested the following possibility: that the acetyl-CoA of these nerve fibers is used to acetylate some compound X (reaction 1) to form acetyl X, which is rate limiting for the extrusion of sodium ions; and that the "condensing enzyme," which catalyzes the reaction acetyl CoA + oxaloacetate to form citrate (reaction 2), reduces the rate of formation of the acetyl X by removing acetyl CoA. The two equations are:

(1) Acetyl CoA + X → acetyl X + Co-A

(2) Acetyl Co-A + oxaloacetate → citrate + Co-A

According to this hypothesis, malonate acts by decreasing the amount of oxaloacetate formed so that more acetyl-Co-A is available for reaction 1. If this idea is correct, oxaloacetate, a normal metabolic intermediate, should cause a decrease in the post-tetanic hyperpolarization, which is indeed found to be the case.[31,45]

These experiments thus suggested to Greengard and Straub[31] that the most immediate demand for glucose of these nonmyelinated fibers may not be as a fuel supply but as a precursor of acetylated compounds. Some experiments of Larrabee and Horowicz[11] on the sympathetic ganglion are of particular interest here. They obtained evidence that the rate of energy production in the ganglion is not reduced enough by withdrawal of glucose to account for failure of transmission, and they suggested that some product of glucose metabolism might be essential to the ganglion cell. However, recent experiments with rabbit cervical vagus nonmyelinated fibers cast some doubt on the hypothesis just discussed.[45] For in these nerves malonate does not increase the post-tetanic hyperpolarization; rather it produces the depression that might have been expected of it as an inhibitor of the Krebs tricarboxylic acid cycle.

Whatever the fate of the glucose might be, these experiments show the very high dependence of nonmyelinated fibers on an adequate glucose supply. This dependence can again be explained in terms of the very high surface to volume ratio of these fibers. For whatever the extrusion method may prove to be by which metallic ions are actively transported through the lipoprotein membrane, the amount of chemical work that has to be done per unit volume of nerve fibers remains, presumably, proportional to the surface to volume ratio and is thus inversely proportional to the fiber diameter. These considerations might well explain why the brain is particularly sensitive to lack of glucose and of oxygen; for in the brain there is a high density of dendritic processes and axon terminals in the synapses, all of which are extremely fine structures and would thus have a very high surface to volume ratio.

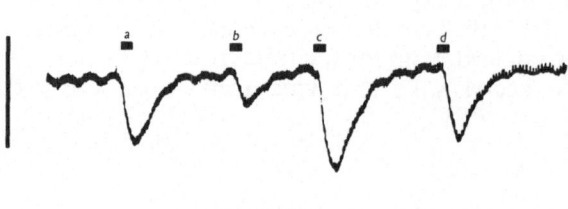

3. Oxygen Consumption

The oxygen consumption of mammalian nonmyelinated fibers is particularly large, again because of their small size. Thus, when the nonmyelinated fibers of a rabbit desheathed vagus nerve contained in a fine capillary tube through which Locke solution flows are stimulated, there is a substantial fall in the oxygen concentration of the effluent (Fig. 5). The results obtained by Ritchie[18] indicate that after stimulation the extra oxygen consumption rises rapidly to the maximum value from which it decays more or less exponentially with time. Clearly the area under the oxygen concentration/time curve, times the flow rate, is directly proportional to the oxygen consumption of the nerve in response to the period of stimulation. As might be expected, procedures that interfere with the sodium pump interfere with this extra oxygen consumption without affecting the action potential. Thus the extra oxygen consumption on stimulation in ordinary Locke solution is abolished by the addition of ouabain to the Locke solution, by replacing the sodium in Locke solution by lithium, or by adding an agent such as salicylate that uncouples oxidative phosphorylation.[18]

Rang and Ritchie[32] showed that the electrogenic pump in mammalian nonmyelinated fibers depends on the presence externally of potassium, or some similar cation such as cesium or rubidium. In keeping with this, the extra oxygen consumption on stimulation similarly depends on the presence of an external cation such as potassium, cesium, or rubidium.[33]

The free energy change involved in the operation of a 1:1 coupled sodium:potassium pump is given by

$$RT\left(\ln \frac{[\text{Na}]_o}{[\text{Na}]_i} + \ln \frac{[\text{K}]_i}{[\text{K}]_o}\right) \text{cal/mole}, \qquad \text{where } [\ ]_o \text{ and } [\ ]_i$$

denote the external and internal concentrations respectively, R is the gas constant, and T is the absolute temperature. Rang and Ritchie[47] found that for C fibers in 5.6 mM potassium-Locke solution $[K]_i = 165$ mM and $[Na]_i = 76.6$ mM. This means that the free energy change is about 2400 cal/mole at 21°C. Taking the amount of sodium and potassium to be transported after a stimulus as 6000 pmole/g,[21] the free energy change becomes 14.5 μcal/g shock; the energy released by the utilization of the approximately 1000 pmole oxygen per gram following each shock,[18] i.e., over 100 μcal/g shock, thus seems quite adequate for active transport of cations.

4. Heat Production

A high degree of metabolic activity of mammalian nonmyelinated fibers is also suggested by experiments on the recovery heat production that follows a period of repetitive stimulation. The thermal changes that accompany electrical activity in nerve fibers are extremely small; however, they are worth studying because they may give some insight into the processes that control the ionic fluxes that generate the action potential. The equipment that is used consists of a very sensitive galvonometer with a photoelectric feedback amplifier similar to that described in detail by Hill,[48] and a thermopile that gives an output of about 5000 μV/°C. Recent experiments by Keynes and Ritchie[49] and by Howarth, Keynes, and Ritchie[35] on nonmyelinated fibers of the rabbit vagus nerve have shown that during the passage of a single impulse there is an initial heat production of about 7 μcal/g impulse of which about 4.9 μcal/g impulse is almost immediately reabsorbed to give a net heat per impulse of about 2 μcal/g. At 20°C this initial heat is followed by a period of at least 5 min after the end of stimulation during which recovery heat is evolved. The recovery heat production can be seen after a single impulse; but it is best studied after a train of impulses in the nerve. Thus, 20 impulses at a frequency of 2.5/sec gives records of the kind shown in Fig. 6. Correcting for the characteristics of the thermopile, one can calculate from such records that the recovery heat production is about 93 μcal/g impulse. Stimulation of the same nonmyelinated fibers of the rabbit cervical vagus nerve at 2.5 shocks/sec causes an extra oxygen consumption of about 900 pmole/g shock[18]; taking the utilization of 1 liter of oxygen to be associated with the production of 5000 cal, so that 1 pmole is equivalent to 0.112 μcal, one would predict from the oxygen experiments that the recovery heat production would be 100 μcal/g shock, which is in excellent agreement with the directly determined value of 93 μcal/g shock.

As can be seen in Fig. 6, the recovery heat production was reduced or abolished by ouabain, and also when the sodium of Locke solution was replaced by lithium. Again, this is in excellent agreement with the electrical and oxygen consumption experiments; for both ouabain and lithium, which are known to impair the activity of the sodium pump in nonmyelinated fibers,[28] abolish the post-tetanic hyperpolarization and the phase of extra oxygen consumption of these nerves associated with stimulation.[18,33]

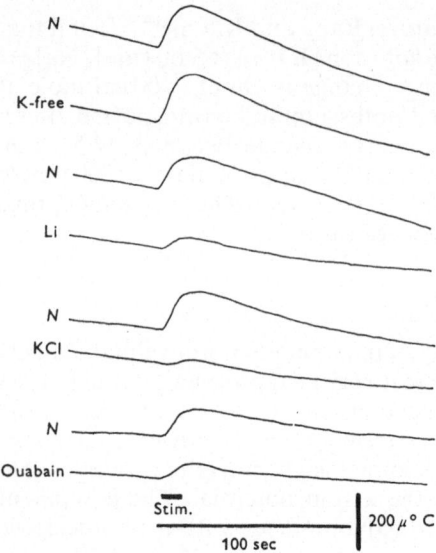

Fig. 6. The total heat production (i.e., net initial heat plus recovery heat) in rabbit desheathed vagal nonmyelinated nerves at 20.8°C after 20 shocks at intervals of 400 msec. The period of stimulation is indicated by the short horizontal bar. The records labeled N were taken in ordinary Locke solution. Interpolated between these records are the responses obtained after the nerve had been equilibrated in a potassium-free Locke solution (K-free); a solution in which all the sodium chloride had been replaced by lithium chloride (Li); a Locke solution in which all the sodium chloride had been replaced by potassium chloride (KCl); and a Locke solution containing 1 mM ouabain. (From Howarth, Keynes, and Ritchie.[35])

D. Density of Pumping Sites

The bulk of the metabolism of nerve axons and quite possibly of all nervous tissue is concerned with maintaining the ionic imbalance between the axoplasm and the external environment. The evidence is now strong that the mechanism involved, the sodium pump, is closely related to the sodium- and potassium-dependent ATPase that is found in membranes, one important piece of evidence being that cardiac glycosides specifically inhibit both the ATPase and the pump. The recent availability of radioactively labeled glycoside has enabled one to further characterize the transport site by allowing the kinetics of uptake and binding of ouabain by tissue to be readily determined. Hoffman and Ingram,[50] Hoffman,[51] and Ellory and Keynes[52] have done this for human red cell ghosts. Recently, Landowne and Ritchie[53] have examined the binding of labeled ouabain to mammalian nonmyelinated nerve fibers. Such studies have suggested that the site density in the nonmyelinated nerve fibers of the rabbit cervical vagus nerve is about 750 per square micron.[53] The density of sites in mammalian nonmyelinated fibers is thus very much higher than that in red blood cells where there is only about one site per square micron. This difference may well be related to the difference in requirements in the two tissues for cationic pumping. It will clearly be of interest to compare the value in mammalian nonmyelinated fibers with those obtained in other nonmyelinated nerves, particularly the squid giant axon where it is known[54] that the number of the tetrodotoxin binding sites, which indicates the number of sodium permeability channels, is fewer than 13 per square micron.

It is interesting to note that although mammalian fibers seem to differ from red blood cells in respect to the density of ouabain binding sites, they do resemble many of the tissues studied by Baker and Willis,[55] who found that in a variety of tissues, such as HeLa cells and skeletal muscle, there are similarly about 500 ouabain binding sites per square micron.

A critical point in comparing the pumping sites in different tissues is how fast they can work. In other words: what is the greatest number of sodium ions that can be extruded by an individual site in unit time? In red blood cells Ellory and Keynes[52] obtained a value of 175 sec^{-1} for the ATPase catalytic center activity (calculated as the ratio (ATPase activity)/ ouabain binding). Direct measurement of the ratio (sodium efflux)/(ouabain binding) in red blood cells at 37° gives a value of 100 sec^{-1} [50] to 150 sec^{-1}.[51] The corresponding value obtained by Landowne and Ritchie[53] in mammalian nonmyelinated fibers at 20°C was 22 sec^{-1}. If this latter value is to be compared with those for red blood cells, it ought probably to be at least doubled because of the difference in temperature, for den Hertog and Ritchie[46] have shown that the rate constant for extrusion of sodium from C fibers is more than doubled when the temperature is increased from 20° to 37°C. When this is done it is clear that the ratio of sodium efflux to ouabain binding in mammalian C fibers is of the same order of size as that obtained for red blood cells. Thus it seems that the same sort of sodium pumping site is involved in the different tissues.

IV. REFERENCES

1. H. McIlwain, *Biochemistry and the Central Nervous System*, 2nd ed., Churchill, London (1959).
2. H. McIlwain, *Chemical Exploration of the Brain* (A study of cerebral excitability and ion movement), Elsevier, Amsterdam (1963).
3. P. C. Caldwell, The phosphorus metabolism of squid axons and its relationship to the active transport of sodium, *J. Physiol.* **152**:545–560 (1960).
4. P. C. Caldwell, A. L. Hodgkin, R. D. Keynes, and T. I. Shaw, The effects of injecting "energy-rich" phosphate compounds on the active transport of ions in the giant axons of *Loligo, J. Physiol.* **152**:561–590 (1960).
5. P. C. Caldwell, A. L. Hodgkin, R. D. Keynes, and T. I. Shaw, The rate of formation and turnover of phosphorus compounds in squid giant axons, *J. Physiol.* **171**:119–131 (1964).
6. A. L. Hodgkin and R. D. Keynes, Active transport of cations in giant axons from *Sepia* and *Loligo, J. Physiol.* **128**:28–60 (1955).
7. A. L. Hodgkin and R. D. Keynes, Experiments on the injection of substances into squid giant axons by means of a microsyringe, *J. Physiol.* **131**:592–616 (1956).
8. S. S. Kety and J. Elkes, eds. *Regional Neurochemistry*, Pergamon Press, New York (1961).
9. S. S. Kety, The general metabolism of the brain in *vivo, in Metabolism of the Nervous System* (D. Richter, ed.), pp. 221–237, Pergamon Press, New York (1957).
10. D. Richter, Brain metabolism and cerebral function, *in Metabolism and Function in Nervous Tissue* (R. T. Williams, ed.), pp. 62–76, Cambridge University Press, Cambridge (1952).
11. M. G. Larrabee and P. Horowicz, Glucose and oxygen utilization in sympathetic ganglia. I. Effects of anesthetics. II. Substrates for oxidation at rest and in activity, *in Molecular Structure and Functional Activity of Nerve Cells* (R. G. Grenell and L. J. Mullins, eds.), pp. 84–122, Publication No. 1 of American Institute of Biological Sciences, Washington (1956).

12. M. G. Larrabee, P. Horowicz, W. Stekiel, and M. Dolivo, Metabolism in relation to function in mammalian sympathetic ganglia, in *Metabolism of the Nervous System* (D. Richter, ed.), pp. 208–220, Pergamon Press, New York (1957).

13. A. den Hertog and J. M. Ritchie, The effect of some quaternary compounds and local anesthetics on the electrogenic component of the sodium pump in mammalian non-myelinated nerve fibers, *Eur. J. Pharmac.* **6**:138–142 (1969).

14. A. L. Hodgkin, The ionic basis of electrical activity in nerve and muscle, *Biol. Rev.* **26**:339–409 (1951).

15. A. L. Hodgkin, *The Conduction of the Nervous Impulse*, Charles C. Thomas, Illinois (1964).

16. R. D. Keynes, The energy source for active transport in nerve and muscle, in *Membrane Transport and Metabolism* (A. Kleinzeller and A. Kotyk, eds.), pp. 131–139, Academic Press, New York (1961).

17. M. G. Larrabee and D. W. Bronk, Metabolic requirements of sympathetic neurons, *Cold Spring Harbor Symp. Quant. Biol.* **17**: 245:266 (1952).

18. J. M. Ritchie, The oxygen consumption of mammalian nonmyelinated nerve fibers at rest and during activity, *J. Physiol.* **188**:309–329 (1967).

19. W. A. H. Rushton, A theory of the effects of fibre size in medullated nerve, *J. Physiol.* **115**:101–122 (1951).

20. B. C. Abbott, A. V. Hill, and J. V. Howarth, The positive and negative heat production associated with a single impulse, *Proc. R. Soc. B.* **148**:149–187 (1958).

21. R. D. Keynes and J. M. Ritchie, The movements of labeled ions in mammalian non-myelinated nerve fibers, *J. Physiol.* **179**:333–367 (1965a).

22. F. Brink, D. W. Bronk, F. D. Carlson, and C. M. Connelly, The oxygen uptake of active axons, *Cold Spring Harbor Symp. Quant. Biol.* **17**:53–67 (1952).

23. F. Brink, Nerve metabolism, in *Metabolism of the Nervous System* (D. Richter, ed.), pp. 187–207, Pergamon Press, New York (1957).

24. R. W. Gerard and N. Tupikova, Nerve and muscle phosphates, *J. Cell. Comp. Physiol.* **13**:1–13 (1939).

25. P. Greengard and R. W. Straub, Effect of frequency of electrical stimulation on the concentration of intermediary metabolites in mammalian nonmyelinated fibers, *J. Physiol.* **148**:353–361 (1959).

26. P. Montant and M. Chmouliovsky, Energy-rich metabolites in stimulated mammalian nonmyelinated nerve fibers, *Experientia* **24**:782–783 (1968).

27. M. Chmouliovsky, M. Schorderet, and R. W. Straub, Effect of electrical activity on the concentration of phosphorylated metabolites and inorganic phosphate in mammalian nonmyelinated nerve fibers, *J. Physiol.* **202**:90–92 (1969).

28. J. M. Ritchie and R. W. Straub, The hyperpolarization which follows activity in mammalian nonmedullated fibers, *J. Physiol.* **136**:80–97 (1957).

29. J. M. Ritchie, Possible mechanism underlying production of after-potential in nerve fibers, in *Biophysics of Physiological and Pharmacological Actions*, pp. 165–182, American Association for the Advancement of Science (1961).

30. O. Holmes, Effects of pH, changes in potassium concentration and metabolic inhibitors on the after-potentials of mammalian nonmedullated nerve fibers, *Arch. Int. Physiol.* **70**:211–245 (1962).

31. P. Greengard and R. W. Straub, Metabolic studies on the hyperpolarization following activity in mammalian nonmyelinated nerve fibers, *J. Physiol.* **161**:414–423 (1962).

32. H. P. Rang and J. M. Ritchie, On the electrogenic sodium pump in mammalian non-myelinated nerve fibers and its activation by various external cations, *J. Physiol.* **196**:183–221 (1968b).

33. H. P. Rang and J. M. Ritchie, The dependence on external cations of the oxygen consumption of mammalian nonmyelinated fibers at rest and during activity, *J. Physiol.* **196**:163–181 (1968a).

34. J. V. Howarth, R. D. Keynes, and J. M. Ritchie, The heat production of mammalian non-myelinated (C) nerve fibers, *J. Physiol.* **186**:60–62 (1966).
35. J. V. Howarth, R. D. Keynes, and J. M. Ritchie, The origin of the initial heat associated with a single impulse in mammalian nonmyelinated nerve fibers, *J. Physiol.* **194**: 745–793 (1968).
36. P. Greengard, Determination of intermediary metabolites by enzymic fluorimetry, *Nature* (*London*), **178**:632–634 (1956).
37. P. Greengard, Adenosin-5'-triphosphat. Bestimmung durch enzymatische fluorimetrie, in *Methoden der Enzymatischen Analyse* (H. U. Bergmeyer, ed.), 551–558, Verlag Chemie (1962).
38. F. Folkow, Impulse frequency in sympathetic vasomotor fibers correlated to the release and elimination of the transmitter, *Acta Physiol. Scand.* **25**:49–76 (1952).
39. W. W. Douglas and J. M. Ritchie, On the frequency of firing of mammalian nonmedullated nerve fibers, *J. Physiol.* **139**:400–407 (1957).
40. R. A. Alberty, Standard Gibbs free energy, enthalpy, and entropy changes as a function of pH and pMg for several reactions involving adenosine phosphates, *J. Biol. Chem.* **244**: 3290–3302 (1969).
41. S. Nakajima and K. Takahashi, Posttetanic hyperpolarization and electrogenic Na pump in stretch receptor neurone of crayfish, *J. Physiol.* **187**:105–127 (1966).
42. R. W. Straub, On the mechanism of posttetanic hyperpolarization in myelinated nerve fibers from the frog, *J. Physiol.* **159**:19–20 (1961).
43. C. M. Connelly, Recovery processes and metabolism in nerve, *Rev. Mod. Phys.* **31**:475–484 (1959).
44. C. M. Connelly, Metabolic and electrochemical events associated with recovery from activity, *Proc. XXII Int. Cong. Physiol. Lectures and Symposia* **1**:600–602 (1962).
45. A. den Hertog, P. Greengard, and J. M. Ritchie, On the metabolic basis of nervous activity, *J. Physiol.* **204**:511–521 (1969).
46. A. den Hertog and J. M. Ritchie, A comparison of the effect of temperature, metabolic inhibitors, and of ouabain on the electrogenic component of the sodium pump in mammalian nonmyelinated nerve fibers, *J. Physiol.* **204**: 523–538 (1969).
47. H. P. Rang and J. M. Ritchie, The ionic content of mammalian nonmyelinated nerve fibers and its alteration as a result of electrical activity, *J. Physiol.* **196**:223–236 (1968).
48. A. V. Hill, Moving-coil galvanometers of short period and their amplification, *J. Scient. Instrum.* **25**:225–229 (1948).
49. R. D. Keynes and J. M. Ritchie, The thermodynamics of nerve and electric organ, in *Nerve as a Tissue*, pp. 293–304, Harper and Row, New York (1965b).
50. J. F. Hoffman and C. J. Ingram, Cation transport and the binding of T-ouabain to intact human red cells, *Proc. 1st int. Symp. on Metabolism and Permeability of Erythrocytes and Thrombocytes*, Thieme, Stuttgart (1969).
51. J. F. Hoffman, The interaction between tritiated ouabain and the Na-K pump in red blood cells, *J. Gen. Physiol.* **54**:343–350 (1969).
52. J. C. Ellory and R. D. Keynes, Binding of tritiated digoxin to human red cell ghosts, *Nature* **211**:776 (1969).
53. D. Landowne and J. M. Ritchie, The binding of tritiated ouabain to mammalian nonmyelinated nerve fibers, *J. Physiol.* **207**: 529–537 (1970).
54. J. W. Moore, T. Narahashi, and T. I. Shaw, An upper limit to the number of sodium channels in nerve membrane, *J. Physiol.* **188**:99–105 (1967).
55. P. F. Baker and J. S. Willis, On the number of sodium pumping sites in cell membranes, *Biochim. Biophys. Acta* **183**:646–649 (1969).

Chapter 11

THE BIOCHEMISTRY OF SINGLE NERVE CELL BODIES*

M. M. Brand† and G. M. Lehrer

The Division of Neurochemistry
Department of Neurology
The Mount Sinai School of Medicine
New York, New York

I. INTRODUCTION

There is no question that functional specificity in the brain is largely centered in the neuron and its processes. It is also highly probable that the structural and functional informational content of the brain resides mainly in specific chemical differences among cells. There is a considerable body of evidence that metabolic, chemical, and immunological differences exist among different neuronal types within any individual. Even superficial observations of differential susceptibility to viral invasion or to toxic or anoxic damage force one to assume such specificity. Except for rudimentary knowledge of differences in the nature of the chemical transmitters produced and the enzyme systems concerned with their metabolism, relatively little is known about the chemistry of neuronal specialization.

Because brain metabolism represents a summation of the activities of component cells, agglomerated in heterogeneous collections and intricately intertwined, studies of whole brain, *in vivo*, in slices or in homogenates, or even of macroscopic domains of the brain can only yield information on the overall capacities or incapacities of all contained components. If we are to ask the question, "What in the chemistry and metabolism of the neuron accounts for its special characteristic functions?", then we must try to sort out the chemistry of the neuron isolated from all other cell types.

It will be the task of this chapter to present a critical overview of what is known about the chemistry of isolated vertebrate neurons. Because this

* Supported by the National Institute of Neurological Diseases and Stroke Grants No. NB05368 and NB05596, National Institute of Mental Health Grant No. MH17592, and by the Benjamin Miller Memorial Grant No. 293 for Research on Multiple Sclerosis, National Multiple Sclerosis Society.
† Dr. Michael M. Brand is a Special Fellow of the National Institute of Neurological Diseases and Stroke.

field has recently been reviewed in great detail,[I,II,III] this chapter will take the liberty of presenting a narrower, and perhaps somewhat biased view. The monumental work of Hydén and his co-workers on isolated, living nerve cell bodies has also been the subject of recent reviews[IV,V] and consequently receives somewhat less attention here.

While the study of the nerve cell body may give us a metabolic view of those processes which are centered in and about the nucleus as they are affected by functional demands made upon them, it should be noted that moment to moment, rapid neural responses probably involve dendrites, endings and axons much more than the perikaryon. Therefore the study of the cell body may give relatively few clues to metabolic capacity as related to synaptic transmission and conduction.

Much of the metabolic machinery of the brain and the neuron is intimately dependent upon the immediate supply of oxygen and metabolites by an intact circulatory system. Consequently, studies of supravitally isolated preparations, be they slices, single cells or homogenates, must always be interpreted in the light of changes induced by the conditions of the study. Ultrarapid freezing of tissues in accurately defined physiologic states may provide precise information about the metabolic states of the intact system at the time of freezing. Ischemic anoxia can convert a tissue into a closed system, entirely dependent upon endogenous substrates for its energy supply. Measurements of metabolite levels after determined intervals of such ischemic anoxia allow extrapolation to steady-state conditions. We are indebted to O. H. Lowry[VI] for the development of these concepts and for the exquisite refinement of the methodology originated by Kaj Linderstrøm-Lang.

II. COMPOSITION OF NEURONS

A. Volume, Mass, and Density

Determinations of the volume, mass, and density of single nerve cell bodies have been obtained in a variety of neurons from several species (Table I). Accurate measurement of the volume of a single cell depends in part on the manner by which the tissue is initially prepared. During fixation tissue may shrink or swell and the cell morphology may be otherwise altered. As cell volume estimates are based largely on planimetric analyses of the cell profile,[1-3] an alteration of cell morphology related to fixation may affect the reliability of results. It follows that adequate controls for dimensional stability must be devised if measures are to reflect *in vivo* conditions. Cell mass has been determined by soft X-ray absorption and densitometry on a photographic plate (X-ray absorption radiomicrography) and comparison with known standards.[4] A scanning and computing cell analyzer capable of analyzing and calculating the X-ray absorption at 12,000 separate cellular loci was developed later.[5,6] This technique showed that cell mass is proportional to cell body volume. Although the average cell density

TABLE I
Volume, Mass, Density, and Content of Protein, Lipids and RNA in Single Neurons

Site	Species	Volume (pl)[a]	Mass (ng)[b]	Density (kg_{DW}/liter)[c]	Protein (ng/cell)	Protein (% of DW)[d]	Lipid (ng/cell)	Lipid (% of DW)	RNA (ng/cell)	RNA (% RNA, w/v)[e]	RNA (% of DW)	Refs.
Deiters' neuron cell body	Rabbit	88–93	20.0– 21.7	0.22– 0.235 0.23–	17.2	79[f]	4.4[g]	17–20[g]	1.54 1.1		7[f] 5.2–[b]	6, 7, 17
cytoplasm			20.8	0.29	16.6	79[f]	4.2	20[f]	1.3		6.2[f]	6, 7, 17
nucleoplasm			0.8	0.23	0.65	81[f]	0.2	25[f]	0.20 0.40		3.7[f] 7	7, 17 18, 19
nucleolus			0.1	0.38					0.56	1.5		7
cell body	Rat	50							0.683			95, 139, 108
nucleus				0.20					0.030			6, 108
Hypoglossal neuron cell body	Rabbit	13	2.7	0.20	2.2	80[f]	0.550[2]	20[f]	0.200	1.5	7.4[f]	7
axon				0.06					0.161			7, 140
Anterior horn cell	Rabbit	22–24							0.555 0.540	2.4 2.5		16, 22
	Man								0.402[h] 0.553 0.640[i] 0.504[j]			43
	Barracuda Cat								3.244			108

TABLE I (continued)

Volume, Mass, Density, and Content of Protein, Lipids and RNA in Single Neurons

Site	Species	Volume (pl)[a]	Mass (ng)[b]	Density (kg_DW/liter)[c]	Protein (ng/cell)	Protein (% of DW)[d]	Lipid (ng/cell)	Lipid (% of DW)	RNA (ng/cell)	RNA (% RNA, w/v)[e]	RNA (% of DW)	Refs.
Anterior horn cell (*cont.*) cytoplasm										1.7[c] 3.7[g]		8, 110
nucleus										0.9[c]		
nucleolus										1.6[c]		
	Frog	11–19		0.205					0.265–0.446	1.9– 3.5		2, 8
Dorsal root ganglion cell	Rabbit	180 (25–500)	13–24	0.23					1.070			3, 6, 20
	Chick embryo							19–28[k]	0.017			141
Sympathetic ganglion cell (cervical) cytoplasm	Cat	5.650	1.080	0.20[m]					0.165			
nucleus		0.590	0.125	0.20[m]							15	21
Cochlear ganglion cell	Guinea pig								0.026–0.036			142
Mauthner neuron cell body	Goldfish[n]	42							2.0	1.0		10, 11
axon									8.0	0.05		

Cortical neuron	Cat	(11–90)			
	Rat		0.022		143 144
Globus pallidus	Man		0.116		108, 145, 146
Reticular neuron	Rat	3.3–34 33–270	0.544		147 130
Purkinje cell	Rabbit	1–5			67
Spiral ganglion	Guinea pig	3	0.400	0.033–	8.2[f] 148
Mesencephalic V neuron	Mouse		0.230		140
Supraoptic neuron	Rat		0.071		136

[a] pl, picaliters = 10^{-12} liters.
[b] ng, nanograms = 10^{-9} grams.
[c] Kilograms dry weight per liter.
[d] Percent of dry weight.
[e] Weight/volume × 100.
[f] Recalculated from authors original data.
[g] Chloroform-extractable lipid.
[h] 0–40 years.
[i] 41–60 years.
[j] 61–80 years.
[k] Hexane and ethanol-extractable lipid.
[l] Negligible contribution from capsule cells.
[m] Fixed, paraffin sectioned tissue.
[n] Carassius auratus.

obtained by these techniques was reported as essentially constant throughout the Deiters' nerve cell body,[7] apparent differences in density between subcellular components such as the nucleolus and the nucleoplasm[7] and the perikaryon and the axon[5] were also reported. It is likely that such differences in density as found between the nucleus and the cytoplasm in the anterior horn cells of the cat[8] and in the spinal ganglion and Deiters' neurons of the rabbit[6] may be largely artifactual, being related to differential shrinkage consequent to enbloc freeze-drying, fixation, and embedding. Hydén and Larsson[6] note that it is inadvisable to use fixed material for gradient determinations or for distribution studies in single neurons. In frozen-dried material Hydén and Pigon[7] found total dry weight per unit volume of 0.23–0.29 kg/liter for Deiters' neurons and about 0.20 kg/liter for surrounding "glial cells." The authors use the term "glial cells" for a ball of cellular material stripped from the surrounds of a Deiters' cell. It is likely that this material also contains a large proportion of synaptic components and the term "glial" is therefore misleading. Our own data on relative densities of cytoplasm and nucleus of supramedullary neurons of puffer fish have been obtained under conditions where volume and mass determinations were performed with internal standards under controlled conditions which assured accuracy within a 2% limit.[9] It is only when such standards can be used that absolute values of density may be meaningful. However, relative values from any one section, obtained under defined conditions may be useful. The appreciably greater densities of the perikaryon as compared with the axon reported for both the rabbit hypoglossal neuron[5] and the Mauthner neuron of the goldfish[10,11] are probably not artifactual.

B. Nucleic Acids

RNA has been studied in single nerve cell bodies and their surrounding neuroglia in various locations in several experimental animals (Table I). A variety of analytical methods have been employed in the analysis of the RNA content of single neurons. The earlier techniques included ultraviolet microspectrophotometry[12] and ultraviolet and X-ray microradiography[8] following Carnoy fixation, or freeze drying and paraffin embedding. The analytical method of Edström[13–16] allows RNA measurement from single isolated neurons after extraction with ribonuclease either on a cellophane strip or in a lens-shaped drop using microspectrophotometric interpretation. The larger Deiters' neuron of the rabbit shows the highest content of RNA.[15]

From measurements from freshly dissected cell bodies, the RNA content, computed originally on the basis of percent of dry protein or as percent of total dry cell weight (Table I) has been relatively uniform. In the Deiters' and hypoglossal neurons of the rabbit these values range from 7.0 to 7.5% of dry mass.[7,17,18] An equal volume and presumably the same dry weight of "neuroglial cells" were reported to have one-tenth the RNA content of adjacent Deiters' neurons.[19] In spinal ganglion cells of the rabbit, the RNA ranged from 4.5% to 8% of the total dry weight.[3,20] More recently, Pevzner[21] using ultraviolet absorption and interference measurement on fixed tissue found the RNA of the nonstimulated cat superior cervical sympathetic neuron to be approximately 13% of the dry cell weight (1205 pg). The average concentration of RNA in cells, recalculated from previous data Table I), is approximately 16 g/liter in the Deiters' neuron, 15 g/liter in the hypoglossal neuron and 5.9 g/liter in spinal ganglion cell of the rabbit. Values reported earlier for the RNA content of anterior horn cells in the rabbit (2 to 3% w/v freeze-dried; 1.7% w/v Carnoy fixed)[8] and the cat (3% w/v freeze-dried; 1.7% w/v Carnoy fixed)[8] may be excessively high due to underestimation of volume because of tissue shrinkage related to paraffin embedding.[9] Edström[22] reported similar values for RNA content (2.5% w/v) in both freeze-dried and freshly fixed rabbit anterior horn cells following ribonuclease extraction.

While Hydén[23] suggested that nucleic acid concentration may vary inversely with cell volume, Edström[22] demonstrated that the RNA content of rabbit anterior horn cells is approximately proportional to surface area. Assuming these neurons have the shape of a regular tetrahedron, the RNA content was found to be 0.095 pg/μ^2 of surface area. A similar proportionality between RNA content and cell surface area was also reported for retinal ganglion cells.[24] Furthermore, as demonstrated in the rabbit hypoglossal nucleus, the anterior horn and the dorsal root ganglion, neurons appear to be arranged in groups with similar values for cell body volume, RNA content, and RNA concentration.[3,22]

RNA is found in axons and their myelin sheaths as well as in the neuronal perikaryon.[10] In the Mauthner neuron of the goldfish the total axonal RNA has been estimated at four times that of the perikaryon (8000 vs. 2000 pg).[10] However, the RNA concentration is significantly lower (1/20th) than in the cell body. There is a proximal to distal diminution in RNA content per unit length along the first two-thirds of the nerve in both the axon and the myelin sheath. However, in the distal axon the RNA concentration is higher than at the proximal end.[11] Axonal RNA with the sedimentation characteristics of ribosomal RNA has also been extracted from the cervical vagus and hypoglossal nerves of the rabbit,[25] and RNA with ribosomal base ratios was found in myelin free axons of the eleventh cranial nerve root of the cat.[26] The role of axonal RNA is yet to be established. Despite the absence of axonal ribosomes,[27] RNA may be of significance in relation to local axonal protein synthesis. Such a role has become credible in light of the demonstration[28,29] that the regeneration of acetylcholinesterase could occur independently in axons following axotomy with total disconnection from the perikaryon. The greater content of RNA in the distal axon under such conditions has been interpreted as a possible response to an increased local demand for the reconstitution of axolemmal protein; as a reflection of continued RNA migration; and as an indication of a possible role for RNA in the presynaptic terminal.[25,26] Edström, et al.[10] demonstrated that the base composition of RNA from the Mauthner axon differed strikingly from the predominantly ribosomal RNA of the cell body and suggested that such axonal RNA may play a basic template role in the local synthesis of axonal protein.[10] Jakoubek and Edström[30] reported a significant decrease in the RNA content in the proximal portion of Mauthner axon following rotational stimulation for up to 90 min.

Microelectrophoretic studies of the nucleic acid base composition in single neurons[31,32] have shown similarity in base composition between nucleolar and cytoplasmic RNA in invertebrate cells.[33–35] Electrophoresis is performed under oil at constant humidity. The carrier is a specially adapted cellulose fiber, 10 to 15 μ thick, and saturated with a highly viscous buffer in which low molecular substances readily diffuse. The separation of RNA bases takes place within 10 min and the results are evaluated microphotometrically. The base ratio composition of RNA in the rabbit Deiters' neuron cytoplasm is predominantly ribosomal and, therefore, is characterized by higher values of guanine and cytosine (62.3) than adenine and uracil (37.7). These values for nuclear RNA are (57.4) and (42.6) respectively. Therefore, the $(A + U)/(G + C)$ ratio for cytoplasmic RNA equals approximately 0.60 and for nuclear RNA equals 0.74. The relatively higher adenine and uracil content found in nuclear RNA is compatible with a more "DNA-like" base composition.[19,36]

Although the total RNA of Deiters' nerve cell body reflects an asymmetric base composition with a high $(G + C)/(A + U)$ ratio of 1.65, typical of ribosomal RNA, microfractionation of RNA with phenol in single neurons has demonstrated that at least three distinct types of RNA are present. These include cytoplasmic RNA (extracted at 0–10°C), nuclear RNA with a GC-rich base composition (extracted at 10–45°C), and nuclear RNA of messenger type with a low $(G + C)/(A + U)$ ratio of 0.85 (extracted at 50–75°C).

C. RNA Synthesis

Spinal motor neurons rapidly accumulate adenine-^{3}H without apparent inhibition by the blood–brain barrier, at a rate which exceeds incorporation by the liver, calculated on the basis of percent of the injected dose per gram of tissue.[37] The uptake of RNA precursors has been shown to be affected by changes in the metabolic state of the animal (hypo-, hyperthyroidism) or changes in a particular neuronal group (regeneration).[38] Egyhazi and Hydén[39] studied the kinetics of RNA synthesis in the lateral vestibular nucleus of the rabbit after orotic acid-^{3}H was administered by cannula into the fourth ventricle. In the nuclei, a low molecular RNA species (8 to 16S) was initially synthesized. After 15 min the sedimentation profile was characterized by larger molecules. A stable profile, typical of ribosomal RNA (18 and 28S) emerged after 180 min. After 30 min the cytoplasmic RNA showed a gradient profile similar to that seen in nuclear RNA after 15 min (radioactivity in the 10 to 12S region). Characteristic cytoplasmic ribosomal sedimentation peaks developed at 180 min. They suggest that RNA synthesis in neurons begins with the initial formation of small, nuclear RNA molecules. They propose that such molecules represent RNA of messenger type which within 15 min are transported from the nucleus to the cytoplasm.

Daneholt and Brattgård[40] observed differences between single neurons and their surrounding glia with respect to the RNA base ratio distribution pattern. They further noted that the incorporation of adenine-^{3}H and cytidine-^{3}H into RNA adenine, guanine, and uracil was about 2.5 times greater in neuroglia than in adjacent single neurons of the rabbit hypoglossal nucleus. Their studies indicated that the conversion relations between the RNA purine precursor pools and between the RNA pyrimidine precursor pools were the same in neurons and neuroglia and were of the same magnitude as in other organs. However, the turnover of RNA in glia expressed as "percent of total RNA newly synthesized" was twice that observed in neurons. They suggest that the central nervous system turns over RNA rapidly (15 to 50 % of total RNA during 4 hr) and that newly synthesized RNA is mainly localized in the nucleolus. This rate of RNA turnover in the central nervous system, therefore, appears in considerable excess of the liver in which new RNA synthesis has been reported as 25 % over 24 hr.[40–42]

In addition to short-lived, reversible fluctuations in cellular RNA content and base composition, cyclic changes in RNA may occur during the lifetime of the neuron. In human anterior horn cells studied postmortem, Hydén[43] observed that the mean cellular RNA content increased from approximately 0.40 ng at age 3–20 years, to 0.65 ng at age 40–60 years and then rapidly fell to approximately 0.42 ng after age 60. Jarlstedt[44] showed that neuronal RNA is comparatively resistant to early postmortem changes.

D. Protein

It has long been known that protein synthesis largely occurs in the neuronal perikaryon and is followed by a proximal to distal flow of proteins and protein precursors along individual axons.[45–47] Protein transported from the nerve cell body may be associated with mitochondria[48] as well as synaptic vesicles[49] and such transport may occur at different rates within similar axons. Grafstein[50] demonstrated that both rapid as well as slower protein transport occurred along optic nerve fibers and suggested that rapid labeling of synaptic endings could be related to a special cellular mechanism for the rapid transport of protein from the cell body to the synaptic terminals. Furthermore, it has been shown that the rates of protein production and transport may in part be age dependent.[51] More recently, Matheson[52] observed that the sciatic nerves of young rats incorporate glycine into their protein fraction at a higher rate than nerves from adult rats *in vitro*, and Barondes[53] showed that the specific activity of labeled soluble and particulate protein fractions isolated from nerve endings following intracerebral injection of precursors tended to be greater in younger as compared with mature mice.

Evidence that axonal protein synthesis may occur independently of or in addition to synthesis in the perikaryon has been offered more recently.[28,54–56] These authors have largely demonstrated that the local synthesis of acetylcholinesterase in the periphery of axons could proceed independently of the cell body. Koenig[55] showed that cats pretreated with the organophosphorus cholinesterase inhibitor diisopropylfluorophosphate (DFP) exhibited acetylcholinesterase activity in both intact and decentralized hypoglossal axons within 24 hr. Enzyme activity at 4 days significantly exceeded that of 1 day following DFP pretreatment and nerve transection. As enzyme restoration in decentralized axons was blocked by puromycin, 5-fluorootic acid, and ribonuclease, such acetylcholinesterase reappearance was interpreted as RNA-dependent synthesis.[54,55] Barondes[48] suggests that it is highly probable that some mitochondrial protein may be synthesized at the nerve ending, possibly by mitochondria already present at that site, possibly directed by mitochondrial DNA. Direct evidence for mammalian axonal protein synthesis has been provided by Koenig in isolated rabbit spinal accessory nerve roots which incorporate leucine-^{3}H into axonal protein *in vitro*.[57]

In the Mauthner axon of the goldfish, Edström *et al.*[10] found local synthesis of protein occurred along the entire length of the axon and the myelin sheath and suggested that both proximal to distal transfer, as well as local protein synthesis, occurred.

Ford *et al.*[58,59] observed that the uptake of intravenously administered, labeled lysine, tyrosine, and glutamine were greatest in single ventral horn cells as compared with muscle, liver, and whole spinal cord. The uptake of lysine was 25 to 30 times higher in anterior horn cells than in ventral horn tissue as a whole. Although the neuron probably is the most active central nervous system cell with regard to protein synthesis,[60] these studies by Ford and colleagues further emphasize the need for the analysis of amino acid uptake and protein synthesis in single neurons in addition to analyses in whole brain or selected topographical zones. The reason for emphasizing single cell analyses lies in the fact that as the more abundant tissue elements surrounding neurons appear to have a less active protein metabolism, the study of heterogeneous tissue may mask the relatively greater protein turnover of the neuron. Furthermore, the turnover of individual brain proteins may vary considerably[60] and a large part of the synthetic capacity of the brain may be devoted to the synthesis of rapidly metabolized acidic brain proteins.[61] Such a protein is the brain-specific S-100 protein which has been localized by macro and microelectrophoretic and immunological techniques mainly in the cytoplasm of oligodendroglial cells and possibly also in the nuclei of large neurons.[61,62] As 15 % of the radioactive-labeled protein synthesized in a cell-free system from rabbit brain can be precipitated by anti-S 100 serum,[63] a considerable part of the protein synthesizing capacity of brain may be devoted to this or similar proteins.

E. Lipids

Data relating to the lipid content of single neurons are given in Table I. Lowry and co-workers[20,64] reported a relatively low proportion of lipid in the cell body layers of Ammon's horn and the retina as well as in single dorsal root ganglion neurons in the rabbit. Cephalins, lecithins, cholesterol, and sphingomyelin accounted for approximately 90% of the total lipid. Cerebrosides were not studied. The cell body layers of Ammon's horn were poor in lipid (total lipid 22%), particularly cholesterol and sphingomyelin. However, the nonmyelinated dendritic synaptic and fiber layers contained considerable quantities of lipid, although in different proportions than in the myelinated layers. Dendritic lipid was composed of equal molar ratios of cephalin and lecithin, respectively, 246–339 mmoles/kg protein and 283–337 mmoles/kg protein. The content of sphingomyelin and presumably of cerebroside was low. Myelinated white matter contained greater quantities of hexane and chloroform-soluble lipids, particularly sphingomyelin, cholesterol, and undoubtedly cerebroside. Lowry has suggested that the high lipid content of dendrites as compared with cell bodies may be related to their greater metabolic activity. However, it must be considered that synaptic areas have much greater concentrations of lipid-rich membranes both by virtue of greatly increased cell surface to volume ratios and transmitter packaging.

Using chloroform extraction, Brattgård and Hydén[65] estimated that the mean lipid concentration of single rabbit Deiters' neurons equaled 560 g/liter, of single spinal ganglion cells 130 g/liter, and of Purkinje cells 210 g/liter. These values are obviously too high, as indicated by a total dry mass of 1040 g/liter in Deiters' cells. However, recalculation of these data as lipid/100 g dry weight would yield 54% for Deiters' cells, 31% for spinal ganglion cells, and about 23% for Purkinje cells. Compared to later data by Hydén and Larsson,[6] Lowry,[20] and Brattgård et al.[5] these values still appear too high. Hydén and Larsson[6] in both fixed and frozen tissue reported that the lipid concentration of Deiters' neurons was greatest at the dendrite, 26% of dry weight. The values for the nucleus and cytoplasm were 20% and 17%, respectively. These values are in good agreement with those of Lowry which ranged from 19–28% of dry weight (hexane and ethanol-extractable lipid). Hydén[7] observed that the mean lipid concentration of "neuroglial cells" is 70–80% compared to 20% for Deiters' cells. Such lipid levels appear excessive since they approach the levels of lipid in heavily myelinated tracts.

Miani[66] reported a primary somatic to axonal shift of phospholipids in the vagus and hypoglossal nerves of rabbits. A slight local synthesis of phospholipid in the axon could not be ruled out. Both swiftly (70 mm/day) and more slowly (40 mm/day) migrating phospholipid fractions were observed. Miani suggested that distally moving phospholipids may exchange with stationary axonal phospholipids. The rate of such exchange appeared greater in the vagus than in the hypoglossal nerve.

F. Enzymes

The activities of a variety of enzymes have been studied in single neurons and their surrounding neuroglia from a number of regions in the central and peripheral nervous system (Table II). Enzyme analyses in single cells

TABLE II

Enzyme Activities in Single Neurons and Neuroglial Cells

Metabolic pathway	Enzyme	Site	Species	Activity			Units	Ref.
				Neuron	Glia	Ratio N/G		
Glycolysis	Hexokinase	Anterior horn	Rabbit	6.3[a]			moles/kg/hr[b]	(67)
	Hexokinase	Anterior horn	Monkey	9.3			moles/kg/hr[b]	(68)
		Spinal ganglion	Rabbit	7.0[a,c]	2.2[d]	3.3	moles/kg/hr[b]	(67)
		nucleus		7.2[e]			moles/kg/hr[b]	(67)
		cytoplasm		1.39			moles/kg/hr[b]	(88)
				2.59			moles/kg/hr[b]	(88)
	Phosphogluco-isomerase	Anterior horn	Rabbit	30[a]			moles/kg/hr[b]	(88)
		Anterior horn	Rabbit	24			moles/kg/hr[b]	(149)
		Anterior horn	Monkey	27			moles/kg/hr[b]	(68)
		Spinal ganglion	Rabbit	36[a,c]	24[d]		moles/kg/hr[b]	(67)
				49[c]				
		Spinal ganglion	Rabbit	29[b]			moles/kg/hr[b]	(149)
		Spinal ganglion	Rabbit	39[g]			moles/kg/hr[b]	(71)
				34				
	Lactic dehydrogenase	Anterior horn	Rabbit	32[a]			moles/kg/hr[b]	(67)
		Anterior horn	Rabbit	21–29			moles/kg/hr[b]	(150)
		Anterior horn	Monkey	60			moles/kg/hr[b]	(68)
		Spinal ganglion	Rabbit	50[a,c]	28[d]		moles/kg/hr[b]	(67)
				60[c]				
		Spinal ganglion	Rabbit	41–57			moles/kg/hr[b]	(150)

							Units	Ref.
Pentose phosphate shunt	Glucose-6-phosphate dehydrogenase	Anterior horn	Rabbit	0.98^a			moles/kg/hr^b	(67)
			Monkey	0.42			moles/kg/hr^b	(68)
		Spinal ganglion	Rabbit	1.6^c / 2.1^e	4.8^d		moles/kg/hr^b	(67)
	6-Phosphogluconic dehydrogenase	Anterior horn	Rabbit	0.99^a			moles/kg/hr^b	(67)
		Anterior horn	Monkey	0.29			moles/kg/hr^b	(68)
		Anterior horn	Rabbit	0.65–0.99			moles/kg/hr^b	(150)
		Spinal ganglion	Rabbit	1.3^a	2.2^d		moles/kg/hr^b	(67)
		Spinal ganglion	Rabbit	0.92–1.25		0.6	moles/kg/hr^b	(150)
Krebs cycle	Isocitric dehydrogenase	Anterior horn	Rabbit	5.1^a			moles/kg/hr^b	(67)
		Anterior horn	Monkey	5.5			moles/kg/hr^b	(68)
		Spinal ganglion	Rabbit	4.5^{c,a} / 6.0^c	6.2^d	0.7	moles/kg/hr^a	(67)
	Malic dehydrogenase	Anterior horn	Rabbit	264^a			moles/kg/hr^b	(67)
		Anterior horn	Monkey	332			moles/kg/hr^b	(68)
		Spinal ganglion	Rabbit	231^{a,c} / 287^c	90.		moles/kg/hr^b	(67)
Respiratory	Cytochrome oxidase	Deiters' nucl.	Rabbit	4.2	11.5	0.4	$\mu l O_2 \cdot 10^{-4}$/hr/cell	(7, 17)
		Spinal ganglion	Rabbit	4.5	1.6^d	2.8	$\mu l O_2 \cdot 10^{-4}$/hr/cell	(96)
	Succinoxidase	Deiters' nucl.	Rabbit	2.1	4.5	0.5	$\mu l O_2 \cdot 10^{-4}$/hr/cell	(7, 17)

TABLE II (continued)
Enzyme Activities in Single Neurons and Neuroglial Cells

Metabolic pathway	Enzyme	Site	Species	Neuron	Glia	Ratio N/G	Units	Ref.
Respiratory (*cont.*)		Deiters' nucl.	Rabbit	2.2			$\mu10_2 \cdot 10^{-4}$/hr/cell	(18)
		Spinal ganglion	Rabbit	3.4	5.1^d		$\mu10_2 \cdot 10^{-4}$/hr/cell	(96)
		Spinal ganglion	Rabbit	1.6	0.5^d	0.14	$\mu10_2 \cdot 10^{-4}$/hr/cell	(96)
		Retinal ganglion cell	Rabbit	4.4			$\mu10_2 \cdot 10^{-4}$/hr/cell	(151)
Amino acid metabolism	Glutamic dehydrogenase	Anterior horn	Rabbit	3.6			moles/kg/hr	(67)
		Anterior horn	Rabbit	2.8			moles/kg/hr	(149)
		Anterior horn	Monkey	5.3			moles/kg/hr	(68)
		Spinal ganglion	Rabbit	2.5^{a,e} / 3.8^c	1.6		moles/kg/hr	(67)
		Spinal ganglion	Rabbit	1.9^g / 3.0^f			moles/kg/hr	(149)
	Glutamic-pyruvic Transaminase	Anterior horn	Rabbit	0.288			moles/kg/hr	(158)
		Spinal ganglion	Rabbit	0.227			moles/kg/hr	(158)
	Glutamic-oxaloacetic transaminase	Anterior horn	Rabbit	26			moles/kg/hr	(149)
		Anterior horn	Rabbit	32^a				(67)
			Monkey	46				(68)
		Spinal ganglion	Rabbit	28–^{g} / 31^f			moles/kg/hr	(149)
				35^c	6.5		moles/kg/hr	(67)
				39^c			moles/kg/hr	(67)
Hydrolases	Alkaline phosphatase	Anterior horn	Rabbit	80^h			mmoles/kg$_{DW}$/hr	(71)
		Spinal ganglion	Rabbit	1324			mmoles/kg$_{DW}$/hr	(71)
	α-Napthyl acid phosphatase	Anterior horn	Man	3.36–4.10			moles/kg$_{DW}$/hr	(86)

	Region	Species	Value		Units	Ref.
	Anterior horn	Monkey	1.12–1.86		moles/kg$_{DW}$/hr	(86)
β-Glucosidase	Anterior horn	Monkey	10.7		mmoles/kg$_{DW}$/hr	(72)
	Spinal ganglion	Monkey	19.2		mmoles/kg$_{DW}$/hr	(72)
β-Galactosidase	Anterior horn	Monkey	150		mmoles/kg$_{DW}$/hr	(87)
	Anterior horn	Monkey	185		mmoles/kg$_{DW}$/hr	(68)
	Anterior horn	Monkey	101		mmoles(kg$_{DW}$/hr	(72)
	Spinal ganglion	Monkey	400		mmoles/kg$_{DW}$/hr	(87)
	Spinal ganglion	Monkey	224		mmoles/kg $_{DW}$/hr	(72)
β-Glucuronidase	Anterior horn	Monkey	37		mmoles/kg$_{DW}$/hr	(68)
	Spinal ganglion	Monkey	23.4		mmoles/kg$_{DW}$/hr	(72)
	Spinal ganglion	Monkey	77		mmoles/kg$_{DW}$/hr	(72)
Phosphatidase A	Spinal ganglion	Rat	0.84		nl CO$_2$/hr/cell	(152)
Peptidase	Anterior horn	Monkey	29.2		moles/kg/hr	(68)
ATPase	Deiter's nucl.	Rabbit	1.0	1.7	pM ATP/hr/cell	(97)
Acetyl-cholinesterase	Anterior horn	Rat	1 – 7^i	0.6	nl CO$_2$/hr/cell	(82)
	Anterior horn	Rat	1.0–5.0			(81)
						(95)
	Nucleus	Rat	5.1		pl CO$_2$/hr/cell	(153)
	Spinal ganglion	Rat	0.2–2.0		nl CO$_2$/hr/cell	(95)
						(95)
	Sympathetic ganglion	Rat	0.2–3.0		nl CO$_2$/hr/cell	(95)
						(78)

TABLE II (continued)
Enzyme Activities in Single Neurons and Neuroglial Cells

Metabolic pathway	Enzyme	Site	Species	Activity			Units	Ref.
				Neuron	Glia	Ratio N/G		
Hydrolases (cont.)		Sympathetic ganglion	Frog	0.2–6.0			nl CO$_2$/hr/cell	(78)(95)
		Sympathetic ganglion	Cat	0.38–5.44			nl CO$_2$/hr/cell	(77)
		Reticular neuron	Rat	0.7–0.8			nl CO$_2$/hr/cell	(147)
		Reticular neuron	Rat	0.44 0.15			nl CO$_2$/hr/cell	(154)
		Diaphragm single end plate	Mouse	3.20–4.50			nl CO$_2$/hr/cell	(156)
		Gastrocnemius single end plate	Mouse	5.00–12.00			nl CO$_2$/hr/cell	(156)
	Butyrylcholine-sterase	Anterior horn	Rat		0.41		nl CO$_2$/hr/cell	(95)(81)
		Spinal ganglion	Rat	0.5	0.7	0.7	nl CO$_2$/hr/cell	(81)(95)
		Sympathetic ganglion	Rat	0.3	0.9	0.3	nl CO$_2$/hr/cell	(78)(95)

		Tissue	Animal					Ref
	Total cholinesterase	Sympathetic ganglion	Cat	0.3–16.0			nl CO_2/hr/cell	(77) (79)
Others	Carbonic anhydrase	Sympathetic ganglion	Rat	3.0	18	0.18	M.10^{-20} CO_2/hr/cell	(139) (95) (155)
	Carbonic anhydrase	Sympathetic ganglion	Rat	6.4	385[j]	0.02	M.10^{-20} CO_2/hr/cell	(139)

[a] Expressed on basis of lipid-free dry weight.
[b] Mole/kg/hr; moles of substrate transformed per kilo dry weight per hour.
[c] Dark appearance in frozen dried state.
[d] "Glial capsule"
[e] Light appearance in frozen dried state.
[f] Four cells used in measurements.
[g] Larger Type A dorsal root ganglion cells which appeared less dense and stained less densely with thionine than Type B (B) cells.
[h] Results believed too low.
[i] Two distinct groups.
[j] Equivalent volume of glia.

have been performed both in experimental animals and humans in the normal (steady) state as well as under conditions of functional and pathological stress. Studies comparing the activities of enzymes involved in carbohydrate and glutamate metabolism from regions rich in cell bodies (pyramidal layer of Ammon's horn; granular layer of the cerebellum) with adjacent regions composed largely of dendrites (molecular layers), as well as the direct analyses of single cells, have shown that the enzyme activities of the major energy-yielding systems are generally at a lower level in nerve cell bodies than in average brain tissue.[64,67-70] As enzymes involved in glycolysis and glutamate oxidation (isomerase, lactic dehydrogenase, and glutamic dehydrogenase) were particularly low in pyramidal cell bodies, it appeared that in the neuronal perikarya of Ammon's horn these processes were relatively less important than in the adjacent neuropil and neuroglia.[64] Similarly, the molecular layer of the cerebellum, composed largely of axons and dendrites, shows higher glycolytic and TCA enzyme activity than the granular layer.[68,70]

Lowry[71] suggested that the metabolism of the brain is chiefly dendritic and that the most metabolically active areas are those containing neural processes, particularly packed dendrites. A relatively smaller percentage of the metabolism of the whole brain was attributed to the nerve cell bodies. The rich content of LDH, malic dehydrogenase, aldolase, and fumarase in the stratum radiatum of Ammon's horn is evidence in support of the high capacity of neuronal processes for both glycolytic and oxidative metabolism.[64,71] Similarly, Robins[68] noted that nerve cell bodies had a lower average enzyme activity than whole brain and that anterior horn cells, in particular, had a lower enzyme concentration than several other areas of gray matter or whole brain. This was particularly true for hexokinase. In contrast, the β-glycosidases are much richer in perikarya. Anterior horn cell bodies have a six to eight-fold greater activity than their surrounding neuropil.[68,72] Other enzymes largely localized to gray matter include those involved in γ-amino butyric acid metabolism. γ-Aminobutyric-α-ketoglutaric transaminase is found in highest concentration in the gray matter of the subcortical nuclei and the brain stem,[73] and glutamic decarboxylase which was virtually absent in white matter exhibited its greatest activity in the globus pallidus.[74] Hirsch and Robins[75] demonstrated that in the cerebellum, the amounts of GABA were similar in cell bodies and their processes and that in the cerebellar and cerebral cortices the levels of γ-amino butyric acid paralleled the distribution of the major enzymes of general energy metabolism.

Although the enzyme activity of single neurons appears lower, in general, than the activity of average brain, the maximal capacities of the enzymes of the glycolytic cycle in single cells nevertheless appears to be in excess of the metabolic demands usually made upon them.[71] This excess of enzyme capacity, Lowry suggests, may be required to keep the intracellular environment constant.

Similarities as well as differences exist with respect to the content and the activity of enzymes from single homologous neurons and homologous areas of brain in different species. The activities of most glycolytic and tri-carboxylic acid cycle enzymes were remarkably uniform in the anterior horn cell bodies of the rabbit and monkey. Only three enzymes, lactic dehydro-genase, glucose-6-phosphate dehydrogenase, and 6-phospho-gluconic dehydrogenase exhibited noteworthy species differences.[67,68] Corres-ponding regions of the cerebellum also demonstrated comparable lactic dehydrogenase and malic dehydrogenase activity.[68,76] In contrast, up to an eight-fold difference in acetylcholinesterase activity was found in the sympathetic ganglion cells of the cat as compared with the rat.[77] Species differences have also been observed in homologous areas of white matter. The malic dehydrogenase and lactic dehydrogenase activities of monkey and rabbit dorsal columns differ significantly and even greater species dif-ferences are found in the optic nerves and tracts where the lactic dehydro-genase activity is three times greater in the rabbit than in the monkey.[76]

Although it might be anticipated that single cell and topographical analyses would demonstrate similar enzyme activity in anatomically or functionally related structures, several studies have shown marked differ-ences in the enzyme activity of anatomically similar or functionally related structures on the one hand and similarities in enzyme activities of anatomic-ally or functionally different structures on the other. In the rabbit dorsal root ganglion, levels of malic dehydrogenase and transaminase activity were found to covary in specific cell bodies and were dissociated from lactic de-hydrogenase and isocitric dehydrogenase.[67,68] Furthermore, in topo-graphical enzyme analyses of the layers of the retina in the rabbit, glycolytic enzymes (LDH) and tricarboxylic acid cycle enzymes (MDH) were distributed in a reciprocal fashion in individual retinal layers. The inner segments of the rods and cones contained 30 times more malic dehydrogenase and trans-aminase than the outer segments. Close association of malic dehydrogenase and glutamic-aspartic transaminase in the same retinal segments is even more striking than the segregation of the dehydrogenases in separate retinal layers[20] and suggests a functional pathway through oxaloacetate.

The nerve cell populations of the spinal sympathetic and dorsal root ganglia and the anterior horns are not pharmacologically homo-geneous.[78-80] A characteristic distribution pattern of cholinesterase activity was not only demonstrated for each ganglion but was also present among histologically similar neurons within the same ganglion. The distri-bution pattern of enzyme activity for a ganglion or nucleus was, however, a constant and typical feature for each species.[78,81] In the rat sympathetic ganglion cells exhibited up to 15-fold differences and anterior horn neurons showed four-fold differences in cholinesterase activity.[82] In a more recent study of sympathetic ganglion neurons in the cat, individual cells showed even wider (50-fold) variation in total cholinesterase activity. There were ten-fold differences in choline acetylase activity among these neurons.[77,79] Large differences were also demonstrated for cytochrome oxidase and suc-cinic dehydrogenase activity among rabbit spinal ganglion cells of the same size.[83]

In topographical enzyme analyses Robins and co-workers[76] showed that glutamic dehydrogenase is twice as active in the molecular layer of the cerebral cortex compared with the molecular layer of the cerebellar cortex in the monkey and rabbit. Such differences in enzyme activity may be related to possible anatomical or functional differences in these regions or different stages in enzyme synthesis.[76,83] Despite the marked anatomical and functional differences between anterior horn and dorsal ganglion cells, levels of glycolytic and tricarboxylic acid cycle enzymes in these cells differed only slightly.[68] Lactic dehydrogenase, an exception, was only half as active in anterior horn cells as in dorsal ganglion cells.

Strominger and Lowry[84] suggested that although in some cases the metabolic activities of axons may in part mirror the activities of their cell body of origin, it is more probable that fibers have a special or independent metabolism which need not be related to that of the cell body. In support of this concept, Lowry[67] cites the striking contrast between the enzyme activities of dorsal root ganglion cell bodies and their surrounding capsules and myelinated fibers. In comparison with the cell bodies, the surrounding capsules and fibers had low hexokinase, malic dehydrogenase, transaminase, and glutamic dehydrogenase activities. The phosphoglucoisomerase, lactic and isocitric-dehydrogenase activities were approximately equal in the cell bodies and capsules and fibers; and the capsules were relatively rich in glucose-6-phosphate and 6-phosphogluconic dehydrogenases.

Characteristic patterns of subcellular enzyme distribution have been demonstrated in neurons. Giacobini and Zajicek[85] showed that the acetylcholinesterase activity was $2\frac{1}{2}$ times greater in spinal sympathetic and parasympathetic cell bodies than in their axons. Giacobini[79] later demonstrated that the intracellular distribution of acetylcholinesterase was generally characteristic for the neurons of spinal and sympathetic ganglia and anterior horn nuclei. Hirsch[86] noted that the activity of α-naphthyl acid phosphatase was many times greater in nerve cell bodies of the cerebellum than in the surrounding neuropil and suggested that lysosomes were much richer in neuronal perikarya than in axons, dendrites, or neuroglial cells. Other (presumably lysosomal) acid hydrolases including β-galactosidase and β-glucuronidase have also been shown to be highly concentrated in nerve cell bodies as compared with the surrounding neuropil.[68,87] Such differences in enzyme distribution between the perikarya and the neuropil are suggestive of considerable segregation of metabolic activity within the neuron.

The enzyme activities in single neuronal nuclei as compared to cytoplasm were examined in quick-frozen-dried sectioned rabbit dorsal root ganglia[67,88] and in puffer supramedullary neurons.[89] Nuclear and cytoplasmic samples were dissected and trimmed so that there was no contamination of nuclear samples by cytoplasm. Evidence for such lack of contamination was the total absence of β-D-glucuronidase from nuclear samples.[89] Both studies revealed approximately equal activities of lactic and malic dehydrogenases and phosphoglucose isomerase in neuronal nuclei and cytoplasm, while hexokinase and glucose-6-phosphate dehydrogenase were present in nuclei at only one-half of the levels found in cytoplasm. The presence of these enzymes in the nucleus would speak for a considerable capacity for local glycolytic, pentose phosphate shunt and perhaps even Krebs cycle metabolism. Slide histochemical methods which purport to localize pyridine nucleotide-dependent dehydrogenases have never demonstrated formazan deposition in neuronal nuclei.[90] This should emphasize the well-documented unreliability of such methods.[157] Large volumes have been written during the past decades using such slide methods to map enzyme localizations and, by implication, their relative activities in areas of the nervous system and in single neurons. Such studies are of doubtful significance and will not be reported here.

Differences in enzyme activity have also been demonstrated among various neuroglial cells. The rate of ATP hydrolysis was 50% greater in "neuronal glia" (richer in satellite oligodendrocytes) compared with "capillary glia" (presumably richer in astrocytes).[91] Hamberger[92] observed similar differences between neuronal and capillary "neuroglia." However, it must be noted that capsular material stripped from neurons is likely to contain synaptic endings which have high ATPase activities. Thus "neuronal glia" may be a misleading term. Hess[93] suggested that the respiratory rate per average neuroglial cell in the white matter was approximately 40% higher than the respiratory rate of neuroglial cells situated in the cerebral cortex.

G. Ions

The Na and K content of vertebrate neurons is still subject to considerable discussion. In many vertebrate species total brain Na is 40–80 mmoles/kg wet weight.[94,95] If a large fraction of this Na were to be considered extracellular, then the extracellular space would contain as much as 50% of tissue water. This is clearly incompatible with electron microscopic estimates and the highest estimates from "tracer" studies are approximately 20%.

Müller,[96] using an integrating micro-flame photometer reported Na in single myelinated frog nerve fibers at 50–80 mM and K at 40–145 mM. Katzman et al.[97] found Na levels of 40–60 mM in mouse sensory and brain stem neurons using both ^{24}Na tracer and flame photometry methods on quick-frozen-dried dissected cells. In a subsequent study of puffer fish supramedullary neurons dissected from supravitally frozen, dried, and sectioned brain, the same authors[95] found cytoplasmic Na to be 102 mM and K 134 mM and nuclear levels of 85 mM Na and 113 M K. Cytoplasmic levels per 100 g of dry weight were 45 mmoles of Na and 61 mmoles of K. These measurements were made with an integrating ultra micro-flame photometer. Keesey[98] using an instrument of similar design on cells dissected from quick-frozen, dried sections of rabbit sensory ganglia reported Na contents of 22–28 mmoles per 100 g dry weight. Assuming that dry weight per liter of tissue volume for these cells is about 0.2, this would yield concentrations of 44–56 mM.

According to the Hodgkin–Huxley–Keynes formulation, Na^+ should be chiefly extraneuronal. Eccles[99] estimated the Na^+ ion concentrations in cat motor neurons at 15 mM intracellular and 150 mM extracellular. From the average spike overshoot of 26 mV (maximum of 42 mV) reported by Bennett et al.[100] for the supramedullary neurons of the puffer, the ratio of extracellular to intracellular Na^+ ion should be 3.5:1. It is therefore likely that a large fraction of neuronal intracellular Na is bound and does not contribute to Na^+ ion activity. The evidence for such binding has been reviewed.[95] It is likely that fixed, intracellular polyanionic substances such as nucleic acids and acidic lipids are principally responsible for such binding. Recent observations with the electron microprobe X-ray spectrometer and with activation analysis[101] show a tendency for Na and P concentrations in various regions of single frozen-dried neurons to covary. This would support the concept that polyanionic phosphate groups are principally responsible for Na binding.

At the present time no direct data are available on neuronal concentrations of other ions or trace metals.

III. NEURONAL-NEUROGLIAL RELATIONSHIPS

Based on topographical analyses, Hess[93] and Pope[102] suggested that neuronal metabolism generally exceeds that of neuroglia and despite the numerical preponderance of neuroglial cells, the respiration of whole brain was largely neuronal. Hess[93] stated that in eulaminate frontal cortex the oxygen consumption per average neuron and its processes was 16 to 50 times that of the average neuroglial cell and that 91 to 95% of cortical respiration was due to neurons and their processes. However, these statements are based on deductive estimates and not on the chemical analyses of single cortical neurons.

In Deiters' nucleus, enzymes localized predominantly in "neuroglia" include cytochrome oxidase, succinoxidase, ATPase, and carbonic anhydrase.[7,103,104] "Neuroglial" ATPase activity is greater than that of adjacent Deiters' neurons[105] and the activity of carbonic anhydrase in "neuroglia" may be up to 120 times that of adjacent rabbit Deiters' neurons.[103] Here again, the method of dissection engenders the inclusion of variable amounts of synaptic ending material, and therefore, what is termed "neuroglia" may metabolically represent "synaptosomal" activity, more characteristic of nerve endings than of glia.

A striking constrast in enzyme activity between neurons and neuroglia has also been demonstrated in the rabbit spinal ganglion where glycolytic enzymes (hexokinase, phosphoglucoisomerase, lactic dehydrogenase) are more active in neurons, and pentose phosphate shunt (glucose-6-phosphate-DH) and certain tricarboxylic acid cycle enzymes (isocitric- and succinic-dehydrogenases) are more active in neuroglia.[103,104]

The activities of several enzymes have been shown to undergo significant changes in neurons and/or their "neuroglia" under a variety of conditions producing enhanced or altered central nervous system activity. During increased functional demand, chemical stimulation, hypoxia, and in pathological states (see below) respiratory enzyme activity may increase in neurons and exceed that of "neuroglia,"[105] may change between astrocytic pericapillary and oligodendroglial neuronal satellite cells[92] and endogenous neuronal substrate may fall to zero.[106,107] For example, reciprocal increase in respiratory enzyme activity in Deiters' neurons and decrease in "neuroglia" and increase in anaerobic glycolysis in neuroglia and a decrease in such activity in adjacent neurons were shown to occur following vestibular stimulation in rabbits.[7] Whether such reciprocal changes reflect a metabolic relationship between the neuron and neighboring neuroglial cells or whether they indicate changes in synaptic material included in the glial "ball" or perhaps changes in the distribution of structures which are separable during dissection cannot be critically stated at this time.

IV. EXPERIMENTALLY INDUCED CHANGES IN THE COMPOSITION OF NEURONS

A. Chemical Agents

Increases have been reported in neuronal nuclear and cytoplasmic RNA following administration of tricyanoaminopropene.[12,19,36,108] The concomitant altered content and base composition of nuclear RNA has been interpreted to reflect a loss to the cytoplasm of nuclear (messenger) RNA as well as a relative increase in nucleolar (ribosomal) RNA. The reciprocal decrease in the RNA content and the base ratio changes in surrounding "neuroglial" samples of equivalent weight and volume were thought to reflect a functional relationship between the neuron and surrounding oligodendroglial cells during chemically induced "stress" in which there is enhanced utilization of neuroglial RNA, possibly by its direct transfer to neurons.[36] However, since the homogeneity of the neuroglial samples varied due to contamination by nerve cell processes and astrocytes,[36] doubt remains as to whether the RNA changes attributed to neuroglial cells reflect these elements alone.

β-β-Imino-dipropionitrile administered to rats (2 g/kg) produces a significant increase in the volume and the mass of anterior horn cells within 12 days.[109] Twelve weeks following administration, the average dry weight of a rat anterior horn cell increased from approximately 0.205 pg/μ^3 to 0.340 pg/μ^3. The axonal balloons which are also induced by this agent have an average density of 0.182 pg/μ^3.[110] Despite the apparent increase in protein synthesis, no significant change in the content or base composition of RNA occurred during 10 weeks in the anterior horn cell bodies or in adjacent "neuroglia." The RNA from the axonal balloons of the IDPN-treated animals differed in base composition from cytoplasmic RNA and resembled the RNA of the Mauthner axon. The finding of RNA in axonal balloons adds further support to the concept that the axon contains RNA with distinctive base ratios[10,26] and that independent, local protein synthesis may occur there.[28,29,54,111,112]

Imipramine (Tofranil) has also been reported to produce a 20% increase in cellular RNA, in Deiters' neurons, accompanied by a 30% decrease in RNA content of the surrounding "neuroglia." The "cytochrome oxidase" activity of single neurons was reported increased 400%[108]; however, it is not clear whether uncoupling of oxidative phosphorylation may have been a factor.

Monoamine oxidase inhibitors are potent stimulators of neuronal RNA production. Phenylcyclopropylamine (0.3 mg/kg) produced a 30% increase in neuronal RNA and a 50% decrease in "neuroglial" RNA accompanied by significant base ratio changes involving guanine and cytosine. Cytochrome oxidase activity increased 250% in neurons and decreased 30% in surrounding "neuroglia." Actinomycin D (0.3 mg i.v.) inhibited these nucleic acid changes and also produced a decrease of RNA per cell, suggesting that chemically induced RNA synthesis is DNA-dependent.[113] Tranylcypromine (0.3 mg/kg) produced a prominent increase in the RNA content of Deiters' neurons within 1 hr. Similar changes occurred earlier and were more pronounced in the surrounding "neuroglia."[91]

Pevzner[21] noted that adrenaline administered to cats (20–25 μg/kg/day) for 2 weeks resulted in a pronounced increase in the mass and nucleic acid content of superior cervical ganglion neuroglial capsule cells and a moderate increase in the cytoplasmic RNA of adjacent neurons.

B. Physiological Stress

Hydén and his co-workers have demonstrated a variety of changes in Deiters' neurons after functional, vestibular afferent stimulation. These include increase in dry weight, RNA and protein and in cytochrome oxidase and succinic dehydrogenase activity. Simultaneously, decreases were observed in RNA content and respiratory enzyme activity of surrounding "oligodendrocytes."[7,17,114] The rate of increase in oxidative enzyme activity could be related to the duration of stimulation. Pevzner[21] demonstrated that electrical stimulation of the superior sympathetic ganglion of the cat resulted in an increase in the nucleic acid and protein content of individual neurons. This was accompanied by a decrease in the RNA content of neuroglial satellite cells. Brattgård[115] observed that the protein content of individual retinal ganglion cells was twice as great in normally stimulated compared with light-deprived 10-week-old rabbits. RNA was often found to be totally lacking in the retinal ganglion cells of nonstimulated animals. Light deprivation for only 3 hr yielded up to 90% decrease in the ribonucleoprotein content of individual ganglion cells. Following stimulation there was a prompt increase.

As suggested by Landauer,[116] the arguments for direct macromolecular transfer of such substances as RNA between glia and neurons are not compelling. Although astrocytes are thought to be involved in the transport of water, electrolytes, and metabolites within the brain and are possibly concerned with neuronal nutrition,[117] their proximity to neurons and blood vessels has also implicated them in blood–brain barrier function where they may be involved in preventing macromolecules from reaching neurons. Oligodendroglial cells are known to be involved in myelination and to bear an intimate anatomical relation to axons but have not been observed to participate in macromolecular transfer *in vivo*. Although the incorporation of RNA has been described both in normal and neoplastic cells *in vivo* and *in vitro*,[118–120] the direct transfer of macromolecules between cells has not been observed *in vivo*.

C. Nerve Regeneration

Brattgård, Edström, and Hydén[5] studied the chemical events in neurons in the rabbit hypoglossal nucleus following nerve crush. During the first 2 days following injury (latent period) the central axon degenerated, the cell body volume remained constant, the RNA content and concentration slightly diminished, the physical state of RNA changed from aggregated to finely dispersed particles (chromatolysis), and the protein content diminished by 30–40%. Between days 3 and 12 thin new axons sprouted (4 mm/day) and contacted motor end plates. The cell volume initially increased (250%) due to water intake, the RNA content remained unchanged (concentration decreased 60%) and the protein and lipid content rapidly increased (100%). Later during this period the RNA content and concentration slowly rose and the cell volume slightly decreased. Between 12 and 90 days (maturation period) the diameters of the newly formed axons increased. There was a second increase (250%) in cell body volume which was maximal at about day 40 and was, in part, the result of increased cell mass. Most conspicuous was a doubling of the RNA content as the protein and lipid content continued to increase. Between 40 and 90 days there was a gradual return of all parameters to normal with reaggregation of RNA into characteristic Nissl bodies. A marked increase in cell volume and protein content with constant or slightly elevated RNA content has also been observed in embryonal neurons.[115,121]

Following sciatic nerve section in the frog, an increase in lumbar anterior horn cell volume and RNA content occurred as early as the fourth day and roughly paralleled the findings in the rabbit hypoglossal nucleus. The maximal increase in the volume and chemical constituents (250% of control) was also achieved at a similar interval following injury (5 weeks). In contrast to chromatolytic hypoglossal neurons, the RNA concentration in cell bodies never decreased.[2]

More recently Watson[122–124] showed that, in the rat, immediately following hypoglossal nerve section, the nucleic acid content of the nucleolus increased first, followed by a subsequent increase in the total cell body nucleic acid content. The uptake of RNA precursors also occurred first in the nucleolus, and decreased in the nucleolus while the cellular content of RNA remained high. Watson suggested that this early increase in nucleolar RNA content represented enhanced ribosomal synthesis, which appeared to be "DNA-primed" as it was altered by actinomycin D. Furthermore, the rate of transfer of nucleic acid from the nucleolus to the cytoplasm appeared closely related to the nucleolar nucleic acid content. The rate of decline of nucleolar nucleic acid could be significantly altered by preventing further RNA synthesis with actinomycin D. The restoration of the normal chemical composition of the neuron appeared in part to be dependent upon changes in the nerve cell volume consequent to axonal disruption and regeneration. In keeping with earlier studies, the closer to the cell body the injury occurred, the earlier the onset and the decline of the nucleolar and cytoplasmic response. During the period of axon outgrowth the rate of transfer of RNA from the nucleus to the cytoplasm decreased below normal and remained low throughout the maturation period. These observations differ from Brattgård and co-workers[5] and Edström,[2] who noted an increase in RNA at this time. Watson[124,125] suggested that both an increased formation and destruction of RNA may accompany the increased protein synthesis observed during axonal outgrowth. Pakkenberg and co-workers[38] noted that the uptake of intravenously administered adenine-^{3}H was greater in the neuropil surrounding nonregenerating as compared with regenerating neurons. They suggested these changes were possibly related to a glial-neuronal interrelationship which could imply a higher turnover or utilization rate of RNA precursors in the regenerating cells.

Watson[126–128] reported that oxygen consumption of isolated hypoglossal neurons remained unchanged for 2 days following nerve section, then diminished to reach a minimum value on the fourth day and gradually returned to normal between the tenth and fifteenth day. Between 3 and 12 days following hypoglossal nerve crush in rats, the activity of succinoxidase and cytochrome oxidase and the rate of oxidation of α-ketoglutarate and glutamate decreased in nerve cell bodies. A marked reduction in acetylcholinesterase activity observed maximally on the tenth day followed the reduction in the oxidative enzyme activities.

D. Learning

Hydén and Lange[129] noted that the quantity and composition of nuclear RNA formed in vestibular and cortical neurons and their neuroglia was altered during a situation in which rats, in order to get food, were required to learn to balance on a thin steel wire. There was an early increase in the nuclear RNA content in Deiters' neurons. In animals who increased their performance linearly with time, the newly formed RNA was rich in adenine. This increase in the A/U ratio in nuclear RNA persisted for at least 48 hr following the completion of the experiment. However, late in the "learning" experience, RNA with a ribosomal base composition was predominantly produced. Animals subjected to rotational stimulation also exhibited a significant increase in the RNA content of Deiters' neurons but without an increase in the adenine and uracil nucleotide fractions. As RNA with an adenine-rich nucleotide fraction was not found in "controls," Hydén and Egyházi[130] postulated that the production of such RNA is not due to increased neuronal function *per se*.

Similarly, in an experiment requiring rats to learn to transfer handedness, there was an early, 25–35%, increase in RNA in Carnoy-fixed neurons of the fifth and sixth nerve cell layers of a cortical region whose destruction had been shown to prohibit the transfer of handedness. On the rising part of the learning curve (3–5 days), as in vestibular neurons, the newly formed RNA was characterized by a DNA-like base ratio composition with respect to adenine and uracil. On the asymptotic part of the learning curve (9–10 days) when learning was completed and the animals performed well, the cellular RNA content was increased further (60–100%) but now the base composition had changed to RNA of ribosomal type.[43]

V. PATHOLOGICAL CHANGES

Hydén and co-workers[91,114,131] showed that hypoxia achieved by administering 8% oxygen and 92% nitrogen to rabbits for 12–15 hr resulted in an approximate 20% increase in the RNA content, and a 30% increase in the cytochrome oxidase activity in Deiters' neurons. Fifteen minutes after occlusion of the abdominal aorta in rabbits, spastic paralysis of the lower extremities appeared. The anterior horn cells showed a decrease in the nucleic acid and protein content of the nucleolus. In cases of mild paralysis the nucleoprotein was reconstituted within $8\frac{1}{2}$ hr of restitution of the circulation.[132] A decrease in cytoplasmic nucleic acid content has also been reported in trauma involving cochlear ganglion cells,[133] in neurotropic viral infection,[134] and in cortical biopsy specimens of patients with schizophrenia.[12]

Urea-induced central nervous system dehydration produced a twofold increase in succinoxidase activity in rabbit Deiters' neurons and in mitochondria isolated from single neurons within 48 hr. The possible contributory effect of altered intracellular osmotic pressure was not ruled out.[135] In rats dehydrated by water deprivation or saline drinking, the consequent increase in the production of antidiuretic hormone was associated with a considerable increase in the RNA content of supraoptic neurons.[136] Conversely, when male rats were castrated or underwent thyroid extirpation to increase the production of gonadotropins, RNA was found to decrease in supraoptic neurons. Generally there was a parallel between the content of cellular RNA and cellular secretory activity. Succinoxidase activity was also found to increase in the Deiters' neurons of rabbits exposed to experimental concussion.[137]

During a 3-week period following a single exposure to 2000–3000 r of X-irradiation increases in the mass, RNA content, respiratory activity, and membrane permeability for potassium were found in Deiters' neurons.[18]

Total mass increased for the first 6 days, RNA content increased transiently at day 5, succinoxidase activity increased within several hours of irradiation and reached a peak at 9 days, and the maximal intracellular loss of potassium occurred between 4 and 7 days. The severity of changes appeared related to the dose administered. A return to normal levels occurred within 2 weeks.

Decreases in the activity of acetylcholinesterase (64%) and choline acetylase (up to 97%) were found in cells dissected supravitally following preganglionic denervation of the spinal sympathetic ganglion of the cat.[77,79] Approximately 35% of individual denervated cells lacked measurable enzyme activity and the ratio, acetylcholine to mecholyl hydrolysis, changed from 1.7 to 1.0. Loss of the adherence of acetylcholine-rich nerve terminals to denervated ganglion cells was offered as a possible mechanism to account in part for the decrease in the enzyme activity.

In alum-induced experimental neurofibrillary degeneration, Embree, Hamberger, and Sjöstrand[138] noted a twofold increase in the cell mass, the RNA content, and the incorporation of isotopically labeled leucine during the early production of neurofibrils. However, in neurons with well advanced neurofibrillary degeneration, total RNA and succinoxidase activity did not differ from controls. The authors suggested that Nissl substance and mitochondria were probably displaced but not destroyed in the neurons exhibiting advanced neurofibrillary degeneration.

VI. REFERENCES

Reviews

I. S. P. R. Rose, in Applied Neurochemistry (A. N. Davison and T. Dobbing, eds.), pp. 332–355, F. A. Davis, Philadelphia (1968).

II. E. Giacobini, in Neurosciences Research (S. Ehrenpreis and O. Solnitzky, eds.), Vol. 1, pp. 1–71, Academic Press, New York (1968).

III. E. Giacobini, in Neurosciences Research (S. Ehrenpreis and O. Solnitzky, eds.), Vol. 2, pp. 111–202, Academic Press, New York (1969).

IV. H. Hydén, Behavior, neural function and RNA, Prog. in Nucleic Acid Res. 6:187:218 (1967).

V. H. Hydén, in The Neuron (H. Hydén, ed.), pp. 179–219, Elsevier, Amsterdam-London-New York (1967).

VI. O. H. Lowry, in The Harvey Lectures, pp. 1–19, Academic Press, New York (1963).

1. J-E. Edström, Ribonucleic acid mass and concentration in individual nerve cells. A new method for quantitative determinations, Biochim. Biophys. Acta 12:361–386 (1953).

2. J-E. Edström, Ribonucleic acid changes in the motoneurons of the frog during axon regeneration, J. Neurochem. 5:43–49 (1959).

3. J-E. Edström and A. Pigon, Relation between surface, ribonucleic acid content and nuclear volume in encapsulated spinal ganglion cells, J. Neurochem. 3:95–99 (1958).

4. H. Hydén, in Biochemistry of the Developing Nervous System (H. Waelsch, ed.), pp. 358–370, Academic Press, New York (1955).

5. S.-O. Brattgård, J.-E. Edström, and H. Hydén, The chemical changes in regenerating neurons, J. Neurochem. 1:316–325 (1957).

6. H. Hydén and S. Larsson, The application of a scanning and computing cell analyzer to neurocytological problems, J. Neurochem. 1:134–144 (1956).

7. H. Hydén and A. Pigon, A cytophysiological study of the functional relationship between oligodendroglial cells and nerve cells of Deiters' nucleus, J. Neurochem. 6:57–72 (1960).

8. J. Nurnberger, A. Engström, and B. Lindström, A study of the ventral horn cells of the adult cat by two independent cytochemical microabsorption techniques, J. Cell. Comp. Physiol. 39:215–251 (1952).

9. G. M. Lehrer, R. Katzman, and C. E. Wilson, Volume and density measurements in subcellular portions of single nerve cell bodies, J. Histochem. Cytochem. 18:44–48 (1970).

10. J-E. Edström, D. Eichner, and A. Edström, The ribonucleic acid of axons and myelin sheaths from Mauthner neurons, Biochim. Biophys. Acta 61:178–184 (1962).

11. A. Edström, The ribonucleic acid in the Mauthner neuron of the goldfish, J. Neurochem. 11:309–314 (1964).

12. H. Hydén and H. Hartelius, Stimulation of the nucleoprotein production of the nerve cells by malononitrile and its effect on psychic functions in mental disorders, Acta Psychiat. Neurol. Suppl. 48 (1948).

13. J-E. Edström, A rapid method for the determination of nucleic acid components by paper partition chromatography, Nature 168:876–877 (1951).

14. J-E. Edström, A chromatographic method for the analysis of nucleic acids in microgram amounts, Biochim. Biophys. Acta 9:528–530 (1952).

15. J-E. Edström, Quantitative determination of ribonucleic acid in the micromicrogram range, J. Neurochem. 3:100–106 (1958).

16. J-E. Edström, Ribonucleic acid mass and concentration in individual nerve cells. A new method for quantitative determinations, Biochim. Biophys. Acta 12:361–386 (1953).

17. H. Hydén, Quantitative assay of compounds in isolated, fresh nerve cells and glial cells from control and stimulated animals, Nature 184:433–435 (1959).

18. O. Hallén. A. Hamberger, B. Rosengren, and H. Röckert, Quantitative response of neurons to x-irradiation: Total organic mass, succinoxidase activity, potassium permeability and RNA content in isolated cells, J. Neuropath. Exp. Neurol. 26:327–334 (1967).

19. H. Hydén and E. Egyházi, Changes in the base composition of nuclear ribonucleic acid of neurons during a short period of enhanced protein production, J. Cell. Biol. 15:37–44 (1962).

20. O. H. Lowry, in Biochemistry of the Developing Nervous System, Proc. 1st Internat. Neurochem. Symposium, pp. 350–357 (1954).

21. L. Z. Pevzner, Topochemical aspects of nucleic acid and protein metabolism within the neuron-neuroglia unit of the superior cervical ganglion, J. Neurochem. 12:993–1002 (1965).

22. J-E. Edström, The content and the concentration of ribonucleic acid in motor anterior horn cells from the rabbit, J. Neurochem. 1:159–165 (1956).

23. H. Hydén, Protein metabolism in the nerve cell during growth and function, Acta Physiol. 'Scandinav. Suppl. 17 (1943).

24. J-E. Edström and D. Eichner, Quantitative unterschungen über den Ribonukleinsaüregehalt der Netzhautganglienzellen bei Rind und Mensch, Zeitsch. f. Mikr-Anat. Forsch. 63:413–421 (1957).

25. N. Miani, A. Di Girolamo, and M. Di Girolamo, Sedimentation characteristics of axonal RNA in rabbit, J. Neurochem. 13:755–759 (1966).

26. E. Koenig, Synthetic mechanisms in the axon-II. RNA in myelin–free axons of the cat, *J. Neurochem.* **12**:357–361 (1965).

27. S. L. Palay and G. E. Palade, The fine structure of neurons, *J. Biophys. Biochem. Cytol.* **1**:69–88 (1955).

28. E. Koenig and G. B. Koelle, Mode of regeneration of acetyl-cholinesterase in cholinergic neurons following irreversible inactivation, *J. Neurochem.* **8**:169–188 (1961).

29. D. H. Clouet and H. Waelsch, Recovery of cholinesterases in the frog nervous system after inhibition, in *Proc. 4th Internat. Neurochem. Symp.* (S. S. Kety and J. M. Elkes, eds.), pp. 243–247, Pergamon Press, New York (1961).

30. B. Jakoubek and J-E. Edström, RNA changes in the Mauthner axon and myelin sheath after increased functional activity, *J. Neurochem.* **12**:845–849 (1965).

31. J-E. Edström, Separation and determination of purines and pyrimidine nucleotides in picogram amounts, *Biochim. Biophys. Acta* **22**:378–388 (1956).

32. J-E. Edström, Extraction, hydrolysis and electrophoretic analysis of ribonucleic acid from microscopic tissue units (microphoresis), *J. Biophys. Biochem. Cytol.* **8**:39–43 (1960).

33. J-E. Edström, Composition of ribonucleic acid from various parts of spider oocytes, *J. Biophys. Biochem. Cytol.* **8**:47–51 (1960).

34. J-E. Edström, W. Grampp, and N. Schor, The intracellular distribution and heterogeneity of ribonucleic acid in starfish oocytes, *J. Biophys. Biochem. Cytol.* **11**:549–557 (1961).

35. J-E. Edström and W. Beermann, The base composition of nucleic acids in chromosomes, puffs, nucleoli and cytoplasm of Cironomus salivary glands, *J. Cell. Biol.* **14**:371–379 (1962).

36. E. Egyházi and H. Hydén, Experimentally induced changes in the base composition of the ribonucleic acids of isolated nerve cells and their oligodendroglial cells, *J. Biophys. Biochem. Cytol.* **10**:403–410 (1961).

37. D. H. Ford and R. Rhines, H^3 accumulation in spinal cord neurons following intravenous injection of adenine-H^3 as determined by liquid scintillation counting procedures on dissected neurons, *Acta Neurol. Scand.* **43**:427–439 (1967).

38. H. Pakkenberg, D. H. Ford, R. Rhines, and R. A. Israely, Adenine-H^3 uptake in nervous tissue, including regenerating nerve cells, as compared with other tissues in euthyroid, hypo- and hyperthyroid male rats, *Acta Neurol. Scand.* **41**:497–512 (1965).

39. E. Egyházi and H. Hydén, Biosynthesis of rapidly labeled RNA in brain cells, *Life Sci.* **5**:1215–1223 (1966).

40. B. Daneholt, and S-O. Brattgård, A comparison between RNA metabolism of nerve cells and glia in the hypoglossal nucleus of the rabbit, *J. Neurochem.* **13**:913–921 (1966).

41. E. L. Bennett and B. J. Krueckel, Renewal of nucleotides and nucleic acids in C57 mice studied with adenine-4, $6^{14}C$, *Biochim. Biophys. Acta* **17**:503–514 (1955).

42. R. W. Swick and A. L. Koch, The measurement of nucleic acid phosphorus turnover in rat liver by the constant exposure technique, *Arch. Biochem. Biophys.* **67**:59–73 (1957).

43. H. Hydén, Behavior, neural function and RNA, *Prog. Nucl. Acid Res.* **6**:187–218 (1967).

44. J. Jarlstedt, The distribution of RNA in the cerebellum of the rabbit, *Exp. Cell. Res.* **28**:501–506 (1962).

45. B. Droz and C. P. LeBlond, Migration of proteins along the axons of the sciatic nerve, *Science* **137**:1047–1048 (1962).

46. A. Lajtha, in *Chemical Pathology of the Nervous System* (J. Folch, ed.), pp. 268–275, Pergamon Press, New York (1961).

47. P. Weiss and H. B. Hiscoe, Experiments on the mechanism of nerve growth, *J. Exper. Zool.* **107**:315–395 (1948).

48. S. H. Barondes, On the site of synthesis of the mitochondrial protein of nerve endings, *J. Neurochem.* **13**:721–727 (1966).

49. A. Dahlstrom and J. Haggendahl, Studies on the transport and life-span of amine storage granules in a peripheral adrenergic neuron system, *Acta Physiol. Scandinav.* **67**:278–288 (1968).

50. B. Grafstein, Transport of protein by goldfish optic nerve fibers, *Science* **157**:196–198 (1967).

51. B. Droz and C. P. LeBlond, Axonal migration of proteins in the central nervous system and peripheral nerves as shown by radioautography, *J. Comp. Neurol.* **121**:325–346 (1963).

52. D. F. Matheson, Incorporation of ^{14}-C-glycine into protein of the adult rat peripheral nerve: Effects of inhibitors, *J. Neurochem.* **15**:179–185 (1968).

53. S. H. Barondes, Further studies of the transport of protein to nerve endings, *J. Neurochem.* **15**:343–350 (1968).

54. E. Koenig, Synthetic mechanisms in the axon-I, local axonal synthesis of acetylcholinesterase, *J. Neurochem.* **12**:343–355 (1965).

55. E. Koenig, Synthetic mechanisms in the axon-III. Stimulation of acetylcholinesterase synthesis by Actinomycin-D in the hypoglossal nerve, *J. Neurochem.* **14**:429–435 (1967).

56. L. Lubińska, S. Niemierko, and B. Oberfeld, Gradient of cholinesterase activity and of choline acetylase activity in nerve fibers, *Nature* **189**:122–123 (1961).

57. E. Koenig, Synthetic mechanisms in the axon-IV. *In Vitro* incorporation of (^{3}H) precursors into axonal protein and RNA, *J. Neurochem.* **14**:437–446 (1967).

58. D. H. Ford, R. Rhines, M. Hartstein, and A. Rhodes, The uptake of DL- lysine-H^3 into the nervous system as compared with other tissues in euthyroid and dysthyroidal male rats, *Acta Neurol. Scand.* **41**:215–232 (1965).

59. H. Ford and R. Rhines, Amino acid uptake compared in motor neurons, spinal cord gray matter, muscle, and liver, *J. Neurol. Sci.* **4**:501–510 (1967).

60. A. Lajtha, Protein metabolism of the nervous system, *Internat. Rev. Neurobiol.* **6**:1–97 (1964).

61. B. S. McEwen and H. Hydén, A study of specific brain proteins on the semi-micro scale, *J. Neurochem.* **13**:823–833 (1966).

62. H. Hydén and B. McEwen, A glial protein specific for the nervous system, *Proc. Nat. Acad. Sci.* **55**:354–358 (1966).

63. A. L. Rubin and K. H. Stenzel, *In vitro* synthesis of brain protein, *Proc. Nat. Acad. Sci.* **53**:963–968 (1965).

64. O. H. Lowry, N. R. Roberts, K. Y. Leiner, M-L. Wu, A. L. Farr, and R. W. Albers, The quantitative histochemistry of brain. III. Ammon's Horn, *J. Biol. Chem.* **207**:39–44 (1954).

65. S.-O. Brattgård and H. Hydén, Mass, lipids, pentose nucleo-proteins and proteins determined in nerve cells by x-ray micro-radiography, *Acta Radiol. Suppl.* **94** (1952).

66. N. Miani, Analysis of the somato-axonal movement of phospholipids in the vagus and hypoglossal nerves, *J. Neurochem.* **10**:859–874 (1963).

67. O. H. Lowry, in *Metabolism of the Nervous System* (D. Richter, ed.), pp. 323–328, Pergamon Press, New York (1957).

68. E. Robins, in Symposium on the histochemistry of the nervous system, *J. Histochem. Cytochem.* **8**:431–436 (1960).

69. E. Robins and D. E. Smith, A quantitative histochemical study of eight enzymes of the cerebellar cortex and subjacent white matter in the monkey, *Research Publ. Ass. Res. Nerv. Ment. Dis.* **32**:305–327 (1953).

70. E. Robins, D. E. Smith, and M. K. Jen, The quantitative distribution of eight enzymes of glucose metabolism and two citric acid cycle enzymes in the cerebellar cortex and its subjacent white matter, *Prog. Neurobiol.* **2**:205–212 (1957).

71. O. H. Lowry, in *Progress in Neurobiology: Ultrastructure and Cellular Chemistry of Neural Tissue* (H. Waelsch, ed.), pp. 69–82, Hoeber-Harper, New York (1957).

72. E. Robins and H. E. Hirsch, Glycosidases in the nervous system: II. Localization of β-galactosidase, β-glucuronidase and β-glucosidase in individual nerve cell bodies, *J. Biol. Chem.* **243**:4253–4257 (1968).

73. R. A. Salvador and R. W. Albers, The distribution of glutamic-γ-aminobutyric transaminase in the nervous system of the rhesus monkey, *J. Biol. Chem.* **234**:922–925 (1959).

74. I. P. Lowe, E. Robins, and G. S. Eyerman, The fluorimetric measurement of glutamic decarboxylase and its distribution in brain, *J. Neurochem.* **3**:8–18 (1958).

75. H. E. Hirsch and E. Robins, Distribution of γ-aminobutyric acid in the layers of the cerebral and cerebellar cortex. Implications for its physiologic role, *J. Neurochem.* **9**: 63–70 (1962).

76. E. Robins, N. R. Roberts, K. M. Eydt, O. H. Lowry, and D. E. Smith, Microdetermination of α-keto acids with special reference to malic, lactic, and glutamic dehydrogenases in brain, *J. Biol. Chem.* **218**:897–909 (1956).

77. E. Giacobini, B. Palmborg, and F. Sjöqvist, Cholinesterase activity in innervated and denervated sympathetic ganglion cells of the cat, *Acta Physiol. Scandinav.* **69**:355–361 (1967).

78. E. Giacobini, Quantitative determination of the cholinesterase in individual sympathetic cells, *J. Neurochem.* **1**:234–244 (1957).

79. E. Giacobini, Cholinergic and adrenergic cells in sympathetic ganglia, *Ann. New York Acad. Sci.* **144**:646–657 (1967).

80. O. Eränkö and M. Härkönen, Noradrenaline and acetylcholinesterase in sympathetic ganglion cells of the rat, *Acta Physiol. Scandinav.* **61**:299–300 (1964).

81. E. Giacobini, The distribution and localization of cholinesterases in nerve cells, *Acta Physiol. Scandinav.* **45** Suppl. 156 (1959).

82. E. Giacobini and B. Holmstedt, Cholinesterase content of certain regions of the spinal cord as judged by histochemical and Cartesian diver technique, *Acta Physiol. Scandinav.* **42**:12–27 (1958).

83. E. Giacobini, *in Drugs and Enzymes* (B. B. Brodie and J. R. Gillette, eds.), pp. 55–63, Pergamon Press, New York (1965).

84. J. L. Strominger and O. H. Lowry, The quantitative histochemistry of brain. IV. Lactic, malic, and glutamic dehydrogenases, *J. Biol. Chem.* **213**:635–646 (1955).

85. E. Giacobini and J. Zajicek, Quantitative determination of acetylcholinesterase activity in individual nerve cells, *Nature* **217**:185–186 (1956).

86. H. E. Hirsch, Acid phosphatase localization in individual neurons by a quantitative histochemical method, *J. Neurochem.* **15**:123–130 (1968).

87. H. E. Hirsch and E. Robins, Localization of β-glycosidases in nerve cell bodies (abstract), *Fed. Proc.* **20**:342 (1961).

88. O. H. Lowry, *in The Harvey Lectures*, pp. 1–19, Academic Press, New York (1963).

89. G. M. Lehrer, C. Weiss, D. J. Silides, C. Lichtman, M. M. Furman, and R. F. Mathewson, The quantitative histochemistry of supramedullary neurons of puffer fishes, *J. Cell. Biol.* **37**:575–579 (1968).

90. G. M. Lehrer, D. J. Silides, and M. F. Farmer, Pyridine nucleotide dependent dehydrogenase: A comparison of quantitative and slide histochemistry, *J. Histochem. Cytochem.* **13**:700–701 (1965).

91. H. Hydén, The neuron and its glia—a biochemical and functional unit, *Endeavour* **21**:144–155 (1962).

92. A. Hamberger, Difference between isolated neuronal and vascular glia with respect to respiratory activity, *Acta Physiol. Scandinav.* **58**:Suppl. 203 (1963).

93. H. H. Hess, *in Proc. 4th International Neurochemistry Symposium* (S. S. Kety and J. M. Elkes, eds.), pp. 200–212, Pergamon, New York (1961).

94. D. B. Tower, in The Handbook of Neurochemistry (A. Lajtha, ed.), Vol. I, pp. 1–24, Plenum Press, New York (1969).

95. R. Katzman, G. M. Lehrer, and C. E. Wilson, Sodium and potassium distribution in puffer fish supramedullary nerve cell bodies, J. Gen. Physiol. 54:232–249 (1969).

96. P. Müller, Experiments on current flow and ionic movements in single myelinated nerve fibers, Exp. Cell. Res. 15:Suppl. 5, pp. 118–152 (1958).

97. R. Katzman, G. M. Lehrer, C. E. Wilson, and A. Yam, The measurement of sodium in mammalian central nervous system single neurons (abstract), J. Cell. Biol. 31:57A (1966).

98. J. C. Keesey, Flame photometric analysis of sodium and potassium in nanogram samples of mammalian nervous tissue, J. Neurochem. 15:547–562 (1968).

99. J. C. Eccles, The Physiology of Nerve Cells, Johns Hopkins Press, Baltimore (1957).

100. M. V. L. Bennett, S. M. Crain, and H. Grundfest, Electrophysiology of supramedullary neurons in Spheroide maculatus. II. Properties of the electrically excitable membrane, J. Gen. Physiol. 43:189–219 (1959).

101. G. M. Lehrer and C. Berkley, Elemental analysis of subcellular portions of single neurons, J. Neurochem. (in press).

102. A. Pope, in Biology of Neuroglia (W. F. Windle, ed.), pp. 211–222, C. C. Thomas, Springfield (1958).

103. E. Giacobini, in Morphological and Biochemical Correlates of Neural Activity (M. M. Cohen and R. S. Snider, eds.), pp. 15–38, Harper and Row, New York (1962).

104. H. Hydén, S. Løvtrup, and A. Pigon, Cytochrome oxidase and succinoxidase activities in spinal ganglion cells and in glial capsule cells, J. Neurochem. 2:304–311 (1958).

105. J. Cummins and H. Hydén, Adenosine triphosphate levels and adenosine triphosphatases in neurons, glia and neuronal membranes of the vestibular nucleus, Biochim. Biophys. Acta 60:271–283 (1962).

106. H. Hydén and P. W. Lange, A kinetic study of the neuron glia relationship, J. Cell. Biol. 13:233–237 (1962).

107. H. Hydén and P. W. Lange, The steady state and endogenous respiration in neuron and glia, Acta Physiol. Scandinav. 64:6–14 (1965).

108. H. Hydén, Biochemical and functional interplay between neuron and glia, Recent Advances Biol. Psychiat. 6:31–54 (1963).

109. H. Hartmann, H. Hydén, and J. Edström, Weight, volume and ribonucleic acid content of nerve cells in rats made hyperactive with iminodipropionitrile, Fed. Proc. 18:1889 (1959).

110. H. A. Hartmann, Quantitative nerve cell hypertrophy after iminodipropionitrile: An x-ray microradiographic study (abstract), J. Neuropath. Exp. Neurol. 25:136–137 (1966).

111. D. H. Clouet and H. Waelsch, Amino acid and protein metabolism of the brain. VIII. The recovery of cholinesterase in the nervous system of the frog after inhibition, J. Neurochem. 8:201–215 (1961).

112. A. Edström, Amino acid incorporation in isolated Mauthner nerve components, J. Neurochem. 13:315–321 (1966).

113. H. Hydén and P. W. Lange, A genetic stimulation with production of adenic-uracil rich RNA in neurons and glia in learning. The question of transfer of RNA from glia to neuron, Naturwissenschaften 53:64–70 (1966).

114. A. Hamberger and H. Hydén, Inverse enzymatic changes in neurons and glia during increased function and hypoxia, J. Cell. Biol. 16:521–539 (1963).

115. S.-O. Brattgård, The importance of adequate stimulation for the chemical composition of retinal ganglion cells during early post-natal development, Acta Radiol. Suppl. 96 (1952).

116. T. K. Landauer, Two hypotheses concerning the biochemical basis of memory, Psychol. Rev. 71:167–179 (1964).

117. E. D. P. DeRobertis, *in Progress in Brain Research*, Vol. 15: *Biology of Neuroglia* (E. D. P. DeRobertis and R. Carrea, eds.), pp. 1–11, Elsevier, Amsterdam (1965).

118. H. Amos, Protamine enhancement of RNA uptake by cultured chick cells, *Biochem. Biophys. Res. Commun.* **5**:1–4 (1961).

119. M. C. Niu, C. C. Cordova, and L. C. Niu, Ribonucleic acid-induced changes in mammalian cells, *Proc. Nat. Acad. Sci.*, **47**:1689–1700 (1961).

120. M. R. Schwartz and W. O. Rieke, Appearance of radioactivity in mouse cells after administration of labeled macromolecular RNA, *Science* **136**:152–153 (1962).

121. P. Sourander, Effect of rabies virus on mass and nucleic acids of embryonic nerve cells, *Acta Path. Microbiol. Scandinav.* Suppl. 95, 3–87 (1953).

122. W. E. Watson, An autoradiographic study of the incorporation of nucleic-acid precursors by neurones and glia during nerve regeneration, *J. Physiol.* **180**:741–753 (1965).

123. W. E. Watson, An autoradiographic study of the incorporation of nucleic-acid precursors by neurones and glia during nerve stimulation, *J. Physiol.* **180**:754–765 (1965).

124. W. E. Watson, The nucleic acid content of neuronal nucleoli and neuronal cell bodies after axotomy, *J. Physiol.* **194**:66P (1968).

125. W. E. Watson, Observations on the nucleolar and total cell body nucleic acid of injured nerve cells, *J. Physiol.* **196**:655–676 (1968).

126. W. E. Watson, Alteration of the adherence of glia to neurons following nerve injury, *J. Neurochem.* **13**:536–537 (1966).

127. W. E. Watson, Some quantitative observations upon the oxidation of substrates of the tricarboxylic acid cycle in injured neurons, *J. Neurochem.* **13**:849–856 (1966).

128. W. E. Watson, Quantitative observations upon acetylcholine hydrolase activity of nerve cells after axotomy (abstract), *J. Physiol.* **189**:38P (1967).

129. H. Hydén and P. W. Lange, A differentiation in RNA response in neurons early and late during learning, *Proc. Nat. Acad. Sci.* **53**:946–952 (1965).

130. H. Hydén and E. Egyházi, Glial RNA changes during a learning experiment in rats, *Proc. Nat. Acad. Sci.* **49**:618–624 (1963).

131. H. Hydén and P. Lange, *in Proc. 4th International Neurochemistry Symp.* (S. S. Kety and J. M. Elkes, eds.), pp. 190–199, Pergamon, New York (1961).

132. I. Hochberg and H. Hydén, The cytochemical correlate of motor nerve cells in spastic paralysis, *Acta Physiol. Scandinav.* **17**:Suppl. 60 (1949).

133. C-A. Hamberger and H. Hydén, Cytochemical changes in the cochlear ganglion caused by acoustic stimulation and trauma, *Acta Oto-laryng.* Suppl. 61 (1945).

134. H. Hydén, The nucleoproteins in virus reproduction, *Cold Spring Harbor Symp. Quant. Biol.* **12**:104–114 (1947).

135. A. Hamberger and S. Løvtrup, The effect of brain dehydration on the activity of respiratory enzymes in isolated neurons, neuroglial cells, and in brain mitochondria, *J. Neurochem.* **11**:687–694 (1964).

136. J-E. Edström, D. Eichner, and N. Schor, *in Proc. 4th International Neurochemistry Symp.* (S. S. Kety and J. M. Elkes, eds.), pp. 274–278, Pergamon, New York (1961).

137. A. Hamberger and L. Rinder, Experimental brain concussion: The early effect of sudden increase in intracranial pressure on the succinoxidase activity of isolated neurons and glial cells from the lateral vestibular nucleus of the rabbit, *J. Neuropath. Exp. Neurol.* **25**:68–75 (1966).

138. L. J. Embree, A. Hamberger, and J. Sjöstrand, Quantitative cytochemical studies and histochemistry in experimental neurofibrillary degeneration, *J. Neuropath. Exper. Neurol.* **26**:427–436 (1967).

139. E. Giacobini, A. cytochemical study of the localization of carbonic anhydrase in the nervous system, *J. Neurochem.* **9**:169–177 (1962).

140. D. E. Slagel, H. A. Hartmann, and J-E. Edström, The effect of iminodipropionitrile on the ribonucleic acid content and composition of mesencephalic V cells, anterior horn cells, glial cells and axonal balloons, *J. Neuropath. Exper. Neurol.* **25**:244–253 (1966).
141. J. A. Burdman, Early effects of a nerve growth factor on the RNA content and base ratios of isolated chick embryo sensory ganglia neuroblasts in tissue culture, *J. Neurochem.* **14**:367–371 (1967).
142. J-E. Edström, *in Methods in Cell Physiology* (D. M. Prescott, ed.), Vol. I, pp. 417–447, Academic Press, New York (1964).
143. M. H. Epstein and J. S. O'Connor, Respiration of single cortical neurons and of surrounding neuropils, *J. Neurochem.* **12**:389–395 (1965).
144. H. Hydén and E. Egyházi, Changes in RNA content and base composition in cortical neurons of rats in a learning experiment involving transfer of handedness, *Proc. Nat. Acad. Sci.* **52**:1030–1035 (1964).
145. H. Hydén, *in The Neuron* (H. Hydén, ed.), pp. 179–219, Elsevier, Amsterdam (1967).
146. G. Gomirato and H. Hydén, A biochemical glia error in the Parkinson disease, *Brain* **86**: 773–780 (1963).
147. R. Pavlin, The effect of lysergic acid diethylamide on the acetylcholinesterase activity of single nerve cells from the reticular formation, *J. Neurochem.* **10**:195–199 (1963).
148. O. Hallén, J-O. Edström, and A. Hamberger, Cytochemical response to acoustic stimuli in the spiral ganglion cells of guinea pigs, *Acta Oto-laryng.* **60**:121–128 (1965).
149. O. H. Lowry, N. R. Roberts, and M-L. W. Chang, The analysis of single cells, *J. Biol. Chem.* **222**:97–107 (1956).
150. O. H. Lowry, N. R. Roberts, and J. I. Kapphahn, The fluorometric measurement of pyridine nucleotides, *J. Biol. Chem.* **224**:1047–1064 (1957).
151. A. Hamberger, B. O. H. Rosengren, and B. Tengroth, Radiation studies on the retina, *Acta Ophth.* **42**:951–956 (1964).
152. E. Giacobini, G. Sedvall, and B. Uvnäs, Phosphatidase A activity of isolated rat mast cells and of red blood cells, white blood cells and spinal ganglion cells, *Exp. Cell Res.* **37**: 368–375 (1965).
153. E. Giacobini, Determination of cholinesterase in the cellular components of neurons, *Acta Physiol. Scandinav.* **45**:311–327 (1959).
154. R. Pavlin, Cholinesterases in reticular nerve cells, *J. Neurochem.* **12**:515–518 (1965).
155. E. Giacobini, Localization of carbonic anhydrase in the nervous system, *Science* **134**: 1524–1525 (1961).
156. M. Brzin and J. Zajicek, Quantitative determination of cholinesterase activity in individual end plates of normal and denervated gastrocnemius muscle, *Nature* **181**:626 (1958).
157. C. W. M. Adams, *in Handbook of Neurochemistry* (A. Lajtha, ed.), pp. 525–538, Plenum Press, New York (1969).
158. G. M. Lehrer, unpublished data (1959).

Chapter 12

NERVOUS TISSUES ISOLATED IN CULTURE*

Margaret R. Murray

Departments of Anatomy and Surgery
College of Physicians and Surgeons
Columbia University
New York, New York

I. INTRODUCTION: METHODOLOGICAL RATIONALE

There would be no need for a chapter such as this—whose subject coverage must be eclectic rather than systematic—were it not that the study of isolated yet intact and functioning areas of the nervous system offers a kind of information that can scarcely be gathered in any other way. To cover with any completeness the areas of overlap between the fields of tissue culture and neurochemistry would be beyond the scope of this volume, and certainly beyond the capacity of this writer. A selection is therefore made of representative studies utilizing tissue culture methods to investigate a number of neurochemical topics.

Considerable information of undisputed value has been obtained through cell and monolayer type cultures, as well as by briefly maintained explants of whole embryos and from organ cultures with maintenance periods of up to 2 weeks. These last are usually of a size and opaqueness which impedes or prevents direct microscopic visualization, and requires preparation by the standard histological procedures. Reference will be made to these methods as results are discussed; the techniques are usually described in fair detail in the published works; however, Willmer's *Cells and Tissues in Culture*, Vol. I,[249] and John Paul's textbook,[190] 1965 or 1970 edition, afford useful sources of general tissue culture methodology. Wilt and Wessels[252] provide a compendium of methods in developmental biology which includes a section covering a variety of culture techniques.

During the past 20 years techniques of long-term organized tissue culture have been developed so as to allow repeated observations, at the

* Recent studies in which the author participated were supported mainly by grants from the NIH (NB-00858, AI-08521 and K6-GM-15, 372) and from the NMSS (Nos. 432, 599 and postdoctoral fellowships to coauthors).

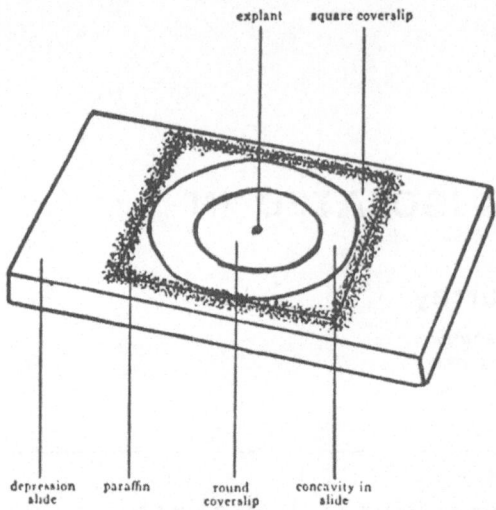

explant   square coverslip

depression   paraffin   round   concavity in
slide   coverslip   slide

Fig. 1. Diagrammatic representation of Maximow-type culture chamber. Actual size of depression slide is $3 \times 1\frac{3}{4}$ inches. This is inverted and culture is incubated in a lying (not hanging) drop of medium.

cellular level, of many neuron types and their associated supporting tissues in stable maintenance, uninjured, for periods of months. Most of these involve the Maximow double-cover slip, depression-slide array described by Murray[169] (Fig. 1). With cultures handled by this basic method, still and time-lapse photographic records can be made sequentially at high magnifications for periods of weeks,[173] bioelectric recordings with microelectrode and modified oscilloscope hookups[46] can be continued for hours; and a variety of chemical agents can be administered directly in physiological concentration to the living cells under continuous observation for days if need be.[167,198] Fluorescence microscopy and electron microscopy are particularly applicable to this type of preparation, in which fixation is immediate and artifacts at a minimum[214,35]; ongoing processes can thus be stopped with precision at a given moment, and embalmed in permanent preparations. The experimental value of the culture's isolation (from systemic factors and topographical relationships) combined with its structural and functional fidelity to the normal uninjured state of nervous tissues *in situ* cannot be overestimated.

On the other hand, substantial disadvantages exist which restrict the uses of this method in chemical investigations: (1) The small size of individual explants ($1–2$ mm^3, maximum); this is compelled by the necessity of achieving gaseous and liquid exchange entirely through diffusion in the absence of vascularity. Microscopic visibility of the living cells (a valuable adjunct) is correspondingly limited to explants $50–100\ \mu$ in thickness. (2) Our inability as yet to utilize adult nervous tissues, which do not long survive explantation in normal states of morphological and functional organization. If a generalization can be made about the most favorable time for explantation of a given area from the neuraxis or its appendages, this would designate a period in development when the overall architecture of the region has become established, (morphogenetic movements having practically ceased) and cytodifferentiation is in its beginnings [e.g. cerebellum of the newborn mouse or rat, spinal cord of 15–17-day murine fetuses; sensory ganglia of 9–10-day chick embryos] (Table I). If explanted during such a period, nervous

TABLE I
Favorable Stages for Explanation[a]

	Otocyst	Spinal cord	Sensory ganglion
Chick	4 day embryo	7–9–10 day embryo	6–9–12 day embryo
⌈Mouse⌉ ⌊Rat ⌋		16–17–18 day fetus	16–18–20 day fetus
	Sympath. ganglion	Hypothalamus	Cerebellum
Chick	13–15–17 day embryo		
⌈Mouse⌉ ⌊Rat ⌋	17–18 day fetus	18–19 day fetus	Newborn
	Cerebral cortex	Retina	
Chick		9–12–15 day embryo	
⌈Rat ⌉ ⌊Mouse⌋	2–3 day postpartum	6–9–12 day postpartum	

[a] The timing is not necessarily exclusive: hypothalamus, for example, is more flexible in its requirements than spinal cord.

tissues recover readily from the incident trauma, and will continue their development *in vitro* to maturity, preserving the vital intercellular relationships of neurons and glia, and some degree of topographical normality (Figs. 2, 3). Myelin sheaths, for example, will be formed *de novo*,[168] as will functional synapses[50] and neurosecretory products.[43] All these developments of course take time, the maturation process advancing in culture at a rate slightly slower than *in vivo*. As a consequence, the material of choice comes from a short-lived and/or rapidly developing species, thus emphasizing the preferability of mouse over man. (3) As will be seen, the overhead is high in time and in skilled labor per culture, especially since all operations must be carried out aseptically and as far as possible in the absence of antibiotics (cf. Section VI, G). Therefore, qualitative determinations are favored over quantitative ones requiring more than micro amounts of tissue. Nevertheless, a number of crucial observations have been made upon organized cultures of the nervous system, as the following text attempts to show; and considerable extension of these is now becoming feasible.

Description of the actual technical procedures will not be undertaken here, since the handling of this type of culture is largely empirical and varies

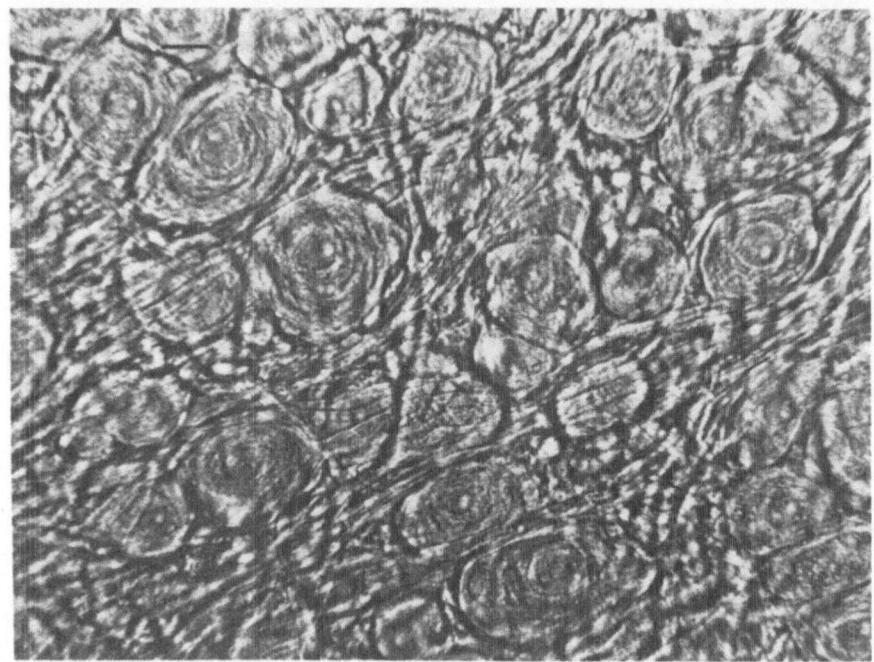

Fig. 2. Living culture of mouse dorsal root ganglion, explant region, 35 days *in vitro*. Nerve fibers run over and between the cell bodies. Bar: 10 μ.

in details according to the nervous region under cultivation, requiring also considerable background knowledge of living cell morphology and behavior (discussed by Murray[169]). It is recommended that those wishing to undertake a program of organized nerve-tissue culture first arrange an apprenticeship for a professional staff member in some established laboratory.

II. EARLY NEURAL DEVELOPMENT IN CULTURE

A. Induction

Since in the earliest stages of vertebrate morphogenesis the dominant tissue is that which will become the nervous system, a word about its induction seems in order. Once the gastrula is formed by invagination of the single cell layer constituting the blastula, morphogenesis begins in the embryonic ectoderm at the dorsal lip of the gastropore. The mechanism of this sudden sally into tissue and organ formation in a spatially restricted area has occupied embryologists for generations. While a wide array of unrelated chemical compounds and tissue extracts (not to mention mechanical procedures such

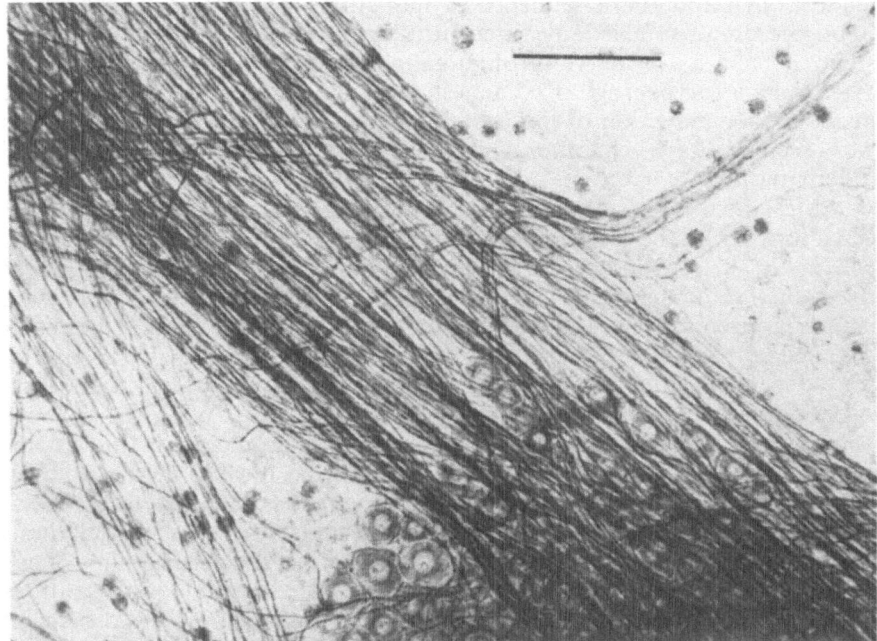

Fig. 3. Rat dorsal root ganglion, 49 days *in vitro*, stained with Sudan black B to illustrate bundles of myelinated fibers. Bar: 100 μ.

as pricking with a needle) have been found to catalyze the induction of new cell types among a population whose genetic equipment is presumably uniform, the inductors' primary site and mechanism of action are still unknown. Current thinking tends to express the problem in terms of regional differential release of factors responsible for gene activation or derepression. Reduced to lowest terms, a search is being directed toward understanding how simple factors such as local change in ion concentration or pH can alter energy-yielding systems and gene function so as to induce tissue differentiation.[7]

Defining induction as the primary stimulus by which cells of a given area are caused to differentiate in a manner different from their apparent destiny, one may investigate (as the Barths have done) the means by which various types of nerve and pigment cells can be evoked from presumptive epidermis (peripheral to the neural plate area). These workers observed that at this stage of development in frog gastrulae, naturally occurring cations such as Ca^{2+} and Mg^{2+} used at concentrations well below the osmotic shock level will induce both ganglionic and motor nerve patterns. Subsequently other natural and "unnatural" cations were likewise employed as tools; anions, viz., Cl^-, Br^-, SO_4^{2+}, were found to be irrelevant. In small explants made from ectoderm, a sequence of inductions was obtained, dependent upon the duration and ionic concentration of the treatments, proceeding stepwise from motor nerve through ganglionic nerve, pigment cells, astrocytes, and other neuroglia.[6] Each step in the sequence was found to be necessary for induction of the next cell type, and each stepwise induction was reversible for 2–3 hr, after which cells were committed to a new type of differentiation.

Motivated by an observation of Ross and Scruggs[211] that the effects of cations upon the electrophoretic mobility of DNA followed the order $Li^+ \gg Na^+ > K^+$, and $Mn^{2+} \gg Mg^{2+} > Ca^{2+}$, the Barths ascertained that the induction of nerve and pigment cells followed the same order and with similar concentrations. Further study revealed that the neural induction initiated by Li^+ and other ions (also by sucrose which was used as a control) were dependent upon an immediate sudden rise in the local concentration of endogenous Na^+ ion, to the equivalent of 88 mM NaCl. With Na^+ at lower concentrations (e.g. 44 mM) other tissue types could be induced in this system; but once differentiation in any direction took place, it could not be modified by Na^+. Since sodium involvement is known to characterize transport at many levels—tissue, cellular, nuclear and subcellular, the working hypothesis was presented that some inductors merely increase the permeability of cell membranes, and the ions in the medium complete the induction—cations possibly combining with nucleic acid PO_4 groups, with the consequent freeing of DNA or RNA from proteins.

B. Morphogenesis, Teratogeny

As development proceeds beyond the stage of primary induction, exposure of the burgeoning neuraxis to various types of inhibitors results in its ablation or teratological modification. The experimental literature is too large and confused to be digested here, beyond the generalization foreshadowed in the above account of induction by cations—namely that the action of experimental agents may be related directly in intensity to the developmental stages at which they are administered, though there is some tissue selectivity reflected in concentration effectiveness. As the embryo develops, centers of morphogenesis shift from neuraxis to somitic (muscle and vertebral) axis, limbs, viscera, special sense organs etc., with susceptibility to injury following the same pattern.

To specify with a few examples other than the most familiar blockers of synthesis or transport: Calf-thymus histone fractions applied to chick embryos explanted at Hamburger and Hamilton's[78] stages 4–6 (head-fold) have a morphostatic effect on CNS and brain development comparable to that of actinomycin D, which acts through inhibition of DNA-dependent RNA synthesis.[142] Histones applied in stages 6–7 (brain development) bring about suppressed or disorganized brain formation (also like actinomycin); but if they are applied at stages 8–10, the result is suppression of the somitic axis.[222] DNP at ∼ millimolar strength produces almost complete absence of neural tissue or severe neural abnormalities in stages 4–7 (head-fold to 2 somites); lower concentrations are ineffective. This action is comparable to that of N-mustards.[24] D-Chloramphenicol at 200–300 μg/ml causes failure of the chick neural-tube to close; the L-compound does not give rise to abnormalities.[15] Chloracetophenone (CAP) at 0.5 mM applied briefly to stage 4 (primitive streak) embryos produces abnormalities predominantly in the nervous system. The effects of CAP can be reversed by subsequent treatment with equimolar SH compounds such as thiomalic acid or GSH; equimolar cystine, but not serine, also achieves reversal.[166] Buckingham and Herrmann,[28] seeking to produce abnormalities in developing chick muscle with 3-acetylpyridine, could do so only at stage 15; protein synthesis, measured by incorporation of glycine-^{14}C, was especially reduced in the nervous system which was, however, not so much distorted as stunted in growth.

Margolis and Kilham[144–146] contribute relevant observations on the action of parvoviruses upon the developing brain of model animal systems. These DNA viruses have an affinity for dividing cells; cells in culture are susceptible to the organisms in direct proportion to their capacity to synthesize DNA. Therefore the same virus produces different lesions in developing individuals, depending on the tissue or system which is growing at the time of infection. Cerebellar hypoplasia has been produced in hamsters by reovirus inoculation at the time of granule-cell multiplication in the external germinal layer; hydrocephalus also has been induced experimentally. If supporting cells are attacked during their phase of multiplication and attachment to large neurons, the neurons will not survive; and since at this stage neuron regeneration cannot take place, lesions may be permanent and crippling.

For selected discussions in depth of developmental chemistry, readers may wish to consult two symposia[129,130] on developmental biology edited by Locke in 1967 and 1968. An early summary (1958) of the then more manageable field was published by McElroy and Glass.[140]

C. Growth

This long-investigated topic is still fragmented in its chemical application to the nervous system, as well as to somatic tissues, except in relation to the nerve growth factor, which is treated in this series as a division in its own right. Since the problems of axon growth and material transport are similarly gathered under special headings, in which such information as can be gleaned from cultures may be included, economy suggests their virtual omission here. Attention is directed to the recent CIBA Symposium[258] on growth of the nervous system.

Perhaps a group of observations made in the writer's laboratory deserve mention because of their novelty. These concern the action of heavy water (D_2O) upon diverse regions of the nervous system in isolation. D_2O in sublethal dosage (up to 25 % of the total ambient medium) exerts a direct stimulatory effect on cellular growth and differentiation, its detailed action varying according to explant source.[170,171] This action is qualitatively and quantitatively different from that of nerve growth factor applied to the same material in culture.

These studies originated from evidence of behavioral disturbances reported by others in adult mice that were subjected to moderate degrees of H_2O replacement by D_2O administered in their drinking water. In mice and rats (Thomson[238]), overt signs of D_2O intoxication are not observed until about 15 % of the body water has been replaced; above this level the animals become hyperexcitable and hard to handle. In the range of 20–25 % D_2O, combativeness is very pronounced and convulsions may occur following even mild stimulation. Mice show increased pilomotor activity and exophthalmos, which have been interpreted as signs of sympathetic erethism, since these symptoms can be simulated by epinephrine and are reversed by ergotoxin. When 30 % of the body water has been replaced, the animals become comatose, and they die before the D_2O concentration reaches 35 %. Mice can be maintained for several months with no apparent ill effects at 20 % saturation, though showing an elevated metabolic rate and body temperature. Notwithstanding the clinical suggestion of autonomic (ANS) involvement in deuterium intoxication, autopsy shows no gross morphological NS abnormalities; kidney and submaxillary salivary glands, however, are selectively damaged. These organs are concerned especially with H_2O transfer.

Nervous tissues in culture respond to D_2O by accelerated growth and maturation; also in some situations by multiplication of differentiated neurons. Submaxillary gland, but not parotid gland, dies rapidly when explanted into D_2O-containing media. Histologically different regions of the neuraxis differ in their response to deuterium. Sympathetic ganglia are the most greatly stimulated: deuteration up to 25 % accelerates and increases growth of neurons and favors their repeated subdivision as an abnormally large size is attained (Fig. 4). The total neuronal population eventually increases twofold or threefold; in turn the progeny—clusters of small, neuroblastlike cells—enlarge and become multipolar. For these light microscopic observations the ganglia of embryonic chicks and rodents were maintained as organized, developing cultures for 2 months or more, during which time they were continuously exposed to D_2O incorporated into their feeding medium as 5, $12\frac{1}{2}$ or 25 %. Cultures exposed to 33 % showed initial stimulation, but died in about 2 weeks. Withdrawal of D_2O after 42 days *in vitro* was not followed by retrogression: the cell population subsequently remained static for at least 2 months.

Electron micrographs reveal an unusual abundance of fibrous and granular elements in the nuclei and cytoplasm of deuterated sympathetic neurons and supporting cells. Neuronal nucleoli are structurally more compact, and intranuclear rodlets (see Section II, D) are more prominent than in controls. Mitochondria may be abnormally dense, and microtubules unusually numerous.

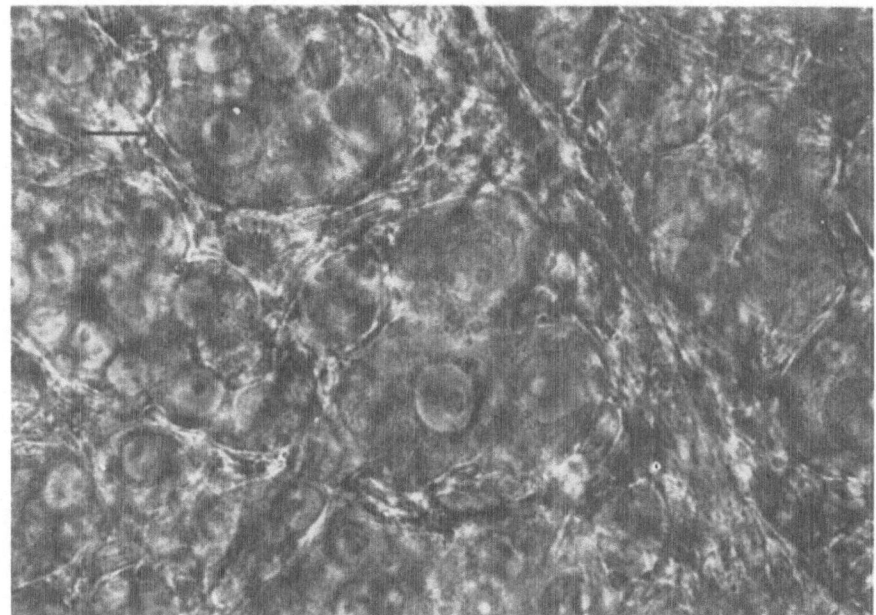

Fig. 4. Sympathetic ganglion from 14-day chick embryo, maintained 54 days in medium containing 25% D_2O. Note clusters of large and small neurons. (After nearly 2 months in culture, control neurons are less numerous and are generally mature.) (Figures 4 and 5 are reproduced from *Growth of the Nervous System*, Churchill, 1968.)[171] Bar: 10 μ.

The response of hypothalamus to D_2O involves both neurons and glia. A neuron (with satellite cell attached) has been followed through mitosis at 16 days *in vitro*. Growth of cerebellar neurons, extension of neurites, migration of glial cells, and myelination are all accelerated by 10% D_2O during early culture stages. Later the controls catch up. In cerebral cortex, 10% D_2O accelerates maturation; some dedifferentiation follows later. In sensory (spinal) ganglia, growth and multiplication of supporting cells is the most conspicuous response to deuteration. Neurons become unusually turgid, and sometimes lobate. They tend to deteriorate as they are abandoned by the activated satellite cells.

In evaluating these observations, one should recall that in the higher vertebrates, neurons once developed beyond the neuroblast stage are not known to divide, with a possible rare exception among sympathetic ganglion cells. Mature neurons almost entirely cease producing chromatin DNA, although they are characteristically prolific in RNA production and utilization. Yet in these deuterated cultures of sympathetic ganglia and hypothalamus we see positive evidence of division in large multipolar cells with which satellite cells have actually established the adult relationship. It may therefore be of particular significance that the nuclei of deuterated neurons are very densely packed with structured particles, many of which conform to the concept of ribosomes. This condition extends also to the specific supporting cells. Such cells *in situ*, though retaining throughout life their capacity for division, normally assume in maturity a stable relationship with neurons, and multiply only in response to trauma or other insult. Schwann and satellite cells continue to divide in maturing deuterated cultures as they do in early stages of development.

In this spectrum of nervous tissues direct application of D_2O induces not only proliferation of normally nondividing cells, but at the same time extensive cell growth with varying degrees of cytodifferentiation. Such a paradoxical combination of developmental processes, registered morphologically, tempts speculation as to possible effects of deuterium upon macromolecules, especially nucleoproteins in transcription and/or translation. Might D_2O whose physical characteristics differ appreciably from those of H_2O, be acting mechanically at the many available membrane surfaces to change sterical configurations, e.g., to prolong or reinstate the early normal activity of DNA polymerase? Deuterium is known to favor the polymerization of microtubular spindle proteins, and will prevent or reverse the action of colchicine in undermining the cytoarchitecture of neurons. Might it inactivate histones, in either case derepressing DNA? And/or might it shift *m*RNA signals for protein synthesis? It seems possible that some of these effects might be related to an exchange of D with readily exchangeable H bonded to N, O, and S in membranes or nucleoproteins.

D. Cytodifferentiation

Though this subject increasingly occupies chemically oriented embryologists, little mileage has accrued so far to the nervous system in culture. However, an outstanding body of work on the circumstances surrounding induction of glutamine synthetase in embryonic neural retina has been put forth by A. Moscona and his associates: these studies deserve some detailed attention here because of their use of a model culture system for investigating control mechanisms in differentiation. As summarized by Moscona[163] and by Moscona et al.[164]:

Glutamine synthetase (GS) has a characteristic and unique developmental pattern in the embryonic chick retina, marked by a very steep rise in activity that begins after the sixteenth day of development and continues until after hatching; it coincides closely with other aspects of retinal differentiation.[201] These workers found that the characteristic sharp rise of retinal GS could be induced precociously, several days ahead of the normal schedule, by growing retinal organ cultures in adult horse serum. Upon fractionating the serum, they discovered that the "inductor" was one of the 11-β-hydroxy corticosteroids, e.g., hydrocortisone.[200] Other hormones, including other types of steroids, as well as cyclic AMP, were ineffective.[165] Both the precocious and the normally timed rises in GS activity require continuous protein synthesis, and immunochemical evidence suggests that the steroid functions by increasing the rate of enzyme synthesis.[1] Inhibition of protein synthesis in the cultured retina at the translational level by cycloheximide completely blocks the steroid-induced increase in GS activity. However, genomic control appears to operate during the first 12 hr at least, since at this time, though not in later stages, actinomycin D also blocks induction.[164] In further study of timing,[162] the relatively long-term organ culture method was exchanged for monolayer culture of trypsin-dissociated retinal cells at various embryonic ages—10, 12, 14, and 16 days. It was found that cells from the younger embryos, which were capable of reaggregating into large clumps, exhibited a very much higher responsiveness to GS induction than did the more differentiated cells from the older embryos (approaching the normal induction time) which were able to re-form only small, first-stage associative groups. Response to the steroid inducer is therefore organization-dependent.

Using as test object dissociated retina from early chick embryos (8 days), Orr and Roseman[186] report the purification from horse serum of a "neural retina-aggregating protein" (NRP) which promotes cell aggregation at 5°C at 1 μg/ml. They suggest that it may be specific for neural retina, since it is ineffective on other tissues tested (e.g., dissociated limb-bud cells). This protein is a congener of horse globulin (IgM), but is not identical with it.

Retinal degeneration in neonatal mice has been reported[185] following parenteral injection of a single large dose (4 mg/g) of sodium glutamate at this critical period in murine retinal differentiation. Olney's observations by electron microscopy that ganglion cells and dendrites of the inner neuronal layer swell and burst, confirm earlier data cited by Lucas[132] and by Lolley (Chapter 20, Vol. 2 of this Handbook) on the selective toxicity of large amounts of glutamate to the synaptic layer of adult mammalian (and reptile) retina. For *in vitro* media containing suitable electrolytes, the toxic level of the directly applied glutamate probably is as low as 0.5–1.0 mM, though there are species differences.

Rod cells in retinas explanted from 1-day-old rats and cultured through the differentiation period display a selective susceptibility (compared to other retinal and brain tissues) to thiol inhibitors at very low concentrations (5×10^{-6}M). SH donors protect the rod cells against the degenerative action of SH blockers, whose effects are most intense at about 8 days *in vitro*, the beginning of outer segment differentiation.[83] This experimental picture is very similar to the course of rod cell degeneration in a heritable type of retinal dystrophy also studied by Hansson.[81] Normal retinas *in vivo* do not respond to the low concentrations of thiol inhibitors which bring about selective rod cell degeneration in culture. For further discussion of inherited retinal dystrophy, see Section IV, B.

The origin and development of the "intranuclear rodlet" long observed in a variety of nerve cells *in situ*, have been followed with the light microscope and electron microscope by Masurovsky *et al.*[149] in long-term organized cultures of sympathetic ganglia (Fig. 5). The rodlet consists of bundles of ~70 Å filaments associated with granular-fibrillar spheroidal bodies ~0.5 μ

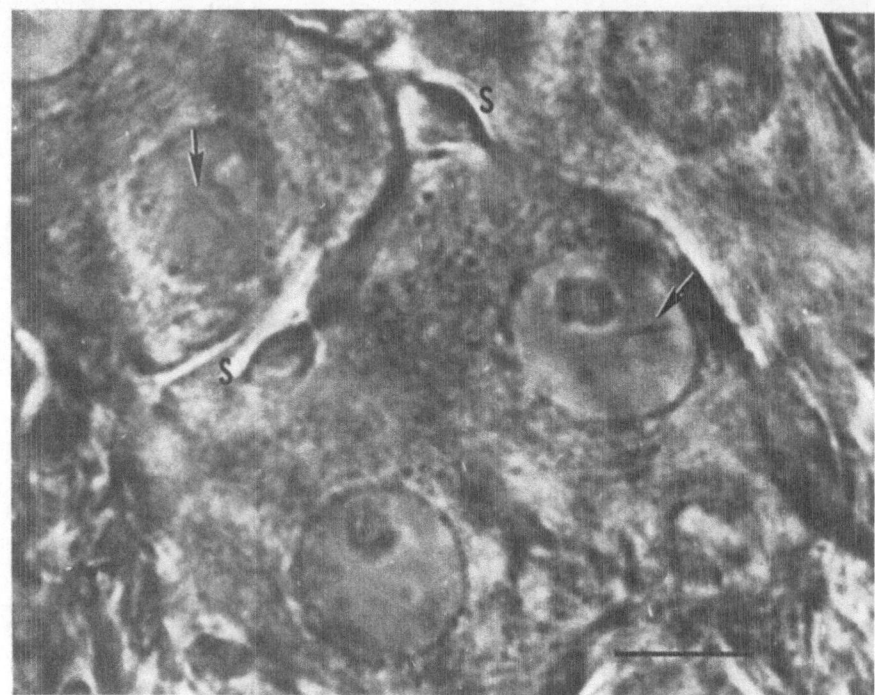

Fig. 5. Sympathetic neurons, 54 days in culture. Note intranuclear rodlets (arrows). S, satellite cells. Bar: 10 μ.

in diameter which are thought to elaborate the rodlet bundle under the influence of the nucleolus. Filaments pass from the rodlet through the nuclear envelope into the cytoplasm of the neuron where it is suggested that they give rise to at least some of the cell's microfibrillar constituents. Histochemically the rodlet bundle appears to consist of nonhistone, nonnucleic acid protein (Kim et al.[111,112]), which includes arginine, lysine, histidine, tyrosine, tryptophan, and possibly aspartic and glutamic acids.

III. METABOLISM (NORMAL) IN CULTURE

Little of a systematic or quantitative nature has been reported on the metabolism of nervous tissues in culture, though there is considerable literature on other types of cells which can be cloned or grown *en masse* as monolayers. For these last, summaries in 1965 of carbohydrate and energy metabolism by J. Paul,[191] and of amino acid and protein metabolism by Eagle and Levintow,[62] also J. A. Lucy,[133] are available; however they probably have limited applicability to our subject since the energetics of rapidly growing and multiplying cells differ substantially from those of relatively stable cell populations engaged in specialized functions.

Introduction of the complex chemically defined media (cf. Paul[190]) which have revolutionized culture methods for somatic cell strains initially raised high hopes among the keepers of nerve cells.[197] Such hopes have remained largely unfulfilled. The defined media will not long support primary explants of any sort without a considerable supplement in "natural" components (generally serum and/or tissue extracts). Thus, since maturing neurons do not multiply, and survive only in primary explants, rigorous investigation of their nutrition via defined media has so far been lacking. Recent reports of improved techniques for dissociation of brain tissue and its subsequent mass cultivation in degrees of cytodifferentiation seem to hold promise.[10,241] In the meantime, histochemical and tracer studies of organized explants of nervous tissues have been feasible and somewhat rewarding.

A. Carbohydrates

The exceptionally high glucose utilization coupled with high O_2 consumption which characterizes nervous tissues, especially CNS, has dictated histochemical studies of oxidative enzymes, whose reaction product can be visualized intracellularly in organized cultures, and related to both developmental stages and topographical distribution among neurons and glia. From the standpoint of nutrition, it was early noted that neuron survival and myelin formation were substantially enhanced by elevation of the culture medium's glucose content to 3–10 times that usually employed for somatic tissues.[204,92,23,177] Myelinating cerebellum and dorsal root ganglion (DRG) cultures were examined by Yonezawa et al.[261,262] before, during, and after myelination, for the reactivity pattern and time sequence presented by succinic dehydrogenase (SD), NAD and NADP diaphorases, and cytochrome oxidase (CO). Cultures were maintained on the optimal glucose concentration of 600 mg% in standard feeding medium.

CNS: During the whole course of cultivation (5 weeks) Purkinje cells and neurons from the dentate and roof nuclei showed high activity of all four enzymes; however neuroglia diverged from this pattern. In astrocytes reaction granules were diffusely distributed in perikarya and processes throughout the cultivation period; but in oligodendroglia granules appeared first in perinuclear array, then spread in increasing numbers through the processes, and at the onset of myelination, aligned themselves along myelinating axons. In older cultures, in which new formation of myelin had decreased, reactivity of SD, NADH, and NADPH subsided coincidently, but CO activity continued at a constant level.

In PNS, the sensory neurons were a little slower to develop SD and NADH reactivity, but maximal product appeared at the time of myelin formation. Supporting cells showed NADH first, the other enzymes appearing later and all exhibiting a remarkable rise within Schwann cells just prior to myelin appearance—when granules were found outlining the segments. Activity of the first three enzymes subsided as myelinization was completed, but that of CO continued as in CNS cultures. Apparently these differences in activity between CO and the dehydrogenase group in culture parallel their *in vivo* accepted roles, CO being concerned more with respiration and maintenance, while the dehydrogenases participate also through the Krebs cycle in phospholipid production. Later, Yonezawa and Iwanami,[265] using cultured cerebellum for a study of thiamine deficiency in the presence of antimetabolites (oxythiamine or pyrithiamine) produced a chronic type of deficiency characterized by degeneration of preformed myelin sheaths; this could be prevented by coincident administration of cocarboxylase. Since in mature myelin the Krebs cycle enzyme activity should have subsided, and have been replaced by activity of G6PDH and 6PGDH,[260] the authors suggested that thiamine deficiency in this material interferes with trans-ketolation in the hexose monophosphate shunt.

Monoamine oxidase activity in mature myelinated cultures of mouse DRG and cerebellar folia was reported by Kim and Murray.[113] Moderate to strong formazan deposits were observed (by the tryptamine-nitro-BT method) in neuron somas and myelin sheaths, but not in glial elements— which showed only weak MAO reactivity. Without reference to myelinization, Kim[110] investigated the activity of eight oxidative enzymes in cerebellar neurons maintained in culture up to 60 days in a 500 mg% glucose medium. All of these (NADH and NADPH tetrazolium reductases; succinate, malate, *iso*-citrate, lactate, α-glycerophosphate, and glucose-6-phosphate dehydrogenases) showed strong reactivity though somewhat different intracellular localizations. A spotty SD distribution at the surface of neuron somas suggested to Kim enzymatic activity of mitochondria in synaptic boutons, whose existence had been demonstrated in similar preparations by silver staining. Kim concluded from the overall reactivity of this array of enzymes that the pentose phosphate pathway, hexose monophosphate shunt, and glycolytic pathway were being utilized by his cultured cerebellar neurons.

However, Hoskin and Allerand,[101] using the same material under improved culture conditions, traced the metabolism of specifically labeled glucose-1-^{14}C and 6-^{14}C to [^{14}C]O$_2$ and since they found no departure from the 1:1 C-6/C-1 ratio, concluded that there was virtually no participation of the pentose phosphate pathway even during active myelination— contrary to the conclusions of Yonezawa et al.,[261] who showed histochemically a selective reactivity of Krebs cycle enzymes in the presumed myelin-forming cell. The metabolic rate in Hoskin and Allerand's cultured material was very much higher than that of cerebellar slices *in vitro*, carried simultaneously. Suyeoka,[237] attempting to grow cerebella from mice 1–2 weeks postpartum, found that the presence of pyruvate in the standard high glucose medium was stimulating to migration and survival.

A method for microdetermination of glucose metabolism in single cerebellar cultures, using antipyrine-^{3}H as marker was recently described by Lehrer and Bornstein,[127] by which measurements could be made in terms of mmoles/kg dry weight of culture per hour. Readings were taken before, during, and at the close of active myelination in Maximow cultures of newborn rat cerebellum. These showed glucose consumption doubling between the first two stages, and leveling off into the third (101, 194, and 213 mmoles/kg/hour); parallel lactate production in the same periods jumped sevenfold and then declined somewhat (1.04, 7.40, and 5.75 mmoles/ kg/hour). Calculated on a wet weight basis this would give ~ 25 mmoles/ kg/hour glucose in stage 1 as against 50–55 in stage 2, which agrees well with *in vivo* measurements by other methods during early postnatal development. Lactate production was extremely low. In earlier work[126] these authors followed seven enzymes of the glycolytic and Krebs cycles and pentose phosphate shunt in the internal granular layer of similar developing cerebellar cultures. G6PDH, 6PGDH, and NADP-dependent isocitric DH activity rose sharply to peak at the beginning of myelination, with continuing high levels in the cultures, though *in vivo* these undergo a gradual decline. The other enzymes followed the *in vivo* trends.

In sum, it appears that both of the principal energy-producing pathways, and the hexose monophosphate shunt are followed in cultured CNS and PNS, the preponderance of a given pathway varying with the stage of maturation. Embryonic tissues *in vivo* are known to rely considerably on anaerobic production of energy from glucose.

B. Proteins and Nucleic Acids

On this subject some suggestive though not definitive observations have emerged from tracer studies of nervous tissues isolated in culture.

Using lysine-^{14}C on cultured dorsal root ganglia (DRG) Bunge and Bunge[32] studied by electron microscopy the distribution of the grains incorporated in neurons, finding 57% concentrated over Nissl bodies, 20% over the "general cytoplasm," 19% over the neuron-satellite-cell boundaries, 3% over the nucleus, and 10% over the Golgi complex. They expressed doubt therefore that this last organelle functions as a packaging center in protein production. Utakoji and Hsu[240] in light microscope studies of several tritiated amino acids in DRG concluded that protein synthesis takes place only in the pyrenophore; by their count, products are transported distally along cultured fibers at the rate of 10 mm/day. Incorporation of leucine-^{3}H into embryonic chick spinal cord[44] occurs first in neuron perikarya; later radioactivity spreads down into the processes.

Burdman[38] using DRG, reported conversion of tritiated adenine to guanine, and cytosine to uracil. The addition of NGF to the medium resulted in increased rate of conversion of the purines to RNA precursors and eventually to RNA. Myelinated cerebellar cultures (of rat) were used by Appel and Silberberg[4] to assess salvage and *de novo* pathways of pyrimidine synthesis, showing that orotic acid-^{14}C and carbonyl aspartic acid were readily incorporated into UMP and into perchloric acid-insoluble RNA; they could be blocked by azauridine. Uridine-^{3}H also rapidly entered the cultures and was similarly incorporated; it was partially inhibited by azauridine. Citrulline was not incorporated into UMP or blocked by azauridine; nor was $NaH[^{14}C]O_3$. The authors conclude that the salvage pathway has potential importance, and the brain depends upon both exogenous and endogenous pyridine precursors.

Utakoji and Hsu[240] and Courtey and Bassleer[44] found that uridine-^{3}H was actively incorporated into their neurons; thymidine-^{3}H was incorporated into spindle cells present in the cultures, but not significantly into neurons; DNA was therefore hardly being synthesized at all. Amaldi et al.,[1a] by applying (^{3}H)uridine to dissociated sensory ganglia, found (by autoradiography) that RNA was present in the naked nerve fibers.

Zanini and Angeletti[268] made an immunochemical assay in chick DRG cultured with and without NGF, for presence of the acidic protein (S-100) that is regarded by Moore et al.[161] as characteristic of the nervous system (cf. Chapter 6, Vol. I, this Handbook). Amounts of NGF which stimulated the typical precocious growth of a halo of neurites with considerable net protein synthesis failed to produce any S-100; they found little in embryonic chick brain; more in adult brain in situ. S-100 appears in highest concentration in the brain of adult mammals; however, Moore reports its presence in rabbit PNS. Pfeiffer et al.[199a] find the S-100 protein appearing in a clonal line of rat glial cells whenever they cease to proliferate and manifest other signs of differentiation.

Yonezawa et al.[266] produced pyridoxine deficiency in myelinating cultures of murine DRG and cerebellum by the use of the analogue deoxypridoxine (DOP) and of isonicotinic acid hydrazide (INH)—considered as causative agents for the deficiency in vivo. High concentrations (10^{-3}M) of these inhibitors were required in culture to decrease substantially the rate of myelin formation, and several weeks' exposure to produce degenerative changes in neurons. If the media contained 25% of serum, cells did not develop deficiency signs even after 2 months. When cultures were maintained on low protein media (less than 15% serum), injury from DOP could be prevented by equimolar pyridoxine, but once moderate damage had occurred, recovery of neurons could take place only in high serum media; seriously damaged cells did not recover at all. INH was more damaging to myelin and supporting cells than to neurons; this injury could not be prevented or counteracted by pyridoxine, and only initial changes were susceptible to reversal by withdrawal of the inhibitor. It would seem that these inhibitors applied directly interfere with protein metabolism, possibly on a competitive basis.

It has been questioned whether neurons are able to take up or utilize exogenous proteins. It is well known that pinocytosis occurs copiously in the growing tips of neurites (e.g., Pomerat et al.[206]) and by inference, that considerable amounts of protein must be swept in from serum-containing media to travel proximad along the fiber. Whether this material ever reaches the pyrenophore, or whether macromolecules may enter either neurite or soma directly from ambient fluids, has not been known, though Klatzo and Miquel showed[117] that fluorescein-labeled protein readily entered glial cells by pinocytosis *in vitro*, and probably protein would do so (from ambient edema) *in vivo*. Recently Holtzman and Peterson[100] reported in electron microscopic studies the uptake of horseradish peroxidase by both autonomic neurons *in vivo* and DRG neurons in culture. First (at 4 hr) the enzyme appeared in unmyelinated axons; by 17–19 hr after administration it was definitely concentrated in the perikarya in a lysosomelike pattern. These workers hold that the neuron soma and fiber are normally pinocytotic,[99] and that the peroxidase is engulfed from the perineuronal space which it has entered through the Schwann and satellite cells; however, they observed very little of it in myelinated nerve fibers. Similarly, Hirano *et al.*[93] working with peroxidase implanted *in vivo* found it to be very nearly excluded from the periaxonal space and the myelinated fibers of rat forebrain, though it filled the extracellular spaces. Smaller molecules such as labeled amino acids do, however, enter into peripheral nerve fibers through the myelin sheath, and are incorporated there into proteins, as reported by Singer and co-workers (cf. this Handbook, Vol. 2, Chapters 17, 18; Caston and Singer[40]). In CNS, Hirano and Dembitzer[94] describe the passage of lanthanum through the transverse bands at the nodes of Ranvier, between myelin lamellae and into the periaxonal space of the myelinated fiber.

Recently a presumptive physiological mechanism for the exchange of materials between ambient fluids and the periaxonal space of myelinated peripheral fibers has been demonstrated cinemicrographically by Murray and Herrmann[173] in the mouse and by Singer and Bryant[233] in the frog. This function involves the Schmidt–Lanterman clefts, which normally stand open, but are periodically squeezed shut by rippling contractions of the Schwannian cytoplasm along the internode. In the mouse, these brief occlusions occur somewhat irregularly but at intervals of the order of 15 min, following which the clefts gape open again like the interstices of a squeezed sponge. The Schwannian contractions travel in both directions— proximad and distad—reversing themselves frequently; they are in no sense "peristaltic waves" which might serve to propel axoplasmic constituents from the pyrenophore toward the periphery. ("Colic" describes them better.) It should be emphasized that these observations were made over a period of 10 days on mature, intact, myelinated fibers which had developed and been maintained in culture without transfer or insult of any kind (except photography) for 4 months.

C. Lipids

Though myelin lipids are a major concern of neurochemistry, and various morphological aspects of myelin are prominent in the tissue culture literature, the latter has afforded few direct studies of lipid metabolism. In a general discussion of glial cells in their relationship to the central myelin sheath, Bunge[31] remarks that the "oligo" is a metabolically active cell and its synthetic machinery seems especially equipped for lipid synthesis.

This judgment holds true even more for the Schwann cell, which because of its greater accessibility for microscopic observation in culture provided material for autoradiographic studies of choline incorporation into peripheral nerve myelin by Hendelman and Bunge.[86] Using the tritiated form of this myelin precursor they examined (with light and electron microscopy) the pattern and degree of labeling in the segment as a whole during myelin development, finding (1) that radioactivity appeared all along the length of the internode without there being a preferential site of initial incorporation, (2) that the myelin–Schwann cell unit in the fully myelinated cultures incorporated choline as actively as it did in newly myelinating cultures. Their observations support the concept that a significant amount of myelin-related metabolic activity occurs in mature tissue, thus indicating a continuous process of myelin maintenance and repair (cf. Chapter 21, Vol. 3, this Handbook).

Synthesis of a myelin lipid, sulfatide, in myelinating CNS cultures was observed by Silberberg et al.[229] with the incorporation of $Na_2{}^{35}SO_4$ into newborn rat cerebellum. Cultures showed a low level of labeled sulfatide in their first 6 days, then a sharply increasing rate of incorporation from the seventh to the twenty-first day in culture, which coincides with the period of active myelination, and is quite similar to the pattern in vivo, where the peak incorporation is around the eighteenth to nineteenth day.

Inferences concerning the role of glucose in myelin synthesis are mentioned under carbohydrates (Section III, B). Observing synthesis of ATP by the myelin of rabbit spinal nerves, Miani et al.[157] report that practically all the glycolytic enzymes (of the Embden–Myerhof pathway) are present in the myelin. Metabolic disorders involving myelin are discussed below.

IV. INBORN ERRORS OF METABOLISM

A. Heritable Human Disease

Metabolic abnormalities which cause disease syndromes are frequently, though not uniformly, reflected in cells isolated from the affected patients. Reference is made below to observations in culture on several diseases which bring about neurological damage, also to attempts to reproduce or further explain the disease signs by experiments with cultured cells.

1. Galactosemia

Abnormal carbohydrate metabolism is represented by galactosemia, now known to be transmitted by a single autosomal recessive; it is characterized by a deficiency of galactose-1-phosphate uridyl transferase.[235] Krooth and Weinberg[122] cultured skin biopsies from homozygous patients and their heterozygous relatives, also normal individuals, in the presence of glucose-1-^{14}C or galactose-1-^{14}C, finding that the galactosemic cells were virtually unable to oxidize galactose, though normal and heterozygous cells could grow equally well in both media. Miller et al.[158] confirmed these results, showing that the galactosemic cells died within 72 hr if cultured in media containing 100 mg% galactose, but survived in equal parts of 50 mg% galactose and glucose. Electron microscopy showed massive dilation of endoplasmic reticulum in the damaged cells.

2. Phenylketonuria

Abnormal amino acid metabolism is represented by several studies: In phenylketonuria (PKU), which is genetically based as an autosomal recessive, phenylalanine hydroxylase fails to differentiate in the liver after birth, and this deficiency allows the accumulation of L-phenylalanine in abnormal amounts, whereas a large part of the compound taken orally should be metabolized to tyrosine. A concatenated dyscrasia follows, affecting particularly the developing brain, in which myelin shows histological lesions. In an effort to use histopathologic features of PKU as an index by which to judge relative toxicity of PKU metabolites, myelinating cultures of newborn rat cerebellum were exposed to various metabolites by Silberberg.[227] Only certain compounds—and not L-phenylalanine—were found to be harmful during the length of the culture period. Indole 3-acetic, -pyruvic, and -lactic acids were most toxic on an equimolar basis. At blood levels and higher (up to 9×10^{-3}M), phenylalanine and closely related congeners were harmless. The indole derivatives (0.85×10^{-3}M) began to exert their effects at 10–14 days i.v.—the period of active myelin formation, first delaying it and eventually bringing about degeneration of such myelin as was formed. Other types of compounds did not affect myelin, and only a high concentrations (up to 14×10^{-3}M) induced granular degeneration of neurons and glia. Applying L-phenylalanine at 50–100 mg/100 ml to newborn canine neocortex, Liss and Grumer[128] reported perikaryal damage during a culture period up to 53 days.

3. Branched-Chain Ketonuria

Maple syrup urine disease (branched-chain ketonuria) involves a deficiency in the oxidative decarboxylation of valine, leucine, and isoleucine. Pathological findings are widespread, showing especially marked alterations in cerebral myelin formation. Using the model system of myelinating cultures from rat cerebellum, Silberberg[228] found that decreased-to-absent and delayed myelination were produced by α-keto isocaproic acid at physiological levels; at higher levels, injury was noted to small cells regarded as glia, but not to neurons. The primary branched-chain amino acids and alloisoleucine had no visible effect in concentrations up to 6.0 mM. α-Keto L-valine and α-keto L-isoleucine did not produce injury at the 3.0 mM level—at which α-keto isocaproic acid was severely damaging. Dancis et al.,[55] working with MSU skin fibroblasts in culture, demonstrated in them considerable reduction of the ability to decarboxylate all three branched-chain keto acids. They suggest that one autosomal gene may control the synthesis of a polypeptide that is common to the three enzymes involved.

4. Refsum's Disease

Refsum's disease (hereditary polyneuritic ataxia) appears to involve an oxidative enzyme defect[236,160] in the first step of phytanic acid degradation which normally leads to formation of α-hydroxyphytanic acid. It is characterized by a hypertrophic neuropathy whose initial site may be the Schwann cell,[64] as a repository of the unmetabolized branched-chain fatty acid. Herdon et al.[88] have shown that though fibroblasts cultured from skin biopsies of patients did not (in a 3-day period) contain elevated levels of phytanate, yet they had rates of phytanate ^{14}C oxidation less than 3% of those from normal subjects. Phytanic acid degradation products were oxidized at normal rates by both Refsum's and normal fibroblasts—thus supporting Steinberg's original hypothesis of a single enzyme defect located at the first step in oxidation. Using normal fetal mouse DRG, Dubois-Dalcq and Gorce [61a] applied phytanic acid to cultures in graded amounts, using palmitic acid as a control. Phytanic acid at concentrations less than 80 mg/liter did not impede myelinization or seem to injure Schwann cells. However, it eventually induced vacuolization of neuron somas, and myelinated fibers underwent Wallerian degeneration. Palmitic acid at the same dosage did not produce neuron vacuolization, but eventually killed DRG cultures after a longer period of exposure.

5. Tay-Sachs Disease

Abnormal lipid metabolism also has been observed in cultured biopsy or autopsy material. In Tay–Sachs disease (ganglioside lipidosis, amaurotic familial idiocy) the ganglion cells are overloaded with monosialogangliosides lacking the terminal galactose; an aminoglycolipid also accumulates in both gray and white matter; glial cells proliferate and become loaded with lipid; myelin degenerates. McKhann et al.[141] cultured (for 7 days) neurons isolated from a fresh Tay–Sachs biopsy, during which time these cells continued to accumulate material with the histological properties of ganglioside (PAS- and oil-red-O-positive granules). Batzdorf et al.[9] with longer-term cultures of cerebral tissue from such a patient, found the predominant cell type to be many-branched and distended with granules. Analysis of the ganglioside fraction of these cells by thin-layer chromatography revealed $G_5(G_{M2})$ ganglioside, the characteristic storage material of Tay–Sachs disease, to be the predominant N-acetylneuraminic acid-containing substance. According to Okada and O'Brien,[183a] there is a generalized absence of a β-D-N-acetyl hexosaminidase component in a variety of non-neural tissue types.

6. Metachromatic Leukodystrophy

Metachromatic leukodystrophy (sulfatide lipidosis), probably based on an autosomal recessive, is characterized by a striking lipid metachromasia in the white matter of CNS and in peripheral nerves, accompanied by disintegration of myelin; there is a marked excess of sulfatide in the white matter. Abnormal lipid inclusions are stored in the lysosomes of the glial cells and phagocytes, which appear to lack aryl sulfatase. Lipids other than sulfatide, which originate from the myelin breakdown, are catabolized normally.[209] These electron microscopic findings are supported by enzymatic studies[208] on a strain of skin fibroblasts from a child dying of MLD, which showed less than 5% of the normal aryl sulfatase-A activity, though other lysosomal enzymes (aryl sulfatase-B, β-galactosidase, β-glucuronidase, and β-N-acetyl glucosaminidase) were comparable in activity to those of control fibroblasts. Cravioto et al.[52] cultured white matter from the same patient for 80 days, during which time the outgrowing (glial?) cells continued to show perinuclear accumulations of granules which stained metachromatically with cresyl violet–acetic acid, histochemical evidence of the persistence of cerebroside sulfate. The MLD white matter grew very much better in culture than did control normal white matter.

7. Gaucher's Disease

In Gaucher's disease (cerebroside lipidosis) an autosomal recessive produces deficiency in a glucocerebroside-cleaving enzyme. Of several clinical pathological patterns for the disease, two at least appear to be distinct entities; an infantile cerebral form, and a chronic (adult) noncerebral form. The latter has been studied in culture by Danes and Bearn[57] using strains of skin fibroblasts from 3 adult patients, 6 of their relatives, and 6 normal individuals. Giant fibroblasts containing metachromatic material were seen in all cultures from affected individuals and the 6 heterozygous "carriers," but not in those derived from normal individuals. Affected cell lines continually produced 20–40% metachromatic fibroblasts which stained with toluidine blue. These cells were morphologically distinct from those seen previously[56] by the same authors in cultures from the mucopolysaccharidoses (Hunter-Hurler syndrome).

8. Diseases of Porphyrin and Heme Metabolism

a. Kernicterus. Kernicterus (bilirubin encephalopathy) associated with neonatal jaundice and hyperbilirubinemia has been approached by Silberberg and associates[231,230] in several elegant experiments utilizing the model system of myelinating cerebellar cultures to assay the neurotoxicity of bilirubin, hemin, and related pigments without regard to permeability of the BBB. Typical morphological changes are induced in cerebellar cultures kept in the dark, by either bilirubin or hemin, in direct proportion to the concentrations administered. Neuronal swelling and vacuolation, incorporation of pigment by macrophages, loss of myelin, and glial swelling, are seen in vitro; these changes are similar to those found in the Gunn rat, a mutant with hyperbilirubinemia and kernicterus devolving from hereditary lack of glucuronyl transferase. Since albumin-bound bilirubin does not pass the BBB, it has been believed that unconjugated bilirubin is responsible for kernicterus; and Cowger et al.[45] reported reversal by albumin of bilirubin toxicity to somatic cells in tissue culture. In this regard the toxicity of bilirubin and also of hemin was found[231] to be pH-dependent: ratios of pigments to albumin in the medium that were innocuous at 7.5 were found toxic to cultures at 7.0, the normal pH for maintenance. Fine structural changes induced by bilirubin or hemin, namely enlargement of mitochondria and accumulation of complex intracytoplasmic inclusions in neurons, were similar to those seen in cerebellar neurons of the Gunn rat. Hemin therefore, as well as bilirubin, may be implicated in kernicterus. Biliverdin was nontoxic at the highest concentration used (15 mg% for 72 hr). Other smaller molecular photodegradation products of bilirubin were later[230] found to be similarly nontoxic to myelinated cultures; this last investigation was undertaken by Silberberg as a check on the phototherapy now widely used in kernicterus.

Haddock and Nader[77a] cultured fibroblasts from normal human skin, finding that bilirubin exerted a toxic effect by inhibiting NADH–cytochrome reductase. This action was prevented by exposing the bilirubin to light; breakdown products had no adverse effects. A clonal strain of rat hepatoma cells, however, can take up and conjugate bilirubin and then excrete the conjugated pigment into the culture medium as products chromatologically identical to those in normal rat bile.[212] Bilirubin UDP glucuronyl transferase activity is high. Flavaspidic acid inhibits bilirubin metabolism in this cell strain as it does in liver *in vivo*.

b. Intermittent Acute Porphyria (Pyrroloporphyria). This is an inherited disturbance of porphyrin metabolism based on a dominant autosomal gene. The metabolic defect is in the liver and leads to overproduction of δ-aminolevulinic acid (ALA), resulting in urinary excretion of this precursor and of porphobilinogen. Clinical manifestations include central autonomic and peripheral neuropathy without photosensitization. Structurally diverse chemicals will induce in animals a disease that resembles the inherited hepatic porphyria of man. Granick[76] showed that the same compounds induce *de novo* synthesis of ALA by chick embryo liver cells in culture, resulting in increased heme formation. Alloisopropylacetamide was one of the best inducers; griseofulvin, hexachlorobenzene, and sulfanol were also active, as were glutethimide, meprobamate, and dilantin sodium. Sterical rather than chemical effects appear to be involved in their activity. It is to be expected that large doses of sedatives or tranquilizers might induce acute episodes of hepatic porphyria in humans genetically disposed to the disease.

B. Genetic Dystrophies

Among the many known animal mutants displaying inherited defects in the nervous system, several have elicited investigation in tissue culture systems. The "jimpy" mouse, with a recessive sex-linked deficiency in CNS myelin (Sidman *et al.*[225]), presents similarities to the human sex-linked sudanophilic leukodystrophy; and an army of neurochemists is now attacking its developmental chemistry. Morphologically, the "jimpy" CNS develops about 10 % of the normal amount of myelin, but peripheral nerves are well myelinated and interneuronal connections are normal. A sudanophilic (nonpolar) lipid is distributed in some white matter tracts. Wolf and Holden[257] undertook an analysis of this dystrophic mechanism in long-term organized cultures of "jimpy," heterozygous and normal cerebellum from newborn animals, and found that the disease is intrinsic to the CNS tissue i.e., not based on a distant or general metabolic defect. Dorsal root ganglia from "jimpy" mice developed and myelinated as well as those from normal and heterozygous controls; but of 23 cerebellar cultures, 21 produced no myelin or very little, while all of the heterozygous and normal sister cultures went on to develop the usual generous amounts. "Jimpy" and normal explants grown on the same cover slip in the same drop of medium had no influence on each other's myelination. Though relatively fewer myelin sheaths were found in "jimpy" cultures than are produced by "jimpy" *in situ*, once the sheath was formed it was stable and structurally normal. For this and other reasons Wolf hypothesized that breakdown and phagocytosis (sudanophilic) generally occur in myelin precursors before these can be incorporated into myelin sheaths, rather than that breakdown occurs in fully formed myelin. Chemical analysis of "jimpy" brain composition by Nussbaum et al.[183] supports the idea that the myelination process in CNS is disturbed at an early stage in lipid synthesis, leading to the destruction of primitive myelin structures, and resulting in a deficiency of mature myelin sheaths. An electron microscopic study of "jimpy" *in vivo* by Hirano et al.[95] indicates that abortive myelination does occur, producing sometimes only one or two lamellae around axons which are normal. There is also myelin degeneration of the "vesicular dissolution" type that is seen in experimental allergic encephalomyelitis.

Inherited retinal dystrophy in mice is a progressive retinal degeneration, which resembles the human disease retinitis pigmentosa, the most common heritable cause of human blindness. The mouse disease is based on a simple autosomal recessive, which exerts its effects during the second week post-partum, when the retina is nearing the end of its developmental period. At this time, when the rod outer segments ought to be differentiating, development ceases in the diseased retinas and the entire photoreceptor cell begins rapidly to melt away, so that by the third week after birth (when the eyes of the normal mouse open and vision begins) most of the photoreceptor cells have disappeared without trace, though the inner retinal layers remain quite normal. (For an outline of retinal histology and metabolism, cf. Lolley, this Handbook, Vol. II, Chapter 20.) Sidman[223,224] further developed an organ-culture technique by which Lucas[132] had studied dystrophic eyes, and cultured normal and diseased eyes from 7–12-day-old mice in isolation from the complexities of the whole organism, finding that all the histological features of the abnormality were reproduced *in vitro* and concluding that the disease process is intrinsic to the eye (not based on circulating toxins, for example, like human kernicterus). He observed that isolated normal eyes did not differentiate fully in tissue culture media without the addition of carotenoids: All *trans*-vitamin A permitted the typical development of retinal components except for the rod outer segments, which were never produced at all. But when 11-*cis*-retinene was added, morphogenesis of an outer layer was regularly obtained in which the rod segments were histochemically typical though somewhat shorter than those developing *in situ*. Dystrophic retinas, however, though offered 11-*cis*-retinene in their media, never developed rod outer segments, and showed a progressive deterioration of photoreceptor nuclei, at the same time bipolar and ganglion cells remained healthy. (In earlier stages of dystrophic retinal development, migration and mitosis occurred normally, and synthetic mechanisms operated similarly.) Since the rod outer segment appears to be the primary site of the disease process, the hypothesis was offered that an error in synthesis of the protein opsin might be the basis of this pathogenesis: viz, production of an opsin of abnormal shape. This then could not combine properly with the 11-*cis*-retinene, which is specifically necessary to produce the sterical configuration of rhodopsin, an essential structural component of the photosensitive rod outer segment. The unstable combination would then break down, followed by the rod outer segment and eventually by the whole photoreceptor cell. This attractive hypothesis has not yet been proven, however.

Inherited retinal dystrophy in the rat also involves progressive postnatal loss of photoreceptor cells, but the course of pathogenesis differs from that in the mouse.[60] In the rat, rod outer segments form normally, but simultaneously an abnormal lamellar tissue component appears between rods and pigment epithelium. Rhodopsin is overproduced, and the excess is located in the lamellar material. The inner segments of the rods and the nuclei are the site of earliest degeneration, then the outer segment and the whole photoreceptor cell is lost, while rhodopsin production declines. Rat retina and pigment epithelium were maintained together and separately in organ culture by Sidman.[224] Neither rods nor extracellular lamellar material were synthesized in layers cultured separately, but more pigment was always formed *in vitro* than *in vivo*.

Pursuing this latter finding, Sidman and Pearlstein[226] cultured eyes from strains of mice and rats homozygous for a recessive gene leading to minimal pigmentation *in vivo*: "pink-eyed dilute." Such eyes made little pigment when cultured in a tyrosine-free synthetic medium.[224] But pigmentation was strong when tyrosine-containing media were offered; the pigment was identified as melanin. Tyrosine failed, however, to modify pigment formation when injected *in vivo*. Competitive inhibition by excess phenylalanine was ruled out, as was deficiency in the copper available for tyrosinase. This melanin deficiency is therefore not resident in the eye tissues; the pink-eyed gene reduces pigmentation *in vivo* despite the presence of an appropriate melanin-synthesizing machinery in the eye. Possibly it causes endogenous tyrosine to be shunted to other pathways, since exogenous tyrosine directly applied to the isolated eye tissues is utilized normally. This gene action may not be dissimilar to that operating in human albinism, where it seems possible that endogenous tyrosine is by some means made unavailable (cf. Stanbury *et al.*[235]).

All these intriguing problems await solution, perhaps by more sophisticated neurochemical methods.

V. NEUROTOXIC AGENTS

A. Exogenous

Organized cultures of nervous tissues have been used to particular advantage in the study of agents which exert a selectively neurotoxic action, since in this type of preparation the initial lesion can be distinguished and the whole course of injury followed in a single tissue complex (by time-lapse or serial photography) from the moment of exposure. The presumptive poison is applied directly to the nervous tissue and is administered in physiological concentration—usually not immediately lethal, but allowing recovery upon withdrawal of the agent. By this means information also can be gathered on neural effects of substances which are metabolized at a distance to produce toxic compounds, or to which the BBB is essentially impermeable (cf. Section IV, A, 7). A brief review of earlier work largely with short-term cultures, is given by May.[154]

1. *Mitotic Inhibitors*

Administration of colchicine or its congeners to humans for therapeutic purposes (in gout or in attempted carcinostasis) and also to experimental animals, produces a well-known peripheral neuritis.[2] A closer evaluation of the drug effects was undertaken by Peterson and Murray[199] using myelinated DRG cultures for light microscopy observations of lesions as they developed in both soma and neurites. During a 3-day exposure to colchicine (0.5 μg/ml) striking changes occur in perikaryal (but not nuclear) architecture. A clear area initially appearing at the periphery is shifted and twisted in tortuous patterns, eventually becoming a juxanuclear crescent or a perinuclear ring. Nissl staining shows, however, that a basophilic nuclear rim is retained throughout the episode. Silver staining reveals[195] that the "clear areas" observed in living neurons are solid neurofibrillar bodies, or "tangles" of neurofibrils. By electron microscopy[33,34] these neurofibrils are shown to be singly ~ 79 Å in diameter, of the same order as those which are normally present as slender bundles in the "roads," or less dense areas of the cytoplasm, running between the Nissl areas, Golgi complex, mitochondria, etc. Under colchicine treatment these small neurofibrillar bundles widen and coalesce into the large masses detectable by light microscopy. Mitochondria congregate around these aggregates and Nissl substance becomes concentrated centrally or peripherally, so that the entire neuronal soma becomes segregated into regions occupied by different organelle systems. Normally, neurotubules are present along with neurofilaments in the cytoplasmic roads; these begin to decrease in number upon colchicine administration, and, at the height of the organelle segregation, have disappeared entirely.

Withdrawal of drug is followed by complete recovery, i.e., restitution of the normal distribution of perikaryal organelles, and disappearance of the clear and vacuolar areas. According to Borisy and Taylor,[17] colchicine (^{3}H) binds to a 6S cellular protein in somatic cells which[221] is a subunit of microtubules. Wisniewski et al.[255] find that a variety of spindle inhibitors including colchicine produce a loss of tubules with coincident proliferation of neurofibrils, in anterior horn neurons in vivo. These changes are reversed in animals allowed to recover. Since microtubules in nervous tissues have as one of their functions the maintenance of an intracellular skeletal structure, their dissolution by colchicine in our cultured neurons probably is a factor in the characteristic perikaryal disorganization. To test this hypothesis, Murray and Herrmann[174] introduced D_2O (which is known to antagonize the colchicine effect on tubules in mitotic cells) into DRG cultures before, and during, colchicine administration. Time-lapse light microscopy films, recording the behavior of DRG exposed to these agents singly and together, show that D_2O in highest tolerable concentration (25%) administered several days in advance, and continued during colchicine exposure, either completely prevents the characteristic drug effects or arrests and reverses them at a very early stage (cf. Section II, C).

In neurites, swellings related to neurofibrillar bodies are often seen adjacent to constrictions in the fiber. Demyelination is segmental in pattern, characterized first by nodal elongation, ballooning of myelin and lamellar splitting; later by a frothy blistering and eventual fragmentation of the sheath within the baglike Schwann cell. After colchicine withdrawal, the intact and many damaged internodes are retained; myelin remnants are resolved within the Schwann cells, and remyelination begins. Electron microscope observations made in DRG cultures by Hendelman and Bunge[85] on the disposition of microtubules in relation to the myelin sheath indicate that these may be involved in myelin formation and stabilization. The shape of the elongated segmental Schwann cell appears to be maintained by microtubules running parallel to the fiber throughout the internode except at the Schmidt–Lanterman clefts and the terminal loops of myelin at the nodes, where they are disposed circumferentially. As this microtubular skeletal structure undergoes sterical disintegration with colchicine binding, one might expect the lamellated myelin structure within the Schwann cell to fragment or collapse. On withdrawal of drug, Schwannian microtubules would be reconstituted, and recovery (i.e., lamellar rewrapping) could take place.

Other mitotic inhibitors, e.g., the vinca alkaloids vinblastine sulfate (VBL) and vincristine sulfate (VCR) which have widespread use in cancer chemotherapy, are similarly neurotoxic and show among their effects the concurrent loss of microtubules and formation of neurofibrillary tangles. Burdman[36] noted with light microscopy a selective toxicity of VCR on DRG cultures, and later described with the electron microscope[105] the production of extensive bands of filaments in sensory neuron somas *in situ.* Journey *et al.*[104] found in ultrastructural studies on somatic cell cultures that the primary effect of VCR was on the mitotic apparatus, with loss of spindle tubules or blocking of spindle protein synthesis, coincident with the appearance of fine filaments; the mitotic arrest was irreversible. In DRG cultures VCR accelerated conversion of neurotubules to neurofilaments. Seil and Lampert[219] applied VCR and VBL to cultures of central and peripheral nervous tissues, finding that the formation of neurofibrillary tangles began with a condensation of argentophilic fibrils around the nucleus; the tangles were seen by electron microscopy to correspond to bundles of closely packed neurofilaments. Cells were able to revert almost completely to the normal pattern after withdrawal of VBL, but recovery from VCR exposure was not noted. Krishan also [121] reported reversal of mitotic arrest in somatic cells after treatment with VBL. Using cord ganglion cultures Peterson[194] reported neurofibrillar bodies produced by exposure to vinca alkaloids and podophyllotoxin, the latter alkaloid being less toxic i.e., more readily reversible than colchicine, which was less toxic than VBL. No production of the microtubular crystals reported[14,216] to be associated *in vivo* with vinca administration was observed in these cultured nervous tissues.

Diwan[59] found that colchicine at 5×10^{-5} mg/ml applied to chick embryos at the primitive streak stage interfered with morphogenesis of the brain region and neural system. More recently Karfunkel[106] in approaching the problem of neurulation mechanics from the standpoint of microtubule participation found that as the cellular components of the neural plate elongated, microtubules became oriented parallel to their long axes. Exposure of the neural anlage to D_2O in very high concentration (70%) stopped development but did not reverse any morphogenesis that had already taken place; microtubules remained in their usual location and orientation. However, when the neural plate was treated with VBL, neurulation was reversed; the neural groove opened up, the folds flattened out, and the component cells lost their elongation, assuming a spherical shape coincidently with the disappearance of microtubules and microfilaments and their replacement with crystalline microtubular subunits. Thus very convincing evidence exists from a variety of sources that the polymerized microtubular protein performs the function of skeletal support in the neuron, and that its subunits can be visualized as crystals or microfibrils. Other possible microtubular functions are mentioned elsewhere in this Handbook (Vol. II, Chapter 18).

2. Bacterial Toxins

In the era preceding mass immunization and antitoxin therapy, a paralytic peripheral neuritis was a common clinical complication of diphtheria. This condition can be produced experimentally as a demyelinating disease in animals (e.g. guinea pig) by administration of diphtherotoxin (cf. Webster *et al.*[245]). Crude toxin (from Lederle Laboratories) administered to rat DRG cultures in concentrations of 1:250 or 1:500 by volume in the feeding medium evokes a segmental demyelination which is in many respects similar to that which is observed *in vivo*.[198] The early changes begin at the nodes of Ranvier, with an initial widening of the nodal gap and a progressive fragmentation of the sheath toward the Schwannian nucleus; myelin subjacent to the nucleus is affected last. The Schwann cell of the damaged segment retracts and thickens as the fragments of myelin pile up inside it. Gradually these fragments become smaller and are completely lysed within the cytoplasm. At the lower level of toxin exposure many segments remain intact for weeks, and most pyrenophores are unaffected. When toxin is withdrawn from the medium, recovery and remyelination can take place. At higher concentrations the culture is killed, but neurons are the last cells to succumb. In spinal and cerebellar cultures myelin is not destroyed at equivalent concentrations of this agent; in the NS the Schwann cell with its myelin membrane system seems to afford the most sensitive acceptor. Since the action of diphtherotoxin in mammalian cells is to interfere with protein metabolism, it may in the Schwann cell produce anabolic block of myelin maintenance, or by inducing secondary lysosomes lead to catabolic processes of demyelination. A number of studies have been made with somatic cells in culture, e.g., Bonventre and Imhoff[16] and Paradisi *et al.*[189]

Diphtherotoxin neutralized by antitoxin produces no lesions in DRG cultures, even when administered in high dosage. Hydrocortisone-Na-hemisuccinate administered at 20 μg/ml simultaneously with active toxin delays the onset and reduces the severity of the demyelinative response.[176] Damage to myelinated fibers from diphtherotoxin is still further reduced by previous conditioning of the culture with hydrocortisone.

3. Metals

The pattern of toxic response by nervous tissues in culture to metallic salts varies widely according to the metal tested. Neurofibrillar aggregations produced experimentally have elicited especial interest because of their not infrequent occurrence in spontaneous human pathological conditions. After it was found that subarachnoid injection of $AlPO_4$ produces, in rabbit cord neurons, neurofibrillary tangles recalling Alzheimer's disease, Seil *et al.*[220] exposed DRG and cord cultures directly to 10^{-2}M suspensions of this compound. Cord neurons were not affected, but in DRG cells, interweaving small bundles of neurofibrils appeared, massing to form spheroids which pushed the nucleus into an eccentric position; the condition somewhat resembled that which characterizes neurons in postencephalitic Parkinsonism. Alzheimer's neurons were more closely simulated in culture by the VCR tangles described above (Section V, A.1).

A distinctive pattern of degeneration is produced by thallium chloride or acetate (Tl $C_2H_3O_2$) applied to DRG cultures, which to some degree explains the *in vivo* findings of myelin degeneration, irregular axon swellings and discontinuities, in the peripheral neuritis that characterizes thallium intoxication. In these cultures the primary damage occurs in portions of the axon—specifically at the nodes of Ranvier, spreading inward along the segment from both ends toward the region subjacent to the Schwann cell nucleus, which is affected last.[198] Within 48 hr of thallium administration, a marked distention and retraction of the myelin sheath is observed at the nodes; this is caused by a vacuolar swelling of the neurite, and does not involve the Schwann cell or the lamellar myelin structure: even the S–L clefts are undisturbed in these early stages of fiber damage—which are reversible upon withdrawal of the agent. Gradually the vacuoles disappear and the distended myelin sheath returns to normal configuration, without suffering fragmentation or dilamellation. However, after longer exposure to thallium, the neuron somas also become granular and vacuolated, Schwann cells succumb, myelin breaks down, and the whole culture deteriorates abruptly. It has been suggested by Pentschew and Garro[192] that thallium may inhibit riboflavin metabolism.

Though it is generally agreed that intoxication by lead brings about nervous disorders, the mechanism of action of lead upon nervous tissues is not well understood. In this connection Sobkowicz and Murray[234] undertook to observe the direct action of lead compounds on PNS (DRG) and CNS (cerebellum) in organized cultures, administering lead acetate for (1) morphological and (2) histochemical studies; also testing (3) its indirect action on cultures through serum from lead-poisoned animals and patients. Results were compared (4) with those of untreated, mercury-, gold-, and triethyl-tin-exposed cultures. Briefly, it was found (1) that the primary action of lead acetate is a destruction of supporting tissues (macrophages and glial tissues) which proceeds slowly (over about 3 weeks) with little regard for concentration (0.0379 to 37.900 μg/ml in feeding medium) and only secondarily leads to degeneration and necrosis of neurons. (2) By means of rhodizonic acid (red) staining for the presence of lead it was found that actual lead uptake by the tissues generally paralleled the morphological evidences of toxicity: lead grains were early found in abundance in macrophages and meningeal cells which chanced to be present; unattached glial cells showed a slower and more delicate uptake, whereas no lead inclusions at all were observed in neurons or their processes, or in supporting cells closely applied to them. (3) Serum from lead-poisoned animals and patients showed demyelinating properties which were not, however, related quantitatively to the actual concentrations of lead in the sera. The pattern of intracellular uptake was similar for both agents. (4) Tin and gold, like lead, provoke patterns of morphological change primarily in cells of mesodermal origin, and then in glial cells. Mercury by comparison is toxic in exceedingly low concentration, and like thallium exerts its primary action on the neuron. The apparent imperviousness of the neuron itself to lead in its environment is noted also by Schlaepfer[215] and Lampert and Schochet[123] who report *in vivo* an experimental lead neuropathy affecting supporting cells. Brun and Brunk[26] report minimal morphological changes (by light microscopy) even in acute lead poisoning, but note with histochemical methods an increased activity and changed distribution of acid phosphatases in the neurons.

It goes without saying that once the toxic substance is present in great enough force to bring about the death of the neuron or the severance of portions of the axon from it, the fiber will undergo a Wallerian type of degeneration. The advantage of the type of study described above is in revealing the exact sequence of events and cellular susceptibilities in the toxic response. Lampert and Schochet[123] also produced in rats *in vivo* a low-grade chronic lead intoxication which allowed them to observe (with electron microscopy) the segmental degeneration of Schwann cells and myelin which was followed repeatedly by repair and renewed breakdown, while the periodically denuded axons remained viable.

4. *Radiation Damage*

In a classic light microscope and electron microscope study utilizing as a model system organized DRG cultures, Masurovsky *et al.*[151,152] described the sequence of cytological changes produced in PNS by acute roentgen irradiation. In the culture situation, unconfused by secondary vascular and other damage to enveloping somatic tissues, direct effects of irradiation on the nervous tissues could be assessed. It was found that Schwann cells investing unmyelinated axons in the fascicle were the first to succumb, followed only after several days by those on myelinated fibers, which with their myelin sheaths degenerated during the ensuing week or 10 days, leaving the associated neurites apparently viable. Satellite cells embracing neurons degenerated *pari passu* with the myelinative Schwann cells, leaving the denuded neurons and their processes as the last to disintegrate (at about 2 weeks).

Exploiting this material further, Masurovsky and Bunge[153] showed that significant radiation protection could be afforded by transguanylated aminoethylthiol (AET) derivatives (2-mercapto-ethylguanidine and bis[2-guanidoethyl] disulfide) to essential cytoplasmic organelles and nuclear structures in neurons as well as to supporting cells in these cultures. Light microscope and electron microscope autoradiography with AET-^{35}S indicated widespread AET binding—especially to mitochondria, Golgi complex, and ergastoplasm.

B. Endogenous

In the section on metabolic disorders reference was made to some indwelling neurotoxic agents of lesser complexity than those which are associated with immune reactivity, and are to be considered here. Among the degenerative diseases of the nervous system whose etiology is little understood, the demyelinative syndromes have become the object of concerted investigative onslaught, in the course of which the concept of disordered immune response as a causative factor has gained considerable ground (cf. H. E. Whipple[248] and Burdzy and Kallós[39a]). Experimentally induced demyelinating diseases in animals have served widely as models for this type of study, especially experimental allergic encephalomyelitis (EAE) evoked by sensitization of several laboratory species with CNS proteins, and experimental allergic neuritis (EAN), evoked by injections of peripheral nervous tissue preparations. During the last decade, model systems in the form of organized, myelinated cultures of rodent CNS and PNS tissues have served as test objects in the assay of myelinotoxic factors present in body fluids and tissue extracts from experimental animals and from humans presenting demyelinative disorders.[18] With the announcement in 1959 that demyelination was brought about in cultures of newborn rat cerebellum by serum from rabbits with EAE,[19] Bornstein and Appel began a series of investigations which have since been widely extended by themselves and others in the effort to identify and characterize endogenous myelinotoxic factors.

1. Demyelinative Diseases of CNS

a. Experimental. Exposed to 20–25% EAE serum, myelinated cultures react with a characteristic pattern of demyelination which begins within a few hours after serum administration, as a swelling of neuroglia or a displacement of myelinated axons, which soon become disfigured by fusiform swellings at irregular intervals along their length. From this point on, the sheath may fragment into droplets, or simply melt away in 12 to 24 hr, leaving the neurites and somas relatively unaffected. On withdrawal of the EAE serum, remyelination will occur after a few days in normal medium.[20] The demyelinating factor in this serum is complement-dependent and was found only in the γ-2-globulin (7 S) fraction by Appel and Bornstein.[3] The serum was selective in its effect on CNS; it did not demyelinate DRG cultures at compararable concentrations; and nephrotoxic serum from immunized rabbits did not produce myelin lesions in cerebellar cultures.[20]

Most recently, Bornstein and his associates[208a] have characterized the fine-structural changes occurring during demyelination from EAE serum as selective damage to oligodendroglia, coincident with myelin dissolution; this is followed by astroglial phagocytosis of the debris. Conversely, if even a small amount of myelinotoxic EAE serum is included in the feeding medium of developing CNS cultures, then myelin is not formed at all. Electron microscopic study shows no myelin destruction; these cultures are "normal" except for a total lack of myelin and an apparent absence of oligodendroglia.

The method of sensitization originally used to produce EAE involved injection of whole brain or cord fortified with Freund's adjuvant into another species. Serum taken from such a sensitized animal after the onset of clinical symptoms was tested for myelinotoxicity in CNS cultures by Bornstein and Appel. Coincidently, attempts were being made to isolate and characterize encephalitogens from myelin; among these a purified basic protein produced by Kies[109] from guinea pig brain has received considerable notice. Assays in culture were undertaken by Seil et al.[218] to compare the myelinotoxic effects of sera from guinea pigs sensitized with whole CNS to those of sera from animals sensitized with Kies' purified basic protein. Though this latter encephalitogen produced clinical symptoms *in vivo*, the serum was not myelinotoxic when applied to CNS cultures; and the authors concluded that the antibodies induced by the purified protein were not the same as those induced by whole spinal cord.[39a] A contrary conclusion was expressed by Yonezawa et al.,[254] who found that sera from animals sensitized with an encephalitogenic basic protein prepared "after Kies' method" demyelinated cerebellar cultures; the experiments have since been repeated with similar results using Kies' own preparations. In this same report the authors distinguished two types of antibodies in sera from sensitized animals: a cytotoxic factor (IgG) and a myelinotoxic factor (IgM) which were complement-dependent, organ-specific and species nonspecific; both were demyelinative in culture.

According to Lumsden et al.[135,136,138] a peptide of low molecular weight (~ 4000) constitutes the common denominator in all encephalitogenic "antigens." It is loosely incorporated in certain basic proteins, and forms the active core of EAE proteolipid. It is completely resistant to carboxypeptidase, which suggests the absence of an N-terminal group, or that lysine is the terminal group. In cytotoxicity it is very specific, damaging only myelin-forming glial cells.

b. Human CNS Disease. The immunologically mediated animal disease EAE was studied especially as a putative model for multiple sclerosis (MS) in man; the next step was therefore to apply human sera from MS and other demyelinating diseases involving possible "delayed-type hypersensitivity" to the myelinated culture model. The first report, by Bornstein,[18] indicated demyelinative activity in sera from a high percentage (66%) of MS patients in active phases of the disease, but none from patients in MS remission. A scattering of positive sera was found in patients with other demyelinative diseases, fewer in patients with nondemyelinative diseases, and one in a normal control. The pattern of demyelination in the human diseases was similar to that produced in EAE where the serum was known to contain circulating antibody.

Subsequent studies by Lumsden[135,136] and Hughes and Field[102] confirmed the presence of myelinotoxic factors in EAE and MS sera. Field and Hughes[66,39a] also noted demyelination of cultures by serum from amyotrophic lateral sclerosis (ALS). Dowling et al.[61] have reported the identification of two complement-dependent antibodies in sera from patients in exacerbative phases of MS. By ultracentrifugation of whole serum, both a 19S IgM and a 7S IgG antibody were found, which separately and together, demyelinated cultures of newborn mouse cerebellum. To characterize these factors further, Kim et al.[115] separated myelinotoxic sera from patients with a variety of disorders into 19S and 7S fractions and assayed these on the CNS tissue cultures. Myelinotoxicity was restricted to the 7S fraction in patients with ALS and viral diseases and in "normal" human controls; but with 50% of MS patients, in addition to a 7S factor, a 19S myelinotoxic factor was present. The authors suggest that the 7S demyelinating factor, being so generally distributed, is probably not disease-specific; the 19S (IgM) factor may be specific to MS patients.

Recently Kim et al.[116] have tested on this same myelinated culture system pooled, concentrated cerebrospinal fluid and brain extracts from

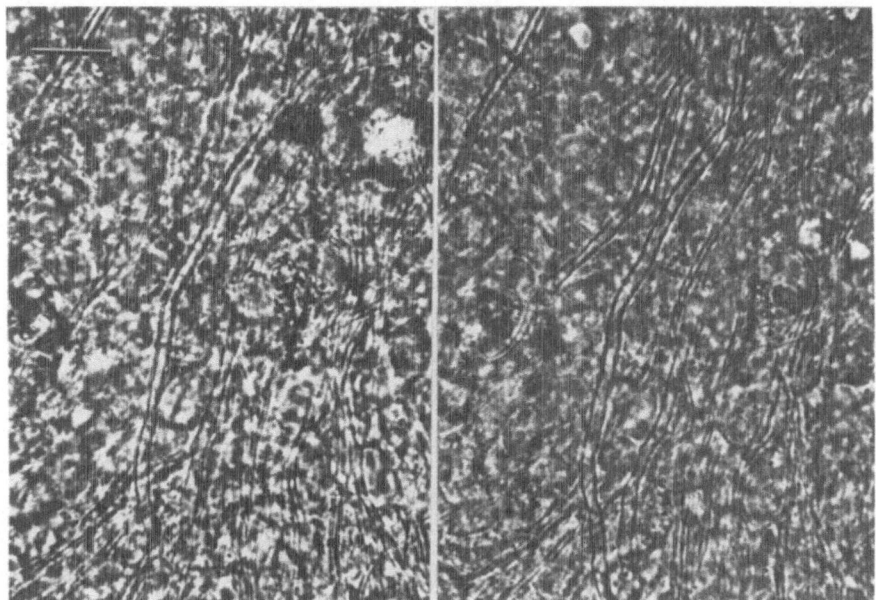

Fig. 6. Myelinated culture of newborn mouse cerebellum used in assay for myelinotoxicity of extract from normal-appearing white matter of MS patient. Left: Before application of sample. Right: 48 hr later. Sample is judged negative. Cf.[116]. Bar: 10 μ.

MS patients. Their CSF (concentrated 200 times) and extracts from the plaque, shadow-plaque, and periplaque areas showed a high degree of thermolabile myelinotoxicity, whereas extracts of normal-appearing white matter of MS patients, normal brain extracts, and concentrated normal CSF registered substantially lower toxicities (Figs. 6, 7). The myelinotoxic activity of these respective samples was correlable with their IgG ratios; however, no attempt was made to separate globulin types.

The possible intervention of an elevated lysolecithin titer in human and experimental demyelinative disease was suggested by Périer[193] and investigated by him in the model system of myelinated CNS cultures. Under certain conditions this substance can destroy all the myelin in a culture, without appreciably altering neuron somas or axons, which are capable of remyelination. Such action is obtained when lysolecithin is dissolved in complete feeding medium at a concentration 5–10 times as great as that in normal human plasma. The morphological pattern of demyelination is similar to that described by Bornstein and Appel,[20] but the effects of lysolecithin at the above concentrations are more severe.

2. Demyelinative Diseases of PNS

a. Experimental. An allergic peripheral neuritis (EAN) comparable to EAE has been produced in guinea pigs, rats, and rabbits by inoculation of

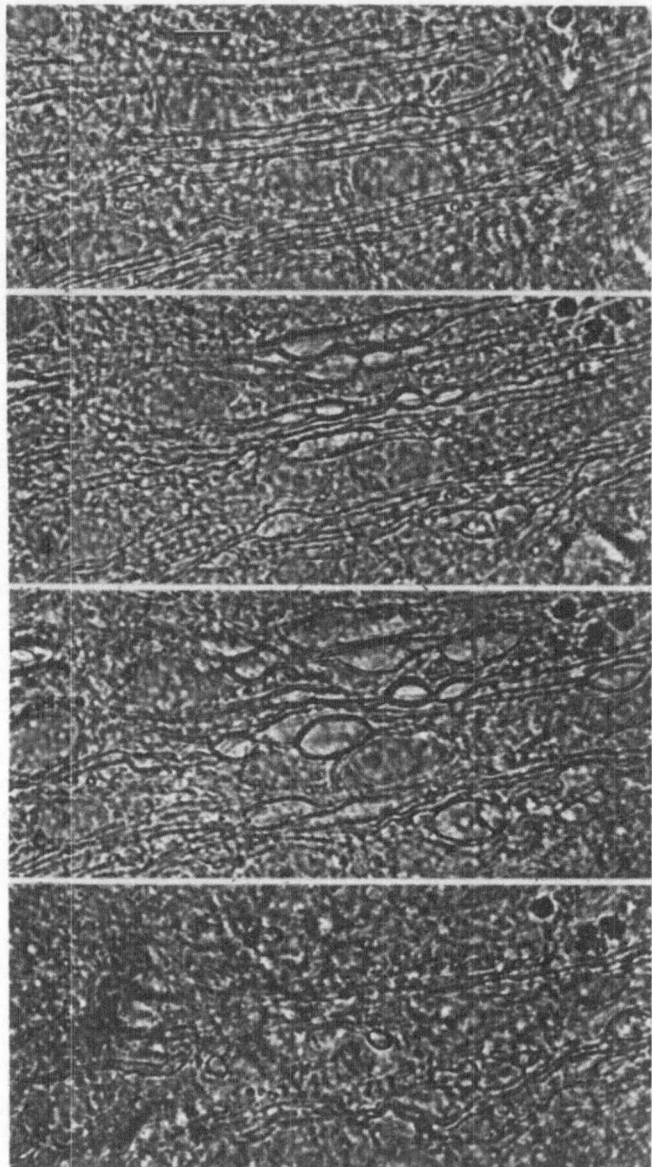

Fig. 7. Sister culture of culture in Fig. 6 exposed to extract from MS plaque of another patient. (a) Before application of sample. (b) 24 hr later. (c) 48 hr later. (d) 72 hr later. Positive sample. Bar: 10 μ.

homologous and heterologous nerve (usually sciatic) homogenates with Freund's complete adjuvant. The disease is characterized by mononuclear cell infiltration and demyelinative changes generally restricted to the spinal ganglia, cranial and spinal nerve roots, and peripheral nerves. *In vitro* study of this syndrome was first reported by Winkler[253] using organized cultures of fetal rat trigeminal ganglia, DRG which follow the same course as the spinal ganglia in development of myelin sheaths.

During the period of clinical latency (about 2 weeks) after injection of antigen into the left hind-paw of rats, their serum was not demyelinative; but from 4 days on, lymphocytes taken from the inguinal nodes on the injected side and seeded into DRG cultures brought about demyelination without showing any particular agglutinative activity. This lymphocytic myelinotoxicity declined somewhat after clinical signs appeared, when it was almost equaled by the activity of serum from the affected animals. Arnason et al.[5] extended these preliminary observations to report that both node cells and peripheral buffy coat cells acquire the capacity to destroy myelin in DRG cultures 5–7 days after immunization. This activity peaks at 7–9 days but remains effective for weeks. Mononuclear cells must be living to exert myelinotoxicity. A rabbit antiserum to rat IgA blocked the cytodestructive effect of EAN lymphocytes, as reported by Winkler and Arnason.[254] Serum, even after the 2-week latent period, was very much less potent than sensitized cells in destroying myelin.

However, Yonezawa et al.[39a,263] obtained strong demyelination of DRG cultures with EAN serum alone, in three patterns: (1) cytotoxic effects upon Schwann cells with segmental demyelination; (2) lytic changes in myelin sheaths; and (3) activation of phagocytes which ingested myelin débris. The demyelinative factors were complement-dependent; they were adsorbed by peripheral nerve, but not by CNS or somatic tissues. Antigenicity of peripheral myelin sheaths was indicated by immunofluorescent analysis. In 1969 Yonezawa et al.[264] reported that basically the same types of myelin alteration were induced by EAE and EAN sera; the cytotoxic factor in each case was identified as IgG, and the myelinolytic one as IgM, in this preliminary account.

b. Human PNS Disease. Like EAE for CNS, EAN has served as a wedge into the study of human peripheral neuritides, chief among these the Landry–Guillain–Barré syndrome (acute idiopathic polyneuropathy) which appears to be a primary demyelinative disease[255] with an unusual pattern of myelin breakdown associated with phagocytosis of myelin by macrophages or large mononuclear cells, which is accompanied by minimal axonal damage. Though the etiology of this disease (LGB) is as yet unknown, evidence is strong that it may be immunologically mediated. An unusual population of primitive basophilic mononuclear cells, many of which are engaged in DNA synthesis (cf. Cook and Dowling[41]), occurs in the peripheral blood of patients with LGB, prior to clinical improvement. Ultrastructurally similar cells have been identified in demyelinated nerve roots in the disease. Short-term cultures of buffy coats containing these reactive leukocytes yield in their supernatant medium a factor which is capable of demyelinating DRG cultures. Production of the demyelinating factor is cycloheximide sensitive and its myelinotoxic activity is complement dependent.

In addition to the buffy coat factor, myelinotoxic factors were identified in the sera of 23 out of 28 patients with LGB, by assay on DRG cultures.[172] During the first 3 weeks of illness, or prior to clinical improvement, myelinotoxicity was most prevalent. It took the form of a segmental swelling or fragmentation of the myelin sheath, or a melting away of the sheath without antecedent irregularities, while the Schwann cell appeared to draw back from close application to the neurite. These changes were often associated with background granularity and the presence of large numbers of mononuclear cells. Despite the extensive involvement of myelin sheaths, neither axons nor somas nor satellite cells showed marked alterations. Fine-structural aspects of this demyelination process[96] involved regular separations of the myelin lamellae, intramyelinic splits, distention of the periaxonal space, and disintegration of Schwann cell cytoplasm. Occasionally, cell processes, apparently viable, were seen invading the damaged myelin sheaths. Axons, sometimes completely denuded, appeared well preserved.

Fractionation of LGB serum by Cook et al. has revealed the presence of at least two myelinotoxic factors, as assayed on DRG cultures.[42] A 19S factor (probably IgM) was found in all (5) sera studied; and a 7S factor was found also in two of these patients. No differences from that of whole serum were observed in these demyelination patterns. Mild demyelinative action was found in whole sera from several other neuropathies and in serum from one normal subject.[172]

VI. PHARMACOLOGICAL ACTIVITY

A. Hormones

Though manifold relationships exist in vivo between the nervous system and products of the endocrine glands, little that is noteworthy has been observed in NS cultures. A possible exception is the work of Hamburgh[79,80] on the influence of thyroid hormone in the development of newborn murine CNS. Here, the most consistent result was an acceleration of myelinogenesis by cerebellar cultures in a medium supplemented with thyroxine in concentrations of 1.5–3.0 μg/ml, and grown at optimal temperatures (~ 36°C). In untreated control cultures grown at 32° and 30° respectively, cytodifferentiation was delayed, but myelin formation was almost completely suppressed. Thyroxine added to the medium increased the incidence of myelin formation at low temperatures but did not restore it to normal. Cultures shifted from low to optimal temperatures recovered in 2–3 days, with or without thyroxine supplementation. Ability of myelinated cultures to maintain their sheaths without refeeding (for 12 days) was decreased by the presence of thyroxine in the medium. Hamburgh suggests that one of the stimuli for myelinogenesis may be a metabolic signal, mediated probably via enzymes sensitive to thyroid hormone for their rate of synthesis. It should be noted here that since the organized NS culture model cannot be supported on a wholly synthetic medium but requires a substantial amount of serum in the feed, these observations relate rather to the developmental response evoked by a degree of hormone excess; a more rigorous study awaits the availability of hormonal inhibitors or antimetabolites comparable to the vitamin analogues. However, the observations do exclude possible effects of endocrine imbalance on distant physiological mechanisms, to which *in vivo* experiments are subject, e.g., a compensation by growth hormone, in thyroid deficiency. Dealing with somatic cells (human kidney) Burki and Tobias[39b] found that the life cycle of this line was reduced from 27 to 20 hr by the presence of thyroxine during exponential growth. This shortening of generation time was due solely to a reduction in the pre-DNA replication period (G-1).

The induction of glutamine synthetase by β-hydroxy corticosteroids during a specific stage of retinal development is discussed in Section II, D. Using a cloned strain of glial cells, induction of GPDH by cortisol (at 1 μg/ml of medium) was obtained by deVellis and Inglish.[242] Besides cortisol, aldosterone, corticosterone and cortisone were active, while progesterone, estradiol, testosterone, insulin, and cyclic AMP were ineffective. Cortisol inhibited nucleic acid synthesis, but incorporation of leucine-^{14}C into proteins remained unchanged, as did levels of LDH, glutamine synthetase, and phosphorylase in this material. The level of LDH, but not GPDH, rose sharply following addition of epinephrine. Cyclic AMP had no effect on LDH activity, and inhibitors of RNA and protein synthesis inhibited LDH induction under epinephrine treatment. The authors suggest that cortisol and epinephrine control respectively GPDH and LDH synthesis at the transcription level.

Reference is made in Section V, A.2 to the effect of hydrocortisone in delaying or reducing the segmental demyelination brought about in DRG cultures by diphtherotoxin. In normal CNS cultures included in this same experimental series[176] the addition of hydrocortisone at 10 μg/ml to the feeding medium enhanced migration and survival of glial cells, though not influencing the amount or quality of myelin produced; this treatment reduces the severity of a transient minor crisis, traceable to unidentified toxic factors present in the normal biological medium, which habitually occurs within the first 2 weeks of cultivation and results in death of varying numbers of glial cells. The explant usually recovers spontaneously; and although cortisone treatment does not entirely avert this episode, it minimizes it. Since hydrocortisone is known to have a stabilizing action on lysosomes, it may have the effect of reducing their release of acid proteases in response to toxic medium. [Cf. Fell et al.[65] and Dingle et al.[58] on response of cultured cartilage and bone to antisera; the role of lysosomes; inhibition by corticosteroid.] Corticosteroids are used to some degree clinically in the treatment of human demyelinative diseases.

B. Neurosecretory Materials

Since the autonomic nervous system (ANS) is the major site of origin for neurosecretory products, observations of their elaboration *in vitro* have lagged, at least partly because of an earlier preoccupation with CNS and PNS culture methodology. The work of Levi-Montalcini and her colleagues on the nerve growth factor utilized short-term (24 hr) and dissociated-cell ANS cultures for assay; and only recently has long-term, organized culture of sympathetic ganglia and hypothalamic and thalamic nuclei become technically satisfactory—this coincidentally with the propagation of histochemical fluorescence methods which permit the direct visualization of transmittor substances in adrenergic and serotonergic neurons.

However, some of the earlier observations deserve mention here— particularly those of Hild[89-91] who provided significant evidence toward the now-established concept that hypophyseal pressor substances (though not the melanophore expander) are produced in the hypothalamus and transmitted along the granule-studded tract leading from supraoptic and paraventricular nuclei through the hypophyseal stalk into the neurohypophysis, where they are stored for release upon occasion (cf. Sachs, this Handbook, Vol. IV, Chapter 17). Hild showed that hormones transferred with isolated pituitary explants into the culture medium are inactivated within 7–10 days, whereas ganglion cells from the hypothalamic nuclei continue to exhibit small amounts of stainable neurosecretory substance after as long as 68 days; and by time-lapse moving pictures he observed a peripheral passage of these materials down the axons. His findings thus indicated (1) that characteristically staining secretory granules were not produced *de novo* by the posterior hypothesis in culture; (2) that under the same conditions they were continuously elaborated by hypothalamic neurons; and (3) that a distally directed traffic existed in these axons.

Recently Wurtman *et al.*[259] maintained rat pineal glands in organ culture, incubating them for 48 hr with tryptophan-^{14}C and observing incorporation of ^{14}C into protein. Since the sympathetic nerve endings leading into the pineal contain large amounts of both serotonin and norepinephrine, the relation of these biogenic amines to neurotransmission there was the point at issue. Finding that incorporation of tryptophan into protein was stimulated by NE and not by 5-HT, they concluded that NE functions as neurotransmitter in the rat pineal gland. According to Klein *et al.*,[119] N^6O^2-dibutyryl AMP stimulates conversion of tryptophan-^{3}H to melatonin-^{3}H in pineal organ cultures, as does L-NE. The dibutyryl analogue also stimulates conversion of serotonin-^{14}C to melatonin-^{14}C without altering HIOMT activity. Klein believes that such pineal organ cultures afford a model for *in vitro* study of the relationship between input and output signals in a neural system, since a neurotransmitter (NE) can influence the metabolism by a neural structure of an output signal (melatonin). The hallucinogen Harmine appears to induce, in pineal structures, the enzyme N-acetyltransferase, which is involved in the metabolism of a biogenic amine; hence, the biochemical actions of drugs also may be studied under defined conditions in this system.

As shown (particularly by Pomerat) in a long series of time-lapse films, explanted nervous tissues habitually exhibit movements, both general (as migratory) and special (as exploratory or contractile). The neuron itself is a relatively quiet and sedentary cell, contented with nuclear rotation and probing pseudopodial movements at its neurites' growing tips. Lumsden and Pomerat[137] and Pomerat[202,203] gave especial attention to pulsatory movements of oligodendroglial cells, which are intermittent in culture; when pulsation does take place it follows a regular rhythm. The beat, which occurs once in about 5 min, consists of a rapid systolelike phase and a more protracted diastolelike phase. It occurred to D. W. Woolley who was then concerned with serotonin as a vasoconstrictor that this slow oligodendroglial contractile rhythm was comparable to that of involuntary muscle, and that these cells too might be expected to respond to serotonin. As then shown by Benitez et al.[13] in time-lapse films, and Murray,[167] cultured oligodendrocytes from corpus callosum after being exposed directly for 2 min to serotonin at 5 μg/ml respond by a prolonged tonic or spasmodic contraction (Fig. 8). The effect becomes apparent a few minutes after exposure, reaches its peak in 40–60 min, and then gradually subsides. It can be prevented or reversed by the 5-HT analogs medmain and 1-methylmedmain. LSD, a congener,

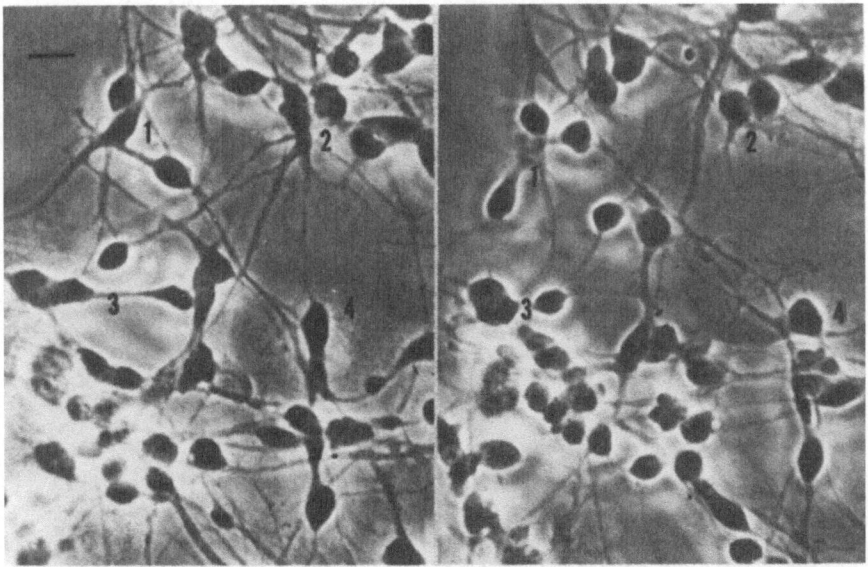

Fig. 8. Effects of serotonin (5 μg/ml for 2 min) on human fetal oligodendrocytes. (a) Shows the culture before application of 5-HT. (b) The same area 35 min later. The same groups of cell are numbered 1, 2, 3, 4 in both photos. (Reproduced from Windle, *The Biology of Neuroglia*, Charles C. Thomas, 1958.)[167] Bar: 10 μ.

is antagonized by 5-HT but does not itself antagonize 5-HT: when administered simultaneously it intensifies or prolongs the action of 5-HT. To investigate the reputed antagonistic effect of chlorpromazine on serotonin, Nakazawa[181] administered these two compounds to cultured oligodendrocytes. He found that each one applied separately increased the contraction rate over the normal; but when run simultaneously through a culture in a perfusion chamber they appeared to cancel each other out. Further studies of drug action on cultured glial cells were published by Nakazawa.[182] The fact that serotonin and various exogenous pharmacologicals induce physiological responses that include cell movements affords another dimension to neurochemical investigation.

Fluorescence histochemistry as outlined by Falck *et al.*[63] permits direct visualization of the transmittor in adrenergic neurons, based on the finding that catecholamines (e.g. NE), also indoleamines (5-HT), in a dry protein layer are converted into strongly fluorescent products when exposed to formaldehyde gas. The method has been employed by A. Dahlstrom[53,54] in her elegant studies outlining the intraneuronal distribution and transport of bioamine storage granules within the sympathetic system *in situ*; these support the theory of perikaryal origin and peripherad transport. In long-term organized cultures of sympathetic chain ganglia, Benitez and colleagues[12,43] have followed the development of green-fluorescing particles (presumably NE) in the somas of immature chick neurons and their later appearance peripherad along the neurites as the cells matured over a month in culture, supporting Dahlstrom's *in vivo* conclusions. An

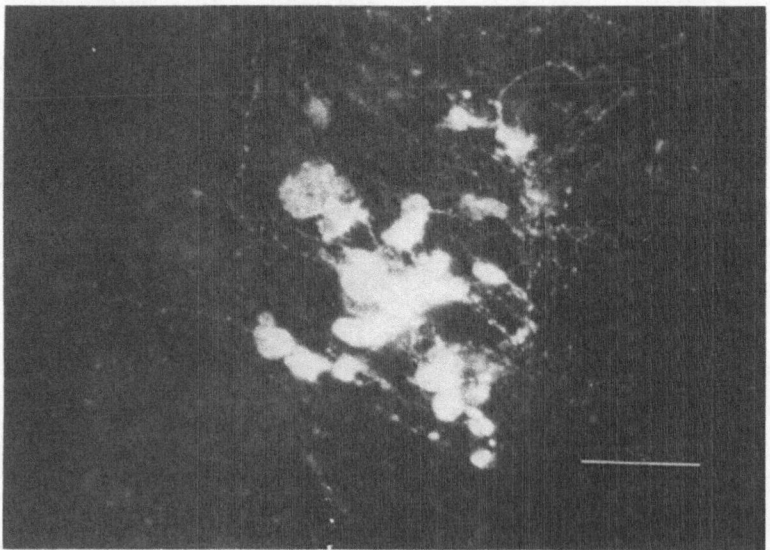

Fig. 9. Sympathetic ganglion from 15-day chick embryo, 22 days *in vitro*. Pattern of catechol fluorescence 6 hr after application of biopterin 10^{-5} M. Bar: 100 μ.

advantage of the culture situation is that the whole neuron with all its extensions can be visualized at one time, and need not be reconstructed from sections. However Sano et al.[214] using similar fluorescence and culture methods with fetal murine chain-ganglia, localized catecholamine granules first in neurite extremities at 8 days in vitro, and only later (after about a month) in neuron somas.

In an attempt to localize and observe directly the biosynthesis of transmittor substances, Coté et al.[43] utilized organized cultures of chick sympathetic chain ganglia to follow the conversion of tyrosine to DOPA by tyrosine hydroxylase. This initial step in the synthesis of NE is the rate-limiting reaction and requires a reduced pterine as cofactor. The addition of biopterin at 10^{-6}M to mature cultures significantly increased the amount of greenish fluorescence between 4 and 20 hr after application, with a peak at approximately 10 hr (Fig. 9). A second application caused a further rise in fluorescence intensity which subsided more gradually. When α-methyl-tyrosine, a potent inhibitor of tyrosine hydroxylase, was added to biopterin-treated cultures, fluorescence was profoundly reduced. The fluorescent results correlated well with the rate of conversion of L-tyrosine-^{14}C to catecholamine in these cultures, using as blanks D-tyrosine and α-CH$_3$-tyrosine. In all these experiments the neuron perikaryon appeared to be the primary site of fluorescent particle emergence.[10a]

Preliminary fluorescence studies of cultured mouse hypothalamus by Benitez et al.[11] and Coté et al.[43] show development of a particulate yellow fluorescence (5-HT) in neuron somas, most pronounced in the preoptic and infundibular areas. Electron micrographs of cultured fetal hypothalamus[148,150] show the development of large numbers of characteristic synaptic profiles during a maturation period of ~3 months in vitro. Some of these terminal structures contain only the small clear vesicles associated with ACh; but especially in cultures from the infundibular region a mixed population of small clear, and large dense-cored (~1200 Å) vesicles is frequently found within irregular calycoid end bulbs which have multiple contact zones. Identification of these vesicles with catecholamines, 5-HT, or possibly even oxytocin must await further study (cf. Figs. 10, 11).

In long-term cultures of sympathetic chain ganglia, synapses also develop prolifically—they are more commonly cholinergic in type, but may show multiple contacts. However, where such cultured ganglia are confronted with populations of heart-muscle cells, they send out bundles of neurites which penetrate the cardiac complex and branch within it. Though these fiber terminals do not exhibit definitive synaptic profiles, they take the form of thickenings and irregular swellings often packed with dense-cored vesicles and other electron-opaque materials,[147] as they lie closely applied to the cardiac muscle cells. Brief repetitive (10/sec) ganglion stimulation with microelectrodes[47] can produce a marked increase in the rate of spontaneous cardiac contractions, from 1–2/sec to more than 5/sec. The effect occurs after a latency of 1 sec and may continue for 10 sec after cessation of stimulation.

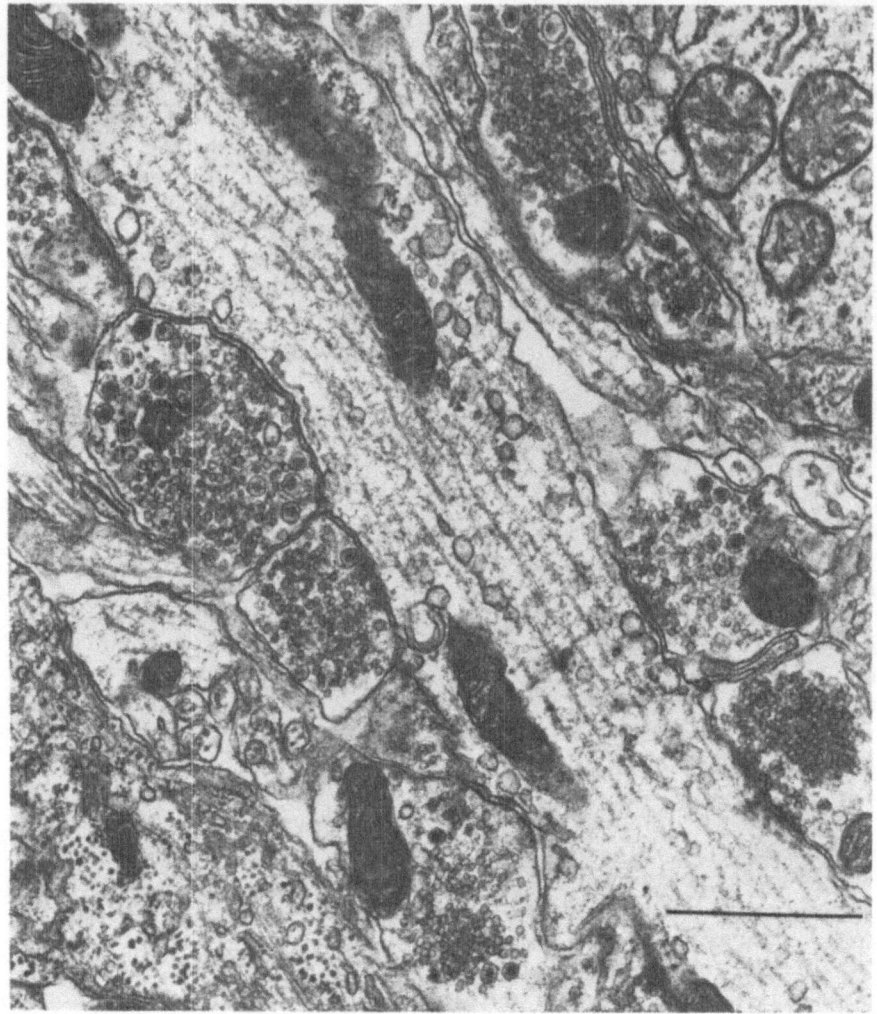

Fig. 10. Electron micrographs (by Dr. E. B. Masurovsky) of newborn mouse hypothalamus, infundibular region 64 days *in vitro*. Multiple contacts of the axodendritic type. Terminals contain many small clear vesicles, with a few larger, dense-cored vesicles interspersed. Bar: 1 μ.

Burdman[37] applied dopamine-^{3}H and NE-^{3}H respectively to cultures of dissociated chick embryo sympathetic ganglion tissue, maintained on 90% synthetic medium (199) and 10% serum, with added NGF. Uptake was very rapid (within 1 min of administration) appearing in neuron somas and processes simultaneously; it was restricted to sympathetic nerve cells, which retained the amines up to 24 hr. Neurons still had the capacity for

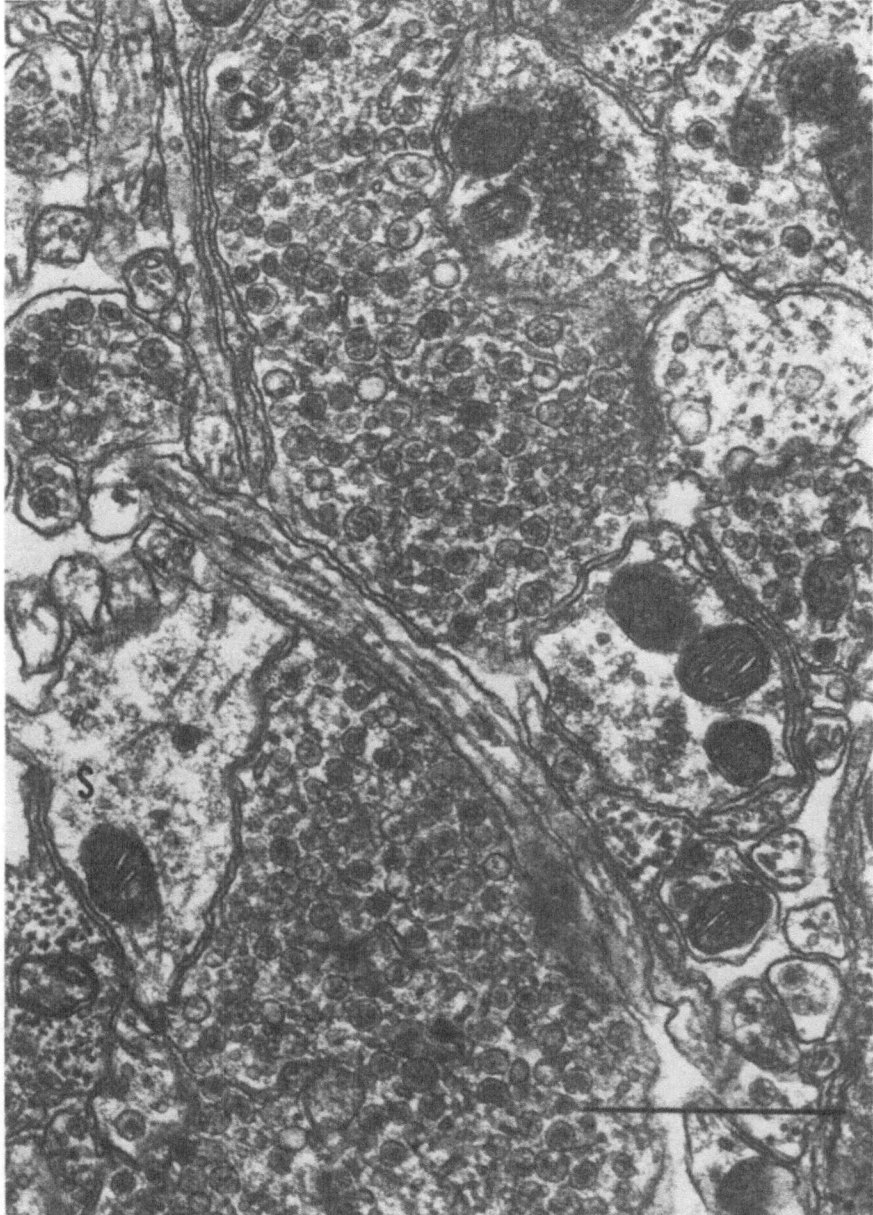

Fig. 11. Electron micrographs (by Dr. E. B. Masurovsky) of newborn mouse hypothalamus, infundibular region 64 days *in vitro*. Two large bodies containing a variety of vesicle types are shown. Both abut on a neurite containing microtubules; one also is closely applied to a spine (S). Bar: 1 μ.

uptake after 1 month in culture. Uptake was inhibited by the presence of nonradioactive DOPA, dopamine, NE, or epinephrine in the pulse media.

C. Acetylcholine Esterase

Before the emergence of fluorescence histochemistry for biogenic mono-amines, students of transmittor substances were of course occupied with the quaternary amine ACh and its biotransformations. The Koelle–Friedenwald thiocholine method[119a] allowing visualization of AChE as CuS was applied in various ways to nervous tissues *in situ*; papers by May *et al.*[155,156] represent its use in detecting the appearance of these hydrolyzing enzymes in cultures of early embryonic (7-day chick) spinal cord developing *in vitro*. The precipitate is already demonstrable as perinuclear granules in the medulloblast at explantation; as these cells differentiate toward the neuro-blast stage, CuS granules progress into the axon from its point of origin and continue with it in the growing tip. During a 48-hr culture period little new enzyme is produced; some types of fibers, however, are completely lacking in enzymatic activity. In 48-hr organ cultures of 15–17-day embryonic chick cord, Turbow and Burkhalter[239] used a modification of the same method to localize AChE in untreated cord cultures and in those which had received supplements of ACh (at 0.02 M) in their culture medium. Enzyme activity was restricted to the gray matter, was especially pronounced in the motor area of the ventral horn, and was substantially increased at 48 hr in the presence of ACh. Veneroni and Murray[243] demonstrated the formation *de novo* and development of neuromuscular junctions in long-term cultures of chick embryo spinal cord confronted with dissociated skeletal muscle cells. As exploring fibers from the cord made contact with the reconstituted sarcoblasts, areas of AChE activity appeared in the sarcolemma. *iso*-OMPA was used to repress nonspecific ChE, after staining by the method of Karnovsky and Roots.[106a] It was suggested that the exploring nerve fiber exerts a triggering action on the myotube membrane, causing it to bulge locally and anchor the neuritic terminal in a condensed area of AChE activity. Peterson and Crain[195a] have extended this type of observation, producing functional neuromyal junctions even across species lines (rat and mouse), as well as more highly differentiated end-plate structures with typically delimited AChE concentrations. They attribute to the exploring nerve fibers a "trophic" influence on the muscle, which is enhanced by the addition of cortisone to the medium.

In a study on the distribution of enzymes in cultured retina from 2-day-old rats and rabbits, Hansson[82] found AChE activity (as distinct from nonspecific esterases) in multipolar neurons and some smaller neurons after morphological differentiation had taken place in culture. Hösli and Hösli[101a] visualized AChE in many of the neurons that they were able to distinguish in short-term (12 day) cultures of neonatal cerebellum and brain-stem, but not in dendrites or glial perikarya. In long-term (3 months) cultures of newborn and late fetal mouse cerebellum and DRG, Kim and Murray[114] demonstrated AChE activity by the Karnovsky and Roots method using *iso*-OMPA as inhibitor for pseudocholinesterase. AChE was localized in neuronal perikarya and dendrites as fine to coarse brown granules. Axons showed a faint diffuse coloration. Glial cells reacted negatively. Purkinje cells and neurons of the cerebellar nuclei, also DRG neurons, invariably displayed activity in their somas, regardless of the time in culture—from 3 to 100 days. A combination of microgasometric analysis and electron microscopic histochemistry was described by Brzin *et al.*[27] for the intracellular localization of AChE in isolated (though not cultured) frog DRG. Fischbach[67a] reports recording synaptic potentials from 2–4 week old cultures of initially dissociated chick cord and muscle which resemble the synaptic responses of adult spinal cord and neuromyal junctions. ACh applied intophoretically to the muscle shows that the synapse is chemically mediated.

D. Exogenous Compounds Affecting Bioelectric Function

Mention was made in the introductory paragraphs of the development of characteristic bioelectric activities in long-term organized nervous tissue cultures. Maturing explants, mainly from CNS, have been exploited as a model system for these studies, especially by Crain and co-workers (cf.[46,51] for reviews of recent work in this area). Development of complex bioelectric function in the cultured material is correlated with the formation of abundant synaptic junctions between neurons in a given explant (e.g. cerebrum) and/or between these and neurons explanted from another area (brain stem) that would normally establish synaptic connections with the first. Companion studies with electron microscope and microelectrode arrays have demonstrated the coincident onset and development of synaptic profiles and of complex functional responses in rat spinal cord-ganglion cultures.[29,50] The techniques used to prepare the cultures have been reported in detail[196] as have the electrophysiological methods.[49]

Cerebral connections with medulla or spinal cord explanted ~ 1 mm away, established *de novo* by outgrowing neurites, have produced complex evoked responses, also spontaneous activities which Crain designates as both excitatory and inhibitory interactions. Application of strychnine (10 μg/ml) facilitates evoked responses; picrotoxin applied to cord-ganglion cultures at similar concentrations also increases amplitude and duration of spike barrages; d-tubocurarine (at 10^{-4} M) provokes a selective block of spike-barrage responses. In cord–muscle hookups this drug rapidly blocks neurally evoked contractions, while electric stimuli applied directly to muscle fibers are still effective. Eserine accelerates recovery of normal function.

Most interesting are recent observations by Crain *et al.*[48] on the functional maturation of cultured CNS tissues during chronic exposure to agents which prevent bioelectric activity. Fetal rodent cerebral cortex and spinal cord tissues explanted prior to the onset of synaptic function were exposed for 5–30 days to xylocaine (50 μg/ml) in the feeding medium, or to a 10-fold increase in Mg^{2+} concentration, which did not interfere with the characteristic morphological development but allowed no evoked or spontaneous biolectric activity at ordinary, or much higher, stimulus intensities. Nevertheless, within minutes after withdrawal of the blocking agent, complex responses characteristic of organized multisynaptic networks could be evoked by the first stimulus.

Sera from EAE animals and acute MS patients, when applied to model cultures of murine cerebrum and cord bring about, in addition to demyelination, rapid, reversible alterations in electrically evoked synaptic responses.[22] The agents, which are complement-dependent, can produce within 30 min a great rise in stimulation threshold coupled with marked decrease in amplitude and duration of slow-wave potentials. Strychnine at 1 μg/ml rapidly lowers this threshold and initiates the return of slow-wave potentials. Replacement of active serum by heated serum or by normal medium restores bioelectric function in about 1 hr. In an electron microscopic study of the same type of culture material exposed to EAE serum, Ross and Bornstein[210] found reversible morphological alterations in synaptic boutons; however these do not occur in synchrony with the functional responses described above. In about 24 hr, synaptic vesicles tend to clump, fuse, and disappear and there seems to be a reduction in the number of synapses; after 6 days there is a 50% numerical reduction, but remaining profiles appear normal. Complete recovery occurs in 24 hr after withdrawal of EAE serum. All these intriguing observations await further explanation.

E. Electrolytes

Following a light microscopic observation that application of the cardiac glycoside ouabain to organized murine DRG cultures resulted in the formation of perikaryal vacuoles, Whetsell and Bunge[247] undertook a fine-structural study of that process. It was found that this inhibitor of active Na and K transport, administered in 10^{-3} to 5×10^{-5} M concentrations for several hours produced selective and progressive swelling of Golgi elements in many, but not all, neurons; other neuronal organelles were not affected, nor were satellite and Schwann cells. Within a few hours after ouabain withdrawal, the neuronal vacuoles disappeared and normal Golgi areas were again seen. These observations suggested to the authors that there is a site for active transport of Na^+ and K^+ on the Golgi membranes of these neurons; and that if this site were directed so as to pump cation from Golgi cisternae into cytoplasm such a pumping mechanism could explain the concentrating function of the Golgi apparatus.

Among the distinguishing properties of the currently accepted Na-pump system based on a membrane-associated ATPase are its specific inhibition by ouabain and its inability to pump Li^+ ions out of the cell efficiently. Incubating DRG cultures in a Li-containing feeding medium, Whetsell[246] found the same cytoplasmic vacuolization of neurons as with ouabain, and this also was revealed by electron microscopy to comprise dilatation of

Golgi components, with little effect on other organelles. Transient focal swellings observed by Mire *et al.*[159] along degenerating unmyelinated fibers in DRG cultures are interpreted by these authors as resulting from failure of active axolemmal ion-pumping mechanisms, preceding the ultimate loss of selective permeability by the axon membrane.

The foot-plates of cultured astrocytes were observed by Friede[68] to be particularly involved in the response of several hydrolyzing enzymes to the environmental concentration of common electrolytes; however only NaCl could elicit an increase in (e.g. NADH, SD) activity at concentrations known to be present in the living organism. Since only astrocytes and not neurons or other glia responded in this way to a rise in ionic concentration, it was concluded that those cells are metabolically involved in the maintenance of the osmotic environment of the CNS, particularly as regards the active transport of Na. Renawek *et al.*[208b] applied ouabain at 10^{-4} M to cultures of rat cerebellum, over periods of 30 min to 24 hr. Oligodendroglia were unaffected, and neurons only very slightly. However, astrocytes displayed a "specific sensitivity," becoming very swollen or bursting in about 12 hr. These workers also conclude that the Na–K pumping mechanism for the brain is localized in the astrocytes.

The effects of different electrolytes, ionic strength $+/+ +$ cation ratio, pH and temperature on the ciliary activity of ependyma from the fourth ventricle of newborn rats were studied by Singer and Goodman.[232] Using a system for rating beat frequency, they found that univalent cations could be ordered in a clear sequence, directly related to the classic lyotropic series for protein macromolecules within the nervous system and in general.

F. Drugs

Though neurotropic drugs have been tested in various tissue culture situations, few observations have actually been made on cultured nerve cells. Reference was made earlier to LSD and to chlorpromazine in relation to the action of serotonin on oligodendroglia; more recently Hendelman and Kelland[87] have published a note indicating that LSD in 10^{-4} or 10^{-5} M concentration had little or no morphological effect on organized synapse-containing cultures of mouse cerebellum, beyond a possible enhancement of lysosomal elements.

A more extended study of chlorpromazine (CPZ) was undertaken by Murray and co-workers,[175,179] based on the circumstance that this phenathiazine derivative is a fluorochrome which can be visualized directly in the living cell. It was hoped that with an elaborate system of spectroscopy the binding sites of the drug could be localized within the neuron, and its metabolic transformations possibly followed.[131] The sensory ganglion was chosen because the disposition of those neurons in culture allows maximum visibility (Fig. 12). A short exposure of DRG cultures to CPZ at 7×10^{-5} M quickly produces a diffuse orthochromatic fluorescence in perikarya. After 24 hr the fluorescent image is no longer diffuse but particulate, corresponding

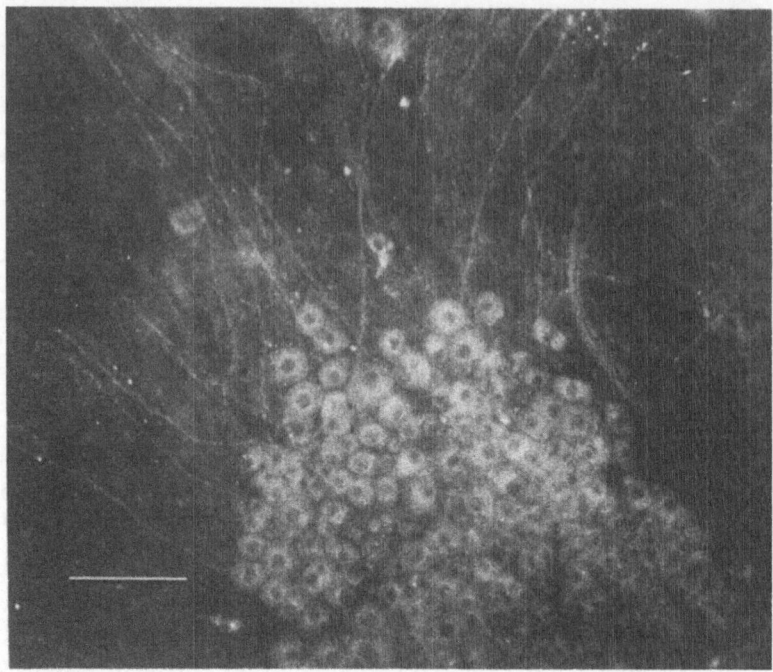

Fig. 12. Fluorescence image, chlorpromazine 7×10^{-5} M visualized in fetal rat DRG cultures 1 hr after exposure. Cf.[179] Bar: 100 μ.

to a somewhat abnormal, transient, granularity seen in the neuron somas at that time, by ordinary light microscopy. Electron microscope study revealed that these fluorescing particles were large, dense, multilaminated bodies associated with the lysosomal system of the cell.[30,67] They become prominent within 4 hr after exposure and at 24 hr have increased considerably in numbers and size. During the early period of their emergence they contain vacuoles of cytoplasm (probably autophagic); and acid phosphatase activity is found within them. At 24 hr, reaction product appears extensively also throughout the perikarya. Within 48 hr the neurons recover uneventfully. It is concluded that CPZ is concentrated within these multilaminated lysosomal bodies, the neurons adopting toward this drug a general pattern of response that is observed in many somatic cell types after exposure to vital dyes (e.g., neutral red and acridine orange).

G. Antibiotics

In monolayer cultures—particularly those concerned with somatic cell lines—antibiotics are used freely, and doubtless the cells involved have adapted to the ambient antibiotic in the course of becoming established. Indeed, antibiotics are routinely incorporated into commercially purveyed synthetic media. Primary cultures, however, may be more fastidious, and nervous tissues from various areas appear to have differential susceptibilities to the spectrum of bactericidal drugs. A good deal of this lore is incorporated into the culture mores on an empirical basis: e.g. penicillin is not introduced into cerebellum cultures, though at 250 units/ml it is acceptable for DRG and spinal cord, as is Achromycin (at 5 μg/ml). Cerebellum also will tolerate Achromycin.

Because of clinical observations incriminating streptomycin, several experimental studies on antibiotics in relation to the otocyst have been pursued in culture; these are summarized and discussed by Friedmann.[69] Friedmann and Bird[70] examined the direct effects of ototoxic antibiotics on the fine structure of the developing chick embryo otocyst isolated in organ culture, finding extensive morphological changes in mitochondria, endoplasmic reticulum in the sensory areas, and nerve tissues. Neomycin, kanamycin, vancomycin, streptomycin, and dimycin (a combination of streptomycin and dihydrostreptomycin) proved highly toxic for the otocyst; in concentrations of 1000–2000 μg/ml, streptomycin, dimycin, and neomycin severely damaged organelles and nerve fibers within 48 hr. In a histochemical study of this material, McAlpine and Friedmann[139] found that their cultured otocysts retained all the enzymes found in the intact ear of the chick except for ChE; treatment with antibiotics did not alter the activity and localization of the dehydrogenases in the damaged area; perhaps these were still active in the damaged mitochondria? Friedmann takes note of the discrepancy between the relatively high dosage required *in vitro* and the much lower doses causing appreciable damage *in vivo*, suggesting that ototoxic antibiotics might be concentrated in the endolymph *in situ*. In long-term cultures of otocyst ganglion cells Orr and Shambaugh[188] similarly found high concentrations of streptomycin sulfate damaging to Nissl substance and neurofibrils. However, in a later study of acoustic ganglion development, Orr[187] routinely incorporated 100 mg/ml penicillin and 100 μg/ml streptomycin into feeding media with favorable results. In an early study quoted by May,[154] Barski and Maurin[5b] reported the effects of streptomycin on the outgrowth from embryonic chick mesencephalon: 100 units/ml (reputed to be comparable to the concentration attained in CSF *in vivo*) was stimulating to neuritic outgrowth; 1000/ml was inhibitory to both neurites and glial processes; and 10,000/ml was frankly toxic to both. Susceptibility of mesenchymal elements was much lower than that of nervous tissues.

VII. NEUROLOGICAL NEOPLASMS

A good deal of work exists on the cultivation *in vitro* of tumors of the nervous system (cf. monographs by Kersting,[108] Lumsden,[134] Sano et al.[213]); most of these studies have been morphological, with the aim of elucidating tumor classification and tissular origin, or aiding in diagnosis.[180] Recently some metabolic investigations, especially in the neuroblastoma family, have produced interesting findings; glial neoplasms have been studied also in relation to chemotherapy.

A. Metabolic Aspects

The neuroblastoma of childhood, believed to arise in embryonal sympathetic neurons, is a widely metastasizing, usually rapidly fatal neoplasm. Nevertheless a good many sporadic observations, well authenticated, indicate that spontaneous regression can take place, and some of these abatements have been correlated with a morphological maturation from neuroblasts to multipolar ganglion cells exhibiting various degrees of cytodifferentiation (cf. Greenfield and Shelley[77]). And in monolayer cultures derived from human neuroblastomas and carried for long periods, Goldstein and co-workers also have observed some morphological maturation.[75]

In children with tumors of neural crest origin there is a well-known raised urinary excretion of NE metabolites and precursors, which apparently is not correlated quantitatively with the degree of malignancy. LaBrosse et al.[25] were the first to investigate this tumor metabolism in living cells in culture; they found that after 2–6 weeks in vitro both neuroblastoma and ganglioneuroma cells produced enzymes capable of degrading NE-^{3}H added to the culture media. Chromatograms of the medium removed after a 48-hr exposure showed areas of radioactivity delineating metabolites resulting from the activity of catechol-O-methyl transferase, monoamine oxidase (MAO), and sulfate conjugase. Later Goldstein et al.[74] compared the incidence of deaminated metabolites in the tumors and in a 2-year old cell line established from a lymph-node metastasis, finding that both exhibited high MAO activity. The major deaminated metabolite of tyramine-^{3}H in both tumor homogenates and cultured tumor cells was p-hydroxyphenylacetaldehyde. Aldehyde dehydrogenase activity was very low, therefore the authors reasoned that the generation of aldehydes and alcohols by the action of MAO might be expected to cause some toxic effects in neuroblastoma patients.

An earlier study by Burdman and Goldstein[39] took note of the presence in sera from children with neuroblastoma (also retinoblastoma and some sarcomas) of a nerve-growth stimulating protein which appeared related to, though not identical with, the NGF obtained by Levi–Montalcini and Cohen from the mouse submaxillary gland (see Schenkein, this Handbook, Vol. V, Chapter 16). It was not found in normal or carcinomatous human sera. This human growth factor could be neutralized by an antiserum produced in goats against the purified mouse NGF; but cross-reactivity between the human NGF and the mouse NGF antisera was low. The human NGF is present in relatively large quantity in fetal sympathetic ganglia, but not in the tumors. Mouse NGF antiserum did not inhibit or destroy neuroblastoma cultures but it did inhibit chick DRG and human fetal DRG and sympathetic ganglion cultures, which were stimulated by the human serum NGF.

Cloned lines of a spontaneous mouse neuroblastoma have been used recently in observations on cytodifferentiation. When these cells are maintained in suspension they take on a round, "anaplastic" form; but when allowed to attach to a glass, plastic, or collagen surface they send out processes up to 3 mm long and assume some of the morphology of neurons. With electron microscopy they exhibit neurotubules, neurofilaments, and some dense bodies which Schubert et al.[217] regard as dense-cored vesicles. These cells have a tyrosine hydroxylase activity, characteristic of ANS neurons, as deduced from their accumulation of dopamine-^{3}H and some NE following exposure to tyrosine-^{3}H.[5a] No morphologically typical synaptic profiles have been observed but some neurites formed club-shaped terminals in close contact with long or wide nerve processes which they tended to indent. Using intracellular recording microelectrodes, and iontophoretic application of ACh, Harris and Dennis[84] found that the somas of these neoplastic neurons were sensitive to ACh though not all the processes were; some cells were electrically coupled via processes, but no chemically mediated synaptic transmission was registered. Klebe and Ruddle,[118] also Olmsted and Rosenbaum[184] attest to stably maintained differentiated functions for this cell line (termed NEURO), as expressed in neuronlike morphology and production of tyrosine hydroxylase and microtubular protein at a level similar to that in brain. The cell line also produces motile amoeboid cells which can be eliminated by treatment with 10^{-4} M FUdR and 10^{-4} M uridine, leaving only a population of large cells which send out arborizing neurites, and can differentiate to this degree without contact with other cells. No supporting cells appear to be present; this condition is (in the writer's observations) characteristic of the human neuroblastoma as contrasted with the ganglioneuroma, in which Schwann and satellite cells are excessively numerous.

This remarkable tumor cell line is now being exploited by numberless investigators, many of them associated with M. Nirenberg, whose reports on various parameters of the work are clustered in the *Proceedings of the National Academy of Sciences* [e.g., **64**: 1004–1010 (1969); **66**: 160–167 (1970); and **67**: 786–792 (1970)]. It is claimed that a considerable number of properties identified with cytodifferentiation in neurons can be regulated in their expression by these neoplastic cells and that the regulatory mechanism is inversely related to the rate of cell division. Among characteristics which are augmented upon restriction to cell division are: ACh activity, amount of protein per cell, axon–dendrite development, formation of electrically excitable membranes, and synthesis of ACh receptors. A very recent and unusually intriguing report by Minna et al.[158a] describes the fusion of neuroblastoma cells with L-cells to form hybrids which still exhibit electrically excitable membranes 10 to 40 generations later. It is stated that at least a part of the genetic information for neuron differentiation can be expressed by N × L hybrid cells, and that evidence for the regulation of action potential components exists in this new neoplastic line. Maintenance of several cell lines arising from neoplastic human glia has been reported[143]—sometimes with companion studies of normal glia, human or animal.[103,207] Enzyme histochemistry of these tumors in culture is described by Kreutzberg and associates[120] with especial reference to oxidoreductases in the glioblastoma multiforme; distribution of acid phosphatase activity in gliomas was studied by Głuszcz et al.[73] Subdivision of glioblastoma multiforme on a cytochemical basis was proposed by Lapham[124]; Lehrer[125] has extended the schema to include a larger spectrum of glial tumors in vivo. Rhythmic contractions of oligodendroglioma cells in vitro were recorded by Pomerat et al.[205] As regards experimental gliomas, Zagon et al.[267] have found that the effects of colcemid and VBL on a methylcholanthrene-induced mouse glioblastoma are morphologically aberrant. Dorfman and Ho[59a] report on a clonal strain of rat glial tumor cells (originally induced by injection of N-nitrosomethylurea) which in culture produces hyaluronic acid, chondroitin-4-sulfate, and heparin sulfate, as well as S-100 protein.

B. Chemotherapy

The especially distressing and invariably fatal course of human glioblastoma multiforme has led to some concentration of chemotherapeutic effort in regard to gliomas of all types. Unfortunately, results have not been encouraging (as summarized by Batzdorf[8]). Problems of availability of the tumoricidal agent to the tumor in vivo, involving metabolism of the agent by the host, filtration by the BBB or utilization of other routes (intrathecal, arterial); untoward local effects, or systemic toxic effects, all render evaluation of potential agents in tissue culture less than satisfactory, as evidenced by early studies. Gellhorn and associates performed a series of screening tests on cultured glioblastoma tissue with potential therapeutic agents which included 8-azaguanine (deaminated by normal but not by neoplastic tissue), 6-mercaptopurine, benzimidazole, miracil-D, thiaxanthenones, aminoquinolines, and many others.[71,72,97,98,178] In some cases experimental gliomas were utilized, and the tumoricidal results of each compound in vitro checked against those in vivo, with poor correlation. Studies of cytostatic effects on human glioblastoma cultures were also reported by Kersting.[107] More recently Wilson and Barker have tested mithramycin and vinblastine sulfate (VBL) on cultured human brain tumors[251] and have shown that in the presence of a CSF-constituted medium the oncolytic effects of these agents were not reduced.[250]

VIII. REFERENCES

1. T. Alescio and A. A. Moscona, Immunochemical evidence for enzyme synthesis in the hormonal induction of glutamine synthetase in embryonic retina in culture, Biochem. Biophys. Res. Commun. 34:176–182 (1969). See also Exper. Cell Res. 61:342–346 (1970).
1a. P. Amaldi and G. Rusca, Autoradiographic study of RNA in nerve fibers of embryonic sensory ganglia cultivated in vitro under NGF stimulation, J. Neurochem. 17:767–771 (1970).
2. J. B. Angevine, Jr., Nerve destruction by colchicine in mice and golden hamsters, J. Exper. Zool. 136:363–371 (1957).
3. S. H. Appel and M. B. Bornstein, Application of tissue culture to the study of experimental allergic encephalomyelitis. II. Serum factors responsible for demyelination, J. Exp. Med. 119:303–312 (1964).
4. S. Appel and D. Silberberg, Pyrimidine synthesis in tissue culture, J. Neurochem. 15:1437–1443 (1968).

5. B. G. W. Arnason, G. F. Winkler, and N. M. Hadler, Cell-mediated demyelination of peripheral nerve in tissue culture, *Lab. Invest.* **21**:1–10 (1969).

5a. G. Augusti-Tocco and G. Sato, Establishment of functional clonal lines of neurons from mouse neuroblastoma, *Proc. Nat. Acad. Sciences* **64**:311–315 (1969).

5b. G. Barski and J. Maurin, Action de la streptomycine sur le tissu nerveux en culture, *Ann. Inst. Past.* **76**:295–302 (1949).

6. L. G. Barth and L. J. Barth, The relation between intensity of lithium treatment and type of cellular differentiation of *Rana pipiens* presumptive epidermis, *Biol. Bull.* **124**:125–140 (1963).

7. L. G. Barth and L. J. Barth, The sodium dependence of embryonic induction, *Develop. Biol.* **20**:236–262 (1969).

8. U. Batzdorf, Chemotherapy of glioblastoma-multiforme: an appraisal, *Bull. Los Angeles Neurol. Soc.* **31**:164–176 (1966).

9. U. Batzdorf, L. L. Sarlieve, V. A. Gold, and J. H. Menkes, Tay-Sachs disease. Demonstration of the stored ganglioside in cultured cells from brain biopsy, *Arch. Neurol.* **20**:650–652 (1969).

10. P. Benda, J. Lightbody, G. Sato, L. Levine, and W. Sweet, Differentiated rat glial cell strain in tissue culture, *Science* **161**: 370–371 (1968).

10a. H. H. Benitez, L. Côté, and M. R. Murray, Regulation of norepinephrine synthesis in organotypic cultures of sympathetic ganglia, *J. Cell Biol.* **47**:18a (1970).

11. H. H. Benitez, E. B. Masurovsky, and M. R. Murray, Hypothalamus: development *in vitro*. Formation of synapses and distribution of monoamines, Excerpta Medical International Congress Series 166, No. 46, p. 38 (1968).

12. H. H. Benitez and M. R. Murray, Circumstances surrounding long-term cultivation of embryonic sympathetic ganglia, *Excerpta Med.* Sec. 1, 19 (No. 10), p. XIV (1965).

13. H. H. Benitez, M. R. Murray, and D. W. Woolley, Effects of serotonin and certain of its antagonists upon oligodendroglial cells *in vitro*. II International Congr. Neuropath. Proc., Pt. II, pp. 423–428, Plates 98–100, Excerpta Med. Foundation (1955).

14. K. G. Bensch, R. Marantz, H. Wisniewski, and M. Shelanski, Induction *in vitro* of microtubular crystals by vinca alkaloids, *Science* **165**:495–496 (1969).

15. F. S. Billett, R. Collini, and L. Hamilton, The effects of D- and L-threochloramphenicol on the early development of the chick embryo, *J. Embryol. Exper. Morphol.* **13**:341–356 (1965).

16. P. Bonventre and J. Imhoff, Studies on the mode of action of diphtheria toxin: protein synthesis in primary heart cell cultures, *J. Exper. Med.* **126**:1079–1086 (1967).

17. G. G. Borisy and E. W. Taylor, The mechanism of action of colchicine. Binding of colchicine-^{3}H to cellular protein, *J. Cell Biol.* **34**:525–534 (1967).

18. M. B. Bornstein, A tissue-culture approach to demyelinative disorders. Symposium on Organ Culture. NCI Monograph No. 11:197–214 (1963).

19. M. B. Bornstein and S. Appel, Demyelination in cultures of rat cerebellum produced by experimental allergic encephalomyelitic serum. Am. Neurol. Assoc. Trans. 84th Ann. Meeting, pp. 165–166 (1959).

20. M. B. Bornstein and S. H. Appel, The application of tissue culture to the study of experimental "allergic" encephalomyelitis. I. Patterns of demyelination, *J. Neuropath. Exper. Neurol.* **20**:141–157 (1961).

21. M. B. Bornstein and S. H. Appel, Tissue culture studies of demyelination, *N.Y. Acad. Sci. Ann.* **122**:280–286 (1965).

22. M. B. Bornstein and S. M. Crain, Functional studies of cultured brain tissues as related to "demyelinative disorders," *Science* **148**:1242–1244 (1965).

23. M. B. Bornstein and M. R. Murray, Serial observations on patterns of growth, myelin formation, maintenance and degeneration in cultures of new-born rat and kitten cerebellum, *J. Biophys. Biochem. Cytol.* **4**:499–504 (1958).

24. P. Bowman, The effect of 2,4-dinitrophenol on the development of early chick embryos, *J. Embryol. Exper. Morphol.* **17**:425–431 (1967).

25. E. H. La Brosse, J. Belehradek, G. Barski, C. Bohuon, and O. Schweisguth, Metabolic activity of neural crest tumors in tissue culture, *Nature* **203**: 195–196 (1964).

26. A. Brun and U. Brunk, Histochemical studies on brain phosphatases in experimental lead poisoning, *Acta Path. Microbiol. Scand.* **70**:531–536 (1967).

27. M. Brzin, V. M. Tennyson, and P. E. Duffy, Acetylcholesterase in frog sympathetic and dorsal root ganglia. A study by electron microscope cytochemistry and microgasometric analysis with the magnetic diver, *J. Cell Biol.* **31**:215–242 (1966).

28. B. J. Buckingham and H. Herrmann, The effect of 3-acetylpyridine on the explanted chick embryo, *J. Embryol. Exper. Morph.* **17**:239–246 (1967).

29. M. B. Bunge, R. P. Bunge, and E. R. Peterson, The onset of synapse-formation in spinal cord cultures as studied by electron microscopy, *Brain Res.* **6**:728–749 (1967).

30. M. B. Bunge, R. P. Bunge, E. R. Peterson, and M. R. Murray, Comparative light and electron microscope studies of chlorpromazine-treated neurons, *Excerpta Med.* **18**, Sec. 1, IX (1964).

31. R. P. Bunge, Glial cells and the central myelin sheath, *Physiol. Rev.* **48**:197–251 (1968).

32. R. P. Bunge and M. B. Bunge, Electron microscopic autoradiographic studies of amino acid incorporation into cultured dorsal root ganglion neurons, *Anat. Rec.* **145**:213 (1963).

33. R. P. Bunge and M. B. Bunge, Electron microscopic observations on colchicine-induced changes in neuronal cytoplasm, *Anat. Rec.* **160**:323 (1968).

34. R. P. Bunge and M. B. Bunge, A comparison of neuronal changes following colchicine treatment, with observations on other conditions involving the accumulation of neuro-filaments, *J. Neuropath. Exper. Neurol.* **28**:169 (1968).

35. R. P. Bunge, M. B. Bunge, and E. R. Peterson, An electron microscope study of cultured rat spinal cord, *J. Cell Biol.* **24**:163–191 (1965).

36. J. A. Burdman, A note on the selective toxicity of vincristine sulfate on chick-embryo sensory ganglia in tissue culture, *U.S. Nat. Cancet Inst. J.* **37**:331–336 (1966).

37. J. A. Burdman, Uptake of [³H]catecholamines by chick embryo sympathetic ganglia in tissue culture, *J. Neurochem.* **15**:1321–1323 (1968).

38. J. A. Burdman, Conversion of adenine to guanine and cytosine to uracil by chick embryo sensory ganglia in tissue culture, *Brain Res.* **15**:515–517 (1969).

39. J. A. Burdman and M. N. Goldstein, Long-term tissue culture of neuroblastomas. III. *In vitro* studies of a nerve growth-stimulating factor in sera of children with neuro-blastoma, *U.S. Nat. Cancer Inst. J.* **33**:123–134 (1964).

39a. K. Burdzy and P. Kallós, eds. *Pathogenesis and Etiology of Demyelinating Diseases*, S. Karger, Basel (1969).

39b. H. J. Burki and C. A. Tobias, Effects of thyroxine on the cell generation cycle parameters of cultured human cells, *Exper. Cell Res.* **60**:445–448 (1970).

40. J. D. Caston and M. Singer, Isolation of radioactive amino acids and derivatives from peripheral nerve, *Exper. Neurol.* **25**:438–446 (1969).

41. S. D. Cook and P. C. Dowling, Neurological disorders associated with increased DNA synthesis in peripheral blood, *Arch. Neurol.* **19**:583–590 (1968).

42. S. D. Cook, P. C. Dowling, M. R. Murray, and J. N. Whitaker, Circulating demyelinating factors in acute idiopathic polyneuropathy (Guillain-Barré), *Arch. Neurol.* **24**:136–144 (1971).

43. L. Coté, H. H. Benitez, and M. R. Murray, Monoamine biosynthesis in isolated regions of autonomic nervous system, in Proc. 2nd Internat'l Meeting Soc. Neurochem (Paoletti, Fumagalli, and Galli, eds.), p. 137, Tamburini, Milan (1969).

44. B. Courtey et R. Bassleer, Etude histoautoradiographique de l'incorporation de thymidine, d'uridine et de leucine tritiées dans des cellules nerveuses d'embryon de Poulet isolées et cultivées *in vitro*, *C. R. Acad. Sci. Paris* **264**:497–499 (1967).

45. M. L. Cowger, R. P. Igo, and R. F. Labbe, The effect of albumin in reversing bilirubin toxicity in tissue culture, 35th Annual Meeting, Soc. Ped. Res. (1965). Quoted by Silberberg and Schutta 1967. See also *Biochem.* **4**:2763 (1965).

46. S. M. Crain, *in International Review of Neurobiology*, Vol. 9, pp. 1–43, Academic Press, New York (1966). See also *J. Exper. Zool.* **173**:353–370 (1970).

47. S. M. Crain, Development of functional neuromuscular connections between separate explants of fetal mammalian tissues after maturation in culture, *Anat. Rec.* **160**:466 (1968). See also *J. Neurobiol.* **1**:471–489 (1970).

48. S. M. Crain, M. B. Bornstein, and E. R. Peterson, Maturation of cultured embryonic CNS tissues during chronic exposure to agents which prevent bioelectric activity, *Brain Res.* **8**:363–372 (1968).

49. S. M. Crain and E. R. Peterson, Complex bioelectric activity in organized tissue cultures of spinal cord (Human, Rat and Chick), *J. Cell. Comp. Physiol.* **64**:1–14 (1964).

50. S. M. Crain and E. R. Peterson, Onset and development of functional interneuronal connections in explants of rat spinal cord-ganglia during maturation in culture, *Brain Res.* **6**:750–762 (1967).

51. S. M. Crain, E. R. Peterson, and M. B. Bornstein, *in Ciba Foundation Symposium on Growth of the Nervous System* (G. E. W. Wolstenholme and M. O'Connor, eds.), pp. 13–40, J. and A. Churchill Ltd., London (1968).

52. H. Cravioto, J. O'Brien, R. Lockwood, F. H. Kasten, and J. Booher, Metachromatic leukodystrophy (sulfatide lipidosis) cultured *in vitro*, *Science* **156**:243–245 (1967).

53. A. Dahlström, The intraneuronal distribution of noradrenaline and the transport and life-span of amine storage granules in the sympathetic adrenergic neuron, Thesis, Dept. Histology, Karolinska Institutet, Stockholm (1966).

54. A. Dahlström, J. Häggendal, and T. Hokfelt, The noradrenaline content of the varicosities of sympathetic adrenergic nerve terminals in the rat, *Acta Physiol. Scand.* **67**:289–294 (1966). See also *Ibid.*, pp. 271–288.

55. J. Dancis, J. Hutzler, and R. P. Cox, Enzyme defect in skin fibroblasts in intermittent branched-chain ketonuria and in maple syrup urine disease, *Biochem. Med.* **2**:407–411 (1969).

56. B. S. Danes and A. Bearn, Hurler's syndrome: a genetic study of clones in cell culture with particular reference to the Lyon hypothesis, *J. Exper. Med.* **126**:509–522 (1967).

57. B. S. Danes and A. G. Bearn, Gaucher's Disease: A genetic disease detected in skin fibroblast cultures, *Science* **161**:1347–1348 (1968).

58. J. T. Dingle, H. B. Fell, and R. R. A. Coombs, The breakdown of embryonic cartilage and bone cultivated in the presence of complement-sufficient antiserum. 2. Biochemical changes and the role of the lysosomal system, *Int. Arch. Allergy* **31**:283–303 (1967).

59. B. A. Diwan, A study of the effects of colchicine on the process of morphogenesis and induction in chick embryos, *J. Embryol. Exper. Morphol.* **16**:245–257 (1966).

59a. A. Dorfman and P.-L. Ho, Synthesis of acid mucopolysaccharides by glial tumor cells in tissue culture, *Proc. Nat. Acad. Sci.* **66**:495–499 (1970).

60. J. E. Dowling and R. L. Sidman, Inherited retinal dystrophy in the rat, *J. Cell Biol.* **14**:73–110 (1962).

61. P. C. Dowling, S. U. Kim, M. R. Murray, and S. D. Cook, Serum 19S and 7S demyelinating antibodies in multiple sclerosis, *J. Immunol.* **101**:1101–1104 (1968).

61a. M. Dubois-Dalcq and F. Gorce, Action de l'acide phytanique sur le système nerveux périphérique in vitro, *C. R. Acad. Sci. Paris, Series D* **270**:2325–2328 (1970).

62. H. Eagle and L. Levintow, *in Cells and Tissues in Culture* (E. N. Willmer, ed.), Vol. I, Ch. 8, pp. 277–295, Academic Press, London (1965).

63. B. Falck, N.-A. Hillarp, G. Thieme, and A. Torp, Fluorescence of catecholamines and related compounds condensed with formaldehyde, *J. Histochem. Cytochem.* **10**:348–354 (1962).

64. M. Fardeau and W. K. Engel, Ultrastructural study of a peripheral nerve biopsy in Refsum's disease, *J. Neuropath. Exper. Neurol.* **28**:287–294 (1969).

65. H. B. Fell, R. R. A. Coombs, and J. T. Dingle, The breakdown of embryonic (chick) cartilage and bone cultivated in the presence of complement-sufficient antiserum. 1. Morphological changes, their reversibility and inhibition, *Int. Arch. Allergy* **30**:146–176 (1966).

66. E. J. Field and D. Hughes, Toxicity of motor neurone disease serum for myelin in tissue culture, *Brit. Med. J.* **2**:1399–1401 (1965).

67. F. C. Fildes-Brosnan, M. B. Bunge, and M. R. Murray, The response of lysosomes in cultured neurons to chlorpromazine, *J. Neuropath. Exper. Neurol.* **29**:337–353 (1970).

67a. G. D. Fischbach, Synaptic potentials recorded in cell cultures of nerve and muscle, *Science* **169**:1331–1333 (1970).

68. R. L. Friede, The enzymatic response of astrocytes to various ions *in vitro, J. Cell Biol.* **20**:5–15 (1964).

69. I. Friedmann, *in The Biology of Cells and Tissues in Culture* (E. N. Willmer, ed.), Vol. II, pp. 521–547, Academic Press, London (1965).

70. I. Friedmann and E. S. Bird, The effect of ototoxic antibiotics and of penicillin on the sensory areas of the isolated fowl embryo otocyst in organ cultures, *J. Path. Bact.* **81**:81–90 (1961).

71. A. Gellhorn, M. R. Murray, E. Hirschberg, and R. Fricker Eising, *In vitro* and *in vivo* effects of chemical agents on human and mouse glioblastoma multiforme, Proc. 2nd Internat. Congr. Neuropath, London, 1955: *Excerpta Med.* May 1957, Pt. 1 pp. 265–271, Pt. III plates 61–62.

72. A. Gellhorn, E. R. Peterson, A. Kells, E. Hirschberg, and M. R. Murray, The effects of 6-Mercaptopurine, 8-Azaguanine, and 1,4-Dimethanesulfonyloxybutane on an experimental brain tumor: Preliminary observations, *Ann. New York Acad. Sci.* **60**:273–282 (1954).

73. A. Głuszcz, K. Sejmicka, A. Drab, and J. Alwasiak, Cytochemical distribution of acid phosphatase activity in gliomas cultured *in vitro, Acta Neuropath.* **11**:113–121 (1968). [*Excerpta Med.* Sec. 16, Cancer 17, No. 2380 (1969.]

74. M. Goldstein, B. Anagnoste, and M. N. Goldstein, Tyramine-H^3: Deaminated metabolites in neuroblastoma tumors and in continuous cell line of a neuroblastoma, *Science* **160**:767–768 (1968).

75. M. N. Goldstein, J. A. Burdman, and L. J. Journey, Long-term tissue culture of neuroblastomas. II. Morphologic evidence for differentiation and maturation, *J. Nat. Cancer Inst.* **32**:165–199 (1964).

76. S. Granick, Hepatic porphyria and drug-induced or chemical porphyria, *Ann. N.Y. Acad. Sci.* **123**:188–197 (1965).

77. L. J. Greenfield and W. M. Shelley, The spectrum of neurogenic tumors of the sympathetic nervous system: Maturation and adrenergic function, *U.S. Nat. Cancer Inst. J.* **35**:215–226 (1965).

77a. J. H. Haddock and H. L. Nader, Bilirubin toxicity in human cultivated fibroblasts and its modification by light treatment, *Proc. Soc. Exper. Biol. Med.* **134**:45–48 (1970).

78. V. Hamburger and H. Hamilton, A series of normal stages in the development of the chick embryo, *J. Morph.* **88**:49–92 (1951).

79. M. Hamburgh, Evidence for a direct effect of temperature and thyroid hormone on myelinogenesis *in vitro*, *Develop. Biol.* **13**:15–30 (1966).

80. M. Hamburgh, *in Current Topics in Developmental Biology* (A. A. Moscona and A. Monroy, eds.), Vol. 4, pp. 109–148, Academic Press, New York (1969).

81. H.-A. Hansson, A tissue culture study of inherited dystrophy of the retina in mice, *Virchows Arch. path. Anat.* **340**:69–83 (1965).

82. H.-A. Hansson, The distribution of acetylcholine esterase and other hydrolytic enzymes in retinal cultures, *Acta physiol. Scand.* **70**:Suppl. 288 (1966).

83. H.-A. Hansson, Selective effects of metabolic inhibitors on retinal cultures, *Exper. Eye Res.* **5**:335–354 (1966).

84. A. J. Harris and M. J. Dennis, Acetylcholine sensitivity and distribution on mouse neuroblastoma cells, *Science* **167**:1253–1255 (1970).

85. W. J. Hendelman and M. B. Bunge, Some observations on the disposition of microtubules in relation to the myelin sheath, *J. Cell Biol.* **31**:46A–47A (1966).

86. W. J. Hendelman and R. P. Bunge, Radioautographic studies of choline incorporation into peripheral nerve myelin, *J. Cell Biol.* **40**:190–208 (1969).

87. W. J. Hendelman and D. M. Kelland, Morphologic observations on the effect of LSD in tissue cultures of cerebellum, *Anat. Rec.* **166**:318 (1970).

88. J. H. Herndon, Jr., D. Steinberg, B. W. Uhlendorf, and H. M. Fales, Refsum's disease: characterization of the enzyme defect in cell culture, *J. Clin. Invest.* **48**:1017–1032 (1969).

89. W. Hild, Elaboration of so-called posterior lobe hormones in the hypothalamus, *Biol. Bull.* **105**:360–361 (1953).

90. W. Hild, Das morphologische, kinetische und endokrinologische Verhalten von hypothalamischen und neurohypophysarem Gewebe *in vitro*, *Z. Zellforsch.* **40**:257–312 (1954).

91. W. Hild, *in Hypothalmic Hypophysial Interrelationships* (W. S. Fields, ed.), p. 17, C. C. Thomas, Springfield (1956).

92. W. Hild, Myelinogenesis in cultures of mammalian central nervous tissue, *Ztschr. Zellforsch.* **46**:71–95 (1957).

93. A. Hirano, N. H. Becker, and H. M. Zimmerman, Isolation of the periaxonal space of the central myelinated nerve fiber with regard to the diffusion of peroxidase, *J. Histochem. Cytochem.* **17**:512–516 (1969).

94. A. Hirano and H. M. Dembitzer, The transverse bands as a means of access to the periaxonal space of the central myelinated nerve fiber, *J. Ultrastruct. Res.* **28**:141–149 (1969).

95. A. Hirano, D. S. Sax, and H. Zimmerman, The fine structure of the cerebella of jimpy mice and other "normal" litter mates, *J. Neuropath. Exper. Neurol.* **28**:388–400 (1969).

96. A. Hirano, J. N. Whitaker, S. D. Cook, P. C. Dowling, and M. R. Murray, Fine structural aspects of demyelination *in vitro*, *J. Neuropath. Exper. Neurol.* **30** (in press Jan. 1971).

97. E. Hirschberg, A. Gellhorn, M. R. Murray, and E. F. Elslager, Effects of miracil D, amodiaquin, and a series of other 10-thiaxanthenones and 4-aminoquinolines against a variety of experimental tumors *in vitro* and *in vivo*, *U.S. Nat. Cancer Inst. J.* **22**:567–579 (1959).

98. E. Hirschberg, M. R. Murray, E. R. Peterson, J. Kream, R. Schafranek, and J. L. Pool, Enzymatic deamination of 8-azaguanine in normal human brain and in glioblastoma multiforme, *Cancer Res.* **13**:153–157 (1953).

99. E. Holtzman and E. R. Peterson, Cytochemical studies of protein uptake, lysosomes and GERL in neurons, *J. Histochem. Cytochem.* **16**:502 (1968).

100. E. Holtzman and E. R. Peterson, Uptake of protein by mammalian neurons, *J. Cell Biol.* **40**:863–869 (1969).

101. F. C. G. Hoskin and C. D. Allerand, Metabolism of specifically labeled glucose by explants of newborn mouse cerebellum, *J. Neurochem.* **15**:427–432 (1968).

101a. E. Hösli and L. Hösli, The presence of acetylcholinesterase in cultures of cerebellum and brain stem. *Brain Res.* **19**:494–496 (1970).

102. D. Hughes and E. J. Field, Myelotoxicity of serum and spinal fluid in multiple sclerosis: a critical assessment, *Clin. Exp. Immunol.* **2**:295–309 (1967).

103. R. Hugosson, B. Källén, and O. Nilsson, Neuroglia proliferation studied in tissue culture, *Acta neuropath. (Berl.)* **11**:210–220 (1968).

104. L. J. Journey, J. Burdman, and P. George, Ultrastructural studies on tissue culture cells treated with vincristine (NSC–67574), *Cancer* Chemother. Rep. **52**:509–517 (1968). [*Excerpta med.* Cancer 17.1, No. 221 (1969).]

105. L. J. Journey, J. Burdman, and A. Whaley, Electron microscopic study of spinal ganglia from vincristine-treated mice, *U.S. Nat. Cancer Inst. J.* **43**:603–619 (1969).

106. P. R. Karfunkel, The role of microtubules and microfilaments in neurulation, *Anat. Rec.* **166**:328 (1970).

106a. M. J. Karnovsky and L. Roots, A "direct-coloring" thiocholine method for cholinesterases, *J. Histochem. Cytochem.* **12**:219–221 (1964).

107. G. Kersting, Cytostatische Effekte in der Glioblastomkultur, *Glioblastoma multiforme, Suppl. Acta Neurochir.* (1959).

108. G. Kersting, Die Gewebszuchtung menschlicher Hirngeschwulste, *in Monographien aus dem Gesamtgebiete der Neurologie und Psychiatrie*, Vol. 90, Lange and Springer, Berlin (1961).

109. M. W. Kies, Chemical studies on an encephalitogenic protein from guinea pig brain, *Ann. N.Y. Acad. Sci.* **122**:161–170 (1965).

110. S-U. Kim, Histochemical demonstration of oxidative enzymes associated with carbohydrate metabolism in cerebellar neurons cultured *in vitro*, *Arch. Histol. Jap.* **27**:465–471 (1966).

111. S-U. Kim, E. B. Masurovsky, H. H. Benitez, and M. R. Murray, Histochemical studies of the intranuclear rodlet in neurons of chicken sympathetic ganglia, *J. Histochem. Cytochem.* **17**:201–202 (1969).

112. S-U. Kim, E. B. Masurovsky, H. H. Benitez, and M. R. Murray, Histochemical studies of the intranuclear rodlet in neurons of chick sympathetic and sensory ganglia, *Histochemie* **24**:33–40 (1970).

113. S-U. Kim and M. R. Murray, Monoamine oxidase activity in central and peripheral myelin, developed in tissue culture, *J. Cell Biol.* **39**:170a (1968).

114. S-U. Kim and M. R. Murray, Histochemical demonstration of acetylcholinesterase in organized cultures of mouse cerebellum and sensory ganglia, *Anat. Rec.* **163**:310 (1969).

115. S-U. Kim, M. R. Murray, P. C. Dowling, and S. D. Cook, Analysis of myelinotoxic factors in human disease, *J. Neuropath. Exper. Neurol.* **29**:147 (1970).

116. S-U. Kim, M. R. Murray, W. W. Tourtellotte, and J. A. Parker, Demonstration in tissue culture of myelinotoxicity in cerebrospinal fluid and brain extracts from multiple sclerosis patients, *J. Neuropath. Exper. Neurol.* **29**:420–431 (1970).

117. I. Klatzo and J. Miquel, Observations on pinocytosis in nervous tissue, *J. Neuropath. Exper. Neurol.* **19**:475–487 (1960).

118. R. J. Klebe and F. H. Ruddle, Neuroblastoma: Cell culture analysis of a differentiating stem cell system, *J. Cell Biol.* **43**:69a (1969).

119. D. C. Klein, G. R. Berg, J. Weller, and W. Glinsmann, Pineal gland: dibutyryl cyclic adenosine monophosphate stimulation of labeled melatonin production, *Science* **167**:1738–1740 (1970). See also *Science* **168**:979–980 (1970) and *In Vitro* **6**:197–204 (1970).

119a. G. B. Koelle and J. S. Friedenwald, A histochemical method for localizing cholinesterase activity, *Proc. Soc. Exper. Biol. Med.* **70**:617–622 (1949).

120. G. W. Kreutzberg, M. Minauf, and F. Gullotta, Enzyme histochemistry of human brain tumors and their tissue cultures with special reference to the oxidoreductases in the glioblastoma multiforme, *Histochemie* **6**:8–16 (1966).

121. A. Krishan, Time-lapse and ultrastructure studies on the reversal of mitotic arrest induced by vinblastine sulfate in Earle's L cells, *U.S. Nat. Cancer Inst. J.* **41**:581–596 (1968).

122. R. S. Krooth and A. N. Weinberg, Properties of galactosemic cells in culture, *Biochem. Biophys. Res. Comm.* **3**:518–524 (1960).

123. P. W. Lampert and S. S. Schochet, Jr., Demyelination and remyelination in lead neuropathy, *J. Neuropath. Exper. Neurol.* **27**:527–543 (1968).

124. L. W. Lapham, Subdivision of glioblastoma multiforme on a cytologic and cytochemical basis, *J. Neuropath. Exper. Neurol.* **18**:244–262 (1959).

125. G. M. Lehrer, in *IV. International Congress of Neuropathology Proc.*, Vol. 1, pp. 66–70, Georg Thieme Verlag, Stuttgart (1962).

126. G. M. Lehrer and M. B. Bornstein, Carbohydrate metabolism of the developing brain *in vivo* and *in vitro*, *Excerpta Medica Int. Cong. Series No. 94*, p. E140 (1965).

127. G. M. Lehrer and M. B. Bornstein, Glucose metabolism in rat cerebellum tissue cultures as a function of age, *Amer. Neurol. Assoc. Trans.* **93**:174–176 (1968).

128. L. Liss and M. D. Grumer, Effect of L-phenylalanine on central nervous system elements in tissue culture, *J. Neurol. Neurosurg. Psychiat.* **29**:371–374 (1966). See also *Proc. VIth International Congress of Neuropath.*, Paris, Masson et Cie, pp. 112–113 (1970).

129. M. Locke (ed.), *Control Mechanisms in Developmental Processes*, Academic Press, New York (1967).

130. M. Locke (ed.), *The Emergence of Order in Developing Systems, Developmental Biology*, Supplement 2, Academic Press, New York (1968).

131. C. N. Loeser, M. R. Murray, E. R. Peterson, and F. F. Barrueto, Intraneuronal fluorescence spectroscopy of chlorpromazine and comparison with standard curves, *Anat. Rec.* **151**:379 (1965).

132. D. R. Lucas, in *Cells and Tissues in Culture, Methods, Biology and Physiology* (E. N. Willmer, ed.), Vol. 2, pp. 457–520, Academic Press, London (1965).

133. J. A. Lucy, in *Cells and Tissues in Culture, Methods, Biology and Physiology* (E. N. Willmer, ed.), Vol. 1, pp. 297–315, Academic Press, London (1965).

134. C. E. Lumsden, in *Pathology of Tumors of the Nervous System* (D. S. Russell, L. J. Rubinstein, and C. E. Lumsden, eds.), 2nd ed., pp. 281–334, Edward Arnold Ltd., London and Williams and Wilkins, Baltimore (1963).

135. C. E. Lumsden, in *Pathogenesis and Etiology of Demyelinating Diseases* (K. Burdzy and P. Kallós, eds.) S. Karger, Basel, pp. 247–275 (1969).

136. C. E. Lumsden, S. Aparicio, and M. Jennings, Electron microscopy of glial reactions observed in CNS cultures, *Brain Res.* **7**:268–285 (1968).

137. C. E. Lumsden and C. M. Pomerat, Normal oligodendrocytes in tissue culture. A preliminary report on the pulsatile glial cells in tissue cultures from the corpus callosum of the normal adult rat brain, *Exper. Cell Res.* **2**:103–114 (1951).

138. C. E. Lumsden, D. M. Robertson, and R. Blight, Chemical studies on experimental allergic encephalitis. Peptide as the common denominator in all encephalitogenic "antigens," *J. Neurochem.* **13**:127–162 (1966).

139. J. C. McAlpine and I. Friedmann, A histochemical study of the effects of ototoxic antibiotics on the isolated embryonic otocyst of the fowl (Gallus domesticus), *J. Path. Bact.* **86**:477–486 (1963).

140. W. D. McElroy and B. Glass, *The Chemical Basis of Development*, John Hopkins Press, Baltimore (1958).

141. G. M. McKhann, W. Ho, Ch. Raiborn, and S. Varon, The isolation of neurons from normal and abnormal human cerebral cortex, *Arch. Neurol.* **20**:542–547 (1969).

142. P. Malpoix and A. Emelinckx, Effect of histones on morphogenesis and differentiation in chick embryos, *J. Embryol. Exper. Morphol.* **18**:143–153 (1967).

143. E. E. Manuelidis and A. R. Pond, Continuous cultivation of cells arising from a human case of glioblastoma multiforme (glioma T. C. 178), *Proc. Soc. Exp. Biol. (N.Y.)* **102**: 693–695 (1959), [*Excerpta med.* Sec. 16, Cancer 9, No. 39 (1961).]

144. G. Margolis, *in Viral Etiology of Congenital Malformations.* Proceedings of a Conference sponsored by the National Heart Institute and the National Institute of Child Health and Human Development, May 1967, pp. 46–66, National Institutes of Health, Bethesda (1968).

145. G. Margolis and L. Kilham, *in The Central Nervous System*, Monograph No. 9, International Academy of Pathology, pp. 157–183, The Williams and Wilkins Co., Baltimore (1968).

146. G. Margolis and L. Kilham, Experimental virus-induced hydrocephalus. Relation to the pathogenesis of the Arnold-Chiari malformations. *J. Neurosurgery* **32**:1–9 (1969).

147. E. B. Masurovsky and H. H. Benitez, Apparent innervation of chick cardiac muscle by sympathetic neurons in organized culture, *Anat. Rec.* **157**:285 (1967).

148. E. B. Masurovsky and H. H. Benitez, Development of synapses in organized cultures of mouse hypothalamus, *Anat. Rec.* **160**:390 (1968).

149. E. B. Masurovsky, H. H. Benitez, S-U. Kim, and M. R. Murray, Origin, development, and nature of intranuclear rodlets and associated bodies in chicken sympathetic neurons, *J. Cell Biol.* **44**:172–191 (1970).

150. E. B. Masurovsky, H. H. Benitez, and M. R. Murray, Synaptic development in long-term organized cultures of murine hypothalamus (in press, 1971).

151. E. B. Masurovsky, M. B. Bunge, and R. P. Bunge, Cytological studies of organotypic cultures of rat dorsal root ganglia following x-irradiation *in vitro*. I. Changes in neurons and satellite cells, *J. Cell Biol.* **32**:467–496 (1967).

152. E. B. Masurovsky, M. B. Bunge, and R. P. Bunge, Cytological studies of organotypic cultures of rat dorsal root ganglia following x-irradiation *in vitro*. II. Changes in Schwann cells, myelin sheaths, and nerve fibers, *J. Cell Biol.* **32**:497–518 (1967).

153. E. B. Masurovsky and R. P. Bunge, Radiation protection in mammalian spinal ganglion cultures treated with AET derivatives. Light- and electron-microscopic autoradiographic studies, *Acta Radiologica* **8**:38–54 (1969).

154. R. M. May, *in Culture et Greffe des Cellules Nerveuses chez les Vertébrés Supérieurs, Monographies de Physiologie Causale* (R. May), Vol. VI, pp. 87–115, Gauthier-Villars, Paris (1966).

155. R. M. May, J. P. Denèfle, and B. Courtey, Comparaison des teneurs en cholinestérases des fibres médullaires *in vitro* et au cours de leur embryogénèse chez le poulet, *J. Physiologie* **55**:167–169 (1963).

156. R. M. May, J. P. Denèfle, and B. Courtey, Distribution différentielle des acétylcholinestérases dans les fibres nerveuses d'embryons de poulet cultivées *in vitro*, *Ann. Histochim.* **11**:19–28 (1966).

157. N. Miani, C. Cavallotti, and A. Caniglia, Synthesis of ATP by myelin of spinal nerves of rabbit, *J. Neurochem.* **16**:249–260 (1969).

158. L. Miller, G. Gordon, and K. Bensch, Cytologic abnormalities in hereditary metabolic disorders: the effects of galactose on galactosemic fibroblasts *in vitro*, *Lab. Invest.* **19**:428–436 (1968).

158a. J. Minna, P. Nelson, J. Peacock, D. Glazer, and M. Nirenberg, Genes for neuronal properties expressed in neuroblastoma × L-cell hybrids, *Proc. Nat. Acad. Sci.* **68**:234–239 (1971).

159. J. J. Mire, W. J. Hendelman, and R. P. Bunge, Observations on a transient phase of focal swelling in degenerating unmyelinated nerve fibers, *J. Cell Biol.* **45**:9–22 (1970).

160. C. E. Mize, J. H. Herndon, Jr., J. P. Blass, G. W. A. Milne, C. Follansbee, P. Laudat, and D. Steinberg, Localization of the oxidative defect in phytanic acid degradation in patients with Refsum's disease, *J. Clin. Invest.* **48**:1033–1040 (1969).

161. B. W. Moore, V. J. Perez, and M. Gehring, Assay and regional distribution of a soluble protein characteristic of the nervous system, *J. Neurochem.* **15**:265–272 (1968).

162. J. E. Morris and A. A. Moscona, Induction of glutamine synthetase in embryonic retina: its dependence on cell interactions, *Science* 1736–1738 (1970). See also *Develop. Biol.* **23**:36–61 (1970).

163. A. A. Moscona, Induction of retinal glutamine synthetase in the embryo and in culture, *Quaderno (Accad. Naz. d. Lincei)* **104**:237–256 (1968).

164. A. A. Moscona, M. H. Moscona, and N. Saenz, Enzyme induction in embryonic retina: the role of transcription and translation, *Nat. Acad. Sci. Wash. Proc.* **61**:160–167 (1968). See also *Exper. Cell Res.* **61**:342–346 (1970).

165. A. A. Moscona and R. Piddington, Enzyme induction by corticosteriods in embryonic cells: steroid structure and inductive effect, *Science* **158**:496–497 (1967).

166. L. Mulherkar, S. S. Joshi, B. A. Diwan, and P. N. Joshi, Reversible effect of chloroacetophenone by sulphhydryl groups on morphogenesis of chick embryos, *J. Embryol. Exper. Morphol.* **17**:263–266 (1967).

167. M. R. Murray, in *Biology of Neuroglia* (W. F. Windle, ed.), pp. 176–190, Charles C Thomas, Springfield, Ill. (1958).

168. M. R. Murray, in *Comparative Neurochemistry* (D. Richter, ed.), pp. 49–58, Pergamon Press, London (1964).

169. M. R. Murray, in *The Biology of Cells and Tissues in Culture* (E. N. Willmer, ed.), Vol. II, pp. 373–455, Academic Press, London (1965).

170. M. R. Murray and H. H. Benitez, Deuterium oxide: direct action on sympathetic ganglia isolated in culture, *Science* **155**:1021–1024 (1967).

171. M. R. Murray and H. H. Benitez, in *CIBA Symposium on Growth of the Nervous System* (G. E. W. Wolstenholme and M. O'Connor, eds.), pp. 148–178, J. and A. Churchill Ltd., London (1968).

172. M. R. Murray, S. D. Cook, and P. C. Dowling, Myelinotoxic activity of Guillain-Barré serum on cultures of peripheral nervous tissue, *Proc. VIth Internat'l Congress of Neuropath., Paris,* Masson et Cie (1970).

173. M. R. Murray and A. Herrmann, Passive movements of Schmidt-Lanterman clefts during continuous observation *in vitro, J. Cell Biol.* **39**:149–150a (1968).

174. M. R. Murray and A. Herrmann, Film: Antagonism of colchicine effect in dorsal root ganglia by D_2O, (unpublished, in preparation).

175. M. R. Murray and E. R. Peterson, in *Comparative Neurochemistry* (D. Richter, ed.), pp. 451–458, Pergamon Press, New York (1964).

176. M. R. Murray and E. R. Peterson, Action of cortisone in myelinated cultures of nervous tissues, *Excerpta Med.* Internat'l Congress Series No. 94:E221–222 (1965).

177. M. R. Murray, E. R. Peterson, and R. P. Bunge, in *International Congress of Neuropathology* (H. Jacob, ed.), Vol. II, pp. 267–272, Georg Thieme Verlag, Stuttgart (1962).

178. M. R. Murray, E. R. Peterson, E. Hirschberg, and J. L. Pool, Metabolic and chemotherapeutic investigation of human glioblastoma *in vitro, Ann. N.Y. Acad. Sci.* **58**:1147–1171 (1954).

179. M. R. Murray, E. R. Peterson, and C. N. Loeser, in *Morphological and Biochemical Correlates of Neural Activity* (M. Cohen and R. Snider, eds.), pp. 225–239, Harper and Row, New York (1964).

180. M. R. Murray and A. P. Stout, Distinctive characteristics of the sympathicoblastoma cultived *in vitro*. A methods for prompt diagnosis, *Am. J. Path.* **23**:429–441 (1947).

181. T. Nakazawa, Effects of chlorpromazine and serotonin on the contractility of oligodendrocytes, *Tex. Rep. Biol. Med.* **18**:52–65 (1960).

182. T. Nakazawa, in *Morphology of Neuroglia* (J. Nakai, ed.), pp. 103–120, Igaku Shoin Ltd., Tokyo (1963).

183. J. L. Nussbaum, N. Neskovic, and P. Mandel, A study of lipid components in brain of the "Jimpy" mouse, a mutant with myelin deficiency, *J. Neurochem.* **16**:927–934 (1969).

183a. S. Okada and J. S. O'Brien, Tay-Sachs disease: generalized absence of a β-D-N-Acetylhexosaminidase component, *Science* **165**:698–700 (1969).

184. J. B. Olmsted, K. Carlson, R. Klebe, F. Ruddle, and J. L. Rosenbaum, Isolation of microtubule protein from cultured mouse neuroblastoma cells, *Nat. Acad. Sci. (Wash.) Proc.* **65**:129–136 (1970).

185. J. W. Olney, Glutamate-induced retinal degeneration in neonatal mice. Electron microscopy of the acutely evolving lesion, *J. Neuropath. Exper. Neurol.* **28**:455–474 (1969).

186. C. W. Orr and S. Roseman, Intercellular adhesion. II. The purification and properties of a horse-serum protein that promotes neural retina cell aggregation, *J. Membr. Biol.* **1**:125–143 (1969).

187. M. F. Orr, Development of acoustic ganglia in tissue cultures of embryonic chick otocysts, *Exper. Cell Res.* **40**:68–77 (1965).

188. M. F. Orr and G. E. Shambaugh, The effect of streptomycin upon ganglion cells in Rose Chamber cultures of chick otocyst. *Excerpta med.* 16, Sec. I, No. 9, Tissue Culture Assoc., XIII Annual Mtg., Washington (1962).

189. F. Paradisi, B. Galanti, and A. Mancini, ATP levels in normal and diphtheria toxin treated cell cultures, *Path. Microbiol.* **32**:24–28 (1968).

190. J. Paul, *Cell and Tissue Culture*, 3rd ed., Williams and Wilkins Co., Baltimore (1965). (4th ed. in press 1970).

191. J. Paul, in *Cells and Tissues in Culture* (E. N. Willmer, ed.), Vol. I, pp. 239–276, Academic Press, London (1965).

192. A. Pentschew and F. Garro, Thallium encephalopathy in monkeys, *J. Neuropath. Exper. Neurol.* **28**:163 (1969).

193. O. Périer, Demyelinization of cultures of central nervous tissue by lysolecithin, *Acta Neurol. Belg.* **65**:78–95 (1965).

194. E. R. Peterson, Neurofibrillar alterations in cord-ganglion cultures exposed to spindle inhibitors, *J. Neuropath. Exper. Neurol.* **28**:168 (1969).

195. E. R. Peterson and M. B. Bornstein, The neurotoxic effects of colchicine on tissue cultures of cord-ganglia, *J. Neuropath. Exper. Neurol.* **27**:121 (1968).

195a. E. R. Peterson and S. M. Crain, Innervation in cultures of fetal rodent skeletal muscle by organotypic explants of spinal cord from different animals, *Zeitschr. Zellforsch.* **106**:1–21 (1970).

196. E. R. Peterson, S. M. Crain, and M. R. Murray, Differentiation and prolonged maintenance of bioelectrically active spinal cord cultures, *Zeitscher. Zellforsch.* **66**:130–154 (1965).

197. E. R. Peterson and M. R. Murray, Modification of development in isolated dorsal root ganglia by nutritional and physical factors, *Develop. Biol.* **2**:461–476 (1960).

198. E. R. Peterson and M. R. Murray, Patterns of peripheral demyelination *in vitro*, *N.Y. Acad. Sci. Ann.* **122**:39–50 (1965).

199. E. R. Peterson and M. R. Murray, Serial observations in tissue cultures on neurotoxic effects of colchicine, *Anat. Rec.* **154**:401 (1966).

199a. S. E. Pfeiffer, H. R. Herschman, J. Lightbody, and G. Sato, Synthesis by a clonal line of rat glial cells of a protein unique in the nervous system, *J. Cell. Physiol.* **75**:329–340 (1970).

200. R. Piddington, Hormonal effects on the development of glutamine synthetase in the embryonic chick retina, *Develop. Biol.* **16**:168–188 (1967).
201. R. Piddington and A. A. Moscona, Correspondence between glutamine synthetase activity and differentiation in the embryonic retina in situ and in culture, *J. Cell Biol.* **27**:247–251 (1965).
202. C. M. Pomerat, Dynamic neurogliology, *Texas Rep. Biol. Med.* **10**:885–913 (1952).
203. C. M. Pomerat, in *Biology of Neuroglia* (W. F. Windle, ed.), pp. 162–175, C. C. Thomas, Springfield (1958).
204. C. M. Pomerat and I. Costero, Tissue cultures of cat cerebellum, *Am. J. Anat.* **99**:211–247 (1956).
205. C. M. Pomerat, B. L. Crue, and F. H. Kasten, Observations on the cytology of an oligodendroglioma cultivated *in vitro*, *U.S. Nat. Cancer Inst. J.* **33**:517–530 (1964).
206. C. M.Pomerat, W. J. Hendelman, C. W. Raiborn, Jr., and J. F. Massey, in *The Neuron* (H. Hydén, ed.), pp. 119–178, Elsevier Publ. Co., Amsterdam (1967).
207. J. Ponten and E. MacIntyre, Long term culture of normal and neoplastic human glia, *Acta Path. Microbiol. Scand.* **74**:465–486 (1968).
208. M. Porter, A. Fluharty, and H. Kihara, Metachromatic leukodystrophy: arylsulfatase-A (AA) deficiency in skin fibroblast cultures, *Proc. Nat. Acad. Sci.* **62**:887–892 (1969). See also *Arch. Biochem.* **138**:646–652 (1970).
208a. C. S. Raine and M. B. Bornstein, Experimental allergic encephalomyelitis: ultrastructural study of experimental demyelination *in vitro*, *J. Neuropath. Exper. Neurol.* **29**:177–191 (1970). See also *Amer. J. Path.* **59**:11 (1970).
208b. K. Renawek, G. Palladini, and L. A. Ieradi, Morphologie de la glie cultivée in vitro après l'ouabaine, *Proc. 6th Intern. Congr. Neuropath., Paris*, Masson et Cie, pp. 458–549 (1970).
209. A. Résibois, Electron microscopic study of metachromatic leucodystrophy. III. Lysosomal nature of the inclusions, *Acta Neuropath. (Berl.)* **13**:149–156 (1969).
210. L. Ross and M. Bornstein, An electron microscopic (EM) study of synaptic alterations in cultured mammalian central nervous tissues exposed to serum from animals with experimental allergic encephalomyelitis, *Lab. Invest.* **20**:26–35 (1969).
211. P. D. Ross and R. L. Scruggs, Electrophoresis of DINA. II. Specific interactions of univalent and divalent cations with DNA, *Biopolymers* **2**:79–89 (1964).
212. H. E. Rugstad, S. H. Robinson, C. Yannoni, and A. H. Tashjian, Jr., Metabolism of bilirubin by a clonal strain of rat hepatoma cells, *J. Cell Biol.* **47**:703–710 (1970).
213. K. Sano, F. Satoh, and H. Chigasaki, in *Morphology of Neuroglia* (J. Nakai, ed.), pp. 133–150, Igaku Shoin Ltd., Tokyo (1963).
214. Y. Sano, G. Odake, and T. Yonezawa, Fluorescence microscopic observations of catecholamines in cultures of the sympathetic chains, *Zeitschr. Zellforsch.* **80**:345–352 (1967).
215. W. W. Schlaepfer, Experimental lead neuropathy: a disease of the supporting cells in the peripheral nervous system, *J. Neuropath. Exper. Neurol.* **28**:401–418 (1969).
216. S. S. Schochet, P. W. Lampert, and K. M. Earle, Oligodendroglial changes induced by intrathecal vincristine sulfate, *Exper. Neurol.* **23**:113–119 (1969).
217. D. Schubert, S. Humphreys, C. Baroni, and M. Cohen, *In vitro* differentiation of a mouse neuroblastoma, *Nat. Acad. Sci., Wash. Proc.* **64**:316–323 (1969).
218. F. J. Seil, G. A. Falk, M. W. Kies, and E. C. Alvord, Jr., The *in vitro* demyelinating activity of sera from guinea pig sensitized with whole CNS and with purified encephalitogen, *Exper. Neurol.* **22**:545–555 (1968).
219. F. J. Seil and P. W. Lampert, Neurofibrillary tangles induced by vincristine and vinblastine sulfate in central and peripheral neurons *in vitro*, *Exper. Neurol.* **21**:219–230 (1968).
220. F. J. Seil, P. W. Lampert, and I. Klatzo, Neurofibrillary spheroids induced by aluminum phosphate in dorsal root ganglia neurons *in vitro*, *J. Neuropath. Exper. Neurol.* **28**:74–85 (1969). See also *Proc. VIth Internat'l Congress Neuropath., Paris*, Masson et Cie, pp. 131–132 (1970).

221. M. L. Shelanski and E. W. Taylor, Isolation of a protein subunit from microtubules, *J. Cell Biol.* **34**:549–554 (1967).
222. G. V. Sherbet, The inhibitory effects of calf-thymus-histone fractions on the development of the chick embryo, *J. Embryol. Exper. Morphol.* **16**:159–170 (1966).
223. R. L. Sidman, Tissue culture studies of inherited retinal dystrophy, in *Diseases of the Nervous System* Monograph Supplement, **22**:14–20 (1961).
224. R. L. Sidman, Organ-culture analysis of inherited retinal degeneration in rodents, *Nat. Cancer Inst. Monograph* **11**:227–246 (1963).
225. R. L. Sidman, M. M. Dickie, and S. H. Appel, Mutant mice (quaking and jimpy) with deficient myelination in the central nervous system, *Science* **144**:309–311 (1964).
226. R. L. Sidman and R. Pearlstein, Pink-eyed dilution (p) gene in rodents: Increased pigmentation in tissue culture, *Develop. Biol.* **12**:93–116 (1965).
227. D. H. Silberberg, Phenylketonuria metabolites in cerebellum culture morphology, *Arch. Neurol.* **17**:524–529 (1967).
228. D. H. Silberberg, Maple syrup urine disease metabolites studied in cerebellum cultures, *J. Neurochem.* **16**:1141–1146 (1969).
229. D. Silberberg, J. Benjamins, N. Herschkowitz, and G. McKhann, Incorporation of $Na^{35}SO_4$ by myelin sulfatide in myelinating tissue cultures, *Amer. Neurol. Assoc., Trans.* **94**:346–347 (1969). [A biochemical parameter of myelination.]
230. D. Silberberg, L. Johnson, H. Schutta, and L. Ritter, Photo-degradation products of bilirubin studied in myelinating cerebellum cultures. Presented at Symposium on Bilirubin Metabolism in the Newborn, University of Chicago, June 6, 1969, Proceedings of the Symposium, National Foundation (in press).
231. D. Silberberg and H. Schutta, The effects of unconjugated bilirubin and related pigments on cultures of rat cerebellum, *J. Neuropath. Exper. Neurol.* **26**:572–583 (1967). [See also *Amer. Neurol. Assoc. Trans.* **92**:284–285 (1967).]
232. I. Singer and S. J. Goodman, Mammalian ependyma: Some physicochemical determinants of ciliary activity, *Exper. Cell Res.* **43**:367–380 (1966).
233. M. Singer and S. V. Bryant, Movements in the myelin Schwann sheath of the vertebrate axon, *Nature* **221**:1148–1150 (1969).
234. H. M. Sobkowicz and M. R. Murray, Effects of lead on organized cultures of nervous tissue (1971, in preparation).
235. J. B. Stanbury, J. B. Wyngaarden, and D. S. Fredrickson (eds.), *The Metabolic Basis of Inherited Disease*, 2nd ed., The Blakiston Div.—McGraw Hill, New York (1966).
236. D. Steinberg, J. H. Herndon, Jr., B. W. Uhlendorf, C. E. Mize, J. Avigan, and G. W. A. Milne, Refsum's disease: nature of the enzyme defect, *Science* **156**:1740–1742 (1967).
237. O. Suyeoka, Studies on the culture media for the central nervous tissue. I. Effects of glucose and related substrates, *Exper. Cell Res.* **47**:22–29 (1967).
238. J. F. Thomson, *Biological Effects on Deuterium*, Macmillan, New York (1963).
239. M. M. Turbow and A. Burkhalter, Acetylcholinesterase activity in the chick embryo spinal cord, *Develop. Biol.* **17**:233–244 (1968).
240. T. Utakoji and T. C. Hsu, Nucleic acids and protein synthesis of isolated cells from chick embryonic spinal ganglia in culture, *J. Exper. Zool.* **158**:181–202 (1965).
241. S. Varon and C. W. Raiborn, Dissociation, fractionation and culture of embryonic brain cells (from 11-day chick embryos), *Brain Res.* **12**:180–199 (1969).
242. J. De Vellis and D. Inglish, in *Proc. 2nd International Meeting Society Neurochemistry* (Paoletti, Fumagalli and Galli, eds.), pp. 151–152, Tamburini, Milan (1969).
243. G. Veneroni and M. R. Murray, Formation *de novo* and development of neuromuscular junctions *in vitro*, *J. Embryol. Exper. Morphol.* **21**:369–382 (1969).
244. C. Waymouth, A chemically defined medium for study of retinal pigmentation, *Develop. Biol.* **12**:115–116 (1965).

245. H. deF. Webster, D. Spiro, B. Waksman, and R. D. Adams, Phase and electron microscopic studies of experimental demyelination. II. Schwann cell changes in guinea-pig sciatic nerves during experimental diphtheritic neuritis, *J. Neuropath. Exper. Neurol.* **20**:5–34 (1961).

246. W. O. Whetsell, Jr., Cytoplasmic vacuole formation in cultured neurons treated with a Na-K-activated-ATPase inhibitor or with Li ions, *Excerpta Med. Internat'l Congr. series* **193**:299 (1969).

247. W. O. Whetsell, Jr. and R. P. Bunge, Reversible alterations in the Golgi complex of cultured neurons treated with an inhibitor of active Na and K transport, *J. Cell Biol.* **42**:490–500 (1969).

248. H. E. Whipple (ed.), Research in demyelinating disease, *N.Y. Acad. Sci. Ann.* **122**:1–570 (1965).

249. E. N. Willmer (ed.), *Cells and Tissues in Culture. Vol. I, Methods*, Academic Press, New York (1965).

250. C. B. Wilson and M. Barker, Chemotherapeutic response of human glial tumor cells cultured in cerebrospinal fluid, *Cancer Chemotherapy Reports* **51**:201–204 (1967).

251. C. B. Wilson and M. Barker, Relative cytotoxicity of mithramycin and vinblastine sulfate in cell cultures of human neural tumors, *U.S. Nat. Cancer Inst. J.* **38**:459–468 (1967).

252. F. H. Wilt and N. K. Wessells (eds.), *Methods in Developmental Biology*, Crowell, New York (1967).

253. G. F. Winkler, *In vitro* demyelination of peripheral nerve induced with sensitized cells, *N.Y. Acad. Sci. Ann.* **122**:287–296 (1965).

254. G. F. Winkler and B. G. Arnason, Antiserum to immunoglobulin A: Inhibition of cell-mediated demyelination in tissue culture, *Science* **153**:75–76 (1966).

255. H. Wiśniewski, M. L. Shelanski, and R. D. Terry, Effects of mitotic spindle inhibitors on neurotubules and neurofilaments in anterior horn cells, *J. Cell Biol.* **38**:224–229 (1968).

256. H. Wiśniewski, R. D. Terry, J. N. Whitaker, S. D. Cook, and P. C. Dowling, Landry-Guillain-Barré syndrome. A primary demyelinating disease, *Arch. Neurol.* **21**:269–276 (1969).

257. M. K. Wolf and A. B. Holden, Tissue culture analysis of the inherited defect of central nervous system myelination in jimpy mice, *J. Neuropath. Exper. Neurol.* **28**:195–213 (1969).

258. G. E. W. Wolstenholme and M. O'Connor (eds.), *Growth of the Nervous System*. A Ciba Foundation Symposium, J. and A. Churchill Ltd., London (1968).

259. R. Wurtman, H. Shein, J. Axelrod, and F. Laren, Incorporation of ^{14}C-tryptophan into ^{14}C-protein by cultured rat pineals: stimulation by 1-norepinephrine (NOR), *Nat. Acad. Sci., Wash. Proc.* **62**:749–755 (1969).

260. T. Yonezawa, The hexose monophosphate shunt—its significance for maintenance of the myelin sheath, *Advances Neurol. Sci. (Tokyo)* **9**:707–716 (1965).

261. T. Yonezawa, M. B. Bornstein, E. R. Peterson, and M. R. Murray, *in IV International Congress Neuropathology Proc.*, Vol. II, pp. 273–274, Georg Thieme, Stuttgart (1962).

262. T. Yonezawa, M. B. Bornstein, E. R. Peterson, and M. R. Murray, A histochemical study of oxidating enzymes in myelinating cultures of central and peripheral nervous tissue, *Ann. Histochim.* **8**:229–238 (1963).

263. T. Yonezawa, Y. Ishihara, and H. Matsuyama, Studies on experimental allergic peripheral neuritis. I. Demyelinating patterns studied *in vitro*, *J. Neuropath. Exper. Neurol.* **27**:453–463 (1968).

264. T. Yonezawa, Y. Ishihara, and Y. Sato, Demyelinating antibodies of experimental allergic encephalitis and peripheral neuritis represented by demyelinating pattern *in vitro*, *J. Neuropath. Exper. Neurol.* **29**:180–181 (1969).

265. T. Yonezawa and H. Iwanami, Experimental study of thiamine deficiency in nervous tissue, using tissue culture technics, *J. Neuropath. Exper. Neurol.* **25**:362–372 (1966).

266. T. Yonezawa, T. Mori, and Y. Nakatani, Effects of pyridoxine deficiency in nervous tissue maintained *in vitro*, *N.Y. Acad. Sci. Ann.* **166**:146–157 (1969).

267. I. S. Zagon, R. Lasher, and G. E. Stone, Some preliminary light and electron microscope observations on colcemid and vinblastine sulfate-treated glioblastoma, *J. Cell Biol.* **47**:234a (1970).

268. A. Zanini and P. Angeletti, Immunochemical assay of a protein characteristic of the nervous system in embryonic sensory ganglia cultured with and without NGF, *Brain Res.* **11**:260–263 (1968).

SUBJECT INDEX

This is a joint index for Volume V, Parts A and B. Pages 1-438 will be found in Part A and pages 439-837 in Part B.